FLOATPLANES OVER THE DESERT

FLOATPLANES OVER THE DESERT

Ian M. Burns

The Adventures of
French & British Naval Airmen
Over Sea & Desert Sand
1914-1918

Cover: *The Bombing of El-Afuleh Railway Junction*
by Major Clifford Roger Fleming-Williams
© Imperial War Museum (Art.IWM ART 2846)

Fleming-Williams was employed by the Northern Aircraft Company at Windermere in early 1915 as ground school instructor and general manager. The company ran a seaplane training school for the RNAS. He also made drawings and assisted with the design of the NAC PB.1 pusher biplane seaplane. He left in August 1916, shortly after the Admiralty took over control of the school, joining the Royal Flying Corps. He taught rigging, theory of flight and general design at the School of Military Aeronautics, Reading. In 1918 he was commissioned by General A.C. MacLean, RAF, to create a series of paintings documenting the activities of the RAF at sea. There was, however, some difference of opinion as to the pictures' merit, as the RAF was looking for technical accuracy, whilst Fleming-Williams' work was more artistic and interpretive. The painting manages to combine both technical accuracy in the Short 184 on the upper right, but the remainder is a breathtaking artistic impression of the raid.

Copyright © 2025 Ian M. Burns

Images copyright expired or no known copyright restrictions

All rights reserved. No part of this book may be reproduced in any form by electronic or mechanical means, including information storage and retrieval systems, without permission in writing from the publisher, except by a reviewer who may quote brief passages in a review

Published July 2025

ISBN 978-1-7636268-4-3 (paperback)
ISBN 978-1-7636268-5-0 (ebook)

Little Gully Publishing
littlegully.com

A catalogue record for this book is available from the National Library of Australia

CONTENTS

	Author's Note	1
Introduction	The British in Egypt	5
Chapter 1	Early Aeronautics in Egypt	9
2	On Land and Sea	29
3	*L'escadrille de Port-Saïd*	41
4	After the Suez Canal Attack	68
5	The East Indies and Egypt Seaplane Squadron	93
6	Early Operations of the EIESS	125
7	Aviation in the Middle East, 1916–1918	148
8	Samson in Command	165
9	Attacking the Railway	191
10	Aden and the Red Sea, 1916	223
11	Castellorizo	255
12	*Empress* with the Eastern Mediterranean Squadron	281
13	*Anne* Carries On	310
14	*Raven*'s East Indies Cruise	329
15	Maintaining a Steady Course	352
16	From Gaza to *Goeben*	379
17	The Final Cruise	409
18	Royal Air Force Days and Ways	431
	Afterword	461
Appendix 1	For the Family	464
2	Samson's Serials for EIESS Floatplanes	472
3	Nieuport *Hydroavions* serving with *l'escadrille de Port-Saïd*, 1914–1916	473
4	Shorts and Sopwiths Serving With EIESS, January 1916–March 1918	476
5	Aircraft Serving with 64 (Naval) Wing, RAF, Egypt, from April 1918	482
6	*L'escadrille de Port-Saïd*, Known Pilots and Observers	486
7	East Indies and Egypt Seaplane Squadron, Known Pilots and Observers	491
8	Summary of Shipboard Operations of *l'escadrille de Port-Saïd*, 1914–1916	499
9	Summary of Shipboard Operations EIESS, 1915–1918	505
10	Some Comparative Ranks	518
	Bibliography	520
	Endnotes	525
	Index	543

MAPS

THEATRE MAPS

I	The Eastern Mediterranean and Red Sea, 1914–1918	vii
II	Northern Sinai	viii
III	Palestine Coast – Gaza to Tripoli	ix
IV	Gulf of Alexandretta and Surrounding Region	x
V	Castellorizo and SW Anatolian Coast	xi
VI	North Aegean and NW Anatolian Coasts	xii

CHAPTER MAPS

Chapter 3	Akaba and Region	53
5	Port Said showing French and British Seaplane Bases	108
10	Aden, 1914–1918	225
11	Battery Locations, Castellorizo, 9 January 1917	267
12	Gulf of Orfano	303
16	Naval Operations, Gaza, 28 October to 7 November 1917	380

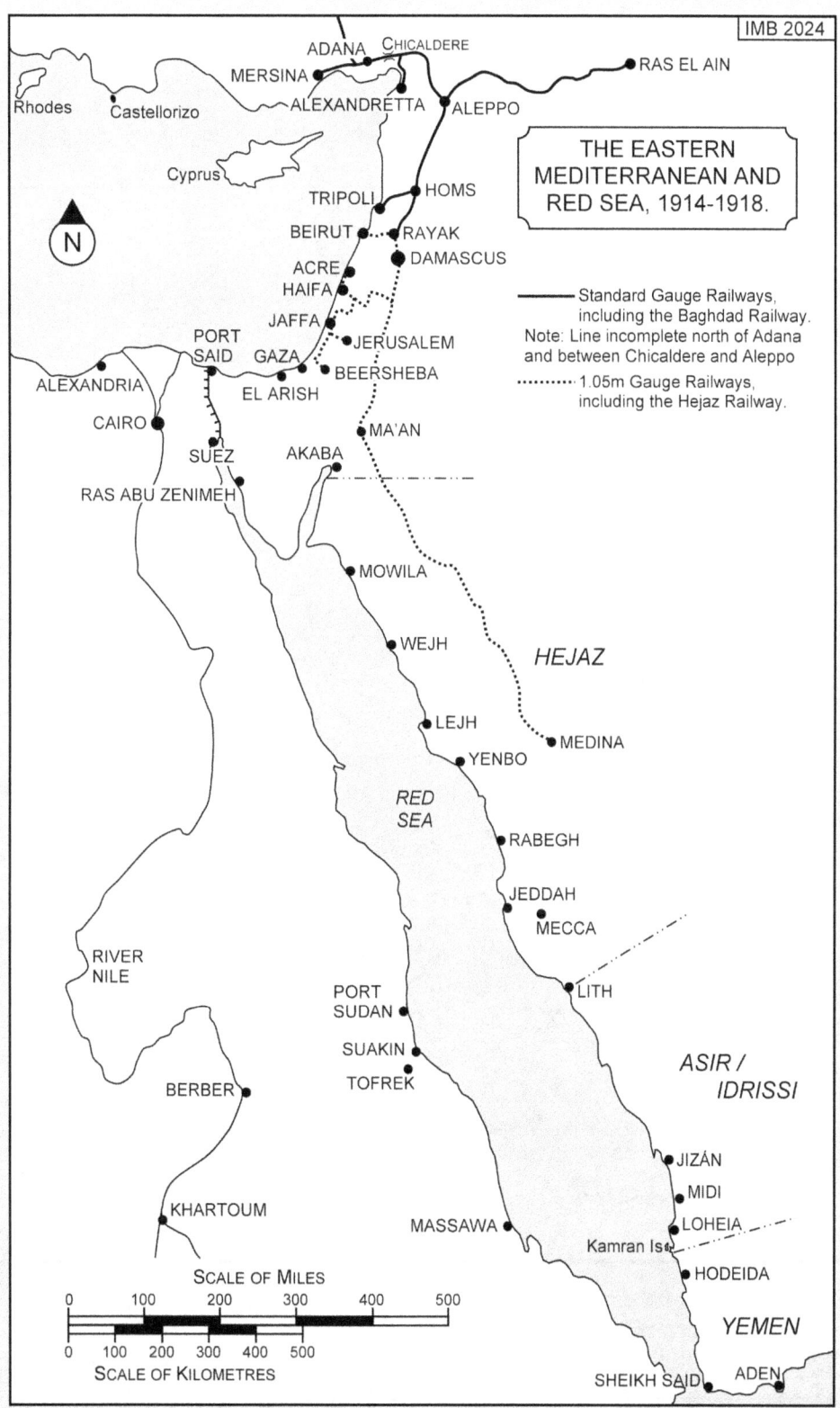

Map I — Eastern Mediterranean and Red Sea, 1914–1918

Map II — Northern Sinai

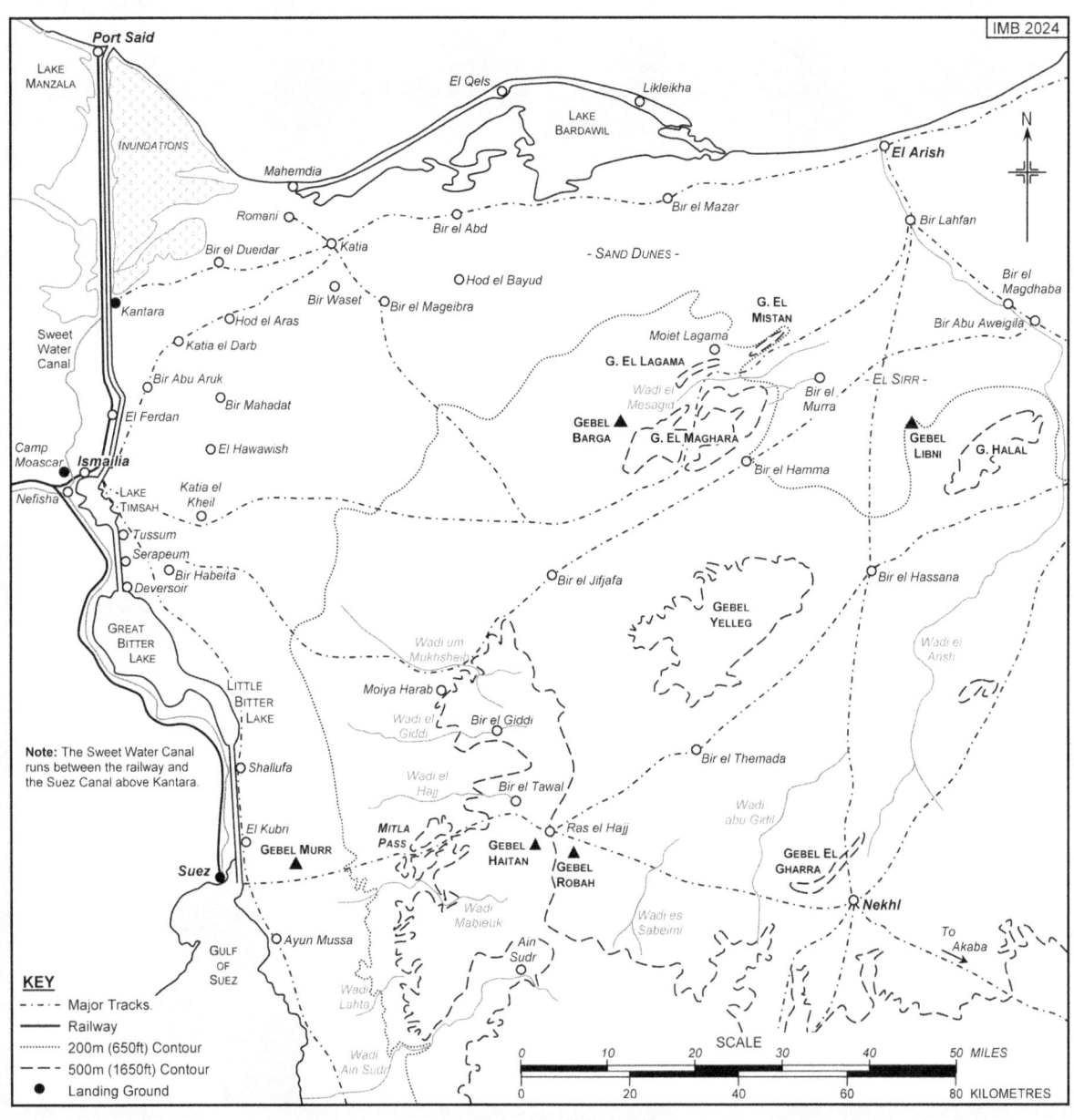

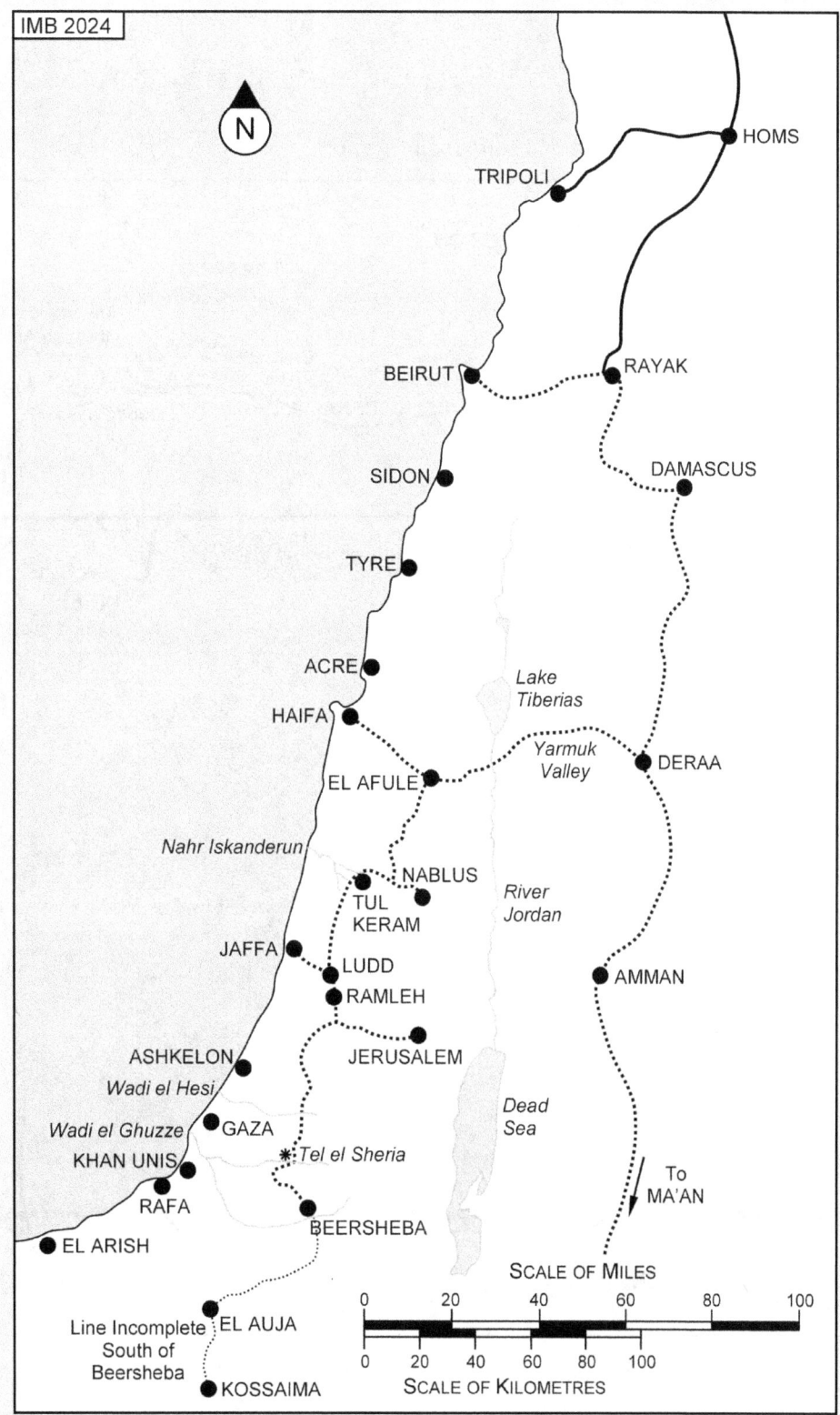

Map III — Palestine Coast Gaza to Tripoli

Map IV — Gulf of Alexandretta and Surrounding Region

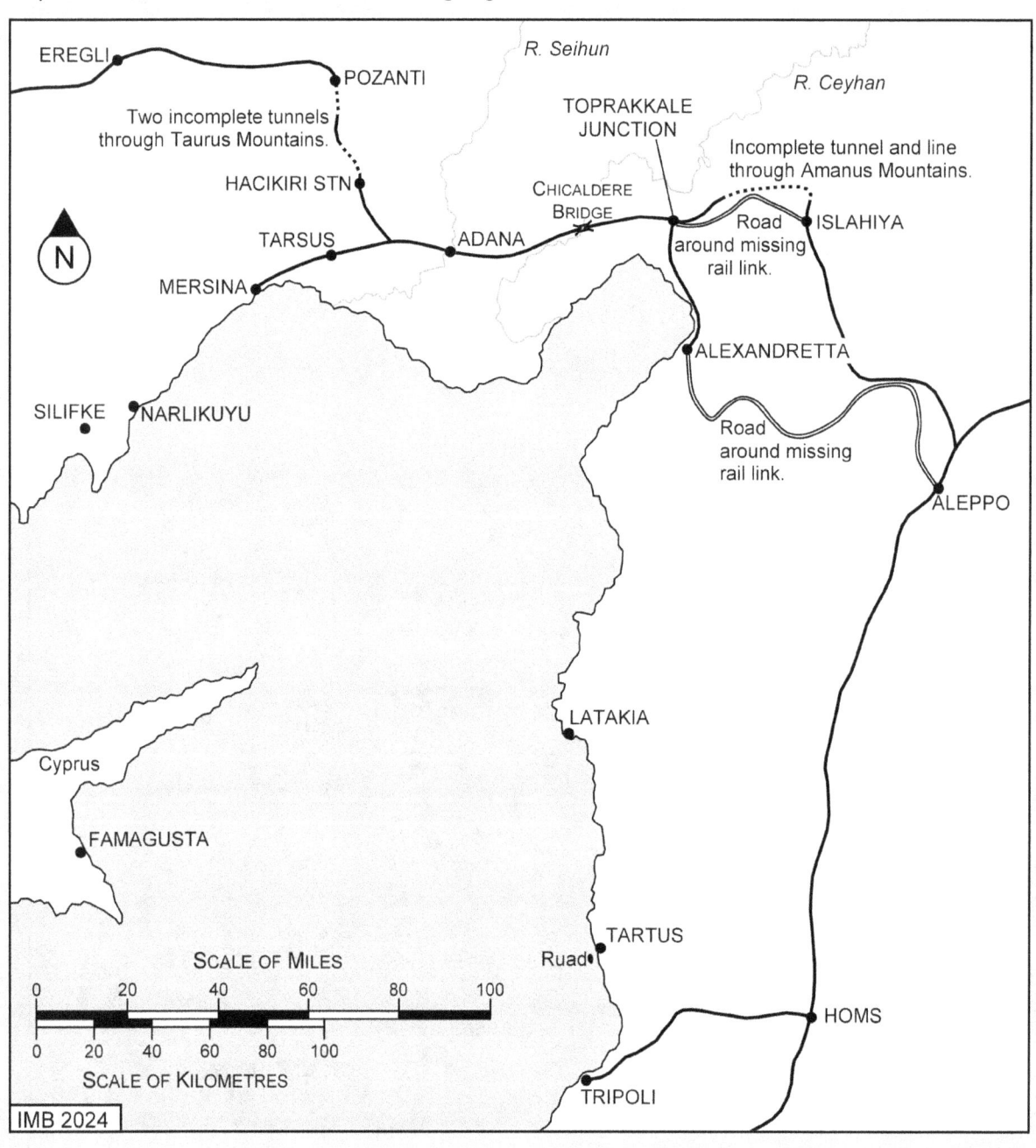

Map V — Castellorizo and SW Anatolian Coast

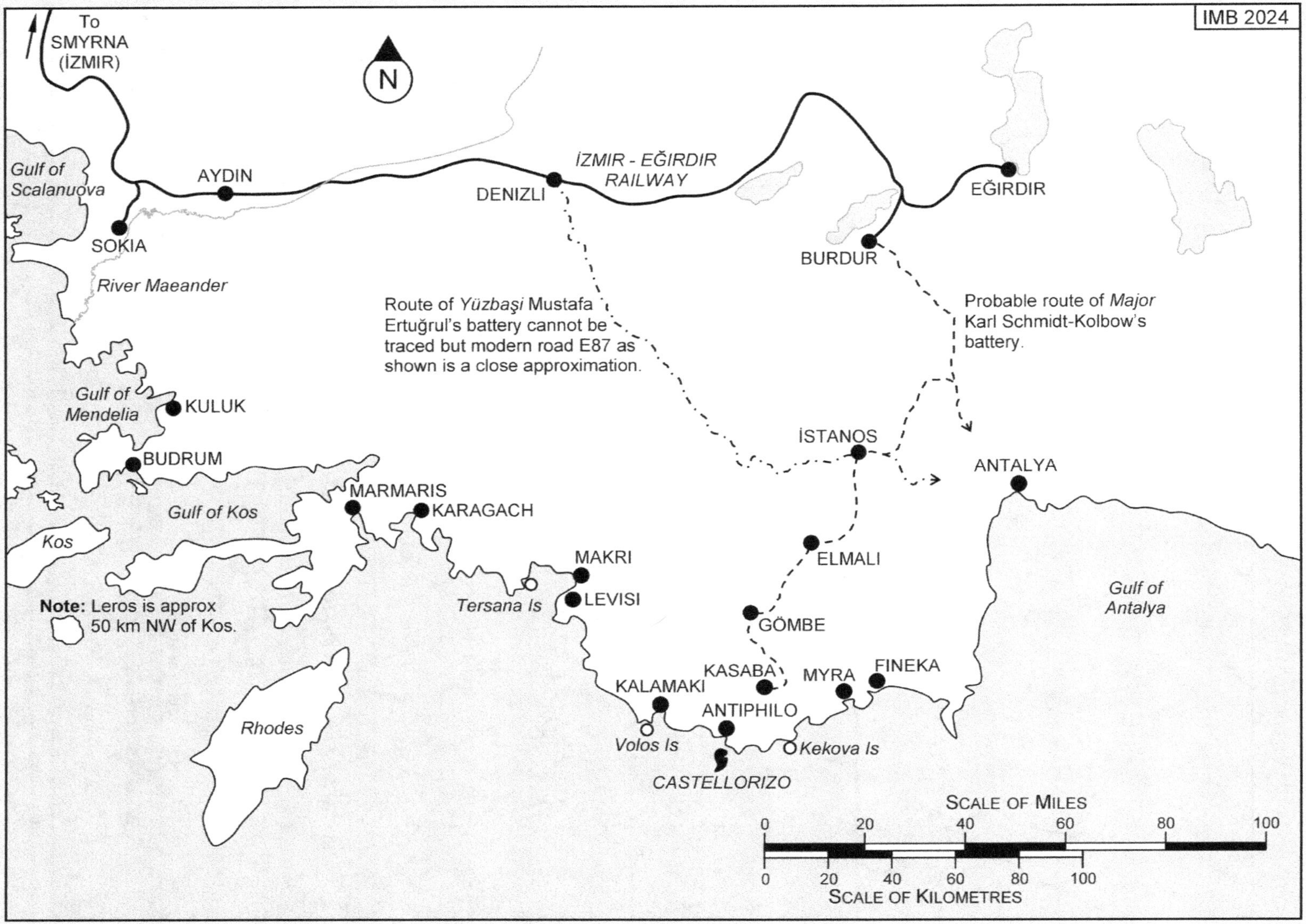

Map VI — North Aegean and NW Anatolian Coasts

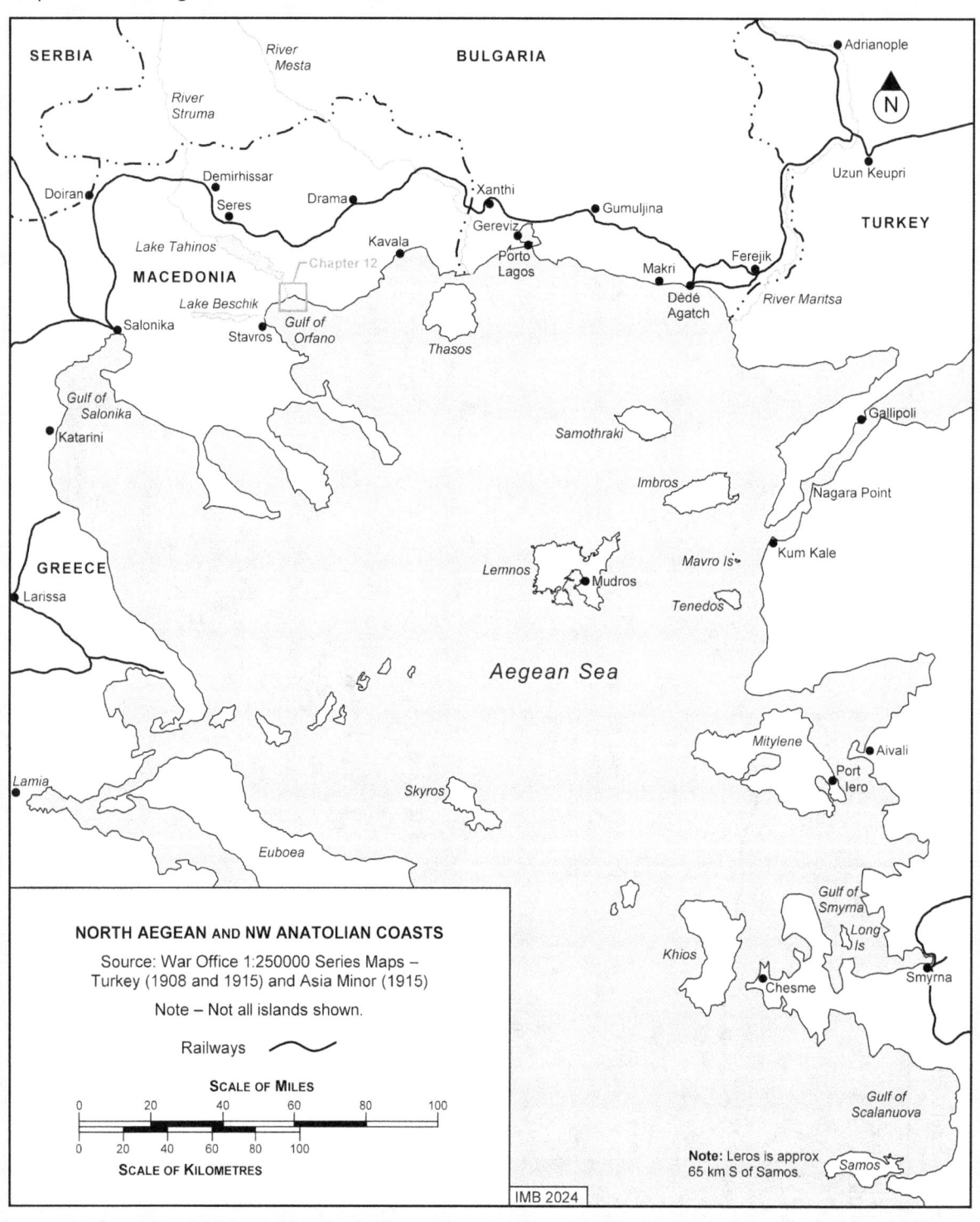

AUTHOR'S NOTE

This is a story of Naval Aviation. It is about the French Navy's *Aéronautique maritime* and the British Royal Naval Air Service (RNAS), and how they came to operate obsolescent floatplanes far behind enemy lines often many kilometres inland from the sea. It is a story set far from the major battlegrounds of the Western Front and the North Sea. It is in fact a story set in a 'sideshow.' Yet the techniques they pioneered of taking naval aircraft to the enemy remain, albeit much refined and powerfully expressed, a prime role of modern naval aviation.

The history of *l'Aéronautique maritime* is briefly detailed in the following chapters, as it is essential to show how and why a French naval aviation squadron came to be based at Port Said, Egypt, from late 1914. For a complete account of early French naval aviation see Robert Feuilloy and Lucien Morareau, *L'aéronautique maritime dans la Grande Guerre* (ARDHAN, 2019).[1] The RNAS, however, only enters our story as a fully formed service in January 1916 with the arrival of the seaplane carrier HMS *Ben-my-Chree* from Gallipoli. For the formation of the RNAS, see David Hobbs, *The Royal Navy's Air Service in the Great War* (Seaforth, 2017); for *Ben-my-Chree*'s story, see the author's *Ben-my-Chree – Woman of My Heart, Isle of Man Packet Steamer and Seaplane Carrier* (Colin Huston, 2008).

Sadly, the sideshow of the last century remains prominent in today's news reporting as the sideshow of which I write ultimately played a prominent role in the formation of today's Middle East. But the Middle East of the Great War period was vastly different from that we know today. Apart from Egypt, the area was an Ottoman fiefdom, ruled with a faltering hand from Constantinople. Egypt at the beginning of the war became a British Protectorate, completing the takeover of what had been a semi-autonomous part of the Ottoman Empire until 1882. How this came about is explained in the Introduction. Assuming it will be as unfamiliar to the reader as it was to the author, I have included brief descriptions of the lands and peoples at appropriate points in the narrative. The existing land and naval forces, and aerial services are covered at greater length with chapters of their own.

I found a lack of consistency in the spelling of place names, especially along the Red Sea coast. In the latter case, many names used cannot be located on contemporary maps or charts, let alone modern gazetteers and maps. In consequence, I have largely adopted the most common contemporary British usage, with modern equivalents provided, when known, within brackets after the first appearance of the name. When quoting original documents, place names are left with their original spelling.

Contemporary records consistently refer to the French Nieuports employed by *l'Aéronautique maritime* and to the British Shorts and Sopwiths of the Royal Naval Air Service as 'seaplanes.' The designation is also employed in the RNAS and RAF unit titles. The term 'seaplane' may date from 1912, Richard Bell Davies recalling in his memoirs that, 'Up to this time, aeroplanes on floats had been called hydroaeroplanes, but I heard Winston Churchill say, "That's a beastly word. Let's give them a better name; let's call them seaplanes." And seaplanes they have been ever since.' However, seaplanes can be subdivided into two main groups, flying boats and floatplanes. Essentially, flying boats combine fuselage and boat hull into one structure, whilst floatplanes have separate flotation devices, floats or pontoons, attached by struts to the fuselage and/or wings. This latter description applied to all the machines employed by *l'Aéronautique maritime* and the Royal Naval Air Service. The author has followed this distinction throughout the text.

Basic measurements are metric, except when quoting or transcribing from contemporary documents and publications, or when original measurements were in imperial units (for example, 6 inch gun, rather than 152 mm, and vice versa). Some units, such as horse power (hp) and tons, are left as is as they remain widely used.

Dates are presented in day-month-year format. Times use the 24 hour clock.

Acknowledgements

I first started researching *Ben-my-Chree*, and the East Indies and Egypt Seaplane Squadron, in the early 1970s. Since then, over many fits and starts, I have been thankful for advice and assistance from many fellow researchers and friends. Too many are no longer with us, but many more remain available to give sound advice and assistance when requested. The following are those I recall:

Jack Bruce and Stuart Leslie, Boris Ciglic, Peter Chapman, Peter Cowlan, Stuart Hadaway, Dick Cronin, Paul Hare, Eric Harlin, Barbara and Colin Huston, Philip Jarrett, R.D. Layman, Elimor Makevet, Errol Martyn, Roger Nailer, Ole Nikolajsen, Colin Owers, Nicholas Pappas, Paschalis Palavouzis, Ray Vann, to name but a few.

Some more I have probably forgotten. If they should happen to see this book please forgive my lapse of memory and get in contact.

L'escadrille de Port-Saïd was in many ways a unique unit, an early example of Franco-British co-operation. Elements of the French and Royal Navies and the British Army came together to perform its duties. As a consequence, records exist in both countries, in both languages. Through the generosity of Bernard Klaeylé and Thierry Le Roy, I have been able to read and translate some of *Lieutenant de vaisseau* de l'Escaille's reports. I have also been assisted by several French historians: Robert Feuilloy and Lucien Morareau (ARDHAN), Michel Bénichou and David Méchin.

L'escadrille de Port-Saïd était à bien des égards une unité unique, un exemple précoce de coopération franco-britannique. Des éléments de la marine française, de la marine royale et de l'armée britannique se sont réunis pour accomplir ses tâches. Par conséquent, des documents existent dans les deux pays, dans les deux langues. Grâce à la générosité de Bernard Klaeylé et de Thierry Le Roy, j'ai pu lire et traduire certains des rapports du lieutenant de vaisseau de l'Escaille. J'ai également bénéficié de l'aide de plusieurs historiens français: Robert Feuilloy et Lucien Morareau (ARDHAN), Michel Bénichou et David Méchin. Je les remercie tous.

Several societies have over the years provided much help and assistance—The Great War Aviation Society (aka Cross and Cockade International), The Australian Society of WW1 Aero Historians, and the Salonika Campaign Society whose Trench Map DVD has been invaluable.

In the UK, the staffs of the Fleet Air Arm Museum, Imperial War Museum, and The National Archives have answered my requests with professionalism and cheerfulness. This publication contains Public Sector information licensed under the Open Government Licence v3.0 and/or a Creative Commons Attribution 4.0 in the UK.

In Canada I have been able to access many of the rarer books listed in the Bibliography courtesy the Naval Marine Archive—*The Canadian Collection* in Picton, Ontario. Their collections contain about half a million items including more than 80,000 books and magazines, and welcomes personal visits or can be accessed online at navalmarinearchive.com.

Despite all the advice and assistance I cannot avoid being responsible for what appears on the following pages. If you, the reader, find any errors or omissions please contact me so that they may be corrected in any future editions.

Photographs

Some of the historic photographs used in this book are showing the ravages of time. Some have been taken from period postcards, magazines or books. Many have been copied from a 'Kodak Story' contact printed at the time and preserved in family albums, or boxes in the attic or basement. However, because of their importance to the story I think they deserve to be seen. We are lucky that they exist at all.

I first started gathering images in the era BC (Before Computer) when printed photographs and albums were photographed, to create new black and white negatives, then returned or passed on to another researcher. Over years of research I have been sent, or collected, different copies of many of the photographs. I have endeavoured to determine the source of the images used, but so many identical prints of wartime photographs found their way into personal albums cared for by families or museums, or have been copied by collectors, that this can be a Sisyphean task. Where known I have acknowledged the original donors, if I have missed anyone I offer my sincere apologies. Any un-credited photographs are from the author's collection or not attributable due to multiple sources. The abbreviation 'LoC' after many photographs show it to be via US Library of Congress collections. Similarly, 'EMK' indicates the Eliot Millar King collection.

All historic photographs used are within the public domain. Any credits should be understood to be *courtesy of* or *via* any credited source.

Family collections form an important part of the images, diaries and papers used in this book.

In the UK the Stone Family Collection and Attrill Family Collection have been generous with their time and making available photographs taken by CPO Charles Henry Stone and CPO Charles (Charlie) Attrill at Port Said during their time with the EIESS. Their photographs help illustrate the experiences of the often overlooked mechanics and aircraft hands.

In particular I wish to acknowledge and thank two New Zealand family collections.

Lieutenant J. L. Kerry's photographs and papers are treasured by his family, and I am very grateful to Kerry Family Collection for making them available for this book.

Eliot Millar King donated his wartime photo albums, log book and papers to the local New Plymouth museum, Puke Ariki. Many of the images credited 'EMK' may be viewed at https://collection.pukeariki.com/objects/35221.

Eliot King's youngest daughter, Adrienne Tatham, recently wrote to me approving my use of these images, also saying 'History is a good teacher, and we need to pass on the information we do have. I have no idea how future generations will learn without printed words!'

It would also be most remiss of me not to thank Bernard de Broglio and Little Gully Publishing (littlegully.com) for the hours spent preparing my manuscript for publication.

* * *

INTRODUCTION

The British in Egypt

In 1517 Egypt became an *Eyālet*, a semi-autonomous part of the Ottoman Empire. In 1867 Egypt was made a *Khedivate*, a fully autonomous tributary state of the Ottoman Empire. It almost immediately fell into debt to European banks.

Construction of the Suez Canal had commenced in 1859, but had been dogged throughout by financial problems. The British Government, and businesses, had declined to take any financial interest in its construction. It was only by the intervention of Muhammad Sa'id Pasha (*Wāli* of Egypt and Sudan from 13 July 1854 – 18 January 1863) who, through a loan from European banks, purchased unsold shares to increase Egypt's holdings to 44% of the Suez Canal Company, that the project was kept afloat. Thus, the canal was controlled by the French owned Suez Canal Company (*Compagnie universelle du canal maritime de Suez*) with a 53% shareholding, 44% being held by the Khedivate of Egypt, and 3% other investors.

By 1875 *Khedive* Isma'il (18 January 1863 – 26 June 1879) through corruption and mishandling of the country's finances had put Egypt on the verge of bankruptcy. To avoid this he sold the 44% holding in the Suez Canal to Great Britain, for just under £4 million (almost £600 million in 2023). With increased British and French control of the country's finances, public unrest led to the Urabi Revolt.[2] *Khedive* Isma'il was deposed in 1879 and replaced by his oldest son, Muhammad Tawfīq Pasha. The revolt continued unabated, leading in 1881 to the Franco-British naval intervention at Alexandria to protect European lives and property. Tawfīq also moved to Alexandria for fear of his own safety.

By June, Egypt was in the hands of nationalists opposed to European domination of the country. Anti-European violence broke out in Alexandria, prompting a British naval bombardment of the city (the French Government had ordered its fleet to withdraw to Port Said). Britain continued to act unilaterally and, fearing the seizure of the canal by the Egyptians, sent an Anglo-Indian expeditionary force to Egypt in August 1882. The combined Anglo-Indian army defeated the Egyptian Army at Tel el Kebir in September and took control of the country putting Tawfīq back in control. *Khedive* Tawfīq, however, was now little more than a puppet controlled by the British Consul-General of Egypt, Sir Evelyn Baring (from 1901, First Earl of Cromer).

The Great Sphinx has been an irresistible draw as a backdrop to souvenir photographs since the 1880s.

Seen here shortly after the Battle of Tel el Kebir in September 1882 is an informal group of 42nd Highlanders (The Black Watch).

(GD483/26/2, National Records of Scotland)

Baring instituted a form of government that came to be called the Veiled Protectorate, whereby he ruled the rulers of Egypt with the assistance of a group of British administrators placed in key positions as advisers to the Egyptian government. In addition, the Egyptian military and police were reorganized and placed under British officers. Baring retired in 1907. His replacements Sir John Eldon Gorst (1907–1911) and Field Marshal The Right Honourable The Viscount Kitchener (1911–1914), largely continued his policies.

In the years immediately prior to the Great War, Egypt had become in essence a British colonial possession. But the French also had interests in Egypt. It still held a majority share in the Suez Canal Company. Consequently, the Suez Canal area and its towns — Port Said, Ismailia and Suez — had a very French atmosphere. Cairo and, to a degree, Alexandria had a very British Colonial feel, but with a strong International community. Cairo's great hotels, notably Shepeard's Hotel, became the social centres of the expatriate population, expanded through the winter season (November to April) by affluent tourists from Britain, Europe and the USA. Outside the cities, Egypt remained the land of the *effendiyya*, or middle class, whose land was worked by *fellahin*, tenant farmers and agricultural workers.

When Tawfīq died on 7 January 1892, he was succeeded by his eldest son, Abbās Ḥilmī. The new *Khedive* Abbās II was initially anti-British but was eventually worn down and, at least publicly, submitted to Baring's rule. However, when Great Britain declared war on Turkey,[3] on 5 November 1914, Abbās was in Constantinople and accused of deserting Egypt by not promptly returning home. The British also believed, correctly, that he was plotting against their rule.

On 18 December 1914, Britain declared a Protectorate over the country.

NOTICE.

His Britannic Majesty's Principal Secretary of State for Foreign Affairs gives notice that, in view of the state of war arising out of the action of Turkey, Egypt is placed under the protection of His Majesty and will henceforth constitute a British Protectorate.

The suzerainty of Turkey over Egypt is thus terminated, and His Majesty's Government will adopt all measures necessary for the defence of Egypt and the protection of its inhabitants and interests.

December 18th, 1914.[4]

The following day:

In view of the action of His Highness Abbas Hilmi Pasha, lately Khedive of Egypt, who has adhered to The King's enemies, His Majesty's Government have seen fit to depose him from the Khediviate, and that high dignity has been offered, with the title of Sultan of Egypt, to His Highness Prince Hussein Kamel Pasha, eldest living Prince of the family of Mahomet Ali, and has been accepted by him.

With these announcements Britain brought to an end Egypt's almost 400 years attachment to the Ottoman Empire.

The Protectorate was placed under a High Commissioner, Lieutenant-Colonel Sir Arthur Henry McMahon, GCVO, KCIE, CSI. Thus effectively bringing the Sultan of Egypt, and all civil and armed services directly under British command.[5]

After the establishment of the British Protectorate in 1882, Shepheard's Hotel became the centre of British social life in Cairo. The Terrace is seen here in 1910. *(Gallica, ark:/12148/btv1b531201421)*

An American lady traveller, Blanche McManus, wrote in 1911, 'As diversified amusement nothing quite takes the place of the Terrace at Shepheard's in the height of the season say about February, when the chairs before the little wicker tables under the gay Oriental hangings are all taken, and a crowd, clothed in all colours, and of all degrees of brilliance, is gathered to hear the band play, gossip and watch the multi-coloured population of this most cosmopolitan of Oriental cities drift ceaselessly past...'

Giza pyramid complex, photographed from Eduard Spelterini's balloon *Wega* at 600 metres on 21 February 1904.

The partially excavated Sphinx is barely visible as a small white shape just above the centre of the image and below the middle pyramid, the Pyramid of Khafre.

The Mena House is to the right of the Great Pyramid of Khufu.

CHAPTER 1

Early Aeronautics in Egypt

Jean-Marie Joseph Coutelle, *commandant du corps des aérostiers*. Coutelle was an experienced balloon observer having spent nine hours in the air during the Battle of Fleurus, 26 June 1794. *(LoC)*

The Aeronauts

Neglecting fantasies of flying Sphinxes and theories surrounding the Saqqara Bird, the earliest recorded flight in Egypt occurred during Napoleon Bonaparte's invasion in 1798. A *compagnie d'aérostiers*, comprising 40 men under the command of *colonel* Jean-Marie Joseph Coutelle, *commandant du corps des aérostiers*, with a single balloon and all necessary equipment, sailed in the fleet. At Toulon their equipment was loaded aboard the transport *Patriote*, the balloon in the holds of the flagship *Orient* (120), whilst the balloonists were aboard *Franklin* (80). Entering the anchorage at Aboukir *Patriote* ran aground on a sandbar and, despite salvage efforts, sank with the balloon equipment still onboard. *Orient*, and the balloon, was destroyed by Rear-Admiral Sir Horatio Nelson's fleet on 1 August 1798. *Franklin*, moored ahead of *Orient*, was severely damaged and captured. However, the balloonists had left the ship shortly after arrival, in early July, to assist with the salvage efforts on *Patriote*.

Nicolas-Jacques Conté was part of the French expedition to Egypt, he may have been one of the party of scientists or, more likely, was part of the *compagnie d'aérostiers*. He was an experienced aeronaut and became the chief engineer with the *compagnie*. When Bonaparte left Egypt to attack gathering Ottoman forces in Syria on 5 February 1799 he left behind the balloon-less *compagnie*. Under Conté's leadership it became a corps of technicians and workers to provide for an army cut off from its supplies. They built furnaces to heat shot for the cannon at Alexandria, helped manufacture gunpowder, weapons, machines for the Cairo mint, even cardboard and canvas. The balloonists were able to manufacture some small unmanned hot-air balloons to demonstrate and entertain in Cairo.[6]

A contemporary illustration of Coutelle and the balloon *l'Entreprenant* at Fleurus. The inflated balloon had to be hauled overland by parties of twenty soldiers for 50 kilometres from its inflation point to the battle site. For the Egyptian Expedition, anticipating problems with the sand, Conté had designed wagons with extra wide rims, but these were lost in the *Patriote*.

> Conté devised and tested a telegraphy system whose stations, it is said, initially consisted of small captive balloons supporting a pulley that could be used to hoist various signals visible from a great distance. However, he soon had to abandon this system and continue his experiments with a mast at the top of which an L-shaped piece of

Nicolas-Jacques Conté. He lost his left eye in his Meudon laboratory following an explosion during the preparation of hydrogen gas. This was a very dangerous operation at the time, as water vapour was passed through a red-hot iron tube, which very often cracked. *(LoC)*

wood assumed different positions. However, he did not give up on balloons, his old experiments and his cherished work at Meudon, and on three occasions launched Montgolfières in Cairo.

The first launch occurred on the *premier vendémiaire de l'an VII*, or 22 September 1798 in the Gregorian calendar.

> Unfortunately, we have very few details about this ascent: what we do know is that the balloon rose very high, then disappeared for good in front of a huge crowd of Egyptians, who, although slightly surprised, remained very calm.
>
> The second experiment was made from Esbekieh Square, on *20 Frimaire de l'an VII* (10 December 1798), at three o'clock in the afternoon in front of nearly forty thousand people.
>
> The 'Machine', as it was then called, was made of paper and was spherical in shape; the gores that made up its surface successively displayed the three national colours and its diameter was twelve metres. It weighed just over one hundred and fifty pounds. The paper that made up the envelope was not strong enough. When it reached a height of about two hundred metres, the balloon broke apart and the debris fell to the ground without causing any accidents.
>
> The last ascent in Cairo during the French occupation took place on the anniversary of the Battle of Rivoli, on *25 Nivôse de l'an VII* (14 January 1799).
>
> The balloon used was thirteen and a half metres in diameter. The inscription Battle of Rivoli with a drawing of a civic crown and palms had been placed on its surface. The envelope was made of canvas and this time the machine supported itself for the entire time that the materials in the stove burned, i.e. for thirty-five minutes.

Picked up near Fort Dupuis, the balloon envelope was recovered 'without having suffered any damage in the fall, in as good a condition as before its removal.'

The French official view of these events is best expressed in a report published by the *Courier de l'Egypte* (a Napoleonic propaganda broadsheet). Of the second ascent it made the following comments.

> The sight of this experiment made the greatest impression on the people of the country; they refused to believe in its possibility; their incredulity lasted all the time that one worked on the preparations; but they were seized with admiration when they saw this large globe moving of itself: when the machine started its movement

those which were in the vicinity of the place where it was charged, fled with the marks of consternation. When they saw the debris of the machine and the stove fall back into the air, they concluded that it was an engine of war, which we knew how to direct as we wished and which we use to burn the cities of our enemies.

The Egyptian view of the same ascent is best expressed by an Egyptian chronicler, Abd al-Rahman al-Jabarti, who wrote:

When the day came the people and many of the French gathered in the afternoon to see this wondrous event and I was among them. I saw a cloth in the form of a large tent upon an erected pole. The cloth was coloured in white, red, and blue. The pole upon which the cloth was suspended was set upon something like the cylindrical form of a sieve in the midst of which there was bowl out of which came a wick immersed in certain oils. This bowl hung from intercrossing iron wires running from it to the cylinder. The cylinder itself was bound with pulleys and ropes which were held by people standing on the roofs of near-by houses. About an hour after the [afternoon prayer] they lit this wick and its smoke rose into the cloth and filled it. The cloth swelled and became like a ball. The smoke sought to rise to its centre but it did not find any exit, so it drew the apparatus aloft with itself. Meanwhile the people pulled it with ropes until it rose from the ground. They cut the ropes and it soared into the air, moving with the wind. Then it began to sail with the wind for a very little while and then its bowl fell with the wick, the cloth following suit. The French were embarrassed at its fall. Their claim that this apparatus is like a vessel in which people sit and travel to other countries in order to discover news and other falsifications did not appear to be true. On the contrary, it turned out that it is like kites which household servants build for festivals.[7]

It would be 86 years before the first confirmed manned flight over Egyptian territory occurred.

Under Muhammad Ali Pasha, governor and effectively ruler of Egypt from 1805 to 1848, the borders of the country greatly extended along the length of the Nile. By 1821 most of Sudan had fallen under Egyptian rule, an occupation fiercely resisted by the Sudanese people. The rise of Muhammad Ahmad bin Abd Allah, the Mahdi, and the Mahdiyya (Madhdist State) led to a series of risings and wars that became known, in European circles, as the Mahdist War (1881–1899). General Charles George Gordon, Governor-General of the Sudan, was under siege in Khartoum from March 1884 until January 1885. A relief expedition was sent up the Nile and from the Red Sea port of Suakin in an attempt to raise the siege. A Royal Engineer balloon detachment was to be part of the Suakin Field Force.

Major J.L.B. Templer and Lieutenant R.J.H.L. Mackenzie, RE, eight NCOs and sappers, departed from the UK in February, 1885. On 24 March a balloon, *Scout* (7000 ft^3), was filled at Suakin, with hydrogen brought in cylinders from the UK. At dawn on the next day the balloon was inflated, the basket attached, and linked to the winch on the wagon. With Major Templer holding the balloon wagon's reins, and Lt Mackenzie in the basket, it took its place in the middle in the middle of a defensive square formation. Also within the square were supply camels and carts, all carrying water destined for Sir John McNeill's zariba at Tore just seven miles from Suakin but in a state of semi-siege. On two previous occasions convoys to the zareba had met with considerable opposition from the Arabs; but this time not a shot was fired nor a single camel's load lost. The balloon was manned for eight hours, Lt Mackenzie being relieved by Sapper Wright at noon. It produced an excellent moral effect, the Arabs being observed to be dispersing in all directions.

On 2 April a small balloon, *Fly*,[8] was transported to Tambuk. An Arab, Ali Kerar, weighing about 84 lb (38 kg), was employed to make ascents, reaching 2000 feet. His observations proved accurate and the balloon was kept at work for more than ten days. As its supply of gas had only taken five camels to bring out, it must be considered a very satisfactory performance. The fact that an earlier balloon detachment was at the same time being employed in Bechuanaland handicapped the Suakim detachment considerably, the available personnel and plant being thus rendered far more limited than was desirable.

However, Gordon had been killed at the fall of Khartoum on 26 January 1885, and the *raison d'etre* for the expedition ended. The campaign gradually wound down mainly for lack of any support from the home government. In May, the entire expeditionary force was withdrawn, and a small Egyptian Army occupying force left at Suakin.

An inflated balloon of the Royal Engineer balloon detachment, Suakin Field Force, March/April 1885. *(Science MuseumGroup Collection)*

In the foreground, from centre left, are several spare baskets, two stacks of hydrogen pressure bottles brought fully charged from the UK, and a hand winch.

Returning to Cairo. Claims by aeronauts, such as 'Professor' Rufus Gibson Wells who claimed to have flown over Cairo in the 1880's, cannot be confirmed. During this period there may also have been visits by other peripatetic aeronauts, none of whom have left any mark on history. The earliest confirmed flight over Cairo, by a gas filled balloon, was that of 'Captain' Eduard Spelterini, a Swiss aeronaut, on Sunday, 2 March 1890. His balloon, *Urania* (1500 m³) made from bright yellow silk, ascended from the Ezbekieh Gardens, Cairo at 4.25 pm. 'The Aeronaut was accompanied by Major General Sir James Dormer, Major Chapman, and two journalists. The wind carried the balloon towards the Desert, and the descent took place at 7.30 p.m. near the village of Chibin-el-Kanater, where the party remained for the night.'⁹ A second ascent was made the following Sunday. Spelterini returned to Cairo in January 1904 with a larger balloon, *Wega* (3260 m³). Over several flights he took the first aerial photographs of the city and, on 21 February, the pyramids at Giza.

Eduard Spelterini ca.1903.
(ETH-Bibliothek Zürich, Bildarchiv)

Top: Spelterini's balloon, *Urania*, Zurich barracks on October 29, 1893.
(ETH-Bibliothek Zürich, Bildarchiv)

The Aviators

Early in the new century, primarily amongst the European community, there was a new interest—the flying machine.

The first aviator to sample the air over Egypt was Baron Pierre de Caters, a Belgian racing driver who had taken to the air. In December 1909 he became the first to be granted a pilot's license from the Belgian Aero-Club, flying a Voisin biplane.

On 2 and 5 December 1909, Baron de Caters made two flights over Constantinople with his yellow painted Voisin aircraft. A flamboyant showman, he wore a yellow beret and jumpsuit to match his aeroplane, and had equipped the Voisin with electric lights on the wings. Both flights ended in crashes with only minor damage to the Voisin and none to the Baron. He made no further flights over Constantinople. As disappointing as his flights undoubtedly were, they were the first powered heavier-than-air manned flights in Turkey. Baron Pierre de Caters was now determined to make the first flight in Egypt. Accordingly, a few days later he loaded the Voisin aboard a ship and sailed off to Egypt.

De Caters arrived in Cairo on 11 December 1909, and was granted permission to use the old parade ground at Abbassia Barracks as his flying field. Abbassia was on north-eastern edge of Cairo on the road to Heliopolis. He was to share part of the gate with the British Military Hospital and the British Soldiers and Sailors Families Association.

AEROPLANE DE CATERS

Top: Baron Pierre de Caters at the controls of his Voisin, as usual with a cigarette in his mouth. *(LoC, G.G. Bain Collection)*

Below: de Cater's Voisin in Europe ca. 1910 from an old postcard.

On Wednesday 15 December de Caters flew his Voisin at Abbassia before a specially invited group of European spectators. At 4.00 pm de Caters was ready to fly, but the wind was a bit too strong. A kite was employed to measure the wind speed, a small square indicator travelling along its line indicating the strength of the wind. Half an hour later the wind had dropped and the Voisin was brought out on to parade ground. After briefly running up the engine, de Caters stopped it explaining that the 'propeller did not work properly'. He then ordered a replacement to be fitted. One wonders if the showman was playing to the crowd. With the new propeller the Baron was able to take off into the fading light and make two tight circuits, landing after a flight of just three minutes. It was the first powered flight in Egypt, but no photographs were taken as it was too dark.

Further flights were advertised for 18, 19, and 20 December between 3.00 pm and 6.00 pm at Abbassia. Tickets ranged from 10 *piastres* for general admission to 400 *piastres* for reserved enclosures. On the first date, once again the winds were strong, but de Caters was eventually able to repeat his earlier flight. On 19 December, in lighter winds, he made two flights beginning at 3.50 pm. He made three circuits of the parade ground in four minutes on the first flight. On the second, he tried to spiral up to 100 metres, but his engine stopped at 40 or 50 metres forcing him to glide to earth. Finally, on 20 December, the Baron made three additional flights. On the first, he remained aloft for six or seven minutes at a height estimated between 10 and 20 metres, flying three circuits.

> He took up a lady with him in each of his two last attempts, the first, who has the honour of being the first lady to fly in Egypt, being *Mlle.* Karri Benedikt and the second Princess Avierino Wizniewski. The former assured us that although it was her first essay at flight she experienced no fear. She described the sensation of gliding through the air as 'beautiful'.[10]

Despite Abbassia being well served by trams from the city, the hoped for crowds failed to attend. On each of the three days there was very meagre attendance and from a commercial point of view the exhibition was a fiasco. Baron de Caters decided to cut his losses, and on Wednesday 22 December sailed from Alexandria for Athens.

As the Baron was leaving Egypt, aviators were beginning to arrive for the Heliopolis Meeting to which we now turn our attention. The Meeting had been announced in the *Egyptian Gazette* late in October 1909, even before the first powered flight in Egypt. Advance notices began appearing in Europe in November. The event was sanctioned by the *Aéro-Club de France*, who sent six observers early in 1910. The organization of the meet itself was entrusted to the *Compagnie Aérienne*, and had a very French flavour, to such an extent that the programme was only available in French. Prizes for distance flown, height and speed achieved, were offered to a total value of 212000 francs (approximately £8400 in 1910, over £1 million in today's money).

The main grandstand at Heliopolis with René Métrot and his Voisin superimposed.

The Oasis of Heliopolis was a specially developed European settlement outside the city of Cairo proper. It had a golf course, amusement park, sports stadium, horse-race course, electric lighting, and two luxurious hotels. It was connected to the city centre by an electric railway and tram line. In 1910 the surrounding area was open desert, ideal for an aviation meeting. A rectangular five-kilometre course was laid out in the desert, two grandstands (*tribunes*) were built, a big one opposite the start/finish line and a smaller one with green silk muslin curtains in front. The latter, referred to as the *tribunes harem*, was intended for women only. On site practice was allowed from 15 December, and all participants were required to arrive by 1 February. The event was to commence on Sunday 6 February 1910, and end the following Sunday.

Twelve pilots and eighteen planes were officially entered:

>   Jacques Balsan (Blériot), #3, #15, and #16
>   Hubert Le Blon (Humber-Blériot), #4
>   *Madame* Raymonde de Laroche (born Élise Raymond Deroche)
>   (Voisin), #18
>   Arthur Duray (Farman), #6
>   Jean Gobron (Voisin), #5, and #17
>   Hans Grade (Grade), #8
>   Gabriel Hauvette-Michelin (Antoinette), #9
>   Hubert Latham (Antoinette), #1, #12, and #13
>   René Métrot (Voisin), #10
>   Adam Mortimer Singer (Farman), #7
>   Frederick van Riemsdijk (Curtiss), #11
>   Henri Rougier (Voisin), #2, and #14

The practice period was not without incident. All three of Latham's

Hans Grade piloting his Monoplane over the course at Heliopolis, where he and his little machine were crowd favourites. Even in the pioneering years this is about as basic an aeroplane possible. It was powered by a 30 hp air-cooled V-4 engine, also built by Grade. *(The Royal Aeronautical Society, National Aerospace Library / Mary Evans Picture Library)*

Antoinettes had been damaged during the transport between Marseilles and Alexandria. Two could be repaired, but a new machine with a big 16-cylinder 100 horsepower engine, had damage to both engine and wings and was not repairable on site. Then, during a test flight on 27 January he lost control of his plane at an altitude of 45 metres. In the crash he was thrown out of the plane, but fortunately escaped with only cuts and bruises. Adam Mortimer Singer, son of the sewing-machine inventor, was not so lucky when he was testing his new Farman on 1 February. He stalled during a turn and side-slipped at a steep angle to the ground from a height of 35 metres. The wings took much of the impact, but he broke his right thigh in three places and injured his back. He did eventually recover, but never piloted an aeroplane again. However, not all was lost, for he married his nurse Miss Madeline Aline Pillavoine on 18 October 1913.

The following is a summary of the flying during the Meeting, a full account can be found elsewhere.[11] Whilst prize winning heights, distance and speeds seem unimpressive today, at the time they were respectable, but not record breaking.

The opening day of the Meeting, Sunday 6 February 1910, was reportedly a perfect day for flying, with a clear sky and no wind. The event was well attended, both *tribunes* were full and the public enclosures packed, the crowd being estimated to be 40000.

> Every balcony and roof over-looking the course was fringed thick with humanity: while the babble that arose from the crowd outside where motors, carriages and people were inextricably herded together, in conjunction with the tuning up of the mammoth machines within the aerodrome produced a volume of sound than can better be imagined than described. The scene of the exodus

along the long dusty roads between the capital and the Oasis reminded one of nothing so much as the rush between London and Epsom on Derby Day. Every motor, carriage and antediluvian bus was in requisition, every tram as it passed reminded one of a swarm of bees, the railway station at Pont Limoun was a seething mass of humanity; and there must have been many hundreds of others who tramped all the way out and all the way back again.[12]

Flying was to start at 2 pm, shortly after which Balsan flew two laps of the course. When he came into land a horse was frightened by the noise and ran over a Mr Tarihaki, who was taken to the hospital by the ambulance service. Following this excitement, the *Khedive* arrived, and joined his family and several government members in the grandstand. The band of the British 7th Dragoon Guards struck up his anthem when he arrived. Flying then resumed. Rougier easily won the three daily prizes for distance, speed and height, the first with 65 kilometres (13 circuits), the second with a height of 195 metres, while his time for the 10 kilometres in the speed test was 9 min 30 sec reaching 60 km/h on occasion. Other good distance flights were made by Balsan, 44 kilometres; van Riemsdijk, 24 kilometres; Métrot, 18 kilometres, who also climbed to 40 metres.

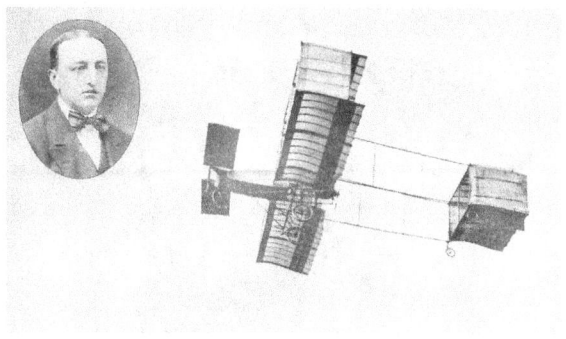

Biggest prize winner (91000 Francs) at Heliopolis was Henri Rougier piloting a Voisin.

> The painful sensation of that afternoon was of course that provided by Mr. Gobron whose Voisin when in the midst of flight, dropped suddenly to earth leaving in its track the bright flash of an explosion and a volume of smoke. Panic was in the air immediately… Motors dashed out across the desert, the ambulance men dashed backwards and forwards in their hopes to render first aid, the crowd stood open mouthed and breathless till there flashed the signal that reassured them. Then after a few seconds came tearing back the motor with the plucky aviator aboard, and instantly hats were waving in the air and shouts of welcome came from every throat as he was piloted up to the Royal Loge where Effendina himself gave him personal congratulations on his safety.[13]

After the flying was over for the day, all flights were to end by 6 pm, the Khedive made a tour of the hangars where the flyers showed their planes. He undoubtedly further commiserated with Gobron whose Voisin was too badly damaged by the engine fire to be repaired. Afterwards the crowd made their way home. The resulting traffic jam lasted until midnight, when trains and trams had cleared the last visitors.

A strong breeze interfered considerably with the flying on Monday, there were no flights until late in the day. The impatient crowd, as big as the day before, were kept amused by an Italian orchestra. Balsan was again first to fly, but crashed after three laps. Rougier and Hauvette-Michelin also flew. Grade, who was becoming a crowd favourite in his little monoplane, won the daily speed and distance prizes with rather modest results, while Rougier reached 219 metres to win the altitude prize.

Tuesday was a blank day, due to high winds causing a minor sand storm, which persisted all day and prevented any flying. In the evening Balsan went up in his racing Blériot, although after one circuit he crashed violently escaping with a few scratches. He was now down to just one remaining Blériot.

Known as *la femme-oiseau* in the French press, *Madame* de Laroche completed the necessary flights at Heliopolis to receive a pilot's license. *(LoC)*

Wednesday brought better weather, sunshine and almost no wind, and another big crowd. During the morning *Madame* de Laroche flew four laps as part of her qualification flights for her pilot's licence. Latham finally managed to get one of his machines in working order and made a short flight. All the other pilots except Le Blon made flights during the day, and at one time there were four planes in the air at the same time. Balsan, recovered from his crash on the day before, won the daily speed prize. Latham won the altitude prize and Métrot the distance prize. There was only one accident, Hauvette-Michelin having to descend suddenly, just missing the hangars.

Thursday was practically a blank competition day, as the wind continued strong. However, *Madame* de Laroche and Le Blon both made their final qualifying flights for their *Aéro Club de France* brevet. Laroche was awarded brevet No. 36, the first female holder, and Le Blon No. 38, both post-dated to 8 March 1910. As an added bonus *Madame* de Laroche won the daily distance prize with four circuits of the course. Rougier continued his winning streak with the daily height prize for an altitude of 48 metres. For the crowd it was a particularly disappointing day. This was the day originally planned for the cross-country *Boghos Pasha Nubar Cup*, worth a total of 12000 francs, a race to the Cheops Pyramid and back. It was called off at a late stage, even after officials were put in place along the course. The wind was strong and the pilots would have had to fly over populated areas with few opportunities for emergency landings. Unfortunately there was no way of informing the public about the cancellation, so huge crowds had gathered at the pyramids and along the course. The prize was not awarded, as there was no opportunity to reschedule it. Perhaps, this was just as well, as the one way straight-line distance to the Great Pyramid from the course was about the same as the greatest distance flown during the Meeting. The prize was not withdrawn, finally being claimed, not without some drama, in December 1913.

Gusty winds also prevented any long flights on Friday. However, in the morning Métrot made a flight with a lady passenger, a *Mlle* Solange Parenty. The winds increased in strength and there was no more flying until later in the afternoon when Le Blon made a four-lap flight. Later he made several flights, breaking both the 10 kilometre and the 5 kilometre speed records. His best time over 5 kilometres was 4 min 2 sec, corresponding to 74.4 km/h, and 8 min 8 sec over 10 kilometres (73 km/h). These won him the daily distance and speed prize. Rougier reached 255 metres, which would be the highest altitude of the Meeting. Latham's run of bad luck continued, when he finally got his engine to run he immediately touched the ground with a wing, breaking it.

On Saturday there was no flying due to strong and gusty winds.

Sunday 13 February was the final day of the Meeting, it brought clear weather, but with strong winds, and huge crowds, larger than those on opening day. Latham had made progress with his engines, and towards the end of the morning he managed to fly 13 kilometres — before crashing again, for the fourth time during the Meeting! Six other pilots flew during the afternoon. Balsan took back the five-kilometre speed record by scoring a time of 4 min 1 sec. At 6 pm Grade tried to take the daily altitude prize and reached 95 metres, but Rougier immediately replied by reaching 116 metres.

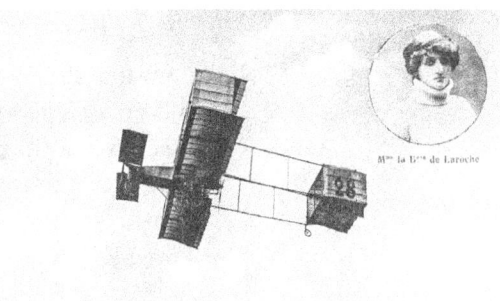

After Heliopolis *Madame* de Laroche returned to France still flying her Voisin. As *la femme-oiseau* she received much publicity, this is one of many postcards bearing her image.

Hubert Latham suffered many problems at Heliopolis but not one to be easily defeated, he took his Antoinettes to the USA later in 1910. He is seen here landing a V-8 Antoinette VII at Tanforan Park, San Francisco, in January 1911.

In summing up Jasper Kemmis wrote:

> Now that the so-called Great Aviation Meeting at Heliopolis is over it is possible to review it in a comprehensive manner and we feel assured that the prevailing note, echoed in the minds of the large majority of those who witnessed the flying, is disappointment. ... Unwholesome rumours have pervaded the air and fantastic excuses have been in circulation to account for the very poor exhibition of flying at Heliopolis but none sufficiently convincing to satisfy the thinking individual who puts his hand deep in his purse and rightly expects something for his money in proportion to the expectations held out to him by advertisement and innuendo.
>
> Admitted that Aeroplaning is in its infancy the general verdict is that, however well intentioned the organizing Committee may have been, no future Aviation Meeting at Heliopolis can hope for success unless an expert Committee can be formed to superintend the details and to frame Bye-laws which will hold professionalism in check and adequately secure the interests of the man who supports the show by paying to witness it.[14]

There were to be no further meetings in Egypt until February 1914, again at Heliopolis. Some winter flying during following seasons may have taken place, but it failed to register in the aviation press. What did catch the press' attention was a 'race' from Paris to Cairo, and two flights up the Nile from Cairo to Khartoum.

Paris — Constantinople — Cairo Flights

Pierre Daucourt, pictured on the cover of *La Revue aérienne*, 10 December 1913.

In the Autumn of 1913 the *Ligue Nationale Aérienne* challenged pilots to attempt to trailblaze the air route from Paris to Cairo, prizes up to 25000 francs were offered. Any flight would require crossing central Europe, an extremely sensitive area politically, and the Balkan mountains before crossing into Asia at Constantinople. After crossing Turkey the 3000 metre Taurus Mountains barred the way to Syria and down to Cairo.

The first pilot to accept the challenge was Pierre Daucourt. With his mechanic J. Roux, he took off from the parade ground at Issy in Paris on 20 October 1913, in a Gnôme powered Morane-Borel monoplane and headed east. They planned to fly via Germany, Austria-Hungary, Rumania, and Bulgaria, to Turkey. At times Roux took to the rails to lighten the machine. After some adventures, they reached Constantinople on 9 November, there taking a rest for both man and machine. On 17 November they crossed the Bosporus into Asia. The route across Turkey followed the Constantinople-Baghdad railway towards Konya in central Turkey. From this point on the terrain became increasingly mountainous, Daucourt leaving Roux to make his way onward by rail. By the end of November Daucourt had reached the Taurus Mountains and, somewhere near Pozantı (alternatively Bozantı) on 26 November, he crashed due to an engine failure. The undercarriage collapsed, and the propeller and engine were damaged. Left under guard over night, the machine somehow caught fire and the left hand wing destroyed. The attempt to reach Cairo was ended. Daucourt returned to France, Roux continuing on to Cairo by rail.

Two more pilots attempted the flight, both reaching Cairo within days of each other. First away on 10 November were Marc Bonnier and his mechanic, Joseph Barnier, flying a 80-hp Gnôme Nieuport monoplane. Following them was Jules Védrines, flying a 80-hp Gnôme Blériot XI-2 solo. Both parties had rather vague plans to fly East but, when they learned of Daucourt's crash, decided to try for Cairo. Making their way across Europe, Bonnier and Barnier arrived in Constantinople on 5 December. Védrines left France on 20 November, whilst nominally under arrest in the French border town of Nancy for flying in a military no-fly zone near the sensitive frontier with Germany, reaching Constantinople shortly after Bonnier. Bonnier was having engine problems and was unable to leave Constantinople before Védrines, who left on 19 December, despite having agreed with Bonnier to continue their flight to Cairo in company 'for the glory of their country'.[15] Bonnier and Barnier were able to get away a day later.

Jules Vedrines.

Védrines continued across central Turkey, following Daucourt's route, reaching Konya around 20 December. His exact route rather depends on which account you read, but from Konya he appears to have reached the coast at Silifke (over 100 kilometres west of Adana). This was a more direct and lower route than through the Taurus passes, but required either a long detour around the coast or a long sea crossing to reach Syria. He decided to fly 300 kilometres across the Mediterranean to either Tartus or Tripoli in Syria, crossing the eastern tip of Cyprus enroute. Even in 1914 such a long over water crossing was risky, accordingly the French Navy had stationed the cruiser *Bruix* to provide assistance if required. Védrines pressed on, reaching Beirut on 25 December and Jaffa two days later, breaking a wheel on landing. This was repaired and at 8 am on 29 December he left for Kantara on the Suez Canal to refuel, then on to Cairo arriving over Heliopolis shortly after 1 pm, where a small crowd had gathered to greet him. Landing on the polo ground, he was greeted by a representative of the *Khedive* and by the French Agent, who placed a laurel wreath bound with the *Tricolore* around his neck.

Meanwhile, Bonnier and Barnier were following close behind. Landing at Eskişehir on 20 December he damaged his propeller, delaying his arrival at Konya by several days. Then on through the Taurus mountain passes to Adana, reaching Beirut on 29 December, and Jerusalem on 31 December.

Marc Bonnier at Jerusalem.

Leaving the Holy City on New Year's Day they landed at Port Said for fuel then on again for Cairo. They were guided to Heliopolis by Marc Pourpe flying his Morane-Saulnier monoplane, landing on the polo ground shortly after 17.00, where they were greeted by Védrines and several other aviators.

Now, Védrines had a reputation for a short temper which led to an incident. *The Aeroplane*'s editor, C.G. Grey, tells the story in his inimitable style.

> Jules Védrines continues to earn the epithet of the "flying apache", which was long ago bestowed upon him. The story of his recent adventures is briefly this: After forsaking their flight across Asia Minor, owing to the destruction of their Borel, MM. Daucourt and Roux, knowing that they were being followed by their fellow-countrymen, left stores of petrol for the use of MM. Bonnier and Védrines. The first to arrive was Védrines on his Bleriot, and in his usual manner he made himself offensive to the ever-courteous Turk. The Turks, somewhat naturally, refused to recognise the fact that Védrines owned the earth as well as the sky, and consequently no petrol was forthcoming and much delay occurred before any arrived. When, however, M. Bonnier, who is one of the most charming of men, arrived on his Nieuport, the petrol was promptly handed over to him, and there is no doubt but for a smashed propeller, he would have arrived at Cairo before Védrines.
>
> Before Védrines arrived at Cairo he had become fully convinced that M. Roux had been the prime cause of his not obtaining petrol, and when that gentleman advanced to greet him on his arrival at Cairo the apache promptly smote him. A somewhat humorous situation then arose, for the undoubtedly aggrieved M. Roux, apparently forgetting that he was a gentleman and could not fight with *canaille*, sent his seconds to Védrines, whereupon Védrines, on the strength of being a member of the Legion of Honour, refused to fight M. Roux, who, he said, was a mere civilian. However, becoming indignant

Marc Bonnier's Nieuport after landing at Jerusalem 31 December 1913, with port wheel deep in a pot hole. The machine, and crew, was undamaged. *(LoC)*

with the behaviour of the *Ligue Nationale Aérienne*, which had finally disowned him because of his conduct, Védrines telegraphed a challenge to Dr. Réné Quinton, the president of the *L.N.A.*, who, curiously enough, seems to have accepted. Dr. Quinton is a specialist who apparently has an idea that subcutaneous injections of sea-water are a cure for all mortal ills, so one might suggest that the duel should be fought with hypodermic syringes at 30 paces. Meantime, Védrines had succeeded in embroiling himself with the French diplomatic circle at Cairo, and as a result the whole of the French colony cancelled all the invitations which he had accepted. Consequently Védrines took up his residence at the Pyramids Hotel, as far as possible from the French official quarters. All of which gives strength to the old proverb that one cannot make a silk purse out of a sow's ear.[16]

The Nieuport on level, if rough and rocky, ground at Jerusalem. It has a competition number 22 beneath the starboard wing and a later photograph in Alexandria shows that there was a crescent and star painted below the port wing. *(LoC)*

Grey's final shot, in the 5 March 1914 edition, 'It is understood that M. Roux, failing to get satisfaction from Jules Vedrines by means of pistols for two, coffee for four, is taking police court proceedings against him for assault.' Védrines had decamped for France, by sea, in early February leaving his Blériot behind.[17]

The original caption: Védrines' special Blériot XI-2, baptized *Nénette* (from the nickname of his eldest daughter Jeanne Noémie), in his "campsite" at the foot of the pyramids, towards the new year 1914. This Blériot XI-2, which Védrines had acquired from Roland Garros was piloted from the rear seat, the front one being usually concealed by a sheet metal plate, absent in the photo. On the back of the photo are inscribed the names "Paillard; Védrines; Jules Munier; Henri Mosseri (child)". Arène Claude (Jules) Munier was a journalist in Cairo. *(via Trois-Chênes-Coupés / Wikipedia)*

The Turkish Constantinople to Cairo Flights

Prior to setting out from Yeşilköy (Constantinople) for Cairo on 8 February 1914 the two Turkish crews are photographed in front of Blériot IX-2 *Muavenet-i Milliye*. Left to right: *Mülâzım-ı Evvel* Sadık Bey (Blériot observer), *Yüzbaşı* İsmail Hakkı Bey (Deperdussin observer), *Yüzbaşı* Fethi Bey (Blériot pilot), and *Mülâzım-ı Evvel* Nuri Bey (Deperdussin pilot).

Turkish army pilots had been very helpful to the French pilots, and were convinced that they too should fly to Cairo. The attempt had political blessing as Turkey had recently lost its large European provinces, and was eager to regain its former influence in the Arab areas and Egypt. A Deperdussin TT, *Prens Celaleddin*,[18] and Blériot IX-2, *Muavenet-i Milliye*, were selected to make the flight. The Deperdussin was piloted by *Mülâzım-ı evvel* (Lieutenant) Nuri Bey with observer *Yüzbaşı* (Captain) Ismail Hakkı Bey. *Yüzbaşı* Fethi Bey was selected to pilot the Blériot with *Mülazım* Sadık Bey as observer. The plan was to follow the route of the French fliers — from Yeşilköy (Constantinople), via Eskisehir to Konya, crossing the Taurus Mountains from Karaman to Adana, and then to Aleppo, thence to Beirut, Damascus, Jerusalem and on to Cairo. Both aircraft were fitted with extra fuel tanks and fuel and spare part depots were placed in Konya and Aleppo. Two navy engineering officers were sent ahead in automobiles to arrange the depots and give assistance if any mishaps occurred. Poor weather delayed the planned departure until 8 February 1914.

Both crews crossed the Taurus range. Fethi arriving at Beirut on 15 February then continuing on to Damascus by the 24th. Nuri, after some delays before the mountains and having to cross them flying solo, reached Tarsus on 20 February waiting there for Hakkı, who had to cross the mountain passes on horseback, to join him. They struggled with engine troubles and storm damage, but reached Damascus on the 27th. Here bad news awaited them.

Yüzbaşı Fethi and *Mülazım* Sadık had left Damascus that morning, but the Bleriot crashed near Lake Tiberius (Sea of Galilee). Both men were killed. The bodies were brought to Damascus by train and buried in the graveyard of the mosque of Salahuddin Ayyubi at Damascus. An estimated 100000 people, including their fellow aviators, followed the funeral procession.

Nuri and Hakkı now flew the Deperdussin down to Beirut then on to Jaffa, arriving on 9 March. Here some engine repairs were carried out and a replacement propeller fitted. Despite a strong gusty wind Nuri insisted on leaving on the 11th, making a successful take-off from the beach. Turning down the coast whilst over the sea, a gust caught the aircraft and it crashed into the sea near the beach. Attending personnel and spectators rushed to help and rescued Ismail Hakkı. Nuri, however was dressed in a heavy leather flying jacket which pulled him down before help could arrive. The next day the body of *Yüzbaşı* Nuri was brought to Damascus and buried alongside the crew of the Blériot. *Yüzbaşı* Hakkı returned overland to Constantinople collecting *en route* enough donations to purchase a new Blériot. Rushed through the factory it reached Yeşilköy on 7 April, where it was named *Edremit*.

Meanwhile a second Bleriot *Ertugrul*, piloted by *Yüzbaşı* Salim Ilkucan with *Yüzbaşı* Kemal Bey as observer, was despatched to continue the flight to Cairo. They left Yeşilköy on 9 March, following a coastal route. The Blériot crashed on 14 March into woods near Küçükkuyu, on the Turkish coast north of the island of Mytilene (Lesbos). Although the aircraft was damaged beyond repair, the crew escaped unhurt. *Edremit* was sent by steamer to Beirut, from where *Yüzbaşı* Salim and Kemal were able to complete the flight to Cairo. Arriving at Cairo on 9 May 1914 to a great welcome. After some demonstration flights over Cairo, they flew to Alexandria on 15 May. Selim, Kemal, and *Edremit*, returned to Constantinople by steamer.

Pilot *Yüzbaşı* Salim Ilkucan (L) and observer *Yüzbaşı* Kemal Bey (R) in front of *Edremit* at Jerusalem judging by the condition of the landing ground. *(LoC)*

Crowds surrounding Turkish Bleriot XI-2 *Edremit* which Salim and Kemal had flown to Jerusalem on 1 May 1914. *(LoC)*

The Final Season

Marc Pourpe flew for the Blériot company in India, Cambodia, and French Indochina. By December 1913 he had left Blériot and was now flying a Morane-Saulnier Type G two-seater monoplane.[19] He arrived in Egypt late in 1913, together with long time mechanic Raoul Lufbery, announcing he would fly to Khartoum, but first he made a flight around the pyramids on 17 December. There was a strong wind blowing, and Pourpe wrote that he had intended only to make a trial flight.

> However, the wind still blew terribly. The shed behind me vibrated like a violin. On the ground everything gave way under the squall, but the motor started first at 800 turns, then boomed to 1200. "Bah!" I said to myself, "let's go on!" And on I went!
>
> I went onward, but not quickly. I left the ground in 15 metres and made a round of the aerodrome. The motor snorted and snorted! At the last turn, from the altitude I perceived away below three sharp pinnacles, which seemed to beckon me. One look at my fuel gauge. I had just 18 litres! But what of that! Ghizeh was only a lap! I had swung around to the south-west. Immediately I felt that I would have to go higher to escape an extremely violent air current. Perhaps it would be calmer higher up. I rose to 1100 metres, but it was useless, the wind was against me and my speed did not increase. I descended to 800 metres, and stayed there.
>
> Although my propeller turned madly it seemed to me that I was not progressing. The gasolene [sic] supply lowered very quickly. Fortunately the immense green plains soon gave way to the large sand plains, and I saw some white tents of English encampments at the north of the 'Mena House.'
>
> The Pyramids! I nearly failed seeing them – they looked like little cones that a child could have let fall from a box of playthings! And the Sphinx? I searched for it — asking myself where on earth it might be. Then I distinguished a vague stony spot on the sand. Evidently it was *HE* — the Sphinx...[20]

Marc Pourpe on the cover of *La Vie au grand air* 17 January 1914.

The flight back, assisted by the wind, took only six minutes, the outward flight had lasted 22 minutes. After landing he immediately submitted his claim for the *Boghos Pasha Nubar Cup*, unclaimed since the Heliopolis Meeting in 1910, and flew directly into a controversy.

Flight in its edition of 3 January 1914, announced that, 'Some time ago the Boghos Nubar prize of £400 was offered for the first aviator who should fly from Heliopolis, round the Ghizeh and Zakkareh Pyramids and back to Heliopolis.

Marc Pourpe's Morane-Saulnier at Cairo. *(AWM C02796)*

This flight was carried out by Olivier on his Farman biplane the other day, and it is announced that he has been awarded the prize.' The only problem was that Olivier's flight was after Pourpe's.

The Aeroplane provided some clarification in its 22 January edition.

> When M. Marc Pourpe made the flight round the Pyramids the other day, he naturally went to the Club and claimed the Bogos prize, but was put off by the statement that as the prize had been unclaimed so long, the donor had given permission for it to be used for other purposes. Apparently M. Pourpe was not satisfied with the explanation, and began to inquire into the matter, whereupon, apparently by some arrangement, Olivier went and did the flight over again, and it was announced that he had won the prize; M. Pourpe's flight being unofficial. Naturally, M. Pourpe is proceeding further with the matter, and the French Aero Club will probably take it up. There is no doubt that he did the flight, as it is sworn to by the whole of the British camp at the Pyramids and by the whole of Cairo, as he flew over the city at about 11 am that morning.

Olivier had arrived in Cairo with a 80-hp Gnôme Henry Farman claiming to have been flying during the recent Balkan Wars. Whilst there is no evidence of his participation, Olivier named his new Farman *Balkanic*, to the ire of the Egyptians who were at the time still supportive of the Ottoman cause. He crashed the Farman whilst taking off at Abbassia, Cairo, on 30 January 1914. Carrying two passengers, having just left the ground the crowd surged onto the field. He deliberately crashed the machine to avoid them. The machine was seriously damaged, but there were no injuries to either pilot, passengers or crowd.[21]

Leaving the controversy behind, Marc Pourpe then set off on his flight to Khartoum. How the problem of the prize was ultimately settled is, regrettably, unknown. His flight to Khartoum was relatively untroubled. Leaving Cairo on 4 January he arrived at Khartoum on 11 January, having flown some 2505 kilometres (1556.5 miles) in five stages, in a total of 25 hours flying time at an average of 62.25 mph (100 km/hr).[22] After some demonstration flights over Khartoum he set out on a leisurely return flight to Heliopolis, arriving on 23 February. By mid-March Pourpe had returned to France and was entered in the Monaco Aerial Rally.

Frank McClean's Short S.80 on the bank of the Nile near Mehaiza, some 80 km south of Luxor, 28 February 1914. (Philip Jarrett)

Not so leisurely was the trouble-plagued flight to Khartoum by Francis (Frank) McClean, and Alec Ogilvie, on the Short S.80 floatplane. McClean had left Luxor on 16 January, following the Nile, he finally arrived in Khartoum on 22 March. During this epic McClean had suffered 13 engine breakdowns, and three heavy landings leading to repairs to the aircraft. His arrival in Khartoum was a testament to perseverance. 'Sir Frances Maclean [sic] flew a seaplane along the Nile to Khartoum in 1914, having literally dozens of forced landings en route, and in many places along the Nile all aeroplanes are still called "Macleans"!'[23] The Short was dismantled and, together with Frank McClean and party, taken by train back to Alexandria, thence by ship to England.[24]

Additional to the Khartoum flyers, Cairo and Heliopolis were host to several more pilots by January 1914. The Paris-Cairo flyers, of course, plus Oswald Watt (an Australian flying an 80-hp Blériot, and a Union Jack from the roof of his hanger), the accident-prone Louis Oliver (Henry Farman), and another Henry Farman pilot/owner, the Italian Leonardi.[25] Then there was Jacques Schneider (of Schneider Cup fame), the aeronaut Maurice Bienaimé, Charles de Lambert, Paul Tissandier (both de Lambert and Tissandier had been taught to fly by Wilber Wright in 1908), and Bernard Dufresne. This party appears to have been involved with a *Hydroglisseur*, or airboat, designed by de Lambert and Tissandier, which they drove up the Nile as far as Khartoum.

Finally, there was a low-key meeting at Heliopolis between 19 and 22 February, the first since the 1910 meeting. This was mainly a showcase for looping flights by Maurice Chevilliard (Henry Farman) and Maurice Guillaux (Blériot). Both pilots made flights, Guillaux making twelve consecutive loops on one flight. Olivier was offering passenger flights, having found another Henry Farman to replace *Balkanic* which was eventually rebuilt at a reported cost of 6000 francs. Towards the end of the meeting, after watching Guillaux make a loop in a strong wind, Olivier attempted a flight but, at 30 feet, was caught between two dust-storms blowing in opposite directions making a hard landing.

Well, that concludes a brief overview of aviation in Egypt prior to the Great War. The next aircraft to fly over the sands of Egypt were from the Royal Flying Corps followed, a few days later, by Nieuport floatplanes of *l'Aéronautique maritime*.

CHAPTER 2

On Land and Sea

This book is essentially about naval aviation in support of the war on land in the Middle East. Hence, at this point, an introduction to land and naval forces also involved is probably a good idea. The aerial services will be dealt with in a subsequent chapter.

Ottoman Forces

The Ottoman Navy of 1914 was in the midst of a modernisation, having ordered two dreadnought battleships, *Sultan Osman-ı Evvel* and *Reşadiye*, from Britain. These were paid for by public donations made to the Ottoman Navy Foundation. The confiscation of the two completed ships by the British government at the outbreak of war led indirectly to a declaration of war.[26] The German Navy's *Mittelmeer-Division*, comprising the battlecruiser SMS *Goeben* and light cruiser SMS *Breslau*, escaped British pursuit in August 1914, passed through the Dardanelles and entered service in the Ottoman Navy as *Yavuz Sultan Selim* and *Midilli*, respectively. *Konteradmiral* Wilhelm Souchon assuming command of the Turkish fleet.

Konteradmiral Wilhelm Anton Souchon, Commander *Kaiserliche Marine Mittelmeerdivision*, later Commander in Chief Ottoman Navy, 1914 to September 1917.

SMS *Goeben* and SMS *Breslau* leaving Messina 6 August 1914 bound for Constantinople. From a German patriotic post card.

İsmail Enver Paşa, Ottoman Minister of War and Chief of the General Staff, 1914 to October 1918.

On 29 October 1914 ships of the Ottoman Navy, under Souchon, shelled Russian ports in the Black Sea. Russia replied by declaring war on 1 November 1914 and Russia's allies, Britain and France, then declared war on the Ottoman Empire on 5 November 1914.

Except for three sorties, the Turkish Navy was restricted to Black Sea operations for the remainder of the war. The first sortie by a single torpedo boat, *Demirhisar*, in March 1915 directly struck against a British seaplane carrier, and will be detailed later. The second was by the Turkish destroyer *Muavenet-i Milliye* over the night of 12/13 May 1915. The destroyer managed to slip past British patrols, torpedo and sink the battleship *Goliath*, then slip back through the Dardanelles to a well deserved triumphal welcome. The final sortie was by *Yavuz* and *Midilli*, on 20 January 1918, resulting in the sinking of several ships but also the loss of *Midilli* and damage to *Yavuz*. Attempts to sink the battlecruiser would involve another British seaplane carrier, again to be related later.

On the outbreak of war the Turkish army was still recovering and re-equipping after the Balkan Wars of 1912–1913. Appointed Minister of War and Chief of the General Staff in January 1914, *İsmail Enver Paşa* (Enver Pasha) controlled the Turkish Army until mid-October 1918 when he was dismissed from his posts.

In 1914 all male subjects of the Ottoman Empire aged between 20 and 45 were liable for military service. They came mostly from rural peasant backgrounds and had little or no formal education and little knowledge of the world beyond the nearest market town. Even many of those recruited from the cities were illiterate. Life in the Ottoman Army was harsh. Ottoman Army officers expected blind obedience from their men and strict discipline was imposed to ensure that they got it. The recruit's background, however, gave them a stoic outlook on life and the ability to endure great hardships. Training was usually limited to whatever the officers in the unit to which recruits were posted were able to provide. Non-commissioned officers were in short supply, and any recruit with a modicum of learning was quickly promoted corporal or sergeant.

The Officer Corps was generally well educated and trained, some officers having been sent to Military Academies, or on Staff courses, in various European countries. Writing of the Palestine campaigns, Dr Erickson notes that, 'The Ottoman and German commanders in Palestine displayed great skill in creating organizations that could manoeuvre rapidly and decisively. At the tactical levels they demonstrated that they knew how to employ effectively the forces under their command in a highly fluid operational and tactical environment. This capability eroded in the final months of the war when logistical support failed to keep up with tactical demands.'[27]

Mehmetçik (Little Mehmet) was an affectionate Turkish nickname for Ottoman (Turkish) soldiers. Nicknames were common in the allied forces as well, two standout for the Turkish soldier — Johnny Turk and Abdul — the latter was a particular favourite amongst Anzac forces at Gallipoli. Neither can be seen as an insult, just an indication of respect from one soldier to another.

> *We* will judge you, Mr. Abdul,
> By the test by which *we* can —
> That with all your breath, in life, in death,
> You've played the gentleman.[28]

Speaking of supply lines, it is necessary to understand that Turkey's armed forces relied on Germany for most of their equipment and military supplies. The land route from Germany to Constantinople was long and subject to interruption due political and military events en route, especially in the early years of the war. From Constantinople, sea transport being unavailable, the Baghdad Railway became perforce the major supply route to Palestine and Mesopotamia.

Unfortunately, there were two uncompleted sections through the Taurus and Amanus mountains requiring transfer of goods on to road transport at each incomplete tunnel. The line from Constantinople (Haidar Pasha) to the Taurus terminated at Pozantı, then a road crossed the mountain range, through the Cilician Gates or Gülek Pass, to rejoin the railway at Tekeli, south of Hacıkırı (Hakkâri) station, a tortuous route of nearly 80 kilometres. A shorter, but more difficult, road paralleled the railway from Pozantı directly to Hacıkırı. Between the Taurus and Amanus mountains the track ran through a relatively flat area bisected by the rivers Seihun and Ceyhan. Both these were bridged; the Seihun at Adana and the Ceyhan at Chicaldere.

Turkish troops passing through a town in Palestine circa 1916. The men are all wearing the khaki woollen kabalak whilst the NCO wears a fez. *(Gunter Hartnagel Collection)*

German trucks, by-passing the Taurus railway tunnels, passing through the Cilician Gates during 1916. Although well-constructed, the road was narrow, winding, subject to rock falls, and often snow-blocked during the winter. *(Gunter Hartnagel Collection)*

From Chicaldere the railway continued to Toprakkale Junction from where it continued into the Amanus mountains until reaching the next incomplete tunnel. A branch line ran south from the junction to Alexandretta (İskenderun). Two possible road routes bypassed the mountain tunnel. One from Toprakkale through the mountains to rejoin the railway at Islahiya. The other ran from Alexandretta through more open country to rejoin the railway at Aleppo, this was a longer but easier route. Work continued on the tunnels throughout the war, using forced labour of PoWs and Armenian prisoners. By mid-1917 the tunnels through both Taurus and Amanus ranges were complete, but only for a 600mm gauge *feldbahn* system. This was then employed to carry war supplies using a fleet of about one hundred narrow gauge engines. The Taurus tunnels were completed for standard gauge in October 1918, too late to be of much benefit to the German and Turkish armies.

Once through the mountains difficulties were not at an end. The main line was built using standard gauge of 1.435 metres (4 feet 8½ inches). At Rayak, north of Damascus, this changed to 1.05 metre (3 feet 5.34 inches) gauge for all lines laid south of the junction, including the Hejaz Railway. Therefore, in addition to having been off loaded onto road transport at each incomplete tunnel on the Baghdad Railway, all stores and equipment had to be transferred again at Rayak. The opportunities for damage, wastage and theft at each of these way points need not be elaborated upon.

Military Operations 1914 to 1917

Palestine, the Red Sea, and Mesopotamia were all considered to be sideshows to the main event, the war in France. Secondary theatres where events would not affect the overall progress of the war. How wrong this judgment was. The three theatres mentioned above proved to be significant in bringing an end to the war as a whole. In bringing about the final collapse of the Ottoman Empire, they hastened the Armistice. They also, for better or worse, brought into being the states of the modern Middle East. Some sideshow!

A glance at the accompanying map will show the political complexity of Ottoman Palestine and Syria. All British accounts of the period simply refer to Sinai, Palestine, and Syria — which convention will be followed in this book. The border between Sinai and Palestine, the Turco-Egyptian Frontier, was established in 1906 as a line running roughly from Rafa to the Gulf of Akaba. But where was the dividing line between Palestine and Syria? Colonel A.P. Wavell in his *The Palestine Campaigns*,[29] describes Palestine thus, 'it will be taken here to include the territory from Dan (Banias) to Beersheba and from the Mediterranean to the Hejaz Railway.' The northern boundary, therefore, runs roughly east-west from the railway through Banias to the coast near Tyre, or approximately along the boundary between the Sanjaks of Acre and Beirut. South and east of Akaba was the ill defined area known as Hejaz.

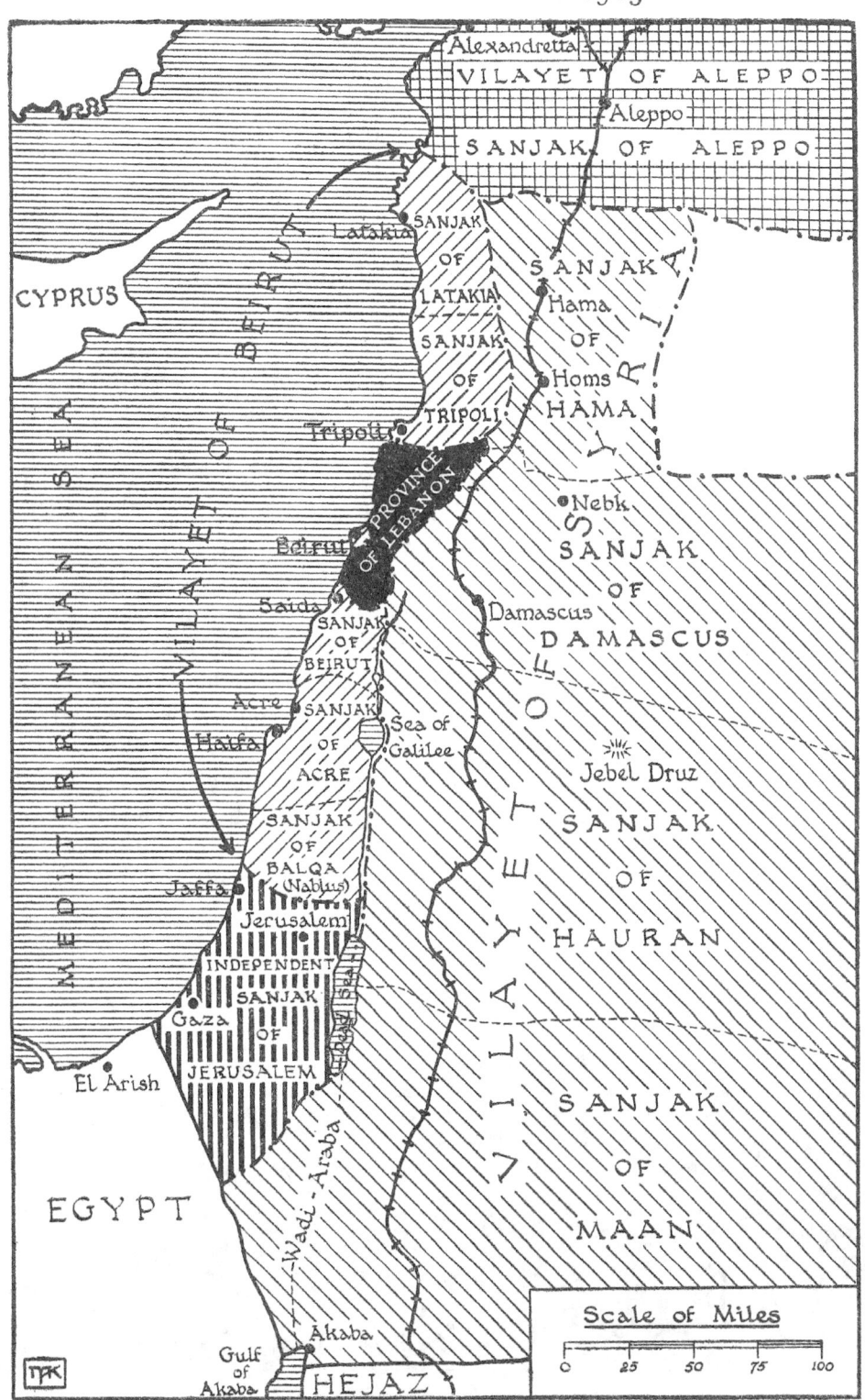

Political map of Palestine and Syria prior to WW1.

Colonel Wavell also divided the campaign into three main phases.

1. Sinai — The Defence of the Suez Canal to the Second Battle of Gaza, April 1917.
2. Palestine — The Third Battle of Gaza to the Taking of Jerusalem, December 1917.
3. Syria — The Final Offensive.

The first Phase might also be divided in two parts — The Defence of Egypt and the Suez Canal, and the Advance across the Sinai Desert to Gaza.

The Suez Canal is one the great short cuts for maritime trade, the second being the Panama Canal. If the Suez Canal were to be closed to allied shipping all supplies and military reinforcements — Imperial Forces from India, and Australia and New Zealand (ANZAC) — would have to sail around Africa, slowing their arrival by several weeks.[30] Not surprisingly therefore, at the outbreak of the war the defence of the Suez Canal was a prime concern for the military forces in Egypt. The British officered Egyptian army was primarily employed to maintain order in the Sudan. Great care was taken not to place it in a position where it might have to confront Ottoman forces, although some volunteer and labour units served in Palestine, Syria and the Hejaz. Therefore, the actual defence devolved on mainly British forces (with French support) — army units from Britain, India, Australia, and New Zealand. Lieutenant General Sir John Maxwell commanded the land forces in Egypt from September 1914.

Throughout 1915, after the defeat of the Turkish attack on the Suez Canal, the Sinai was the scene of numerous small patrol actions. The Turkish forces sent several small units to infiltrate the British patrols and to attempt to disrupt traffic on the Canal, as in the following example.

Turkish infantry muster on the Plain of Esdraelon, north of Jerusalem, prior to the attack on the Suez Canal. *(LoC)*

Ottoman 1st Camel Regiment, Beersheba ca.1915 — note Bedouin attire. *(LoC)*

Between 17 June to 23 July 1915 mobile Turkish forces, mainly camel cavalry,[31] led by *Mülazım-ı evvel* (Lieutenant) Sırrı operated in the northern Sinai. Their intent was to carry out small nuisance raids and to plant mines in the canal. One force was discovered at Bir Waset, south of Katia, on 17 June by Lt Murray and Capt Conran of the Egypt Detachment, RFC,[32] on BE2c 1757. Conran reporting a mixed camel and infantry column two hundred strong. Over the following days the largest group seen was just 35 men. Lieutenant Sırrı probably splitting his force into small groups who were able to blend with the local Bedouin. Their main success came on 30 June when the British steamer *Teiresias* was damaged by a mine at the southern exit of the Little Bitter Lake, causing the canal to be closed for fourteen hours whilst it was swept for mines.

Otherwise the British force fell into a defensive state of mind. So much so that Lord Kitchener, visiting Egypt in November 1915, famously asked, 'Are you defending the Canal, or is the Canal defending you?' He was not being entirely fair in his criticism as the army in Egypt was viewed as a ready source of trained manpower for the trenches. As divisions, mainly Indian or ANZAC, were trained they became more valuable as replacements for the constant drain of manpower in France, than sitting in apparent idleness waiting for something to happen in a sideshow.

In January 1916, Lieutenant General Sir Archibald Murray was given command of the Mediterranean Expeditionary Force, lately evacuated from Gallipoli, and now holding the Canal Zone. This reduced Maxwell's sphere of responsibility to the troops in the Nile Delta, the Western Desert, and the Sudan, but he continued to administer martial law over the whole country including the Canal Zone. A third independent command, the Levant Base, was under the command of Lieutenant General Edward Altham. It was formed at the end of 1915 as a pool of stores of for the Gallipoli, Salonika, and Egyptian theatres. This system quickly proved unwieldy and, in March 1916,

General Murray was appointed to command of what now became the Egyptian Expeditionary Force (EEF). General Maxwell returned to the UK to be thrown into the unforgiving situation in Ireland following the Easter Rising of April 1916.

With the formation and expansion of the EEF a slow advance across the Sinai towards El Arish commenced. Initially held at Katia on 24 April, the British advance continued to the decisive Battle of Romani, 3–5 August, a British success against a strong and well led Turco-German force. The advance continued through El Arish, and Rafa, until halted by the fierce defence of Gaza during the First and Second Battles of Gaza, 26 March 1917 and 19 April 1917. Murray was replaced as commander of the EEF by Lieutenant General Sir Edmund Allenby on 28 June 1917. Allenby was ultimately able to break the Gaza defences in November and advance steadily northward, capturing Jerusalem at the end of 1917.

The Arab Revolt

For this book we also need to consider the Red Sea and Indian Ocean, including the Arab Revolt in the Hejaz and the Defence of Aden. So, it is appropriate at this point to briefly describe the geography and organisation of the Arabian coast of the Red Sea, relying once more on Colonel Wavell.

> There is room here only for the briefest possible sketch of the geography and politics of that portion of the Arabian Peninsula which borders the Red Sea. It comprises three regions: the Hejaz, Asir, and the Yemen. The Hejaz, the holy land of the Mohammedan religion, stretches roughly from just south of the Gulf of Akaba to below Jeddah, the port of Mecca. The Hejaz railway from Damascus ends at Medina, some 250 miles north of Mecca. To the south lies the barren and rocky province of Asir, whose chieftain, the Idrisi Sayyid, was an inveterate enemy of the Turks. Further south, the Yemen stretched down to British territory at Aden, with Sana'a as its principal town and Hodeidah as its port. The ruler of the Yemen, the Imam Yahya of Sana'a, was a man of parts; he had been such a thorn in the side of the Turks that, after several costly campaigns, they had granted him a treaty on exceptionally favourable terms shortly before the outbreak of the Great War. During the war the Imam abided loyally by this treaty, and, though not actively assisting the Turks, fed and harboured the Turkish garrisons in the Yemen.

> Over all three provinces the Turks maintained a nominal but unsubstantial rule. They in fact exercised little authority outside the principal towns where their troops were quartered. At the outbreak of war their garrisons consisted of one division in the Hejaz, one in Asir,

and two in the Yemen.³³ The British blockade of the Red Sea practically isolated the troops in Asir and the Yemen. Those in the Yemen made an attack on the Aden protectorate in July, 1915, and at one time threatened the fortress, necessitating the despatch of a brigade from Egypt to drive them off. After this no events of importance took place in the southern portion of the insula during the remainder of the war. We contented ourselves with holding the fortress and made no attempt to throw the Turks out of the hinterland which they had occupied.

Emir Hussein ibn Ali al-Hashimi, Sharif of the Holy Cities of Mecca and Medina, had impeccable ancestry traceable back to the Prophet Mohammed. Selected in 1908 to take over the Sharifate, he succeeded in suppressing the anarchy then reigning in the Hejaz and bringing order and peace to the Vilayet. Emir Hussein had four sons; Ali, Abdullah, Faisal, and Zeid. Although initially supportive of the Ottomans Hussein had commenced secret negotiations with Sir Henry McMahon, High Commissioner for Egypt early in the war. He was eventually persuaded into open revolt by the arrival in Medina at the end of May 1916, of 3000 additional Turkish troops, intended to pass through the Hejaz and Asir to reinforce the Yemen. When the revolt broke out on 5 June 1916, the greater part of the Turkish garrison of Mecca was in summer quarters at Taif, in the hills south-east of Mecca. The few troops actually in the town were soon disposed of, and those at the sea port of Jeddah surrendered later in June. The force at Taif surrendering on 22 September, when under attack by Arab forces with Egyptian support.

Emir Hussein ibn Ali al-Hashimi, Sharif of the Holy Cities of Mecca and Medina.

All efforts of the Arabs against Medina failed. Thanks to the railway the forces there were reinforced towards the end of 1916. Medina held out throughout the remainder of the war. On 10 January 1919, Fakhri Pasha, under duress from his own men, surrendered the city and its garrison of over 8000 men to Abdullah, Hussein's second son. The Turkish soldiers were sent to camps in Egypt before being repatriated to Turkey. Fakhri Pasha was transferred to Malta, where he was a prisoner of war until 1921. On 13 January 1919, Abdullah's men entered the city, officially ending Turkish control of Arabia and the holy sites of Islam.

Ömer Fahrettin Türkkan (Fakhri Pasha), Commander of the Hejaz Expeditionary Forces and Defender of Medina.

The early moves against Jeddah, the port of Mecca, were supported by the Royal Navy, including floatplanes from *Ben-my-Chree*. Throughout the remainder of the war the Royal Navy maintained a small squadron of older ships, the Red Sea Patrol, in support of the Arab forces. But its influence, even with periodic support from the East Indies and Egypt Seaplane Squadron, did not extend far beyond the coast. Both the British and French sent military missions to the Hejaz to support the revolt. Reinforcements and supplies for the Arab cause had to be brought by ship and off loaded at various ports along the Red Sea as they were captured by Arab forces. Further details will be found in a later chapter.

The Naval Situation 1914 to 1918

The naval situation was quite different from the hard fought campaigns on land, the French and British had almost total command of the sea. This meant that the Turkish right flank in Palestine and Syria, resting on the Mediterranean coast, was wide open to raiding, bombardment and, by using floatplanes launched from seaplane carriers, observation far inland. Command of the sea enabled both navies to employ older ships, freeing up modern warships for more active theatres. This continued until 8 February 1916, when the old French armoured cruiser *Amiral Charner* (1894) was torpedoed off the coast of Syria with the loss of all but one of the crew. The casual use of unescorted ships ceased almost overnight. In the Red Sea and Indian Ocean there were no enemy vessels to consider, once the raiders had been accounted for.

Command in the Eastern Mediterranean, following the Anglo-French Naval Convention of 6 August 1914, was a French responsibility. The Red Sea and Indian Ocean were looked after by the Royal Navy. However, there was much inter-operability, both navies showing a high degree of co-operation. For example, when *vice-amiral* Boué de Lapeyrère was appointed Allied Commander-in-Chief of the Mediterranean navies on 22 August 1914,[34] he requested that the Royal Navy should be responsible for protection of trade between Malta and Port Said. This eventually came under the purview of Vice Admiral Sackville Hamilton Carden commanding the Royal Navy's Eastern Mediterranean Squadron, based at the Dardanelles but also looking after Egyptian naval affairs.

French naval command of the Eastern Mediterranean was vested in *contre-amiral* Henri de Spitz, *Commandante Division Navale de Syrie*, from 10 November 1915 to 4 March 1917. He was succeeded by *contre-amiral* Georges Varney, from 4 March 1917 to January 1919. The principle force based on Port Said was the *3ème Escadre*,[35] comprised mainly of old battleships and cruisers. With the arrival of the U-boats these were increasingly harbour-bound and inactive, several ships being disarmed. The active force was made up of older destroyers and torpedo boats, some of the latter refitted for anti-submarine work, supported by armed trawlers and tugs.

Throughout November and December 1914 French and British ships had, in the spirit of the Convention, been co-operating in operations along the Palestine and Syrian coasts from El Arish to the Gulf of Alexandretta. As the Dardanelles situation developed, requiring more attention by Admiral Carden, the Egypt based operations were added to the responsibilities of Vice Admiral Sir Richard H. Peirse, Commander-in-Chief, East Indies and Egypt Station, who moved his flag from Singapore to Port Said, where he arrived aboard *Swiftsure* on 17 November 1914.[36] Rear Admiral Sir Rosslyn Erskine Wemyss was appointed to succeed Admiral Peirse on 1 January 1916,[37] but did not arrive in Egypt until 23 January. He was in turn succeeded by Rear Admiral Ernest F.A. Gaunt on 23 June 1917, who served until 1919. The Red Sea was considered important enough to warrant the creation of the Red Sea Patrol, dividing the long narrow waterway into two distinct zones, North and South of Jeddah. The Red Sea Patrol was commanded from the old (1893) protected cruiser *Fox* by Captain W.H.D. Boyle until October 1917 when he was succeeded by Captain H.A. Buchanan-Wollaston. The work of the Patrol will feature later in this book.

Along the Palestine and Syrian coast there were few good harbours. Haifa was an open anchorage with piers from shore, the largest being the Hejaz railway pier. Acre and Jaffa both had small boat harbours within the remains of ancient ports. Beirut was the most developed harbour on the coast sheltered by two moles leaving a narrow entrance about 150 metres wide. Unfortunately, the middle of the harbour was occupied by the half sunken wreck of the Turkish Ironclad, *Avnillah* (1870). She was moored in Beirut as a local defence ship from 1910, and sunk by two Italian armoured cruisers, *Giuseppe Garibaldi* and *Francesco Ferruccio*, on 24 February 1912 during the Italo-Turkish war of that year. *Avnillah* was finally removed by the French in the 1920's. Beirut had a good supply of fresh water and was connected to the railway system by a 1.05 metre gauge branch line. With the appearance of enemy submarines, Beirut was considered a likely harbour for them to refuel and replenish stores. The port was occasionally visited by U-Boats, but the visits were rare.[38]

Amiral Charner (armoured cruiser, 1894) with reduced weight pole masts replacing the original military masts. *Amiral Charner* was torpedoed by the German submarine *U-21* on the morning of 8 February 1916. Sinking in only two minutes, there was only a single survivor rescued five days later.

The limited capacity of the railway system required the Turkish army employ local dhows and schooners to run additional supplies along the coast. These were frequently intercepted by patrolling allied warships and, to prevent capture, were often run on to the shore. Abandoned by their crews they would be shelled or set on fire by a landing party. Sometimes they were captured on the open sea and taken to Port Said. A little piratical activity thoroughly enjoyed by the British and French crews.

One further allied naval force was active in the Mediterranean during 1917 and 1918, the Imperial Japanese Navy. From April 1917 a squadron comprising a cruiser and two modern destroyer divisions was based at Malta and Port Said. It was later joined by a second cruiser and a third destroyer division. They were mainly engaged in convoy escort between French and Italian ports and Egypt.

April 1914 in Monaco, probably during the *Rallye aerien de Monaco* which preceded the second race for the *Coupe d'Aviation Maritime Jacques Schneider* (Schneider Trophy). From left to right, four pilot officers all *Lieutenant de vaisseau*: Henry de L'Escaille, Charles Marie Henri Dutertre, André Marie Emile Nové Josserand and Antoine Valentin Marcel Destrem. At the time of the photograph de l'Escaille, Nové Josserand and Destrem were serving on *Foudre*, Dutertre was in command of *CAM* Fréjus – St. Raphaël. *(ARDHAN)*

Group at the Nieuport School, Villacoublay, 16 April 1914. Seated L-R, *Quartier-maître* Jean-Marie Le Gall, *Quartier-maître* Georges Trouillet, *Quartier-maître* Paul Poggi, *Quartier-maître* Hervé Grall. Standing L-R, *Enseigne de vaisseau* André Lorfèvre, unknown, Gabriel Espanet, unknown. They are standing in front of a Nieuport Type II from the school. *(ARDHAN)*

CHAPTER 3

L'escadrille de Port-Saïd

French naval aviation, *l'Aéronautique maritime*, had a head start of almost a year on British naval aviation. For manifold reasons, beyond the scope of this chapter, it failed to prosper. One advantage *l'Aéronautique maritime* held was a year round warm water base, the *Centre d'aviation maritime (CAM)* at Fréjus – Saint Raphaël, on the Mediterranean coast. It also had the use of a seaplane tender *Foudre*, originally designed as a torpedo boat tender, then converted to a repair ship, she was considerably more practical than the Royal Navy's converted cruiser *Hermes*. *L'Aéronautique maritime* had, however, been unable to form the same close relationship with aircraft designers and manufacturers willing to develop aircraft specific to naval requirements that the RNAS had achieved.

On the outbreak of war *l'Aéronautique maritime* was beginning to prosper, but entered the war ill-equipped for the immediate future. In August 1914 it had just 27 pilots (less than half on active duty), 100 men and 14 floatplanes (six each Nieuport monoplane and Voisin *13,5 Mètres*; single examples of the Caudron Type J and Breguet Canton-Unné). By comparison the RNAS had about 100 pilots, 700 men, 39 aeroplanes and 52 seaplanes (about half ready for active service), and seven airships.[39]

Almost by default, *l'Aéronautique maritime* had a preference for the Nieuport floatplanes. Although already outmoded the Nieuport Type XH (80-hp Le Rhône or Clerget) was most practical available floatplane, despite being fitted with Nieuport's unique control system.[40] But obtaining more of them was a challenge. Nieuport was winding up production of the monoplane in favour of a new biplane, somewhat confusingly the Nieuport XB. Fortunately for *l'Aéronautique maritime* a few Type XH were still being completed — some to outstanding French Navy orders, plus at least three from a now cancelled Turkish order. A final French naval order and an order for twelve Nieuports for the RNAS were placed in 1915. The French order was for six machines to be equipped with a 100-hp Clerget engine and being fitted for *TSF* (*télégraphie sans fil* – wireless telegraphy) equipment. These last six also appear to have been fitted with conventional controls, as were the RNAS machines, and were delivered in late 1915.[41]

Enseigne de vaisseau Gustav Delage in 1910 at Châlons-sur-Marne.

One of the leading lights in *l'Aeronautique maritime* was *Lieutenant de vaisseau* Henry Julien Paul de l'Escaille,[42] who entered the French Naval Academy in October 1898. Qualifying as an *aspirant de 1ère classe* in 1901, then spending several years in the Far East and Mediterranean until entering *l'École de canonnage* in 1908. Later serving aboard the battleships *Charlemagne* and *Démocratie* in Toulon, he was promoted *Lieutenant de vaisseau* in October 1910. Bitten by the flying bug, learning to fly at Buc, probably on a Blériot, gaining *Aero Club de France* certificate 791 on 9 March 1912, then gaining his Military certificate 133 on 22 July 1912. De l'Escaille was the twelfth French naval officer to qualify as a pilot. He briefly commanded the *Centre d'Aviation militaire* at Villacoublay before joining the seaplane tender *Foudre* in October 1912. Over the following two years he was involved in testing and trials of new machines for *l'Aéronautique maritime*.

A pre-war photograph, probably a montage, of *Foudre*, showing the large hangar aft of the funnels, and a Nieuport taxiing toward the camera. *Foudre* was attached to *CAM* Fréjus – St. Raphaël for trials and experimental work from 1912 to 1914.

At the outbreak of war *l'Aéronautique maritime* was looking for a role to play with its limited resources.

> On the day of mobilization, Saint-Raphaël was home to a host of elite pilots; aircraft of the most diverse types were being tested on a daily basis, but no military squadron trained for wartime operations had been formed. *L'Aviation maritime* went to war with the elements of its laboratory and research center. After a Mediterranean tour that took them to Nice, Bizerte, Malta, Antivari and Albania's Lake Scutari, a French seaplane squadron arrived in Port Said on December 1, 1914, and was commissioned by the British authorities to help defend the Suez Canal.[43]

On 24 November de l'Escaille had prepared a report for *vice-amiral* Boué de Lapeyrère, Commander in Chief of the allied Mediterranean navies, in it he made recommendations for the future employment of the Nieuports.

Group of pilots and mechanics of the second *détachement au Monténégro*, in front of Nieuports N.6 and N.14, aboard *Foudre* in October 1914. Front row L-R: Parcelet, Sauzet, *Matelot Mécanicien d'air* Jean le Corf, LV Destrem (seated), LV Cintré, *Matelot* (*Mot*) Xavier, *Mot* Julien Levasseur (pilot), *Mot* Abgrall, *Mot* Lejeune, *Mot* Victor Marchal, *Mot* Maurice Duval. Back row L-R: unknown, *Mot* Deluy, *Mot* Edouard Duffaud. This is possibly the earliest datable use of the large red-white-blue cockade on any *Aéronautique maritime* Nieuport. *(ARDHAN)*

Two possible solutions:

1. — Send to the Dardanelles: the *escadrille* would stay on *Foudre*. Only two aircraft would be able to fly, the others serving as a replacements. For reconnaissance, the distances to be travelled would be such that airplanes could only carry petrol and oil, and could only be used for observation. Finally, the weather conditions are such that until at least March, one could count on a limited number of days of use of the machines.

2. — Send to the Suez Canal. The *escadrille* could be based on land. All machines could be used, and *Foudre* released for other duties. The escadrille could make reconnaissances to prevent troop movements at least one day before their arrival on the canal, and also troop attacks.

In summary, the first is an uncertain solution with poor results, the other a sure solution with very good results.[44]

Concurrently, the British Admiral Eastern Mediterranean, Admiral Carden, had been requesting use of the French seaplanes if they were not to be employed by the French Navy. A few days after de l'Escaille delivered his report, Admiral Carden gratefully accepted them for use at Port Said.

A Base at Port Said

Port Said was a company town, built by and for the Suez Canal Company.[45] Located at the northern entrance to the canal it provided ships using the canal with docking, supplies, fresh water and, especially, coal. Coal islands were created on the east side of the port where local workers loaded lighters which were towed out to ships requiring coal, the same workers then transferred the coal into the ship's bunkers. It was a back breaking and dirty business, and passengers always left their ship whilst coaling to enjoy the various enticements of Port Said. On the west side of the canal lay the city and the business side of the port. Several basins, or docks, had been cut and lined with quays. The third from the north, and largest, of these was the Sherif Basin. On its south side was a large, broad quay, it was here *l'escadrille de Port-Saïd* made its base for the next eighteen months. On 2 December *l'escadrille*, initially known as *l'escadrille d'avions du Canal*, commenced landing and setting up its base on the quayside. 'In time that had elapsed since the declaration of war, favourable circumstances had made it possible to form a homogeneous squadron and the Foudre landed six Nieuport single-seater seaplanes with 80 HP engines.'[46] The Nieuports and workshops were housed in a mix of Hervieu and Bessonneau canvas hangars.[47] Within two days the hangars and Nieuports had been erected. De l'Escaille recorded the difficulties facing the unit.[48]

The Port Said base in 1915. At the top of the photograph are existing warehouses and storage. The first hangar is a Bessonneau, flying the tricolour, with seven single aircraft Hervieu tent hangars. *(ARDHAN)*

> Only its homogeneity enabled this squadron to live and fight for over a year and a half in particularly difficult conditions, with precarious supplies due in part to the remoteness of bases, but above all to the cessation of seaplane production by the Nieuport company, which devoted all its resources to the needs of land aviation.
>
> On the other hand, this squadron benefited from the powerful resources of the *Compagnie Universelle du Canal de Suez*, whose entire staff, at all levels of the hierarchy, demonstrated the greatest interest and support, much appreciated by the personnel of *l'escadrille de Port-Saïd*.
>
> A severely damaged aircraft was carefully preserved and maintained until another accident allowed a new aircraft to be rebuilt from old parts. The need to make use of leftovers was so compelling that small expeditions were sometimes mounted to recover aircraft abandoned after breakdowns between the lines in the desert.

The Nieuport aircraft that equipped the squadron was a remarkable machine for its time; its only weakness was its engine, which after development was to show much improved qualities. But the first engines built, in the rather harsh conditions of use at the squadron due mainly to the existence of sand-laden winds, had to be overhauled after less than 20 hours of operation, and were prone to frequent breakdowns.

The floats proved to be very seaworthy, as demonstrated by constant use on the open sea for take-offs and landings. Apart from engine failures, there were few flying accidents over two years of intensive wartime service, despite very difficult circumstances. In particular take-off and landing manoeuvres in the crowded waters of the harbour were difficult. While the squadron's losses meant that pilots sent from France, with barely any experience of land-based aircraft, had to be trained on the spot.

No military equipment had been provided for the Nieuports, which placed a heavy burden on the squadron, which sought to be ready for any eventuality. For reconnaissance missions, the wings had to be modified to improve visibility. [The inboard leading edges were removed back to the main spar.] The squadron also worked on the development and installation of bomb launchers and aiming devices [see below], though these yielded poor results in terms of accuracy — an inevitable outcome given the complexity of the problem, which requires high-precision solutions that were not perfected at the time.

Foudre had brought five Nieuports (N.7, N.11 to N.14) from Malta to Port Said. All were powered by the 80hp Le Rhône, except N.13 which had an 80hp Clerget. Also aboard were *LV* de l'Escaille who remained in command until the unit left Egypt in April 1916, and four pilots, *LV* Destrem, *LV* Delage, *QM* Grall and *Mot* Levasseur, two observers, *LV* Louis Barthelemy de Saizieu and *LV* Alfred Cintré, with 38 mechanics, carpenters and aircraft hands. Of the five Nieuports the oldest, N.7, was in poor condition and had to be completely overhauled by *l'escadrille*. As a single-seater its operational use was limited but it may have been useful converting de Saizieu and Cintré, who were qualified landplane pilots, to the Nieuport.

Within a few days de l'Escaille was requesting additional Nieuports and pilots.[49] Three more Nieuports (N.15, N.16 and N.17), each fitted with an 80-hp Clerget engine, were delivered on 3 January 1915. *QM* pilots Georges Marius Etienne Trouillet and Jean-Marie le Gall arrived in Port Said late in December 1914, there were to be no more pilot reinforcements until October 1915. With the pilots came two old Nieuports, N.3*bis* and N.5, neither of which was suitable for service and were returned to France at the first opportunity.

The Clerget engined Nieuports were particularly welcome, as *l'escadrille*

Nieuport N.22 being hoisted from *Anne* in November 1915, with pilot *LV* de Saizieu and observer 2Lt H.M.C. Ledger *(see Chapter 4)*. A bomb is strapped to the wing and an example of the 'prehistoric' bomb sight (as described by 2Lt Williams) is attached to the fuselage ahead of the wing leading edge. The bomb tube is clearly visible within the web of undercarriage struts. *(Philip Jarrett)*

had from the outset been experiencing problems with the 80-hp le Rhône engines. De l'Escaille in one report commenting, 'We have once more to complain about the "Le Rhône" engines. The quality, perhaps because it was originally intended for the Turks, is deplorable. We constantly have damage resulting from the poor quality of metal, magnetos… The only Clerget engine we have had so far [N.13] has been very satisfactory.'

Even with the more reliable Clerget the performance of a fully loaded Nieuport was not sparkling as 2Lt Kenneth L. Williams recalled.

> The machines had by no means a good performance even in those days. Speed about 52 mph [84kph] at sea level, with a ceiling of about 5000 ft, and the climb to 1,000 metres (about 3,300 feet), with three hours' fuel, pilot, observer, a rifle and forty rounds of ammunition, and one 10 kg bomb, or a camera, took 30–35 minutes. We were not allowed to fly inland under 1,000 metres, and so our radius of action was a good deal handicapped by this slow climb.[50]

The 1000 metre requirement was, perforce, more honoured in the breach than the observance.

Armaments supplied to *l'escadrille* were somewhat makeshift. Other than a carbine for the observer the Nieuports were unarmed. De l'Escaille requested two *mitrailleuses* in October 1915 when agents reported aircraft hangars at Beersheba. However, he was instructed to find them locally and, by

February 1916, he was reporting, 'tests of Lewis guns, given to us by the English. Our tests were satisfactory but obviously our planes are poor fighter aircraft. It is difficult to adapt them as machines for all services'. Details of the installation are unknown.

Early in January 1915 500 boxes of flechettes, each box holding 100 darts, and 200 105 mm shells were provided.[51] The shells were to be converted into bombs by fitting sheet metal fins. One or two of the bombs could be carried strapped to the upper surface of the wing adjacent to the observer in the forward cockpit. Both bombs and flechettes were dropped by means of a tube fitted through the floor on the starboard side of the observer's cockpit to carry them clear of struts, wires and floats. The tube can be seen in many photographs of *l'escadrille*'s Nieuports. 2Lt Williams has provided a detailed account of how a bomb was prepared, aimed and dropped.

> The bomb-sight was a prehistoric edition of the equal distance sight, and consisted of a piece of three-ply wood attached to the starboard side of the machine, into which three nails had been driven in the form of a triangle. A sand-glass was used in conjunction with the sight, in place of the reversible watch at present used. Before starting off on a bombing stunt the observer was supplied with the sand-glass, on which the height in metres was marked on the stem, the striker of the bomb, a piece of string and a knife.
>
> Before dropping, the bomb had to be prepared—quite a long and complicated business. First of all, the bomb had to be unstrapped from the wing, and placed on the observer's knees. A cork was then drawn from the nose, and the striker screwed in its place. One end of the string was then attached to a ring at the tail of the bomb, and the other end to a cross strut in the fuselage. One had then to remove the safety-pin from the striker and manoeuvre the bomb into the bomb-tube, a dangerous proceeding, as the bomb with the striker in it was almost the same width as the fuselage.
>
> Presuming that all this has been safely accomplished, and that we are now approaching our target, we stand up, and, holding the sand-glass in one hand, and the knife in the other, we take our sights. The sand-glass is started when the top and forward pins are in line with the target, and is reversed when the top and rear pins are in line with it. We can now leave our sight, and trusting to our pilot to keep on the line, which incidentally he cannot see from his seat at this point, we note our height on the altimeter. When the sand has risen to the correct height in the stem of the sand-glass we cut the string, and the bomb falls.
>
> OK's by this method are not very common.

Manhandling a box of flechettes through the same tube is best left to the imagination.

Before looking at their operations, the subject of operational control must be considered. *L'escadrille* was somewhat of an anomaly at Port Said. The pilots and ground crew were French, the observers mostly seconded British army officers. Whilst the unit was administered by the French navy, its orders came from both navies and the British army, often directly from Lt General Sir John Maxwell, the British C-in-C Egypt.

Lieutenant Colonel Percival Elgood was Chief Intelligence Officer, HQ Canal Defence Zone, Port Said, representing the Department of Military Intelligence in Cairo. For some time Elgood had been gathering information from inhabitants and tribesmen within Turkish territory. He was looked on as the ideal man on the spot to manage this new source of information. In his memoir he has the following to say on *l'escadrille*.

> The French sea-plane squadron had arrived in Port Said unexpectedly. In itself, the squadron was one of the most complete units ever seen on the Canal, and the personnel the most daring. Operating from Port Said, [it] watched the Turkish concentration at Bir Saba [Beersheba], El Arish, and other frontier towns. Like the RFC, the squadron had brought its own pilots, but both units required observers. There was no difficulty in obtaining volunteers for this perilous service from local sources…[52]

Elgood, however, remains a shadowy figure with most of the work falling to his man with *l'escadrille*, Captain L.B. Weldon, of whom more anon.

Early Operations from Port Said

During its time at Port Said *l'escadrille de Port-Saïd* carried out over 1000 flights, the majority of these being from the base itself. Regular patrols along the Suez Canal and over the approaches to Port Said itself accounted for many of these. Reconnaissances over the Northern Sinai were made regularly. Some of these followed the coast down to Tineh then over Lake Bardawil towards Bir El Abd before returning, others started off down the canal to Kantara then followed the caravan road across the desert to Bir El Abd before returning along the coast. Many of these employed *matelot* from *l'escadrille* itself as observers, but few records remain of their work. Flying from the base at Port Said severely limited the usefulness of the unit. Their machines were all floatplanes. So, the obvious solution was to transport them up and down the coastline to where they were needed, then they could be lifted on and off the sea for take-off and return. Thus greatly increasing the areas they could cover. As had already been discovered, in pre-war experiments, warships did not make ideal aircraft transports. Initially, they were all that was available.

Their limitations would soon be explored.

L'escadrille's first reconnaissance set out from Port Said on 5 December 1914. Piloted by *LV* Destrem with *LV* Cintré as observer a Nieuport set out to fly along the road to El Arish, reaching a point close to Bir El Abd. They encountered a Turkish cavalry patrol of 30 riders, receiving an early lesson in the accuracy of Turkish rifle fire, returning with multiple bullet holes in the fuselage and wings. This was also the first indication that the Turks were exploring and pushing out patrols towards the Suez Canal. General Maxwell immediately requested reconnaissances of El Arish, Beersheba and Akaba, all seen as being possible assembly points for any attack on the Suez Canal. These were all beyond the range of the Nieuports from Port Said, but if carried aboard ships they could be reached. So, two Royal Navy cruisers, *Doris* and *Minerva*, were instructed to each carry a Nieuport along the Palestine/Syrian coast and into the Gulf of Akaba respectively. De l'Escaille was also instructed to use two British army officers, with experience of the Sinai, as observers.

Doris (Capt Frank Larken) and *Minerva* (Capt Percival H. Warleigh) were both *Eclipse*-Class cruisers dating from 1896. Their decks were crowded with guns (each had eleven 6-inch guns in single mounts), boats, etc, with very little space to safely manoeuvre a delicate floatplane. Both ships had been advised that two floatplanes were to be carried but, in the event, only one could be accommodated. *Doris* located hers on the quarterdeck aft of the two 6-inch guns already mounted there, one each to port and starboard. Using a boom rigged to the main mast the Nieuport could be lifted to and from the sea to either port or starboard. *Minerva* adopted a different solution, Capt Warleigh deciding to locate the Nieuport between the ship's aft funnel and mainmast. Midshipman Parkes-Buchanan noting, 'the hoods were removed from the after set of engine room cowls so as to facilitate the building of a platform for the aeroplane.'[53] Once again, a boom mounted on the main mast was used to handle the floatplane.

HMS *Doris* with the handling boom for Nieuports rigged over the quarterdeck. Space for the floatplane was found in the limited area aft of the two side mounted 6-inch guns. Although sister ships, *Doris* and *Minerva* each found a different location for the Nieuport floatplane they carried. *(CCI Archive)*

Doris was instructed by Vice Admiral R.H. Peirse, SNO, Egypt, to 'proceed to Ghaza and reconnoitre by seaplane as far as Beersheba. Having accomplished this, to commence patrolling the Syrian Coast as far as Alexandretta. … The seaplane should prove useful.'[54] After taking onboard a Nieuport, pilot *LV* Delage and Capt James R. Herbert (from the Survey of Egypt Department of the civil service) as observer, the cruiser sailed from Port Said on 11 December. The following day, the Nieuport set off at 14.00 and headed inland but had to return with engine trouble, apparently a fractured oil pipe. A float strut was damaged on landing and the ship returned to Port Said that evening.

Doris' Nieuport somewhere off the Syrian coast mid-December 1914. The pilot is *LV* Destrem and observer Capt James R. Herbert, the identity of the Nieuport is not known. The floatplane is being towed away from the ship by *Doris'* cutter prior to starting up. *(JMB/GSL – CCI Archive)*

A replacement Nieuport with *LV* Destrem as pilot and Capt Herbert as observer, plus two mechanics, were received aboard on the afternoon of 13 December. *Doris* sailing at 06.00 the following morning. A few hours after sailing, the ship was approaching El Qels, located halfway along the sand spit separating Lake Bardawil from the Mediterranean. The water was shallow and poorly charted, *Doris* touching bottom on two occasions without damage. As there were no submarines to worry about, *Doris* anchored, and the Nieuport was sent off for a local reconnaissance finding nothing.

Off Gaza the following morning, and once the sea mist had cleared, Destrem and Herbert set off for Beersheba. They returned after a couple of hours having had an untroubled flight, approximately 150 kilometres mostly over land. At Beersheba they reported 80 tents but otherwise no signs of troops. At the time this flight remained uncommented upon, meriting just a single sentence in *Doris'* report, but it was the first of many *l'escadrille* made to Beersheba. Few were as uneventful as this one. In the afternoon Larken headed *Doris* north along the coast. Approaching Ashkelon a new earth work was observed and shelled. A small landing party was then sent ashore covered by the cruiser's guns. A few rounds of shrapnel chased off a force of Turkish infantry and the landing party was free to examine the earth work, finding some rifles, ammunition and a copy of the Koran, all of which were brought aboard *Doris*. This was the first of many landings made on this cruise, most were to be much more eventful.

On the following days, 16 and 17 December, the Nieuport made local flights over Jaffa then Haifa and Mount Carmel. On both flights riflemen fired on the machine but no hits were recorded. Overnight the wind strengthened, remaining strong and squally for the next few days. At times the wind was trying to lift the Nieuport off the deck. By the end of the day of the 19th the wings had been damaged to such an extent that no further flights were possible. The airmen remained passengers for the rest of *Doris*' cruise although, as a naval officer, Destrem seems to have helped out with watch keeping.

At this point, we must leave *Doris*; suffice to say that the days until her return to Port Said were busy. Cutting telegraph lines, wrecking trains and bridges, and a little diplomatic blackmail, are just part of the story.[55]

Doris returned to Port Said a few days after Christmas 1914. On 1 January 1915, she sailed once more, this time with *Mot* Levasseur, an unnamed observer and a Nieuport floatplane. The same afternoon the Nieuport was over El Arish, but little was to be seen. On 2 January an ambitious programme was planned. In the morning an attempt to fly inland to Hebron was defeated when the Nieuport's engine failed to provide sufficient power to climb the Judean Hills to reach the town. In the afternoon a flight to Beersheba had to turn back when the propeller was hit by a rifle bullet. The Nieuport was able to reach the ship, with a failing, vibrating engine, but the propeller was not repairable. Later that night, at 02.00, the Nieuport and the detachment were transferred to Royal Navy *TB.63* for return to Port Said.[56] *Doris* sailed on to the Gulf of Alexandretta and out of our story.

HMS *Minerva*, every bit the smart pre-war cruiser. Her Nieuport was housed between the aft funnel and main mast, where it would have been partially sheltered by the ship's boats. Several pairs of engine room ventilator intakes can be seen around the funnels, usually cowled at sea they are seen here with hoists rigged for air scoops. The cowls for the aft pair were not installed and a platform built over them for the Nieuport. A boom was rigged from the lower main mast to hoist the floatplane on and off the ship.

Akaba

Akaba, at the head of the eponymous gulf, was an important stop on the ancient trade and pilgrim routes from Egypt into Arabia and north to Syria. It was much shrunken from its glory days, but, with its small fort and garrison, remained a depot for caravans. Akaba is located at the start of the Wadi Araba, the southern extension of the Jordan Rift Valley. From Akaba the wadi rises slowly over 77 kilometres to a height of 230 metres at the watershed between the Dead Sea and the Red Sea. Numerous side wadis ascend out of Wadi Araba leading to the Sinai on the west and northern Hejaz to the east. Of particular interest to the British military were Wadis Ithm and Gharandal leading to Ma'an on the Hejaz Railway, a vital supply route for any attack on the Suez Canal through southern Sinai. Wadi Ithm headed north-east from Akaba towards Ma'an some 105 kilometres distant. The mouth of Wadi Gharandal lies some 65 kilometres up Wadi Araba from Akaba, leading towards Ma'an itself another 50 kilometres away. For both approaches the land rises to over 1500 metres. Either route was a tremendous challenge to a Nieuport floatplane.

De l'Escaille flew a Nieuport down the canal to Suez on 7 December 1914 to join *Minerva*. Midshipman Parkes-Buchanan recording that, 'About 10 am our hydro aeroplane arrived; she was flying very low but she did not recognise us and made straight for the "Askold"'. The Russian cruiser *Askold* stood out because of her five tall thin funnels, she was known as the 'Pack of Woodbines' to the British bluejackets after W.D. & H.O. Wills 'Wild Woodbines' a popular brand of cheap cigarettes available in packs of five. About the same time *LV* Destrem and two mechanics came aboard, having travelled down from Port Said by train, de l'Escaille later returning by the same method. The Nieuport was towed over to the cruiser and hoisted aboard at 13.30. Finally, the British observer Capt Walter Francis Stirling joined ship at 18.40. Stirling, another officer familiar with Egypt, made his first flight as a passenger in Marc Bonnier's Nieuport IVG whilst he was secretary of the Alexandria Sporting Club in January 1914. *Minerva* sailed at 21.45 on 7 December, arriving off Akaba at 06.00 on the 9th.

This was not *Minerva*'s first visit to Akaba. Together with the destroyers *Scourge* and *Savage* the cruiser had bombarded the village and its fort on 1 November 1914, before the official British declaration of war with Turkey on 5 November. In Constantinople on 31 October the Allied Ambassadors had been instructed to leave Turkey. At the same time Admiral Carden was ordered to commence hostilities against Turkey, one of the immediate consequences being the bombardment of Akaba. Following the bombardment the village and fort were abandoned. The Turkish garrison, about 100 strong with a few field guns, retired into Wadi Ithm to keep watch from behind stone sangars overlooking the head of the gulf. Regular infantry and cavalry patrols were also sent down to Akaba and into Wadi Araba.

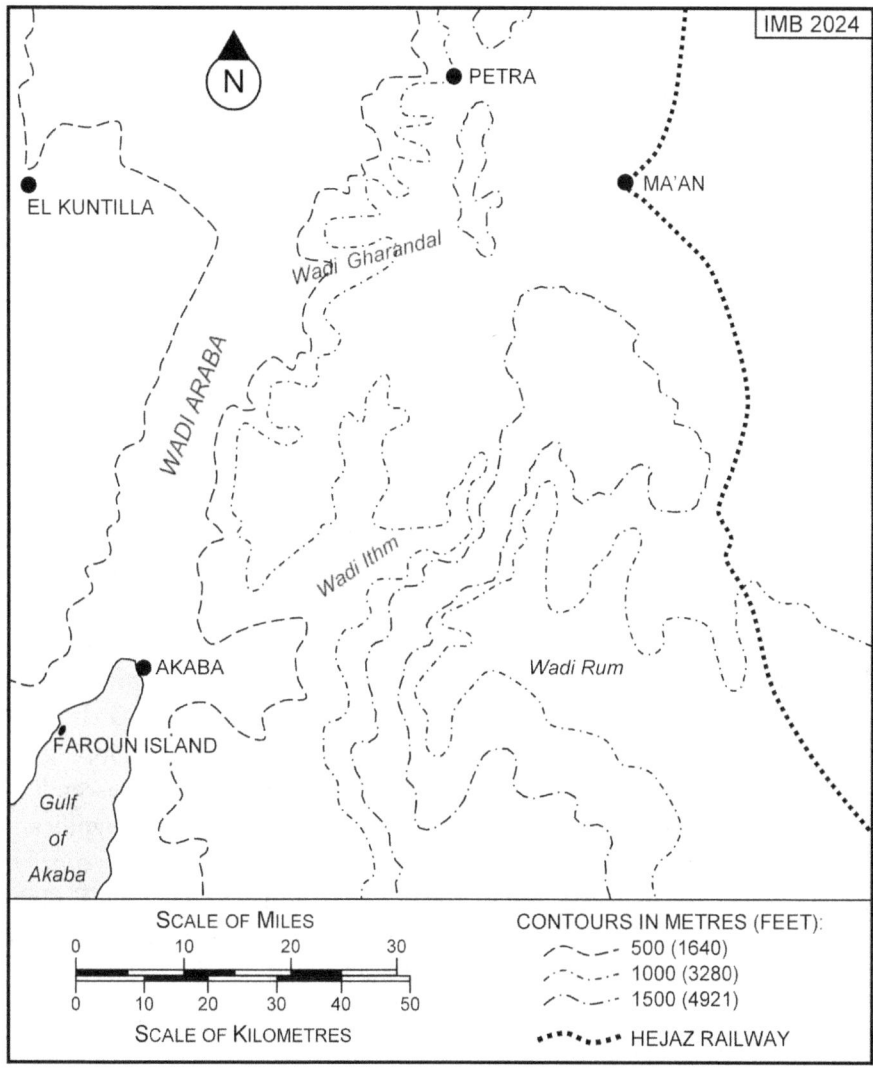

The Nieuport was hoisted out at 08.30 and attempted to take-off. After skipping over the surface several times the machine finally rose to about 15 metres then came down hard, breaking a rigging wire. Towed back to the ship, hoisted in and repaired, a second attempt was made. For this the amount of petrol was reduced and the observer's carbine and ammunition left behind. At 12.30 Destrem succeeded in getting the Nieuport into the air, climbing to around 300 metres but unable to rise any higher. It returned and was hoisted aboard. By then it was too late to make another attempt to reach Ma'an. That evening strong winds and a thunderstorm swept the area. The Nieuport was protected and kept dry by spreading awnings over the boom to tent the machine. The weather remained unsuitable for flying throughout 10 December, and it was the following day before another attempt could be made. The Nieuport set off at 06.30, reaching 750 metres but no higher. Returning at 08.00 it was made fast to the ship's lower boat boom whilst the mechanics attempted to correct an engine fault. Unable to get the engine running properly the machine was hoisted aboard at 10.00. Shortly afterwards *Minerva* sailed for Suez.

Stirling felt that the failure to reach Ma'an was due, in part, to 'my pilot, who was known as Gingembre, a fat French naval officer with an enormous red and spade-shaped beard.'[57] This was a little harsh on Destrem, who was undeniably a well built man, as the Nieuports would have to be working at maximum efficiency to climb over the mountains to Ma'an. De l'Escaille, warned of the returning cruiser sent down to Suez a replacement Nieuport that, 'had risen in Port Said to 2,500 metres.' He also replaced Destrem with *l'escadrille's* lightweight *Mot* Julien Levasseur. Stirling, although no heavyweight, was replaced with Capt A.J. Ross, RE, a temporary loan from the RFC.

Returning to Akaba, the lightweight crew and high climbing Nieuport set off on 16 December shortly after 06.30, first circling and climbing over Akaba then turning into Wadi Ithm. When they returned, after a flight of almost 3 hrs, the expectation was that they had reached Ma'an. Sadly, they failed to climb high enough to pass over the mountain ridges. Instead, Levasseur flew north along the mountains to investigate the upper reaches of Wadi Gharandal, possibly hoping to find a lower ridge to cross. Here, for a heart stopping six seconds, their engine failed. Levasseur was able to restart it, whilst descending down Gharandal to Wadi Araba before returning to the ship. No significant groups of Turkish solders had been seen and Wadi Gharandal was reported to be empty. The following morning the Nieuport made a short local flight trying to locate Turkish troops that had been sniping at the ship overnight. Nothing unusual was seen and the machine hoisted aboard. *Minerva* then sailed once more for Suez. On arrival, the aviation party left the ship and the Nieuport was flown up to Port Said.

De l'Escaille was particularly impressed by Levasseur's flight:

> The fact is that *Mot* Levasseur's attempt was one of the greatest feats yet accomplished in aviation. He made it after two failed attempts by *LV* Destrem, knowing all the difficulties and dangers of a flight in the mountains with a seaplane. He fought hard against the mountain for almost 2 hours and only gave up after an engine failure. I believe that *Mot* Levasseur showed the best qualities of military pilot in wartime on this occasion, and I ask you to grant him the Médaille militaire.

Levasseur was awarded a *Croix de guerre avec palme en bronze* for having, 'Participated in numerous and perilous reconnaissances under enemy fire. He was particularly distinguished in a bold attempt to cross a mountain 2200 meters high by hydro-plane.'

On 22 December, Stirling returned aboard with pilot *QM* Hervé Grall and two mechanics, their Nieuport having been flown down from Port Said earlier. *Minerva* sailed again for Akaba. General Maxwell, deeming the risk too great and could cost the lives of irreplaceable pilots and observers, prohibited any further attempts to reach the railway town. Stirling proposed instead 'to fly up the Wadi Araba as far as the Dead Sea, to see if I could spot any trace of troops passing over the wadi into Sinai.' The Dead Sea lies almost 170 kilometres north of Akaba, but could be observed from the watershed between the Dead Sea and the Red Sea. But before anything could be attempted the Nieuport had to cooperate.

Minerva dropped anchor off Akaba during the early evening of 23 December. All seemed quiet. At 03.00 Turkish field guns opened fire on the cruiser. There were no direct hits but shells passed close overhead, shrapnel scarred the ship's paintwork, leaving the Nieuport undamaged. *Minerva*'s forward 6-inch gun fired shrapnel towards the estimated location of the field guns which, after three rounds, ceased firing. The ship then weighed anchor proceeding 4 miles down the gulf before re-anchoring. At 07.00 Grall and Stirling set off towards Akaba, returning almost immediately with a misfiring engine. The mechanics found a cracked inlet pipe and a broken wing stay. A new engine was requested and *Minerva* turned back towards Suez, arriving at 09.00 on 26 December, the new engine was delivered half an hour later. It was fitted and tested but failed to perform. Port Said was asked to send a replacement machine (N.13) which arrived the following morning. With the Nieuports exchanged, *Minerva* sailed yet again for Akaba at 11.30 on 27 December.

At 07.00 on 29 December, 12 miles south of Akaba, *Minerva* came to a stop and hoisted out the Nieuport which set out on a reconnaissance. It quickly returned with sparks emanating from the magneto. This was corrected and the following morning Grall and Stirling made an attempt to fly up Wadi Araba, but the machine was unable to rise above 300 metres and, as visibility in the wadi was poor, the attempt was abandoned. Finally, early in the morning of 31 December *Minerva* anchored on the western side of the gulf close to Umrashash a deserted settlement of four mud huts.[58] Overnight Grall and Stirling had decided to reduce weight to a minimum, they carried no weapons and only one water bottle between the two of them. The Nieuport took off around 09.45, circled and gained some height before setting out to follow Wadi Araba.

Whilst the Nieuport was away, firing was heard from the shore and the ship's marines landed to assist the Captain of Marines, Capt J.W. Snepp, and his escort who had landed earlier to examine some earthworks. The party

Quartier-maître Hervé Grall. The shape of the fuselage patch, in itself quite common, suggests that he is standing alongside N.13, if so this would be at Port Said in 1914. *(ARDHAN)*

was withdrawn under fire having lost one marine killed and Capt Snepp seriously wounded but survived. Several Turks were killed or wounded by fire from *Minerva*'s 12-pdr guns. All this drew attention away from the Nieuport which, by 13.00, was overdue. At 16.00 the ship weighed anchor and proceeded down the gulf to bury Private Frank Ward, RMLI, at sea. Shortly afterwards *Minerva* reversed course to return to the head of the gulf.

At 20.00 the ship was moving slowly close to the shore, keeping a sharp watch for any movement or sounds from the shore. Someone was heard shouting from the shore and a searchlight directed towards the spot, quickly illuminating a waving figure. Armed boats were sent in, Captain Stirling found and brought out to the ship. Revived by a half bottle of champagne, Stirling told his tale.

They had reached the watershed and, having been able to climb to around 1500 metres, looked on towards the Dead Sea. With petrol reserves getting low they could go no further, Stirling telling Grall to turn back towards the ship. Just over halfway back, south of Ein Ghadian (south of Yotvata, Israel) some 30 kilometres from Akaba, the engine cut out and a crash was inevitable.

Captain Walter Francis Stirling, time and place unknown. However, the head dress suggests that it was later in the war when he was serving as Chief Staff Officer to Lawrence in the desert. His observer's wing is partially hidden by the folds of his *keffiyeh*.

Grall was thrown out of the Nieuport when it somersaulted as the floats hit the ground. Stirling hung onto the cabane struts and remained in the machine, having to dig himself out of the wreck. He was shaken but uninjured. Grall, however, broke his collarbone and several ribs then drank most of their only bottle of water.

The pair set out to walk to the coast, Grall was leaning heavily on Stirling. After three hours they had barely covered 6 kilometres. At that point they agreed that Stirling should head off solo and try to get help. Grall would follow up as best he could.

> In my whole life I have never walked as fast as I did then. From time to time I had to dodge parties of Bedouin on camels; luckily none saw me, but in the middle of the afternoon I heard guns firing in the distance from the direction of the Gulf, and was not at all reassured by the sound. [This was *Minerva* firing on a party of Turks observed in Wadi Ithm at 15.30. Some hours later, after the sun had set, Stirling heard horsemen approaching.]

> In the whole length of the Wadi Araba there is only one tree, one single tree in all those miles between the Dead Sea and Akaba, and providence dictated that at this critical moment I should find myself standing next to it. It was a gnarled old thorn tree, and in its protective shadow I thankfully lay down, burying my head in

my arms, as a whole squadron of cavalry, all lined out at ten paces' interval, moved up across the valley. It was clearly a search party.

The officer in command rode past me quite close, then halted his men just short of the tree. The squadron was accompanied by a dog, which I feared would be bound to betray me. It came sniffing round the tree and even cocked its leg against it, but miraculously paid no attention to me at all. I can only suppose that the castor oil in which I was soused made me smell like nothing human.

[Eventually], to my great relief the officer gave the equivalent order to 'Walk March' and the squadron moved off. When they had got clear I started off again, but twenty minutes later heard some more horsemen approaching, and this time, with no tree to protect me, I had to sprint to get clear of their outer flank. As I came over the last sandy ridge before the head of the Gulf I was amazed to see a long line of Turkish camp fires strung out in front of me: a most unexpected development, for the Turks had never so far left the Wadi Ithm. Somehow I had to get through. From the marshy ground nearer the sea I plucked some reeds which I stuck into my flying helmet for camouflage; then, on all fours, I began to crawl. It was a long way to march, quite literally, on one's stomach, but at last I was clear of the enemy line and was presently in sight of the sea. The ship was not there! This was about the last straw and so I curled up and went to sleep.

I woke up sooner than I expected: another of those cursed cavalrymen was jangling past, so close to me that I could have stretched out and touched him. I looked in the other direction, and could hardly believe my eyes. There was Minerva steaming up the Gulf!

Here was a fresh problem. Should I go down to the water's edge and hail her and risk getting shot by the Turks? Or should I slip quietly into the water and swim, taking the risk of being eaten by a shark? I chose the former, but need not have worried, as the Turks were in such a state of panic at the sight of the cruiser returning that they were far too busy packing up to stop to worry about me.

A searchlight was put on from the ship and slowly traversed the beach; when it suddenly shut off, I knew they had seen me.[59]

Meanwhile, Grall was struggling gamely on towards the sea. After Stirling left he fell asleep until the cold of the desert night woke him. He slowly moved forward through the growing heat of New Year's Day 1915. At one point he went into hiding in the shadows of some rocks when he saw an armed party clearly searching for something or someone. This was almost certainly a

Nieuport N.13 at Port Said in December 1914. The patch on the fuselage under the tailplane is visible, also seen in the photograph of QM Grall. The hand-worked crane appears in several photos at the base, it is difficult to see it as a practical method of hoisting the floatplanes on to the water. Possibly that is de l'Escaille sat on the bollard keeping a careful eye on things. *(ARDHAN)*

landing party from *Minerva* which had set out shortly after 07.00 to search for the missing pilot. It comprised the ship's Royal Marine detachment, a maxim gun party, several 'sections' of seamen and a supply party, in total probably close to 80 officers and men. They cautiously penetrated around 10 kilometres up the wadi before being ordered to return. Some small Turkish outposts were driven out, but there was no sign of Grall. A somewhat despondent party returned to the ship by 13.30.

Grall finally reached the coast around 17.30 only to see *Minerva* weigh and head off down the gulf. He collapsed on the beach, exhausted, and fell asleep.

As was her normal routine *Minerva* was heading slowly south for the night, something the Turks would have been expecting, and something she had done before returning for Stirling the previous night. At 22.30 she turned back to Akaba and, moving at full speed, reached the head of the gulf around midnight. Immediately the shore was swept by her searchlights and a man was spotted just east of the huts at Umrashash. Two of the ship's cutters were manned, armed and rowed into shore where the man was taken aboard. On their way back to the ship the Captain hailed them to know if the missing pilot had been found. Being answered in the affirmative the news quickly spread through the ship and the crew manned the side to welcome Grall with cheers. *Minerva* then returned down the gulf heading for Suez.

For his exploits QM Hervé Grall was recommended for the French *Médaille militaire* and the British DSM.[60] Both were granted. Stirling received no award other than, perhaps, permission to rejoin his regiment the Royal Dublin Fusiliers then serving at Gallipoli, where they had had heavy casualties and lost all but one officer landing from the *River Clyde*. He later served in the desert with Lawrence, as his Chief Staff Officer.

In early January 1915, *Minerva*, having just returned from the Gulf of Akaba with Grall and Stirling, was at Suez preparing to return to Akaba. A Nieuport could not be provided before she sailed and, like *Doris*, *Minerva* sails out of our story.

Aenne Rickmers and *Rabenfels*

Back at Port Said the British were daily expecting an attack from the Turkish army on the Suez Canal. Both *l'escadrille* and the Egypt Detachment, RFC, began providing frequent enemy contact reports. As both units continued to work in concert throughout 1915, a few words about the latter are called for.

Commanded by Captain S.D. Massy, the Egypt Detachment, RFC (EDRFC) arrived at Alexandria on 17 November 1914. Ismailia, centrally located on the Suez Canal, was chosen to be its main base, making its first reconnaissance over the Sinai Desert on 27 November. It was not well equipped, arriving with just three Maurice Farmans, one Shorthorn and two Longhorns, to which were added two Henry Farmans, found at Heliopolis and needing reconstruction, later a single BE2a from India joined the flight line. Using temporary advanced landing grounds (ALG) flights deep into the desert could be made. The establishment of these usually required a significant military operation, so most flights kept close to the Suez Canal. It was also required to provide a basic training in observation to the army officer volunteers, many of whom transferred to *l'escadrille*.[61]

Henry Farman F.20 Type (80-hp Gnome), located at Heliopolis and extensively reconstructed by the Egypt Detachment, RFC and serialled HF.2. The wings, and nacelle were original; the engine was from a Bleriot; everything aft of the wings appears to have been assembled from Maurice Farman Longhorn parts—including a single upper tailplane and elevator, and two rudders. The loaded bomb rack carries three Hales 20-lb bombs. *(Ray Vann, via CCI Archive)*

Whilst the EDRFC could extend their range through the use of ALGs, to extend *l'escadrille*'s work further up the Palestine and Syrian coast two German tramp steamers were provided by the British, *Aenne Rickmers* and *Rabenfels*.

Located within the Suez Canal there were a number of German and Austrian merchant ships nominally protected from seizure by the 1888 Constantinople Convention which stated that the canal was open for use by all shipping. Clearly, this was an unacceptable situation for the British, and a gentleman's agreement appears to have been reached between the ship's captains and the authorities. In the forenoon of 15 October 1914 the German merchant ships commenced sailing from the canal at approximately 30 minute intervals, to be stopped and seized by the British armoured cruiser *Warrior* waiting just outside the three mile limit, then sent on to Alexandria. All ships were formally taken in charge by the British Prize Court in Egypt. Two of these ships, *Aenne Rickmers* and *Rabenfels*, were selected for conversion to seaplane carriers and made available to the *l'escadrille de Port-Saïd*. Both were cargo steamers with holds fore and aft of the centre island containing the bridge, officers' and some passenger accommodation and the engine and boiler rooms. *Aenne Rickmers* (4083 tons gross;[62] length 367 feet; 11 knots) was owned and operated by Rickmers-Linie, had been built in 1911 at Bremerhaven in the company's own shipyard. *Rabenfels* (4706 tons gross; length 394 feet; 10 knots) was Tyne-built by Swan, Hunter & Wigham Richardson Ltd, launched in 1903 and owned by DDG Hansa of Bremen.[63]

Initially both ships were placed under the orders of GOC Egypt (Maxwell), to be manned by the Egyptian Ports and Lights Administration, using the prefix SS and flying the Red Ensign. By the end of March, the prefix HMFA (HM Fleet Auxiliary) was being used. The latter would require flying the Blue Ensign and an RNR officer as captain. On *Aenne Rickmers* the master was quickly commissioned as Lieutenant Robert A. Gaskell, RNR (seniority 12 Jan 1915), from the outset *Rabenfels* had Lt John Jenkins, RNR, as master. They were rushed into service with heterogeneous crews, as Captain Lewen Barrington Weldon explains.[64]

Captain Lewen Barrington Weldon, Royal Dublin Fusiliers, whose stories form an important part of this book. Here he is on the bridge of *Anne* (aka *Aenne Rickmers*) sometime in 1916. *(EMK)*

The instructions I had received in Cairo had prepared me for something rather out of the ordinary, but it was not until I had talked matters over with Colonel Elgood that I began to realise the complexity of the duties which I was expected to perform. I was to be a kind of mixture of Liaison, Intelligence and Commanding Officer rolled into one, and that the seaplanes with which I was to work were French, but it soon appeared that this was not all. Someone was wanted to distribute spies, or more politely "agents", behind the Turkish lines, and this little job also fell to my lot. At that time we knew that there were many people in Palestine and Syria who were willing to help us with information of enemy movements, etc., if we could arrange some system for collecting their news. The only way of doing this was to land agents on the coast behind Turkish positions, and to pick them up again when they had found out all our friends had to tell them.

As soon as I gathered exactly what I was in for, I realised that to land my agents at all I should need a boat fit for surf work, and boatmen who knew the coast to man it: so I spent most of my first day in Port Said hunting for recruits. Luckily they were not hard to find, for there happened to be four Syrian-Christian boatmen in the town who had been stranded there on the outbreak of war and were then at a loose end. These agreed to work with me — at a price. So the next day I was able to join the ship.

The *Aenne Rickmers* was a German cargo boat, steel built and single screw. Formerly she was owned by the Rickmers family of Hamburg, but had been commandeered and was now attached to the French seaplane squadron. Her accommodation being much the same as is found on all cargo boats of her size, a saloon to sit eight, a couple of

Rabenfels at Port Said probably shortly after having been converted. Canvas shelters for the Nieuports are rigged on the forward and aft well decks. At this time only the white upper works have been over painted grey, the pre-war black hull remains untouched. Also visible is an identifying feature of *Rabenfels / Raven*, on the poop is a tall ventilator/ kingpost for a derrick unique to the ship and her sisters.

two-berth cabins and a single bath-room. In the saloon was a portrait of Aenne, the daughter of Rickmers, and there it remained throughout all the cruises on board.

The personnel was nearly as mixed as my job, and rather more cosmopolitan. The captain, the chief engineer, the observers and the bluejackets and marines were English: the pilots and mechanics French: the mates and the crew Greek: one of the engineers was Maltese: and I myself, the O.C., Irish. Moreover, the captain was not then holding a commission, and the crew mostly belonged to a country — Greece — which had not come into the war. Also, we flew the Red Ensign: and the original cargo, worth about £250,000 [over £30m in today's money], was still on board. Yet two aeroplanes rested each on a hatch cover on the after well deck, and the uniformed sailors and marines — lent by HMS *Swiftsure* — were obviously not men of peace. Taken all together, a regular Harry Tate shipload, reminiscent of the London Hippodrome at its best![65]

Defence of the Suez Canal

To attack the Suez Canal the Turkish forces had first to cross the Sinai, a formidable task in its own right. The Sinai Peninsula comprises a narrow sandy coastal plain of the north, then a wide area of sand dunes, a limestone plateau in the middle, and mountains to the south. In 1914 it was crossed only by camel tracks. An attacking force needed water. There were a few wells and stone cisterns, remnants of bygone civilizations, in which winter rainwater was collected by the Bedouin. After the winter rains large pools often existed for short periods.

Through the winter of 1914 there were some heavy storms, providing a good, short term, supply of water. Exploiting these resources Djemal Pasha, Turkish Governor and C-in-C Syria, with his German Chief of Staff Colonel Kress von Kressenstein, planned a three pronged attack on the Suez Canal. A northern route following the coast road from El Arish, the ancient invasion route to and from Egypt. Through the centre of Sinai from Beersheba to Ismailia and, finally, a southern route from Ma'an and Akaba to Nekhl and ultimately to Suez. The Turkish advance commenced on 14 January 1915.

However, the British were well informed of the approach. Good information came from Elgood's Bedouin informants and other in country sources. Also, both *l'escadrille* and RFC began providing frequent enemy reports. The EDRFC, based on Ismailia but with semi-permanent bases at Kantara and Suez, probed the Sinai from Bir el Abd on the El Arish coast road, along the escarpment east of Ismailia, south to Bir Mabieuk and the Mitla Pass. The flights began to see elements of the Turkish advance guard from 17 January 1915.

On 14 January, Captain Massy from the EDRFC went to sea with *l'escadrille de Port-Saïd* aboard the *Aenne Rickmers*. The EDRFC were planning an ambitious reconnaissance to Kossaima, some 80 kilometres inland from El Arish, for which a temporary ALG on the sand spit separating Lake Bardawil from the Mediterranean Sea was proposed. Also aboard the ship was a section of two Nieuport floatplanes from *l'escadrille*, a detachment of the 128th Pioneers, Indian Army, and a RN landing party from *Swiftsure*. *Aenne Rickmers* arrived off the eastern end of the lake, adjacent to Likleikha, the following morning. Prior to the attempted landing Massy was able to observe first-hand the operations of the French floatplanes when *LV* de Saizieu and his observer Capt R.E. Todd made a flight over the sand spit, as far west as El Qels, finding no activity. There was a sea running when the Nieuport was recovered, resulting in minor damage when it was hoisted aboard. Capt J.R. Herbert reported, 'At 1.30 I went away with Capt. Massey [sic], Lieut. Birkbeck and an Indian officer [both from 128th Pioneers] in a boat towed by the steam pinnace to examine the sand spit to see if it was suitable for an aeroplane to alight on. It proved unsuitable, so Capt. Massey decided to abandon the project. We returned on board about 4 pm.'[66] The ship returned to Port Said early in the morning of 17 January.

Aenne Rickmers sailed from Port Said once more in the afternoon of 17 January, having two Nieuports, N.15 and N.16, aboard with pilots de Saizieu and Levasseur and observers Captains J.R. Herbert and R.E. Todd, RAMC. During the afternoon of 18 January, de Saizieu and Herbert on N.16 flew to Kossaima and El Auja. Effectively negating the need for any further RFC plans. Levasseur and Todd, N.15, were over El Arish, Rafa and Khan Yunis (midway between Gaza and Rafa). Both flights returned after sighting assembling Turkish forces.

The next day de Saizieu and Herbert took N.16 on a three hour flight to Beersheba. Herbert, reporting a very different scene from his earlier flight with Destrem in December. Taking off at 14.10 they crossed the coast 40 minutes later having only climbed to 700 m. Continuing to climb, eventually to 1800 m, they reached Beersheba at 15.30. Herbert reporting 'I saw several long strings of Camels here, and I should estimate the number of troops to be from 8000 to 10000. I could see no sign of a railway, but the road in the direction of Anjer appeared to be well worn.[67] There appeared to me to be more buildings to the SE of the town than I saw during my previous reconnaissance on Dec 16th'. They returned to the ship, landing alongside at 16.10.

The RFC meanwhile had sighted groups of infantry and cavalry along the coast on 17 January. Then, on 25 January, troops were sighted exiting the Mitla Pass and at the Bir Mabieuk wells, also advancing along the tracks from Katia towards Kantara. These columns were intended as distractions. The main Turkish force was to advance through the middle, from Beersheba using the track between Gebels El Maghara and Yelleg, passing through Bir el Jifjafa, intending to launch an attack on the canal south of Ismailia. Leaving Beersheba on 14 January 1915, the Turkish 25th Division, 12000 strong, marched into the Sinai. The Turks brought with them pontoons, with which to bridge the Suez Canal. These forces were detected by the EDRFC on 26 January and in much greater strength the following day. By the end of the month the main force had been clearly identified as being located near Wadi Mukhsheib, around Moiya Harab and Khabret Zohra. A strong, but immobile, force was also being observed between Katia and Kantara, and a much smaller force around Bir Mabieuk to the south.

On 22 January, Delage and Cintré from Port Said visited Bir El Abd, on the coast road 60 kilometres from the canal, carrying two bombs just in case they saw anything. This was possibly a follow up to a visit by the EDRFC on the 20th which had reported, and bombed, 1000 infantry and 200 cavalry. Cloud prevented observation and engine failure forced them down on the sea. A naval patrol rescued the crew and towed the Nieuport back to Port Said.

On 23 January Sir Henry McMahon, the newly appointed High Commissioner, accompanied by General Maxwell and Admiral Peirse, paid an official visit to *l'escadrille de Port-Saïd* and was taken for his first flight by Delage to examine the defence preparations along the canal. Admiral Peirse, thanked de l'Escaille for the work they were doing.

QM Le Gall and 2Lt Basil G.N.B. Partridge left Port Said in Nieuport N.14 at 14.13 on 27 January. Le Gall had arrived at Port Said in December, whilst Partridge had only been with the EDRFC for three days when he transferred to *l'escadrille* on 23 January. This was their first operational flight. They flew down to Kantara then followed the track to Katia and Bir El Adb. On reaching Katia, their troubles started. With the engine failing Le Gall turned back

The abandoned N.14 as seen from the deck of *Aenne Rickmers* on 28 January 1915. The floats appear to be waterlogged, the Nieuport would probably have sunk within a few hours, but it appears otherwise undamaged. *(EMK)*

towards Port Said. They landed on the lake twice attempting to repair the engine, but were over the sea when it finally failed.

When they were overdue de l'Escaille informed Admiral Peirse's staff aboard *Swiftsure*. They immediately sent a torpedo boat to search the coast, and sent a message to *Aenne Rickmers* about the missing floatplane and crew. She was out at El Arish from 26–29 January, but a frustrating series of engine failures to N.11 and N.15 had prevented any useful work. On 28 January, around noon, they sighted N.14 near El Qels and hoisted it aboard. On examination, an oil pipe was found to have broken and the both floats were damaged. Within the cockpit the pilot's helmet and observer's map and notes were discovered. *Aenne Rickmers* then returned to Port Said.

Both Le Gall and Partridge were able to reach dry land and decided to follow the coast line back to Port Said, approximately 35 kilometres away. Shortly after setting out they encountered the inundations, an area of salt marsh that had been flooded as part of the canal defences. From there they were restricted to a very narrow stretch of land between the floods and the sea. At a point some 10 kilometres from Port Said, in the middle of a windy and cloudy night they encountered a patrol of the 1/6 Ghurkha Rifles. When challenged for the night's password, which they did not know, perhaps unwisely they continued to advance with their hands raised and Le Gall whistling a British march. Both men were shot and killed.

De l'Escaille commented that 'It was two o'clock in the morning (January 28) and the patrol was especially vigilant, as the advance of the Turkish advance guard was general on all other roads towards the canal.' He also added that 'All the English authorities have sent their regrets about the deplorable accident.'

Aenne Rickmers was out again between 1–8 February, covering the period of the Turkish attack on the canal. Aboard were two Nieuports, N.15 and N.17, with pilots *QM* Grall and Trouillet and observers 2Lts Harry G. Hillas and Sir Robert Paul. Typically, for most of *l'escadrille*'s observers, both had been transferred from the EDRFC on 28 January following minimal training. Bad weather and engine problems prevented any useful flights until 7 February, by which date the Turkish attack had been repulsed. However, on that day a fine flight to Beersheba was made by Grall and Paul on N.17. They took off at 08.40, crossed the coast near Gaza ten minutes later and reached Beersheba at 09.25. In the town and immediate vicinity they reported approximately 800 tents and 20000 men. These were probably an intended second wave if the canal attack had been successful. Flying north-east towards Hebron, road improvements and baggage trains were observed. Unable to climb above 1400 metres, insufficient to reach Hebron, they turned towards Gaza and the ship, landing at 10.43.

The following morning an attempt to reach El Auja was brought to a quick halt by a damaged float. The weather not improving, the attempt was abandoned and the ship turned towards Port Said. Enroute *Rabenfels* was encountered on her first cruise along the coast with just one Nieuport, N.12, aboard. She had evidently rushed out to sea as a boat was sent to *Aenne Rickmers* to collect maps, medicine chest and binoculars. *Rabenfels* had a steep learning curve ahead of her.

Nieuport N.15 undergoing maintenance at the Port Said base 1915. The two open boats, with an Indian sowar on guard, look like examples of the landing boats brought across the Sinai for the Turkish attack on the Suez Canal. The harbour itself is typically crowded and busy, always making flight operations difficult and hazardous. *(ARDHAN)*

The Turkish attack on the Suez Canal commenced at 03.00 on 3 February, along a 25 mile wide front centred on Ismailia. It was repulsed by the evening of the following day. The Turkish forces commencing a long withdrawal to their starting points. The EDRFC was busy throughout the days of the attack. On 1 February there were four flights, on the next two days three morning flights each day, and on 4 February two morning flights. Afternoon flights between 2 and 4 February were reportedly prevented by dust storms in the Ismailia area. Observation was their only role, no bombs were dropped and no directed shoots were carried out; army and naval gunners relying solely their own observations.

Whilst the EDRFC was kept busy flying observations missions over the attacking forces, the role of *l'escadrille* during the actual attacks is not clear. De l'Escaille's report for 11 February noting that 'During the attack on the canal on February 3, the British, their Ismailia based aircraft being busy, requested a plane be sent to Ismailia for fire control. At that time there were three on reconnaissance and I could not meet this request.'

Nieuport N.12 after a landing accident in the Sherif Basin, Port Said, 13 February 1915. One of the floats has been torn apart, the outer right hand wing also torn and smashed. The basic strength of the Nieuport fuselage is demonstrated by its relatively undamaged condition. The Nieuport was written off but parts from it were used to rebuild N.16. *(ARDHAN)*

Major-General Alex Wilson, GOC Canal Defences, in his report was very complimentary of the work done by the flying services before and during the attack.

> In conclusion I desire to express my high appreciation of the valuable work done by the pilots and observers of the French hydroaeroplane squadron and the detachment Royal Flying Corps in the numerous reconnaissances carried out by them previous to and during the advance of the enemy. They were constantly under shrapnel and rifle fire and carried out their difficult and dangerous duties with courage, resourcefulness and success.

A view echoed by General Maxwell, in his Official Despatch.

> The French Hydroplane Squadron and the detachment Royal Flying Corps have rendered very valuable services. The former, equipped with hydroplanes with floats, ran great risks in undertaking land reconnaissance, whilst the latter were much handicapped by inferior types of machines. Notwithstanding these drawbacks, they furnished me regularly with all information regarding the movements of the enemy.[68]

Following the maximum effort during the Turkish attack *l'escadrille* was feeling the strain. Only five Nieuports remained available for operations; N.7 the single-seater, being returned to France sometime in March. Fitted with unreliable 80-hp Le Rhône engines, N.11 and N.14, were used as little as possible but could not be retired due to the lack of machines. Most of the work fell to the 80-hp Clerget machines, N.15, N.16 and N.17. There would have been six Nieuports, but a local patrol led to the loss of Nieuport N.12 (80-hp Le Rhône) on 13 February when the Nieuport capsized on landing at Port Said. The crew were unhurt but the floatplane was a wreck. Parts were used to rebuild N.16 which was severely damaged whilst operating from *Rabenfels* on 16 February. N.16 was back in service early in March.

L'escadrille also lost two of its pilots. In February 1915 *LV* Delage was recalled to France to resume his position as Technical Director of the Nieuport Company. Here he took over the design of the famous sesquiplane Nieuport biplane series. He continued with the company, from 1926 known as *Nieuport-Delage*, until retiring in 1932. Gustave Delage, *Officier de la Légion d'honneur*, died in Paris on 20 April 1946. Then in March *Mot* Levasseur was repatriated to France in March 1915 for medical reasons. He spent most of 1915 convalescing at Saint-Raphaël, before a posting to *l'escadrille de Brindisi* from October 1915 to February 1916. He spent the remainder of the war as an instructor at *l'Ecole de pilotage* at Saint-Raphaël and Hourtin. Demobilized after the war, no further details are known, although he is thought to have died before 1945.

The remaining pilots, however, had to continue without reinforcement until October 1915. With *LV* de l'Escaille essentially stuck at the base, although he often flew patrols from Port Said, the work load devolved on to *LV*s Destrem, Cintré, and de Saizieu, both the latter having recently qualified as Nieuport pilots, and *QM*'s Grall and Trouillet (both would be promoted *Second maître—SM—* during May). The British observers, in contrast, were quite numerous. At least ten, following minimal training with the EDRFC, are known to have flown with *l'escadrille* during 1915. They would all be kept busy over the ensuing months.

Nieuport N.16 being hoisted aboard Rabenfels 16 February 1915 following a capsize landing after a mission. Just visible is the hand hold on the upper port longeron, forward of the wing trailing edge. With damage to both wings, tailplane and fuselage, not to forget the floats, it was fortunate that N.12 was available to provide parts for the rebuild. Once back in the air N.16 saw much further service. *(ARDHAN)*

Nieuport N.16, following its rebuild in February, sports a different style of serial and lacks the fuselage footstep. Seen here on *Rabenfels*, with a wooden divider separating the two aft deck floatplane spots. Pilot is *LV* Cintré and observer Captain R.E. Todd, 11/12 April 1915. *(ARDHAN)*

CHAPTER 4

After the Suez Canal Attack

The principle task of *l'escadrille de Port-Saïd* for the remainder of 1915 and into 1916 was to keep an eye on Turkish troop movements and concentrations in Sinai, Palestine and Syria, and to cooperate with naval forces as required. Most flights were at the request of General Maxwell, and his successors, or French *vice-amiral* Louis Dartige du Fournet commanding the *3ème Escadre* at Port Said. *L'escadrille*, whilst continuing local patrols and reconnaissances from Port Said, frequently operated from the two seaplane carriers.

Depending upon availability of Nieuports and pilots, flights from Port Said were made as often as possible. As mentioned earlier, Port Said flights were more numerous than those from ships. What little evidence there is suggests that these included regular observation flights along the coast to watch the area around Lake Bardawil and approaches to Port Said, convoy escort, anti-submarine patrols, and mine spotting. Details of two strangely similar observation flights along the coast to Lake Bardawil can be extracted from the official records.

A Nieuport of *l'escadrille de Port-Saïd* at Port Said in 1915. A French tricolour flies from the Bessonneau hangar; all the remaining hangars are by Hervieu. The two British officers and a lady friend make one wonder about security. *(Michel Benichou)*

QM Grall and 2Lt Ledger took off from Port Said on N.16 at 10.00 on 21 April 1915 to reconnoitre Bir El Abd, Katia, and Romani. Grall later reported that they had an engine problem half an hour after leaving Port Said. Landing safely on the sea, the wind blew them towards the land. After five hours the Nieuport beached on the narrow sandbar separating Lake Bardawil from the sea. They were about 60 kilometres from Port Said, and began to walk back along the beach.

When they failed to return de l'Escaille had taken a Nieuport to search for them, also asking the Suez Canal Company to send one of its tugboats, *Hardi*, to explore the coast. Spotting *Hardi* at sea Grall and Ledger returned as quickly as possible to the floatplane, where they found three armed Bedouin examining the machine. Both parties were equally surprised, recovering quickly Grall began shouting and brandishing his water bottle. Shocked, two of the Bedouin ran away, the third raised his arms in surrender. *Hardi* sent a boat for the men, returning to Port Said at 16.30 with both flyers, and their prisoner. Provided with a naval armed party, *Hardi* sailed again at 21.30 to recover the Nieuport, returning the following morning with N.16 in tow.

SM Grall and 2Lt Ledger took off from Port Said on N.11 at 08.39 on 4 December 1915, once again to reconnoitre the coast and Lake Bardawil. Again, they failed to return. At first everything proceeded well, a few boats were seen drawn up on the coast of the lake but nothing of importance. Then at 09.54 their engine started to fail and they turned back towards Port Said. At 10.41 they crashed into the sea two miles off the fishing village of Tineh. Holding on to the remains of the Nieuport, an onshore wind slowly drifted them inshore until they were able to swim to the beach. The machine drifted ashore, and they were able to recover their clothes.

When de l'Escaille realised that they were not returning he again set out to look for them, with *SM mec* Victor Rigobert Marchal as observer.

> I pushed up to Bardawil, but did not see anything, despite my low altitude (200m). Knowing from previous experience that an aircraft is difficult to discover, I flew back to the sea and dropped to 50m in height. I then discovered the aircraft washed up on the coast, completely broken by the sea. Having been able to make sure, by descending to 10m and seeing their tracks, that the pilots had taken the road to Port Said, I returned.

The two shipwrecked aviators returned to Port Said on foot, a 35 kilometre march from Tineh where they had come ashore.

Difficult flights could be balanced with celebratory occasions. On 18 August 1915, at a parade attended by French and British naval and military personnel, *LV* Henry de l'Escaille was presented with the British DSC by Vice Admiral Richard H. Peirse, Commander in Chief, East Indies and Egypt Station.[69]

Another overlooked aspect of the base was the hard work put in by the *mécaniciens* of *l'escadrille*. From the outset they had demonstrated their skills by rebuilding the Nieuport *monoplace* N.7. Lack of replacement machines forced the rebuilding of several Nieuports damaged or worn out in service, including N.11, N.14, N.15, N.16 (more than once), and possibly others. Their workshop was a single Bessonneau hangar at the base, it was only space where work tables and some machine tools could be set up and complete aircraft and engines worked on. De l'Escaille recognised the importance of their hard work, 'The intense use of the machines was possible thanks to the professional skills of the repair teams and their hard work.' Some relief in the form of three new 80-hp Clerget-engined Nieuports, N.18, N.19 and N.20, came in mid-April, there would be no more until October.

Group of sailors and airmen aboard *Aenne Rickmers*, 17–22 January 1915 (ARDHAN).

Front row standing L-R: unknown, Lt Robert Gaskell, RNR (ship's master), *Mot* Julien Levasseur, an Egyptian officer, Capt James Herbert, *LV* de Saizieu, Capt R.E. Todd, remaining are unknown. *Mot* Fernand Baille is standing behind Levasseur, between the RN ratings is *Mot* Edouard Duffaud. Group also includes several RN and RM (white and dark uniforms) junior ranks, the Egyptian officer, three Syrian boatmen, and possibly two Greeks. All very typical of *Aenne Rickmers'* heterogeneous crew.

A Typical Cruise

Most of the operations and flights made by *l'escadrille* throughout the remainder of its time at Port Said were very similar. Reports surviving for most of the flights from the seaplane carriers. Therefore, we will look at just one typical cruise in detail and pick some highlights from the remainder.

Typical is perhaps a misnomer, none of the flights were exactly the same but, for all of them, the same risks were always present. The Turkish infantry proved time again to be excellent marksmen. It is perhaps fortunate that the Nieuports never had to contend with enemy aircraft, any of which would have been far superior to the obsolescent, and underpowered floatplanes. One constant concern for the crews was engine failure overland but sea take-offs and landings could provide their own excitements, as described by 2Lt Williams.

When at sea, after the machine had been hoisted out of the ship, itself a slow and difficult job with the rough winches, booms and tackle used for cargo lifting, began business of getting the engine started. The starting handle was carried in the observer's cockpit, which, being very small, made it very difficult for anybody larger than a small boy to swing it from anywhere inside the machine. One usually wound violently for about five minutes, at the end of which the engine would fire once, backfire twice, and catch fire, which meant hurling oneself out of the machine on to the undercarriage, and plugging the air intake pipe with the nearest thing at hand, usually one's own flying cap.

Having at last got the engine started, we commenced a game of ping-pong with the ocean, and on each bump the observer's head came in contact with the forward part of the cabane. An 80 h.p. machine has not much power with which to get "unstuck", and so, if there were no waves, we sometimes got off, but usually did not, and if there were any waves, we certainly never did, but buzzed about like a million infuriated bees, covered in spray.

During the forenoon of 7 April 1915 at *l'escadrille*'s base a *section d'aviation* was being prepared, comprising two Nieuports, with a pilot and observer for each, and four *mécanicien d'air*. Final checks and maintenance on Nieuports N.15 and N.16 were completed; the two floatplanes towed out and hoisted aboard *Rabenfels* between 15.00 and 16.00. The pilot, observers and mechanics arriving aboard shortly afterwards. Unusually only one pilot was aboard, *LV* Cintré, the observers were Capt Todd and 2Lt Paul. The observers were responsible for assembling the required maps and charts, and note book, and writing the post flight reports. The full cruise report was prepared by the OC of the ship, on *Aenne Rickmers* this was usually Weldon, on *Rabenfels* the duty fell to the senior observer until Capt R.E. Todd was appointed Intelligence Officer in June 1915.

The ship sailed shortly after midnight, arriving off El Arish the following morning. *Rabenfels* remained in the vicinity for next few days, frequently at anchor, using bearings from the town mosque's minaret to confirm its position. Anchoring, or even stopping for any length of time, would be unthinkable a few months later once the German U-boats made their appearance but at this time was obviously thought acceptable.

Not until the morning of 10 April could any flights be attempted. The weather had been reported as 'unfavourable for flight' on the two previous days. Cintré and Paul set out on N.15 at 07.35 and returned at 09.32. They had flown over El Arish, then 20 kilometres inland to Bir Lahfan and to Bir el Murra, in the plain of El Sirr, another 30 kilometres. At Bir el Murra they found a large camp site capable of housing up to 7000 men, including a hospital. They also reported a 'Large wooden shed, shaped as if to hold one aeroplane, situated south of the camp.' This report was one of the reasons the EDRFC made a bombing raid on Bir el Murra on 16 April. In the afternoon Cintré and N.15 were off again, this time with Todd as observer. Over a two hour flight they covered the coast from El Arish up to Gaza. Until Gaza only a few small camps were seen. Inland from Gaza three large camps, each holding a brigade of 5–8000 men with artillery were reported. Todd considered it worthwhile to attempt shelling the camps from the sea, suitable arrangements being put in hand.

Off Gaza the following morning the coastal clouds were too low for a proposed Gaza – Jaffa (now part of Tel Aviv) flight but, on the advice of Cintré, an attempt to reach Beersheba was decided upon. Leaving the ship at 08.50, Cintré with Paul on N.15, crossed the coast 15 minutes later, and followed the main road straight to Beersheba some 55 kilometres inland, arriving there at 09.38 having climbed to 900 metres. The garrison had shrunk considerably since the attack on the Suez Canal, Paul reporting a few tents and 2–3000 men. At 09.50 they turned back towards the coast.

When approximately 25 kilometres from the coast an exhaust valve rod broke, causing an explosion blowing a hole in the cowling. Engine revolutions fell to 1000, 100 to 150 below normal, and the floatplane began to lose altitude. Closing the coast Paul reported a 'body of 30 men with Maxim gun apparently waiting for seaplane's return, as not noticed on outward journey. They did not hit us though they fired at us while within range, although at the time the plane was only 200 m.' Cintré landed the damaged floatplane short of the ship, which quickly steamed up and hoisted them aboard.

Capt Todd, as senior observer, commented in his final report, 'One cannot but comment on the coolness and skill of the pilot *Lieut de V* Cintré. Indeed, on this his maiden voyage as a pilot, the only one on board, his flying has been admirable. This with his cheeriness, which is unfailing, made him a most desirable companion.' Cintré was subsequently awarded the British DSC for this flight.[70]

Shortly before the return of the Beersheba flight, *Rabenfels* was joined by French pre-dreadnought *Saint Louis*.[71] In the afternoon Cintré and Todd made two attempts to direct the fire of the ship on to the camps reported the previous day. Both N.16 and, a quickly repaired, N.15 had engine problems and were unable gain sufficient height. An attempt the following morning was more successful. Lacking wireless, smoke signals were used to direct naval gunfire.

> For co-operation with artillery the squadron was provided with an infernal type of firework, called a "smoke-bomb". These were issued in four forms — "Trainit", which made a long trail of smoke, and three

other kinds forming one, two, and three balls of smoke respectively. All these four types were bottle shaped, fitted with a ring at the top, and in order to ignite them, this ring, and about a yard of string attached to it, had to be pulled slowly out. The time between the beginning of the pull and the burst being only four seconds, one had to be pretty quick getting them down the bomb-tube.[72]

Unfortunately, smoke signals were often confused with the bursts of Turkish anti-aircraft fire.

Setting off on N.15 Cintré and Todd directed a successful morning shoot by *Saint Louis*' broadside 138.6 mm guns on the camps at Gaza. 'I was able to confirm the observations made during my reconnaissance [on 10 April]. After the ship first fired, the camps swarmed with men scattering for cover in all directions. There could not have been less than 12000 to 15000.' An attempted afternoon shoot failed due to a problem with the timing and ship did not fire. N.16, having engine problems, then failure and had to volplane down to the sea. *Rabenfels* sailed for Port Said that evening, arriving just after dawn on 13 April. The end of a fairly 'typical' cruise.

The French armoured cruiser *Montcalm* (1900) seen here at Sydney, Australia in 1911. Although most of the mid-ships is covered the sun shade, space to carry a Nieuport was even more limited than on the British cruisers. The best location would have been between the second and third funnels, but hoisting the machine would have had to involve modifying some boat davits. *(State Library of New South Wales)*

A Return to Akaba and the Red Sea

Throughout 1915, the Royal Navy was mostly operating in the southern Red Sea, the French Navy patrolling the northern part of the Red Sea. This resulted in a call for *l'escadrille* to provide a Nieuport for the French Navy in May 1915, for a return to Akaba.

The French cruiser *Montcalm* had been on the South Pacific Station pre-war and assisted the Royal Australian Navy Squadron in the capture of German Samoa and New Guinea.[73] In February 1915, *Montcalm* sailed for Suez, arriving after the Turkish attack had been repulsed. The cruiser remained as part of the *3ème Escadre* for the remainder of the year. In May she was tasked to revisit Akaba.

Nieuport N.20 with pilot LV Destrem and observer 2Lt F.O. Baxter, joined the ship at Suez and sailed 8 May for the head of the Gulf of Akaba. This was Baxter's first operational trip following transfer from the EDRFC, although Destrem was quite familiar with the area having been aboard *Minerva* in December 1914. Their experiences this time around were to be all too familiar as well, as Baxter's report relates.

Between 11–14 May, at Akaba, 'Four trials were made here to attain sufficient altitude to proceed up Wadie Yetham [Wadi Ithm]. Owing to the rarity [sic] of the atmosphere, & the intense dry heat carburation was not good, &

further very strong gusts, and down draughts made it impossible to mount beyond 500 meters. The trials were made each at a different time and conditions.' On the morning of 14 May the Nieuport suffered an engine failure and the plane had to glide back to the ship. The problem was repaired whilst *Montcalm* proceeded from Akaba into the Red Sea to visit Yenbo. On 16 May a morning flight was made over Yenbo, noting a few trenches but no signs of troops. Baxter did note that 'The largest house in the town had a large red crescent flying, outside a horse & carriage standing the only sign of life in the town.'

Proceeding next to Port Sudan, where they remained for two days probably coaling. *Montcalm* then returned to Akaba arriving on the morning of 23 May. Two flights were made looking for mines. Finding three newly placed mines close to their previous anchorage, *Montcalm* anchored some distance away. During an early morning flight the next day it took over an hour to climb to 600 meters and they returned to the ship. An afternoon flight was more successful. Taking off at 16.20 they reached 700 metres and were able to proceed 20 kilometres up Wadi Araba before returning to make observations around Akaba and the entrance to Wadi Ithm. Then down the gulf to Wadi Taba, mid way between Akaba and Faroun Island on the west coast, before returning to the ship at 18.00. *Montcalm* sailed from Akaba two hours later, arriving at Suez on 27 May.

Whilst this brought visits to Akaba to an end, *l'escadrille* still had work to do in the Red Sea. This time working with the Royal Navy in the south.

Hardinge (1900), an armed troop transport built for the Royal Indian Marine, could carry up to 1400 troops and was armed with six 4.7 inch guns. During the defence of the Suez Canal *Hardinge* was hit several times by Turkish guns, having to move out of canal into Lake Timsah to avoid sinking in the channel. The French ship *Requin* located the guns which were quickly silenced by her single forward 274 mm gun.[74] *Hardinge* was lucky to avoid any casualties and was quickly repaired.

RIMS/HMS *Hardinge*. Throughout June 1915 *Hardinge* had mixed success operating Nieuports N.16 and N.18 in the Red Sea. The Nieuports were probably carried on the upper deck aft of the funnels, the handling boom is seen here rigged from the lower main mast. *(EMK)*

On 4 June, at Suez, *Hardinge* took onboard a section comprising N.16 with LV Cintré and Major H.P. Fletcher, plus 2 mechanics. Ship's boats were removed to clear space on the upper spar deck aft of the funnels, and a boom fitted to the main mast capable of handling the floatplane. Sailing the following morning, the ship arrived off Mowila (Al Muwaileh) around 06.00 on 6 May. A successful flight was made that morning, departing at 07.50 and returning at 09.30. Flying a few miles up the coast to Wadi Said and back to Mowila, dropping 'pamphlets & newspapers' at both places. Whilst little activity was seen at Wadi Said, returning to Mowila they were 'repeatedly under fire from the Fort.

The pilot and self both of opinion that only rifles employed. No guns were observed. 75–100 soldiers seen.'

An overnight inspection discovered engine damage in N.16's Clerget. *Hardinge* signalled for a replacement Nieuport whilst returning to Suez, arriving on the morning of 8 June. N.18 arrived and was taken aboard at 17.30. Somehow, *Hardinge* found deck space to accommodate both Nieuports, before sailing once more for Mowila. The next few days are best described as frustrating for all concerned, and a brief summary will suffice.

One of the two Nieuports carried on *Hardinge* was N.18 seen here at Suez on 4 July 1915. *LV* Cintré is giving a flight to one of the ship's officers, Lt Thomas Milne-Henderson, Royal Indian Marine. *(ARDHAN)*

On 10 June, N.16 burnt out a cylinder, and inspections found the wings to be in such poor condition that they were unfit for flight. Then N.18 attempting to take-off had engine problems and nearly caught fire from leaking petrol. The ship headed south towards Jeddah whilst repairs were carried out. On the morning of 13 June, whilst still 15 miles from Jeddah, N.18 was hoisted out. The engine ran well but the sea was too rough to allow take-off without risking serious damage to the floats. The following day the wind strengthened, so again no flying. It is not difficult to imagine the reaction of Commander T.J. Linberry, commanding *Hardinge*, to all these delays. But, knowing the weather in the Red Sea at this time of year was mercurial, and hoping for better weather in a few days time, he decided to return to Suez to take the time to thoroughly overhaul the machines and his ship, which was suffering from 'condenseritis'[75] and in need of coal. On arrival at Suez the two Nieuports were off-loaded and sent to Port Said for overhaul. They returned a week later, joined by *LV* Cintre, observer 2Lt Hillas, and four mechanics. After taking all on board, *Hardinge* sailed for the south on the morning of 24 June, planning an ambitious aerial survey of the coast from Mowila to Jeddah, a distance of some 750 kilometres.

In comparison to the previous cruise the next few days were almost trouble free, both weather and Nieuports cooperating. Between 27 and 30 June two daily flights, one morning and one afternoon, except for a single morning flight on the 29th, were made whilst proceeding down the Arabian coast from Mowila to Jeddah. Most of these were made on N.16, on only one flight did the engine run rough but cleared itself and did not repeat, probably a little sand in the fuel line. However, the landing after the morning flight on the 29th in a heavy swell but no wind resulted in damage, and N.18 completed the final two flights. No photographs were taken and no maps have survived. On completion of the survey *Hardinge* returned to Suez and off loaded the section. Successfully completing *l'escadrille*'s final visit to the Red Sea.

Gallipoli

Aenne Rickmers sailed from Port Said on 24 February 1915 on another trip to Gaza. Aboard were two Nieuports, N.11 and N.17, with pilots *LV* Destrem and *QM* Grall and observers Capt Todd, 2Lt Paul, and 2Lt Williams. They made three unsuccessful flights, one to Beersheba and two to Jaffa, before being ordered to the Gulf of Smyrna (İzmir) on 3 March. At maximum speed, all of 10 knots, *Aenne Rickmers* arrived off the entrance to the Gulf during the morning of 6 March. They were to join a small bombarding force under the command of Admiral Peirse. The Nieuports were to spot the fire of two pre-dreadnoughts *Swiftsure* and *Triumph* and the cruiser *Euryalus* (flag) on to the port of Smyrna with the intent of making it incapable of being used as a base by submarines.[76] However, ongoing engine problems prevented either Nieuport from spotting. On the night of 10–11 March *Aenne Rickmers* was lying at anchor off Chustan (or Long) Island when, at 02.05, she was hit by a torpedo.

Turkish torpedo boat *Sultanhisar*, seen here pre-war, was one of four ships identical to *Demirhisar*. They were fitted with one bow torpedo tube and two on a revolving mounting aft of the funnels. They also had a light 37mm gun each side aft of the bridge. Its already low profile could be reduced further by lowering the mast.

In early March 1915 *Demirhisar*, a small Turkish torpedo-boat, slipped out of the Dardanelles and into the Aegean.[77] When new she could steam at 26 knots, now just 16 knots was her maximum speed, but her three torpedo tubes still worked. Her Turkish commander *Yüzbaşi* Lütfi Talat and crew of five officers and 20 men were subordinate to German *Kapitänleutnant* Freiherr von Fricks, another officer and five petty officers. She attempted to attack the fleet at Tenedos but was detected and was lucky to escape. The small ship sheltered during daylight hours in the small port of Chesme (Çeşme) on the mainland across from the island of Khios. Two of the torpedoes were expended in an unsuccessful but undetected attack on two ships, possibly *Triumph* and a collier, just off Chustan Island on the night 9–10 March. The following night, leaving Chesme for Smyrna to refuel and rearm Fricks turned *Demirhisar* into the Gulf of Smyrna and sighted a ship at anchor. As the last torpedo hit *Aenne Rickmers*, *Demirhisar* slipped away into the night, entering Smyrna harbour before dawn.

Aenne Rickmers at Mudros awaiting repairs. The torpedo hit on the starboard side and the hole cannot be seen in this view. However, the cargo removed from the forward holds to lift the hole clear of the water is piled on the aft well deck and deck house. *(EMK)*

The torpedo struck on the starboard side of *Aenne Rickmers'* forward hold. The original cargo probably saved the ship. The hold was filled with baulks of white oak which both diminished the force of the explosion and provided vital floatation over the next few days. The adjacent hold was filled with sacks of antimony ore which buttressed the bulkhead. In the afternoon, with a collision mat covering the hole, *Aenne Rickmers* sailed to Mudros, Lemnos island, escorted by *Swiftsure*. The collision mat did not long survive the open sea, fortunately the shored up bulkhead between the two holds held. Nevertheless, it was a long, cold night passage to Mudros, which was reached with relief the following day.

Many misadventures were to befall *Aenne Rickmers* before she could be repaired. For details of these you must look elsewhere,[78] but one story must be told. Shortly after their arrival it became known that *QM* Hervé Grall had been awarded the *Medaille Militaire* for his exploits at Akaba. Weldon recalled, 'We were all delighted. No man deserved it more. We all, officers and men, both French and British, donned our best uniforms and fell in on deck, and Destrem, after a short speech in French, pinned the medal on Grall. I made a short (very) speech also in French (bad), and then called for three cheers for Grall.' Grall also received the British DSM from Admiral Peirse on 14 April 1915 during a special Anglo-French parade at Port Said.[79]

Eventually, *Aenne Rickmers* was beached at Mudros but, being a low priority, it was 12 May before a temporary repair was finished. Whilst marooned at Mudros her original crew were gradually taken away for more important tasks, never to return. So, a new crew had to be found to sail her to Alexandria for permanent repair. John Kerr, previously master of the *River Clyde* prior to it being converted to a landing ship, was appointed master and a raggle-taggle crew put together. They arrived at Alexandria on 16 May with 600 tons of water in the fore hold due to failure of the temporary patch. For now we leave her at Alexandria to be repaired and refitted.

De l'Escaille remained in the dark about *Aenne Rickmers* until 14 March when he was informed that the ship had been torpedoed. At the same time he received an order to send *Rabenfels* with reinforcements to the Dardanelles. *Rabenfels* was immediately recalled from the Palestine coast, and prepared for Gallipoli. She sailed on the evening of 18 March, arriving at Mudros three days later. Aboard she had three Nieuports N.14, N.15 and N.16, *LV*s de l'Escaille, de Saizieu and *QM* Trouillet with two British observers

Tour boats were regular visitors to *Aenne Rickmers*, although the attraction was probably not the Better 'Ole but the baulks of white oak it revealed. The wood was invaluable in shoring up the trenches on Gallipoli.

QM Hervé Grall is rarely seen with a smile on his face. He has a good reason to smile in this photograph having just been awarded the *Medaille Militaire* at a parade on the deck of *Aenne Rickmers*. (ARDHAN)

Capt Herbert and 2Lt Hillas, and a maintenance party from *l'escadrille*. There were no machines remaining at Port Said, the base being left in the hands of a small party commanded by Cintré. This fact, once communicated to General Maxwell, quickly resulted in an urgent request for *l'escadrille*, or at least a portion of it, to be returned to Port Said.

Before being recalled, *Rabenfels* sent several flights over Gallipoli. Most notable was a visit to the Gulf of Smyrna where, on 26 March, a flight by Nieuport N.16 (pilot and observer are unknown) over Smyrna spotted 'a two-funnelled torpedo-boat with two torpedo tubes had come out from near Fort No.1 and had passed through a passage about 100 yds. wide between three sunken ships at the entrance.' *Demirhisar* had been found. Although rearmed, the torpedo-boat scored no further successes. She was eventually trapped by superior forces on 16 April, after an attempt to sink the troop transport SS *Manitou*, run aground and destroyed on the Greek island of Khios.

One of the Nieuports, probably N.17, being brought ashore at Tenedos Island on 8 April 1915. A mainly French landing party with a Royal Navy petty officer and rating standing watching. *(Album of Cmdr James Plumpton RNR)*

Returning to Mudros *Rabenfels* spent the next few days establishing a French base on Tenedos. When she sailed for Port Said on 31 March two Nieuports, N.14 and N.17, with pilots de Saizieu and Trouillet, the British observers Herbert and Hillas, a French petty officer and 10 *matelot* were left on Tenedos. Nominally, according to de l'Escaille they were '*sous les ordres du* Commander Samson.' As Samson and his squadron had only begun landing on Tenedos on 26 March, it is difficult to see what orders he could have provided, even if he was aware of their existence.

The small, under equipped unit did the best it could. They were tasked to support the French forces during the upcoming landings. Prior to the landings they made several flights, although details of only a single flight have been located. On 16 April 'A French seaplane flew as far as Paleo Kastro reporting troops and trenches and making a sketch of positions.' There is no shortage of paleo kastros, or ancient castles, along the adjacent Asiatic coast ranging from the ruins of Troy to abandoned Ottoman watch towers. The most likely candidate was on the shores of Eren Keui Bay, approximately 40 kilometres from Tenedos, well within the Dardanelles. It would have been a noteworthy flight.

The French landed on the Asian shore at Kum Kale on the morning of 25 April. It was intended both as a diversionary attack and to protect the Helles landings from Ottoman artillery firing across the straits. They quickly occupied the fort and village. Ottoman counter-attacks were held with great gallantry by the outnumbered French.[80] The detachment made several flights over Kum Kale, again only a single flight can be detailed. In the evening of the first day, a flight by de Saizieu on N.14 spotted and reported, by his unnamed

Taken in late 1915 a view showing *Raven* and *Anne* together at Port Said. *(Robert Cyril Vickers Photographs, John Oxley Library, State Library of Queensland)*

The first ship on the left is a standard merchant ship, then comes *Raven* with canvas aircraft shelters rigged, and *Anne* with the wooden bulkhead along the port side of the aft well deck. Side by side the two ships are very similar, but they do have individual features to help identify them. *Raven*'s is mentioned in an earlier photograph, whilst *Anne*'s was the bowed out centre to the wheelhouse and bridge structure, a feature typical of Rickmer's built ships of the period.

observer dropping a message bag, Turkish reinforcements marching up from the south. This timely information prevented the French from over extending themselves. They were withdrawn on the night of 26/27 April.

The two Nieuports continued operations, mainly over the Asian shore, from 26 April to 14 May. These included reconnaissance and bombing flights and attempts to control naval gunfire. At some point in their exile at least one of the floatplanes was converted to a landplane.[81] Finally, on 25 May, they were loaded on to the cargo ship *Crosshill* for return to Port Said. Both Nieuports, probably the aircrew and mechanics as well, were now in poor condition requiring major overhauls.

At Port Said, in May 1915, *Rabenfels* was the only available seaplane carrier until *Aenne Rickmers* returned from repairs at Alexandria in July. She was fully employed on regular trips along the Palestine coast through May and into August.

Aenne Rickmers returned to Port Said on 1 July. Whilst Kerr and some of his officers remained with the ship, a new crew had to be found. Weldon was informed that he would have to raise a crew himself.

> Well, we set about it. We got firemen all right. And at last we even got a crew of sorts. They were the sweepings of Port Said, and I don't think any of them — with one exception — had ever been on a ship before, except perhaps to carry coal on board. The bo'sun had been to sea years ago on a Khedival boat — that's why he was made bo'sun — but as he had been working as *bash suffragi* (head waiter) in the Eastern Exchange Hotel for the past four years he was a bit out of practice. I doubt if any ship ever sailed with such a crew since the days of the Ark.

Rabenfels became HMS *Raven II* on 12 June 1915,[82] and *Aenne Rickmers* became HMS *Anne* from 5 August 1915. As HM Ships they were part of the Royal Navy with naval crews, and flying the White Ensign. Lt John Jenkins, RNR, remained as captain of HMS *Raven* and HMS *Anne* retained Lt John Kerr, RNR.[83] Both ships also received a vintage 12-pdr gun on a high angle mounting on their stern. Now carried aboard Royal Navy seaplane carriers, the French naval *l'escadrille de Port-Saïd* embarked on several longer ranged cruises, deep desert flights and other expeditions. Some of these will be examined below, starting with a visit to the Gulf of Alexandretta.

The Gulf of Alexandretta and the Railway

Both British and French navies had regularly visited the Gulf of Alexandretta since the beginning of the war, *Doris*' December 1914 visits have already been mentioned. The main reason for this ongoing interest was the Baghdad Railway which ran close to the coast around the top end of the Gulf. It was the main route for the Turkish and German forces in Palestine, and Mesopotamia, as detailed earlier.[84]

On the 20 April *Rabenfels*, just returned from spotting at El Arish and Gaza for *Saint Louis*, welcomed aboard a replacement section comprising Nieuports N.18 and N.20 with pilot LV Cintré, observers Capt Todd and 2Lt Paul with four mechanics. Sailing the following day *Rabenfels* arrived off Mersina (often Mersin in contemporary reports) on 24 April, there joining two French cruisers, *d'Entrecasteaux* and *d'Estrées*,[85] who would keep company over the next week. The intention was to perform a complete survey of the coast and inland areas from Mersina to Alexandretta. This was accomplished, with few problems, over seven flights.

The highlight of this visit was a successful bombing attack on the railway yards at Tarsus. On 29 April Nieuport N.18, with pilot Cintré and observer Todd, left *Rabenfels* at 09.30 to reconnoitre the railway between Mersina and Adana. There was little change to report from a flight made five days earlier other than a considerable number of trucks in the sidings at Tarsus. Arriving over Tarsus flying at 850 metres, approximately 90 railway trucks (box cars) were observed on sidings near the station. Todd dropped a single bomb on the target 'producing an explosion among the trucks out of all proportion to the size of the bomb used. It must be concluded that the wagons contained either ammunition or some explosive. The wreckage among the trucks and the station was considerable.' The sound of the explosion was heard aboard the ships some 35 kilometres distant likened to distant thunder.

Rabenfels arrived back at Port Said during the afternoon of 1 May, returning to the Gulf of Alexandretta just over a month later, this time in company with the cruiser *Jeanne d'Arc*,[86] to launch the first of many bombing attacks on the Chicaldere bridge.[87] Located about 20 kilometres from the coast its four

Lieutenant John Jenkins, RNR, captain of *Rabenfels*/*Raven* throughout her time as a seaplane carrier. Jenkins was an experienced RNR officer having served since 1911. *(EMK)*

Lieutenant John Kerr, RNR, captain of HMS *Anne*. He assumed command of *Aenne Rickmers* in May 1915 at Gallipoli, and remaining in command of *Anne* when she was formally commissioned as a ship of the Royal Navy on 5 August 1915. *(S/Lt H B Buck album, Australian National Maritime Museum)*

steel truss spans carried a single track across the meandering River Ceyhan, some 50 kilometres east of Adana. With the small bombs available, a direct hit could, at best, delay deliveries of supplies for a few days. However, at 08.10 on 7 June, SM Grall and 2Lt Paul set out on N.11 to bomb the bridge.

> At the French Admiral's order a flight was made starting from five miles north of Port Ayas to Chikadir [sic] the object being to destroy the railway bridge over the river JEIHAN IRMAK.[88] Two bombs were taken but no damage was effected from either. In the pilot's opinion the first bomb fell within a few metres of the abutments of the eastern end of the bridge; the second probably fell in the water as no smoke could be seen anywhere. No troops were seen in the neighbourhood of the bridge but after returning to the ship a bullet hole was found in the wing of the plane.

In August both seaplane carriers, now renamed *Anne* and *Raven*, briefly operated together in the Gulf of Alexandretta. On the morning of 18 August, *Anne* rendezvoused with *Raven* and the cruiser *Jeanne d'Arc* at the entrance to the Gulf of Alexandretta. Between them the seaplane carriers had four Nieuports and crews. Weldon noted, 'It was too stormy for a flight, so we had to wait until the next morning, when we went into the Gulf of Tarsus [Mersina] and hoisted out both our 'planes. Grall and Fletcher [N.11], Herbert and Destrem [N.20] flew in ours: and de Saizieu and Ledger [N.17], Trouillet and Paul [N.14] flew in the *Raven*'s.'

Anne's Nieuports had mixed results. In the morning Grall and Fletcher reached Adana and dropped a bomb on the station then proceeded to Tarsus, dropping their second bomb on that station and returned safely to the ship. At the same time Destrem and Herbert, also with two bombs, attempted to get off but, as Weldon recalled:

> ... after they had made several unsuccessful attempts to get off the water, we had to hoist them on board again. Destrem was much disgusted and excited. He rushed up to me and said, "Weldon, I do not like going round and round like what you call dog with rabies, but I do try to conserve my hair!" The fact was that he was a heavy man and, the bombs being also weighty, the 80-h.p. engine was not powerful enough to lift the machine.

In the afternoon, without bombs, they were able to get off and take photographs of Adana town and station.[89]

Raven's morning flights had returned after a few minutes reporting poor visibility overland. In the afternoon an attempt to reach the railway ended very quickly when both machines had to return with broken 'stays'. N.17 was otherwise undamaged and repaired overnight. But N.14 'broke a stay trying

Nieuport N.14 alongside *Raven* on 19 August 1915. Attempting to take off in rough seas in the Gulf of Tarsus one of the port float struts splintered. Pilot and observer, *SM* Trouillet and 2Lt Paul, were unhurt, the Nieuport was repaired back at Port Said. The Nieuport crashed during a reconnaissance to Beersheba on 9 October 1915, the machine was destroyed and its crew, once again *SM* Trouillet and 2Lt Paul, taken prisoner. *(ARDHAN)*

to get up with the result that propeller struck float heavily & smashed float and propeller, strained engine & framework, & is now out of action for rest of the trip.'⁹⁰

On 20 August it was intended that a floatplane from *Anne* would make a reconnaissance of the railway from Osmaniye (10 kilometres east of Toprakkale junction) east to the tunnel construction site. If successful, in the afternoon the three floatplanes would bomb any locomotives found along the line. Unfortunately, Destrem and Herbert were unable to take-off in the morning owing to a complete lack of wind. As clouds were seen to be accumulating over the mountains in the direction of the proposed flight it was called off. Instead an afternoon attack on the bridge at Chicaldere was substituted, the ships making for Ayas Bay at the mouth of the Ceyhan river.

Between 15.15 and 15.30 the three available Nieuports were able to take-off, each carrying two bombs. First off, from *Raven*, was N.17 with de Saizieu and Ledger.

On *Raven*'s bridge wing July 1915. L-R: not known, 2Lt Horace Ledger, *LV* de Saizieu, Capt J. Jenkins, not known. *(ARDHAN)*

They dropped their bombs, apparently without result, and returned to the ship shortly before 17.00. *Anne*'s N.11, with Grall and Herbert, left the water at 15.24 and after climbing for half an hour headed inland at 600m. They reached the bridge at 16.07 and dropped a bomb from 800m, no smoke was seen so it probably dropped in the river. The second bomb was dropped from 900m, it fell 150m NW of the bridge. On returning to the ship bullet holes were found through the tailplane and left-hand wing. Destrem and Herbert, N.20, were able to get off the water but, unable to climb above 300m, had to return to the ship. This time, Destrem may well have been pulling his hair out.

All ships now turned down the coast, returning to Port Said on 23 August. *Anne* would return briefly at the end of January 1916, otherwise this ended the *escadrille*'s direct involvement in the Gulf of Alexandretta. Immediately after returning to Port Said, *Raven* sailed for an extended refit at Alexandria. Returning to Port Said in October she would not be used again by the *escadrille* until December. *Anne*, however, was fully employed over the following months.

Anne at Ruad and Musa Dagh

On 30 August, *Anne* sailed once more from Port Said with a section comprising N.16 and N.17 with pilots *LV* de Saizieu and *SM* Trouillet, observers Major Fletcher and 2Lt Hillas with five mechanics. Also on board were 23 Turkish prisoners, with one sergeant and ten men from the Glasgow Yeomanry as guards. All the prisoners were over military age, mostly sailors of captured schooners, several coming from the island of Ruad, just off the Syrian coast; Weldon's orders were to repatriate them. First stop was a routine reconnaissance the following day over El Arish by Trouillet and Fletcher on N.16 to check for any changes since the previous visit on 13 August, no reinforcements were seen. Then on to Ruad.

Ruad (Arwad), is just over two kilometres from the coast near Tartus. A small flat rock, no higher than 25 metres above the sea and barely 800 × 400 metres in size. It is surrounded by the remains of a Phoenician curtain wall, with an almost equally ancient harbour facing the mainland. Its principal fortifications date from the Crusading era, when the island was their last stronghold in the Holy Land, finally falling in 1302. In 1915 it had a population of around 3000 eking a living from trading, fishing and sponge diving, and dependant on rainfall, or mainland springs, for water.[91]

Amiral Dartigue du Fournet was instructed to occupy the island, taking possession of the island in the name of France on 1 September 1915. *LV* Albert Trabaud, from *Jeanne d'Arc*, was appointed governor, and was provided with a force of three *Enseigne de vaisseau*, a doctor, an interpreter and 80 *matelot*. In addition to the party's rifles, it was provided with two *65mm canon de débarquement*, and two machine guns. The force was accommodated in two houses on a small islet linked to the parade ground by a footbridge. The houses were contiguous to a small octagonal battery, of indeterminate age, protecting the harbour. Also based at Ruad were two armed tugs, *Laborieux* and *Cydnus*, both with a single 47 mm gun. The continued survival of Ruad, and Castellorizo similarly located off the Turkish coast, are a testament to sea power. Without control of the sea they could not have survived as French enclaves.

Anne arrived off the island just as the ceremony of occupation was concluding. Weldon provided each prisoner with fifteen days rations and handed them over to the new *gouverneur*. Trabaud immediately released the natives of Ruad to their homes. The remainder, from Tartus and Beirut, were instructed to find accommodation on the island until they could be returned to the mainland. Following a brief tour of the island, Weldon returned to *Anne* which sailed south for Haifa at 17.00. At 22.00, whilst still north of Beirut, three agents were landed with instructions to make their way to Ruad and report to Trabaud once they had completed their missions – the start of Ruad's role in intelligence gathering.

Later in the year Weldon and *Anne* revisited Ruad. On his earlier visit to the island Weldon had noted a Crusader era plaque on the wall alongside the gate to Ruad's citadel. Taking this as inspiration, he presented Trabaud 'with a flag we had made for him on board. It was a fine flag—with the ancient arms of Ruad (a lion chained to a palm tree) tastefully embroidered. Trabaud was tickled to death.'

In the early hours of 2 September, from a position mid-way between Acre and Haifa and five kilometres off the coast, *Anne* prepared to send off two flights; first de Saizieu and Fletcher on N.17 who were to fly to Nazareth. De Saizieu got away at 09.45, circling for 45 minutes until they reached 1100 metres and headed inland some 40 kilometres towards Nazareth, which lies at 350 metres above sea level, but is surrounded by mountains up to 1850 metres in height. Slowly climbing to 1225 metres they had the town in clear view by 10.47, then turned back towards the coast keeping some distance south of their outward route headed for the Haifa-El Afule railway. Over Haifa they saw a few carriages near the station but no trains on the line, circling to drop pamphlets and letters written by 'Turkish prisoners of war in Egypt to their friends, saying how happy and how well treated they were.' They recovered to the ship at 11.20.

Whilst de Saizieu was still circling, Trouillet with Hillas on N.16 got away at 10.02 and quickly climbing to 300m headed off on their mission, *amiral* Dartige had requested that they search Acre Bay for mines. Heading for Acre they then flew up and down the coast several times in a zig-zag, keeping to the 10 fathom line. No mines were observed. Having now attained 1100m, Trouillet turned over land to check the Haifa to Acre railway, observing little activity. He then headed further inland to get a good view of the railway bridge at Kishon between Haifa and El Afule. Trouillet returned to the ship at 11.00. A wireless message was sent to the *amiral* with the results of the flight.

Anne now headed further down the coast to Jaffa. Where the last flight of the trip was on 3 September by de Saizieu and Hillas on N.17. Leaving the ship south of Jaffa, they flew inland to Ramleh then followed the railway south past the junction to Jerusalem, then returned to the coast at Nahr Suqrier, halfway between Jaffa and Ashkelon. They were brought aboard two hours after leaving the ship, half the time having been spent circled over the water until clouds over the land had cleared sufficiently to permit observation from a safe height. *Anne* then returned to Port Said and a brief rest before being called out again.

Musa Dagh is a coastal mountain located just south of the entrance to the Gulf of Alexandretta. In July 1915 the Armenian villagers close to the mountain received word that they were to be deported. As the Turkish forces approached the villagers retreated to the mountain and resisted all attacks for fifty-three days (21 July to 12 September). By early September they were reaching the end of their resistance, supplies and ammunition were running short. One of the leaders was able to contact patrolling French ships and an emergency evacuation quickly arranged. On September 12 and 13, despite a strong swell, the *Desaix, Guichen, Amiral Charner, Foudre* and *d'Estrées* using their boats and rafts managed to take on board all the refugees from the beach. Assisted by naval gunfire the rear guard, having retreated from crest to crest, were the last to leave. They were mostly taken aboard the sole British ship to be involved, *Anne.*

> One morning — it was in early September — we received orders to sail to assist the French blockading squadron in bringing away about six thousand Armenians who were fighting for their lives somewhere near Alexandretta. So off we pushed. I had with me Herbert, Paul, and eight of my faithful Glasgow yeomen. Next day we received a wireless from the French cruiser *Desaix*, telling us to take-off the rearguard of these Armenians, who were fighting in the Bay of Antioch.
>
> We arrived there at daylight, and brought off three hundred, including two poor old women who must each have been well over eighty years of age, but had both had their legs broken. When the first boat reached the shore – the Turks were gaily firing away all the time – it was met by the Armenian Commandant, who was carrying something wrapped up in a cloth, which he asked should be taken to the Admiral. He said that it was a proof that he and his friends were really fighting the Turks. A proof it was – for on opening the parcel we found a Turk's head freshly hacked away from its body.

> We got our refugees on board without much incident. As they came up the gangway, I had their arms taken from them — and what a collection these were! Every kind of gun, rifle and revolver that was ever made. Many of them were charged, so I put half a dozen Armenians to unload them. In some cases the only way to do this was to fire them, and, as I thought the Armenians knew more about antiquities than our men did, I gave the former the job. Incidentally, one nearly blew my head off, the bullet touching my hair just above my ear, and the explosion completely deafening me. One man arrived on board in a frock coat, carrying two rifles slung on his back and a Singer's sewing machine. He turned out to be the village tailor.[92]

Over 4000 women, children and men were rescued, approximately 1000 having died during the siege. The survivors were accommodated in a camp adjacent to the canal at Port Said, provided with a communal kitchen, water and baths, a hospital, clothing and workshops. The children were provided with an education by the local Armenian community. British attempts to recruit the men into military service failed, although a few were employed by the intelligence department. The French were more successful, around 600 men providing the backbone of *La Légion d'Orient*, later renamed *La Légion Arménienne*, serving with great courage in the Palestine and Syrian campaigns.

Not strictly part of *l'escadrille* story, although two British observers were aboard, but nevertheless a story that sadly still resonates a century later. However, the rescue redounds greatly to the credit of *la Marine Nationale*.

Back to Beersheba

What was so important about Beersheba that caused *l'escadrille* to make over twenty attempts to reach the town? Location, water, and the railway made Beersheba important during the First World War. The main attack on the Suez Canal in February 1915 was launched from the town and it became the inland anchor of the Gaza defences in 1917. The wells have been exploited since biblical times, but it was not until October 1915 that the railway arrived. An extension of the railway from Beersheba to El Auja and Kossaima was in work throughout 1916 and into 1917 until the battles for Gaza terminated construction. A direct flight from the coast at Gaza to Beersheba was some 55 kilometres one way. However, the most common approach was by way of Wadi el Ghuzze (Wadi Gaza) to a point someway south of the town, then following the railway construction to Beersheba. The return by a more direct line to the coast via Gaza then to the ship 8 to 15 kilometres out to sea. This way was closer to 150 kilometres for the round trip.

Early flights have already been mentioned. Between May and October 1915 there were no flights to Beersheba. During this period the concentration was

Nieuport N.11 being brought aboard *Aenne Rickmers* in March 1915, with LV Destrem and Lt Sir Robert Paul. The Nieuport led a long operational life taking part in eight trips aboard the seaplane carriers. The last being from *Anne*, a flight to Beersheba with SM Grall and Major Fletcher on 6 November 1915. Nieuport N.11 came to an end crashing into the sea close to west end of Lake Bardawil on 4 December 1915. The crew, Grall and 2Lt Ledger returned to Port Said on foot, and N.11's engine and instruments later salvaged. *(David Mechin)*

towards the northern section of coast and Red Sea, only from early October did attention shift once more to the area around Gaza. The first of the new series, on 10 October, ended in disaster.

Anne sailed on the 9 October and was off Gaza the following morning. At 10.10 Nieuport N.14, with SM Trouillet and 2Lt Paul as pilot and observer, set off for Beersheba. With petrol for three hours flight, they were not expected back for at least two hours. When last seen from the ship they had climbed to altitude and were headed inland.

When, after three hours they had not returned, Weldon sent off LV de Saizieu and 2Lt Ledger on N.17 to retrace their track. Leaving the water at 14.47, and climbing for height, they crossed the coast at the mouth of Wadi Gaza, seven miles south of Gaza, at 15.49. They then followed branches of the wadi to reach three miles west of Beersheba at 16.18. Unable to climb higher, and having reached the limit of their petrol, they had to turn back towards Gaza following the railway to Tel el Sheria. At this point they turned west directly to Gaza, crossing the coast at 16.51 and taking the water close to *Anne* at 16.58. Although not spotting the downed Nieuport they were able to complete most of the reconnaissance the Trouillet and Paul had been sent on.

Weldon always provided flares for the inland flights 'and instructed them that if they had to make a forced landing they were first to destroy their machines and then to try to make their way to the shore, where if they burnt their flare at night we might have a chance of taking them off'. For the next two nights *Anne* cruised off the coast; but with no result. A few days later a German wireless message was intercepted by *Montcalm* saying 'East of El-Arish an enemy reconnaissance plane has been shot down — machine captured, crew prisoners.' Some months later Weldon received a letter from Sir Robert Paul, dated 26 November from his prison camp near Constantinople, giving some details of what had happened.

> I was taken prisoner owing to the engine of the aeroplane failing on the way back from following a line to a town in Syria, where I had gone to do reconnaissance. Everything had gone well, and we had just turned for home when the engine started to weaken, though nothing to frighten us at first. However, it soon got worse, and when we were about twelve miles from the coast we had to come down. My pilot, a Frenchman named Trouillet, made a magnificent landing without upsetting. The only pity was we came down among a crowd of Arabs, who were on to us in a minute, before we had time to set light to the machine. However, the Arabs proceeded to smash it up pretty effectively, as they explained afterwards to the Turkish officers, they were afraid of it getting up and flying off on its own!!
>
> The Arabs gave us a bad time for a bit, and took our clothing, watches and cigarette-cases, etc. I offered them £200 to help us to escape, but it was no use. In the afternoon another aeroplane from the ship flew over us, no doubt coming to look for us. We spent the day with the Arabs. Trouillet had been taken in one direction and I in another.
>
> About eight in the evening three Turkish officers, two lieutenants and an aviation officer, arrived from the town over which we had flown in the morning to take us back there. I cannot say too much for the kindness of these officers. They returned all our property that the Arabs took from us, and gave us coats and food, mounted us on fine riding camels, and about 10 p.m. we started off southwards. Towards daybreak we arrived at the town, and just as we were arriving my camel took fright at some trucks and I fell off, but was not hurt a bit.
>
> When we arrived the officers took us to their tents, gave us hot tea, whisky and food, and afterwards a suit of pyjamas and camp-beds. In fact they did all they could for us while we were with them. Three days later we were taken by carriage to Jerusalem and thence by rail to Damascus and Constantinople. Trouillet and I have not been separated, and we have all we need, and are being very well treated.

The mention of the aviation officer was, of course, a way of informing *l'escadrille* that there were now some aeroplanes in Palestine. He was a little premature. A German NCO had been sent to Palestine where he organised an airfield near Beersheba late in 1915, but it would be April 1916 before any aeroplanes operated from the airfield.[93]

Both Paul and Trouillet survived captivity. Sir Robert Paul (5th Baronet of Paulville in the County of Carlow, Ireland.) returned to his home, where he died on 9 July 1955. Trouillet, who had been awarded the British DSM in December 1915,[94] returned to Saint-Raphaël but left the navy in July 1919. He died on 12 June 1948 in Constantine, Algeria.

Anne continued her cruise, but bad weather and heavy seas prevented most flights. One attempt damaging the propeller and, after that had been replaced, another flight failed when one of the bolts attaching a float strut fractured and N.17 had to be quickly hoisted aboard. *Anne* arrived back in Port Said on 16 October. Her next two trips, 19–24 October and 5–13 November, also included flights to Beersheba, five in total. The October attempt got within six miles, of the four attempts in November only one reached Beersheba.

Anne left Port Said on the evening of 5 November with a section of two Nieuports, N.11 and N.22, pilots LV de Saizieu and *SM* Grall, and observers Major Fletcher and 2Lt Ledger. The following morning, off Wadi Gaza, *SM* Grall and Major Fletcher on N.11 left *Anne* at 09.20, they reached a point ten miles south of Beersheba on the railway line to El Auja, then turned up towards the town itself. They noted construction on the El Auja line, 'Fifty men working on what appears to be a Decauville Railway S. of Wadi Gaza on Auja Road. About 3 miles of this appears to be completed but does not cross Wadi Gaza to Bir Saba.'[95] Reaching Beersheba at 10.20 they spent twenty minutes circling and mapping the area, noting '18 guns, 300 tents, 60 camels and 300 horses, 17 piles of stores, 1 engine and 6 trucks in the station, 4 huts 50′ × 20′ and 1 Hangar or Garage 100′ × 40′.' All the time over the town they reported being shelled and fired on by rifles, without being hit. Having seen enough, Grall turned towards Gaza, returning to the ship at 11.45.

An afternoon attempt by de Saizieu and Ledger, N.22, failed to cross the coast returning with engine problems. However, the engine was running well enough the following day to enable the Nieuport to carry de Saizieu and Fletcher along the railway from Ramleh north to Tul Keram. A bomb was dropped on the camp at Ramleh and one on Tul Keram station. Short of petrol on return, they had to land 3 miles from ship in heavy seas — damaging a float and one wing on landing.

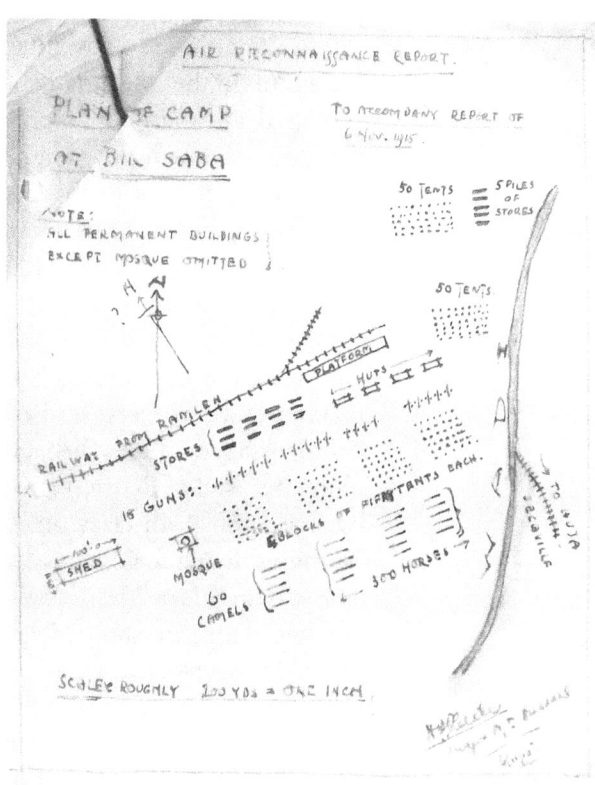

Major Fletcher's map of Beersheba following his reconnaissance flight of 6 November 1915 with *SM* Grall on Nieuport N.11.

Two attempts on 11 November, using N.11, to reach Beersheba were not so successful. In the morning two cylinders burnt out, a quick engine change permitted an afternoon flight which also had to return with damaged cylinders. Following overnight engine repairs an attempt to bomb Gaza was made, but another burnt out cylinder forced a return after just 10 minutes. Flying in air with a constant suspension of fine sand was hard on the engines, both the Le Rhônes and Clergets could take only so much mistreatment of this kind.

Anne sailed once more from Port Said on the evening of 20 December, with a section comprising N.17 and N.22, two pilots *LV* de Saizieu and *SM* Grall and observers 2Lts Ledger and Williams. Their orders were to make reconnaissances of Beersheba, Khan Yunis and Rafa, Abu Aweigila (some 30 kilometres west of El Auja) and finally, Lake Bardawil and Bir el Abd.

Arriving off Wadi Gaza on the morning of 21 December, rough seas and fog caused flying for the day to be cancelled. The following morning, with calm seas and clear skies, de Saizieu and Ledger on N.17 left the water at 09.07 with petrol for three hours flight. Forty minutes later they crossed the coast headed for Beersheba. When they failed to return by 12.30 Weldon 'gave them up as lost. At 2 pm I intended sending in the second machine with Grall Pilot & Williams Observer to search for them but as a fog was overland & coming up from SW so I decided it was not advisable.' After a short delay, as the fog was no worse, Weldon ordered *Anne* to within three miles of Khan Yunis and sent off Grall and Williams to reconnoitre Khan Yunis and Rafa, with strict instructions to return if fog returned. They made a useful flight over the two towns and returned after an hour and a half. *Anne* cruised close to the coast all night with a special watch being kept for flares from the missing airmen. The only lights seen were Turkish watch fires along the coast at mile intervals between Khan Yunis and Gaza.

The weather on the morning of 23 December being favourable, Grall and Williams set off on N.22 to fly to Beersheba, to complete the reconnaissance and look for signs of the downed machine. The left the water at 08.43 and passed over the coast at 09.01 just north of Gaza, following Wadi Imleih towards the railway at Tel el Sheria. At 09.30, over the railway, they turned south to follow the line to Beersheba, arriving at 09.48. Williams reported that 'There have been considerable alterations in the dispositions of the camp since the last reconnaissance.' This was on 6 November as noted above. Williams reported at various locations around the railway: 9 zaribas, 390 tents, and 27 piles of stores. The animal lines now held more than 2000 camels and 950 horses, 22 field guns. In the station were three goods trains. They then flew south along the railway towards El Auja. The light railway was no further advanced but work continued on bridges, embankments and the track bed. Reaching the head of Wadi Gaza they turned back to the coast, returning to *Anne* at 10.38, having seen no signs of the missing machine. Ordered back to Port Said, *Anne* arrived at 09.15 on 24 December.

The first news of the missing airmen was again thanks to a radio message picked up by the cruiser *Montcalm,* 'We shot down a plane near Beersheba and made the pilot prisoner, the observer, an English lieutenant, died.' A brief note was received in early January 1916 from de Saizieu, that had been carried and delivered by an agent. After a brief account of the flight, de Saizieu said that Ledger had been killed. Over Beersheba, a rifle bullet struck the petrol tank causing an irreparable leak. De Saizieu was able to coax the machine to within 20 kilometres of the coast before being forced down. The undercarriage collapsed on hitting the ground but both airmen were able to climb out of the wreckage, where they were quickly surrounded by a troop of Bedouin horsemen. Whilst, in accordance with standing orders, de Saizieu attempted to set fire to the machine, Ledger fought off the Bedouin with his carbine and a revolver. Both flyers were then involved in a desperate hand to hand fight before Ledger was killed and de Saizieu captured. Later that evening de Saizieu was handed over to the Turkish military authorities.

LV Louis Marie Jules Barthélémy de Saizieu was not released until the end of 1918. He was awarded the *Croix de Guerre* in April 1915 and February 1916, and the British DSC in December 1915.[96] After his release and re-training at Saint-Raphaël, he commanded *CAM de Toulon* (June 1919 to July 1921), then the *aviso Bapaume*, which had been fitted with a flying off deck forward of the bridge, during flying trials in 1921–22. He was made *Officier de la Légion d'Honneur*, 1 January 1921. De Saizieu left the navy on 1 October 1925. He died in Paris on 22 June 1951.

Two more attempts were made to reach Beersheba in 1916, both from *Anne* — on 11 February and 16 April — only the first was successful. In the end, *l'escadrille* reached Beersheba on 14 flights and failed on eight occasions. Their observations were always important and welcomed. Further visits to Beersheba would be made later in 1916 by machines of the East Indies and Egypt Seaplane Squadron.

Reinforcements

At the end of 1915 the original pilots remaining active were *LV* de l'Escaille and Destrem, and *SM* Grall. *LV* Cintré was posted to Brindisi as second in command of a new unit forming there. Two new pilots were posted to Port Said late in the year to complete their training, *QM* Raymond P. Bourgeois and *SM* Charles Albert Gramont.[97] Bourgeois may have learnt to fly using the conventional control system and had to be re-trained on the old Nieuport system, to which he quickly adapted. Gramont started to learn to fly in April at the Nieuport School in France, when posted to Port Said he was not fully qualified. His training took several months, not flying his first operation until March 1916.

In May 1915 eight Nieuports had been available for operations, by year's end just three remained operational; three (N.11, N.14, and N.17) were lost on operations and N.15 and N.16 considered only suitable for local patrols and training; leaving just N.18, N.19 and N.20 available for operations. Fortunately, three new Nieuports, N.21, N.22 and N.23, had been delivered in October, with three more arriving early in the New Year.

At the request of the Italian government, three Nieuports intended for Port Said had been deployed as an *escadrille* based at Brindisi. De l'Escaille was understandably unhappy at having three new Nieuports taken from him. He forwarded a report in mid-October to the *ministre de la Marine* pointing out his need for additional aircraft. In this he was supported by *amiral* Frédéric Paul Moreau, commander of the *3ème Escadre*, who also pointed out the need to strengthen *l'escadrille de Port-Saïd*, whose role was of extreme importance. General Maxwell added his weight to the request, saying that the French airmen were rendering great services with the only hydroplanes he had at his disposal. Both officers emphasising the importance of *Anne* and *Raven*, pointing out that without the French floatplanes they would be useless. On 5 December the *Ministre* approved the transfer of *l'escadrille de Brindisi* to Port Said.

The aircraft (NB.1, NB.2, and NB.3) with pilots *SM* Jeanblanc, and *QM* Roussillon, and some of the mechanics departed for Port Said by month's end. Although damaged by weather during transport, two of the Nieuports, NB.1 and NB.2, were quickly repaired and put to use, possibly by robbing the third machine for spare parts. Included in the package was the sole Caudron Type R, which was poorly received, de l'Escaille commenting on 24 February, 'The Caudron aircraft arriving from Brindisi cannot be carried on our aircraft carriers. Given its poor manoeuvrability on the water and the arrangements at Port Said, its use as a hydroplane on site would present great difficulties.' The unloved Caudron was sent to *CAM de Salonique* on 14 March 1916, it was employed by the *CAM* until February 1917, when it was returned to France.

The six new Nieuports had 100-hp Clerget engines but were not immediately welcomed at Port Said. They were different to the ones the pilots were used to as the Nieuport style of controls had been replaced with conventional controls. Recapping, the Nieuport controls comprised foot pedals for wing warping and a control column to operate the elevators and rudder. In the conventional (or Deperdussin) style, still used to this day, the wing warping or ailerons, and elevator respond to the control column and the rudder is worked by the foot pedals. It required a mental reset to fly the new machines.

With the New Year, 1916, came changes and fresh challenges.

CHAPTER 5

The East Indies and Egypt Seaplane Squadron

The formation of the East Indies and Egypt Seaplane Squadron (EIESS) may be dated from the arrival of HMS *Ben-my-Chree* in Port Said on 12 January 1916. But the origins of the EIESS can be traced back to early December 1915.[98]

On 7 December 1915, whilst planning for the evacuation of the Dardanelles was ongoing, Lieutenant-General Charles Munro, commanding the Mediterranean Expeditionary Force telegraphed the War Office.

> Excellent work is being done for the Army by Royal Naval Air Service. But it is essential, from the military point of view, that the establishment of Royal Naval Air Service should be increased with the least possible delay. The increase should comprise two additional wings, including as many fighting machines as possible, also two seaplane carriers. The Admiralty have been informed by telegram from Wemyss of the necessity of this increase.[99] A statement is needed urgently saying whether these additions will be made.

The War Office forwarded this request to the Admiralty on 10 December 1915, adding:

> In this connection the (Army) Council are also of the opinion that the Turkish Lines of Communication are specially vulnerable to attack by aircraft, and that valuable assistance could be given by seaplanes under Naval escort operating off the coast of Southern Syria. The Council would be glad to learn if Their Lordships will be able to make provision for such Naval cooperation, and also whether it will be possible to comply with the requirements set forth on the accompanying telegram from [General Munro].

On instruction from the Chief of the Admiralty War Staff, Vice Admiral H.F. Oliver, the Admiralty replied on 15 December that the requested reinforcements were, subject to availability, being sent and that, 'On arrival of the *Empress* in the Eastern Mediterranean there will then be 3 Seaplane Carriers there,[100] one of which it should be possible to detach to Egypt when required.

This would provide for the naval co-operation desired by the Army Council in operations off the coast of Southern Syria.'

The successful evacuation of the Dardanelles, the last troops left Anzac and Suvla on 20 December 1915 and Helles at 04.00 8 January 1916, changed the dynamics of the Eastern Mediterranean campaigns. In consequence of which, *Ben-my-Chree* was ordered to Port Said, and *Empress* also directed to Egypt.

Ben-my-Chree at Gallipoli. Two folded Shorts, 841 and 842, are visible in the entrance to the hangar, both left the ship before it sailed for Egypt. Short 842 made the first and second successful torpedo attacks in August 1915.

The East Indies and Egypt Seaplane Squadron

In January 1916 Vice Admiral Sir Rosslyn Wemyss was appointed Commander-in-Chief, East Indies and Egypt Station, relieving Vice Admiral Peirse. With Wemyss came a fresh outlook and changes that would directly affect *l'escadrille de Port-Saïd*. The first sign of change was the arrival in Port Said on 12 January 1916 of *Ben-my-Chree*. Her commanding officer, Squadron Commander Cecil John L'Estrange Malone, was immediately involved in discussions with Admiral Wemyss and his staff regarding future employment and organisation of the seaplanes and ships at Port Said. Malone clearly understood that their prime role would be that of intelligence gathering, starting his report of 21 January 1915 with the following:

> It appears that the only sources of intelligence available for the whole country between say, Alexandretta and Kossaima, are:—
>
> (a) Agents
>
> (b) Seaplanes
>
> Of the two, (b) is probably more reliable, when it is feasible, as the observations are made by British Officers instead of presumably foreign civilians.[101]

He then proceeded to use the above argument to support a claim for additional machines, materiel, men and ships. These are summarised in a telegram forwarded to the Admiralty by Admiral Wemyss.

> To carry out efficiently work required by the Army in Egypt to be done by Naval Air Service following are requirement of material and personnel in addition to those already available.
>
>> Seaplanes 225hp Sunbeam Short or 200hp Canton Unne Short, 11 in number.
>>
>> Pilots with experience of above machines 15 in number.
>>
>> Mechanics including proportion of Petty Officers and trained men 20 in number.
>
> If the above total is not available request information as to what establishment is to be supplied in the first instance what periodical supply may be expected and at what intervals in order that arrangements may be made.

Rear Admiral Sir Rosslyn Wemyss, Commander-in-Chief, East Indies and Egypt Station. Seen here as a Rear Admiral, he was promoted Vice Admiral on 7 December 1916. *(LoC)*

To which the Admiralty replied on 26 January.

> Short seaplanes are being sent out by first available ship.
>
> Further seaplanes can be supplied at the rate of one per fortnight.
>
> 225 white tractor seaplanes [presumably, Wight Admiralty Type 840] can be supplied for instructing new pilots if required.
>
> 15 pilots asked for are not available, four will be sent shortly and some others as they become available.
>
> 20 ratings will be sent.

Not everything Malone requested, there was for example no mention of spare engines which he made a point of requesting in his reply to the above on 31 January, but overall not a bad result.

Also on 26 January, Squadron Commander Malone was appointed to command the seaplane carrier squadron, *Ben-my-Chree, Empress, Anne* and *Raven*. Admiral Wemyss also informed the Admiralty that he had granted Malone 'Rank as Acting Commander [ie; Wing Commander, RNAS], placed in command of Squadron of 4 Seaplane Carriers'. However, the Admiralty quickly replied it was 'Unable to confirm Commission as Acting Comdr on account of Seniority. If necessary, more Senior Lt or Comdr can be sent out from England.'[102] So, for now, Malone continued in command of the EIESS but with the substantive rank of Squadron Commander.

Initially, the Squadron was a Franco-British unit, de l'Escaille and Malone seemingly working together in a spirit of co-operation. The major difference would be the source of their orders. To some extent *l'escadrille* had been General Maxwell's private air force. Whatever the General wanted was passed on to Major Elgood through the Intelligence Office in Cairo, who then wrote the orders to be passed on through Weldon. However, Maxwell had been replaced by Lieutenant General Sir Archibald Murray in January. Murray was not as interested in naval aviation and, to cut a long story short, on 12 February 1916, Royal Navy Egypt Order No.39 was promulgated.[103]

39. ORDERS FOR SEAPLANE SHIPS ON THE EAST INDIES STATION

All seaplane ships are under the orders of the Naval Commander-in-Chief, East Indies, and henceforward the Naval Service will take over the internal administration of H.M. Ships "Anne" and "Raven II."

2. The Ships will form a squadron under the command of an officer appointed by the Naval Commander-in-Chief, and his address will be:–

> The Officer Commanding Seaplane Ships,
> Navy House,
> Port Said.

3. General Head Quarters, Ismailia, General Head Quarters, Cairo, and the Admiral Commanding French 3rd Squadron are requested to forward their requirements for flights to the Commander-in-Chief, East Indies, or in his absence to the Senior Naval Officer, Egypt, who will arrange.

4. The departure of seaplane ships and their objective will always be intimated to those concerned, and any requests for flights in the neighbourhood of ships can be sent by W/T and will be carried out if feasible.

5. Flights will be discussed and arranged by:–

~~(1) Lieut.-Colonel Elgood, C.M.G., or his representative.~~ [Crossed out on original.]
(2) The Officer Commanding Seaplane Squadron or in his absence the Senior Flying Officer in the harbour.
(3) The Officer Commanding French Seaplanes, under the orders of the Senior Naval Officer.

6. The Officer Commanding the Seaplane Squadron will acquaint the Senior Naval Officer at Port Said of the proposed departure of any seaplane ship or ships, explaining their mission, on obtaining his approval, and that of Admiral Commanding French 3rd Squadron or his representative when flights are in the French area. Sailing orders

for all seaplane ships are to be made out by the Officer Commanding Seaplane Squadron or his representative, and a copy sent to the Senior Naval Officer, Port Said, for information.

7. Reports of flights will at once be communicated by W/T to the Commander-in-Chief, East Indies, and to the Admiral Commanding French 3rd Squadron. The former will forward them to Headquarters, Ismailia and Cairo.

The full written reports of aerial reconnaissances are to be forwarded as soon as prepared by the Officer Commanding Seaplane Squadron or his representative to:–

(1) Commander-in-Chief, East Indies.
(2) General Headquarters, Ismailia.
(3) General Headquarters, Cairo.
(4) Admiral Commanding French 3rd Squadron.

8. The positions of seaplane ships at sea will be made daily at 10 p.m. to Naval Commander-in-Chief and Vice Admiral Commanding French 3rd Squadron.

9. Seaplane ships whilst in the zone of operations of the French Fleet are to consider themselves under the orders of Vice Admiral Commanding French 3rd Squadron, and will carry out the Sailing Orders given them before leaving, unless the French Vice Admiral considers it imperative they should be changed or modified, when he will inform the Naval Commander-in-Chief.

> R. E. WEMYSS
> Vice Admiral
> Commander-in-Chief

Admiral Wemyss clearly agreed with Malone's assessment of the value of reliable intelligence, as the order cut Lt Col Elgood and Captain Weldon out of the link, although the latter continued to serve aboard *Anne* until late January 1917. Elgood, perhaps coloured by this experience, had little respect for the new arrangements, writing in his memoir,

> His [de l'Escaille's] little unit of six Nieuports got through a wonderful amount of work, and the contrast between the elaborate organization of the British Naval Seaplane Detachment (which replaced de l'Escaille's command in January 1916), and the modest equipment of the French squadron was very striking. Incidentally, the British, unfortunate in their choice of engine, flew no farther or no more frequently over enemy territory than their predecessors had done.[104]

The order also effectively formalised the East Indies and Egypt Seaplane Squadron (EIESS), bringing together *Ben-my-Chree* and *Empress* with their RNAS detachments, *Anne*, *Raven* and *l'escadrille de Port-Saïd*. Soon to join them was a French seaplane carrier, *Campinas*. East Indies & Egypt Station — Seaplane Squadron printed forms were in use by 11 February 1916 with a Report of Flight from *Anne*. Although an identical typed heading was in use as early as 27 January. The printed form was in use until April 1918, and continued in use well into the RAF period.

Sqn Cdr Malone spent much of his time, including several meetings with the army and C-in-C's staffs, defining the uses to which the EIESS could be put and could hope to achieve with the equipment available. The emphasis was changing from defence of the Suez Canal, the paramount role of *l'escadrille de Port-Saïd* and RFC throughout 1915, into preparation for offensive action. The prime need now was to keep the army informed of activity along the Egypt–Palestine border, involving long inland flights from off El Arish to Kossaima, El Auja and Beersheba, monitoring troop movements and traffic on the railway leading to Beersheba, plus other long range reconnaissances along the open flank of the Palestine and Syrian coast as required. Next in order of importance, a visit to the railway bridge at Chicaldere, at the head of the Gulf of Alexandretta, with a view to its destruction by bombing. Additional bombing raids on camps and railway junctions in the Haifa–Ramleh area were to be made whenever possible. Finally, cruises into the Red Sea to reconnoitre the Hejaz Railway, and along the eastern coast of the Red Sea from Akaba to Aden, and beyond.

Having observed the formation of the EIESS, it is time to examine the ships, men and machines.

Ben-my-Chree and *Empress*

All the seaplane carriers employed by the Royal Navy during the Great War were converted merchant ships, most of them passenger ships. The first three, *Empress, Engadine,* and *Riviera*, were cross-channel ferries owned and run by the South Eastern & Chatham Railway Company (SECR), mostly on their Folkestone-Boulogne and Dover-Calais routes. The fourth, *Ben-my-Chree*, was owned by the Isle of Man Steam Packet Company and ran the summer holiday service between Liverpool and Douglas.[105]

Ben-my-Chree (2651 tons gross, 390 feet oa.) entered service in 1908. She was built by Vickers Sons & Maxim at Barrow-in-Furness. A turbine steamer, on trials she made 26.83 knots. The fourth ship to be requisitioned, on 2 January 1915, she was taken from her lay-up berth in Liverpool, across the river Mersey to the wet basin at Cammell, Laird and Company, Birkenhead, to be converted into a seaplane carrier. The aft passenger accommodations were replaced by a large slab-sided hangar capable of housing up to five floatplanes, usually a mix of Short 184 and Sopwith Schneider/Baby, with all necessary maintenance and handling facilities. An armament of four 12-pdr low-angle guns, two forward and two aft, and two 3-pdr anti aircraft guns mounted on the forward corners of the hangar roof, was provided. *Ben-my-Chree* was commissioned on 23 March 1915.

Appointed to command *Ben-my-Chree* was Squadron Commander Cecil John L'Estrange Malone, RN. He entered the Royal Navy in 1905, passing through the Royal Naval College at Dartmouth, then followed the typical career path of a Royal Navy officer spending time in battleships and cruisers followed by the Lieutenants course at the RN College, Greenwich. Promoted Lieutenant on 15 December 1911 he was selected to take a course of flying training at the Naval School, Eastchurch, following which he became the third man to fly from the deck of a ship, flying Short S.38, Sommer Pusher Biplane, RNAS No.2/T2, from a flying-off platform installed on the the pre-dreadnought *London*. On 4 July 1912, when approximately 19 miles from Portsmouth, Malone took off whilst the ship was steaming at 12 knots into a 20 knot wind, reportedly the 'Machine apparently lifted without run.'[106] He flew on ahead to land at the Royal Marines' Eastney Barracks. Since that time Malone had been actively involved with many aspects of naval aviation, including the development of torpedo aircraft. He was appointed to command HMS *Engadine* in August 1914, shortly after his promotion to Squadron Commander, and was in overall command of the seaplane carriers of the Harwich Force during the Cuxhaven Raid of 12 December 1914.

Squadron Commander Cecil John L'Estrange Malone, RN, commanding officer *Ben-my-Chree* and East Indies and Egypt Seaplane Squadron. He is seen here at Mytilene inspecting some Greek irregulars during 1915. *(Attrill Family Collection)*

Photographs of Erskine Childers in RNAS service are very rare. He is seen here, to the right of Malone, in the middle of the back row, whilst on board *Engadine* for the Cuxhaven Raid of 25 December 1914. *(YEORN Personnel 0297. Courtesy of The National Museum of the Royal Navy)*

99

Ben-my-Chree was sent to Harwich for a short while, but at the beginning of June was ordered to Dardanelles, arriving at Iero Bay, Lesbos, on 12 June. She operated Short 184s and Sopwith Schneiders. Highlights of her time at Gallipoli include; torpedo attacks, the Short flown without an observer could just lift a 14-inch torpedo, on Turkish ships in the Dardanelles; bombing attacks on the railway near Dedeagatch (Dedeağaç) in Bulgaria, modern Alexandroúpolis, Greece; spotting the fire of the big gun monitors. On 2 September 1915 she rescued survivors from the troopship SS *Southland*, torpedoed south of the island of Strati, just over 30 miles from Mudros. In total she took aboard 815 Anzac troops and crew members, among whom were nearly 200 injured, 15 seriously.

On the day that Cape Helles was evacuated, 27 December 1915, *Ben-my-Chree* should have received a signal ordering her to proceed to Port Said to join the flag of the C-in-C East Indies. Somehow the signal went astray and the first *Ben-my-Chree* knew of it was a testy inquiry from the Flag asking what had become of her. She sailed for Port Said, two weeks later than intended, on 10 January 1916, making landfall in Egypt at the Damietta Light, some 50 kilometres north west of Port Said, at 07.00 on the morning of 12 January.[107]

Empress (1695 tons gross, 320 feet oa.) entered service in 1907, she was built by William Denny & Brothers at Dumbarton on the Clyde for the South Eastern & Chatham Railway Company. A turbine steamer, on trials she made 22.25 knots. She and her half-sisters, *Engadine* and *Riviera*, were requisitioned in August 1914. Initially used as aircraft transports, they were fitted with temporary wooden platforms surrounded by canvas shelters fore and aft for the Cuxhaven Raid on Christmas Day 1914. This was the first carrier strike in history and, although a failure, provided many lessons for future operations. Six months after entering service, *Empress* received a more permanent conversion, in which a steel hangar, equipped with shutter doors

In this snapshot at Port Said *Empress* is being assisted to her mooring by two of the Canal Company's tugs. *Raven* lies at anchor at the left of the photograph, with the Suez Canal Company office building framed between the two ships. The lighthouse towers over Port Said in the distance. *Empress* arrived at Port Said on 23 January 1916. (FAAM, 1994-243-0016. Courtesy of The National Museum of the Royal Navy)

and heating arrangements, was constructed aft. The hangar had curved upper edges, making her the easiest of the three very similar SECR ships to identify. At each aft corner of the hangar an electric crane was fitted.

Her captain at Harwich had been Sqn Cdr Frederick W. Bowhill, a pre-war RNAS pilot. Since August 1915 she had been commanded by two non-flyers Lt Cdr Valentine D. English and Lt Cdr Reginald E. Marcon. The expansion of the German submarine offensive into unrestricted warfare in the late summer of 1915 was the cause of her detachment to Queenstown in southern Ireland, so that her seaplanes could patrol the sea lanes converging there. Like most anti-submarine patrols without the focus of a convoy it was ineffective and after four months *Empress* was released. Too slow to operate with the Fleet, *Empress* was sent to the Mediterranean to join the East Indies and Egypt Seaplane Squadron, arriving at Port Said on 23 January 1916. Marcon was succeeded by Lt Cdr Edward D. Drury, RNR, on 4 February, although another non-flyer he remained in command for the remainder of the war.[108]

Pilots and Observers

With regard to pilots, *Ben-my-Chree* was well provided. Squadron Commander Cecil John L'Estrange Malone, her commanding officer, was a qualified pilot. Chief pilot was Flt Cdr Charles Humphrey Kingsman Edmonds, the other pilots were Flt Lts John Thearsby Bankes-Price, George Bentley Dacre, and Maurice Edward Arthur Wright. All had served aboard *Ben-my-Chree* throughout the Dardanelles campaign. There was a fifth pilot, Flt Lt Archibald Spencer Maskell, he had broken his wrist during earlier North Sea operations and had spent most of his time as equipment and divisional officer. At Gallipoli, due to a chronic shortage of trained personnel, he often flew as an observer. He resumed piloting after arrival in Egypt.

Ben-my-Chree's intelligence officer and chief observer was Lt Robert Erskine Childers, RNVR, reprising his role on *Engadine*. Childers was an unlikely airman — lightly built, he was at 44 considerably older than his peers, needed glasses and walked with a limp due to sciatica contracted in his late teens. Of Anglo-Irish stock, he was a supporter of Irish Home Rule and had been involved in a gun running trip to Howth, a small harbour five miles north of Dublin, with his yacht *Asgard* immediately prior to the war. This incident was either unknown to, or overlooked by, the Director of Naval Intelligence, Rear Admiral Henry Oliver, when he made Childers the offer of a commission in the RNVR soon after the outbreak of war.[109] Prior to joining *Engadine* he had been briefly appointed to the air station at Felixstowe, then to the Admiralty, the intention being to put his intimate knowledge of the German North Sea coast and islands to good use. At some point, the charts and logs he kept during an epic 1897 voyage in the yacht *Vixen* through these very waters were used to prepare an updated and corrected chart of the area.

Privately printed during the war, and known as Childers' Charts, these were prized possessions of later RNAS pilots and Coastal Motor Boat crews who rated them better than the standard Admiralty charts. He had also used this knowledge in his book *The Riddle of the Sands*, the progenitor of the modern spy thriller. First published in 1903, it has seldom been out of print since.

Empress had arrived at Port Said on 23 January, but was completely unprepared and unequipped for Egyptian operations. Childers commenting,

> The *Empress*' Flight Commander is one [Flt Lt R.M.] Field who seems very capable & energetic. They have 2 225HP Sunbeam Shorts, like ours, & seem fairly content with them. Also 4 Monosoupape Schneiders which they dislike. The strange thing is that they have been using 135HP Canton-Shorts like those the *Ark Royal* had. We also had two decayed specimens which would not climb & were destroyed last August. Field says that they succeeded perfectly during six months of cruising & flying, mainly in the Atlantic, off Ireland, looking for submarines. No breakdowns of any consequence, climb to 7500 feet with passenger & bombs in an hour. Reliable. Easily controllable, etc, etc.

> The unfortunate Empresses have no photographic gear, or camera, having never even heard that photography was now a necessary part of our work! This gives one a glimpse into the condition of the Air Department. Field was much impressed by our photos. I visited *Empress* & gave the RNVR observer, a sub [Lt], hints about maps & work. He is quite ignorant of observation except looking for submarines. Asked if one took a map up with one. He has no office, & no maps yet.[110]

Also, aboard *Empress* when she arrived in Port Said were pilots FSLs C.V. Arnold and R.M. Clifford, the observer has not been identified. Flight Commander William G. Sitwell followed *Empress* out to take over as Chief Pilot. Flt Cdr Sitwell had joined the Royal Navy in 1905 following a typical career path. He learnt to fly in 1913, taking his 'ticket' (RAeC 505) on 2 June on a Maurice Farman at CFS Upavon. He served at the Isle of Grain, Yarmouth, and Redcar before being appointed to *Empress*. Flt Lt Field also remained in the ship, the two making a good team through the vicissitudes ahead. Not until 16 February did *Empress* venture a flight, which ended with the Short 8031 wrecked after side-slipping in from 100 feet. The crew FSL R.M. Clifford and 2Lt E. Williams (*le petit*) were apparently uninjured. Much hard work was necessary to bring her up to standard in time for the first Squadron operation in March.

We should now consider the subject of observers. By the end of January the lack of experienced observers was being felt throughout the EIESS. *Ben-my-Chree* only had one regular observer, Lt Childers, *Empress*' observer was lacking any training or experience, and many of *l'escadrille*'s experienced observers were being called back to their units as they were despatched to different fronts. To resolve the shortage, early in January 1916 Admiral Peirse had requested twelve army officers to be trained as observers.[111] Childers was charged with the creation of a training syllabus for the volunteers. He began 'writing a series of 'Practical Hints' under various heads: Navigation, Photography, etc & having them typed.' Once the school was established Childers' lectures covered the whole gamut of an observer's duties—Lewis guns and bombs, photography and cameras, wireless, navigation, basic seamanship and airmanship, spotting and intelligence reports. De l'Escaille reported in March that the course 'lasted three months, the French squadron providing practical instruction and some lectures. The results look very good.' He also expressed the hope that the new observers would remain with the *escadrille*. In the event, whilst some would serve briefly with the Nieuports, most would fly in the RNAS's Short 184s.

The lack of trained observers was felt throughout the RNAS at this time. Not until October 1915 was a school for observers established in the UK. The first graduates, early March 1916, were all posted to RNAS units in the UK or France.[112] Childers' school at Port Said was a stand alone achievement. Commenting on the situation, shortly after his return to the UK, Childers wrote:

> It appeared to be necessary to obtain observers from
> the Army, not only because Naval Officers were lacking
> but because military knowledge and experience were
> needed for the reconnaissance work of the Squadron.
>
> The officers in question (all Lieutenants or 2nd Lieutenants)
> made a favourable impression. They were well chosen by the
> Military authorities, who had evidently taken trouble to select
> young men who would do the R.N.A.S. credit; They were keen
> on their work and were learning it rapidly when I left.[113]

Shorts and Sopwiths

Returning to January 1916, it was also time for the French and British to visit and take stock of their new comrades. De l'Escaille reporting, in quite envious terms, the arrival of *Ben-my-Chree* and her floatplanes, in particular mentioning 'a robust fully-formed hangar, she carries 5 aircraft with wings that fold along the fuselage.' Of the Short 184 he wrote, 'These machines can carry a torpedo without an observer and then climb to 1000m in 50 minutes. There are five 250 kg bombs on board that they never used.'

The Short brothers Horace, Eustace and Oswald, began constructing balloons in 1902. Their first heavier-than-air products were the 1909 Short S.1 and S.2, copies of the Wright 1908 biplane. They followed the copies with six licence-built Wright biplanes. From there the brothers went on to specialise in designing and constructing floatplanes for the Royal Navy.[114] Their Type 184 was a product of a process of evolution over several years, benefiting from the first lessons of war, it was a large single-engined two-seater and was the most widely produced and used floatplane of the war. Close to 1000 examples were built by several manufacturers, of which over 300 remained in service at the end of the war. The Short 184 soldiered on long after the war in several smaller air arms. The last service use was by the Estonian Air Force as late as November 1933.

The basic design remained little altered throughout its long production run. The only major changes made were installation of different engines of increasing power, from 225 hp in the prototype to 275 hp at the end, and upgrades in armament. Early models were powered by a 225 hp side-valve V-12 Sunbeam Mohawk water cooled engine. A large box shaped radiator was located on top of the fuselage ahead of the wing, obstructing the pilot's forward vision. The observer's cockpit, behind the pilot, had no provision for armament in the early production machines, but it did provide storage for all his paraphernalia — navigational board and charts, W/T transmitter, flares, Very pistol, sea anchor and, on some North Sea operations, a basket of carrier pigeons. Later versions provided a mounting for a single Lewis gun, either a pillar mounting or a Short designed ring mounting, finally a Scarff Ring was installed in the observers cockpit. The EIESS often flew the Short 184 with two single Lewis guns, as will be discussed later. A torpedo could be carried, on two arched spreader bars, between the forward thrusting, flat bottomed, floats. Alternatively, racks for bombs could be installed below the fuselage, between the floats. A single-seat bomber variant will be discussed in a later chapter.

Port Said 16 February 1916, Short 8031 taxying from *Empress*. (Australian War Memorial H13641)

A tail-float with water rudder was installed on struts under the rear fuselage, small cylindrical air bags were also fitted to protect the lower wing tips. Spanning 63½ feet the Short's wings could be folded back alongside the 40 foot 7 inch long fuselage, reducing the width to 16½ feet. The prototype, *184*, had ailerons fitted to the upper wings only, initially unbalanced they were soon fitted with rubber springs to hold them in flying position. With these roll control was marginal, and a second set of ailerons were added to the lower wings on all subsequent machines. The Type 184 had a large, shapely fixed fin with a cut away ahead of the rudder hinge line to accommodate an aerodynamic balance surface on the rudder itself. A conventional tailplane with hinged, rubber sprung elevators completed the assemblage.

A few moments later the Short crashed, the crew FSL R.M. Clifford and 2Lt E. Williams were uninjured. Not so the Short, seen here being towed away by boats from *Empress*. (Attrill Family Collection)

Flt Cdr A.H. Sandwell, writing long after the war,[115] had the following opinion of the Short 184.

> It was a physical impossibility to fly a Short at much more than 75 miles an hour. If you tried to dive it steeply it would start taking the control away from you at, say, 65 mph, and would have flattened itself out before it picked up another ten miles an hour. No pilot was strong enough to hold the wheel forward so that it would continue to dive, and if he had been he would probably have broken the control wires, or the horns on the elevators themselves. Consequently, even if you had the height to spare, you could not get anywhere in a hurry on a Short by stuffing its nose down as you could on most land machines. It was, however, the pilot's dream for putting in hours — docile, stable, obedient, and thoroughly deserving its affectionate nickname Home From Home.

The 'two-two-five' Shorts, as the early 184's were known from their engine power, were fitted with a compressed air starting system. A contemporary writer described the preparations and take-off of a shore based Short floatplane in the following terms:

> [The pilot] switched on the magnetos and opened the cock in an air-bottle. A stream of compressed air hissed into the cylinders of the engine and turned it over, the pistons sucked in the petrol mixture, a spark fired it, and the high-speed engine began to run smoothly.

> He warmed up the oil, tested the engine full out, and then gave the signal for the chocks to be knocked away. The working party ran the seaplane down into the water. It floated clear of the trolley.
>
> When the engine was opened out the tail of the seaplane came up to the horizontal. It leaped forward, planing along the top of the water on the two floats. As the pilot pulled back the controls it skipped along with only the rear edges of the floats touching, taking little jumps off the surface as it encountered the tiny waves. And then it was in the air.[116]

The great weakness of the Short lay in its engine. Childers referred despairingly to 'flying 50 miles across a desert behind that engineering manual the Sunbeam,' an opinion shared by Flt Lt Dacre in his diary,[117] who several times writes of 'our unreliable Sunbeam engines.' Even de l'Escaille, in his brief contact with the Short, noted that 'The officers complain about their [Gnome] Monosoupape and Sunbeam engines. Particularly with the Sunbeam which breaks in pieces after 15 hours flying time and [they have] gone through 8 engines in 8 months for 3 aircraft.' On the same theme, *Ben-my-Chree*'s medical officer, Surgeon H.R.B. Hull, RN, who had a keen eye for what went on aboard ship, made the following comments on the reliability of the two types of floatplane.[118] Of the Short, 'They will carry fuel for 5 hours but in practice it is found that the engine will not last more than 3 hours before something goes wrong with it, they are most unreliable', and of the Sopwith, 'They cannot be relied upon for a flight of more than one hour's duration.' In defence of both engines and machines it should be noted that both had been designed for a temperate northern climate, not a hot Mediterranean summer. It became normal practice for the Shorts to be flown without engine cowlings in order to improve cooling.

Sopwith's Schneider was from thoroughbred stock. A keen flyer himself, 'Tommy' Sopwith had put all his company's skills into designing a small two seat racing biplane the Type St.B, later called the Tabloid. One was completed as a single-seat floatplane, Type HS, to compete in the 1914 Schneider Trophy race at Monaco, which it won convincingly at an average speed 85.8 mph. The Tabloid floatplane caught the attention of the RNAS, and a slightly improved version was ordered as the Sopwith Schneider Trophy Seaplane. This mouthful was commonly shortened to Sopwith Schneider as the machine began to enter service in the spring of 1915.

Of wood and fabric construction the Schneider spanned 25 ft 8 in and had a length of 22 ft 10 in. The fuselage was designed to separate aft of the cockpit for ease of stowage. In practice this feature does not seem to have been much used, the machine being small enough to fit comfortably into most seaplane carrier hangars. Power was provided by a 100 hp Gnome rotary engine.

The engine could be started from the cockpit by the pilot. He was provided with a geared handle with which to rotate the engine, priming the cylinders with a petroleum/castor oil mix, prior to applying a spark from the magneto and, hopefully, commencing ignition. The handle was fitted with a simple mechanical disconnect when the engine fired, however, if the engine backfired the handle would reverse and keep turning. This is what had happened to Maskell breaking his wrist, the handle then smashed all the instruments in the cockpit as it continued to rotate with the engine. Conventional fixed and moveable tail surfaces were fitted. On early production aircraft roll control was via wing warping, this unsatisfactory method was replaced by ailerons on later production machines, although the steering wheel was retained. Its weakest point was the design and strength of its float undercarriage, which frequently failed during operations.

The Schneider had the reputation of being a 'hot ship', it was very light and sensitive to control, and difficult to land. Gerry Livock recalled that they 'were considered to be very difficult to handle', particularly on the water. However, he also noted that, 'They were lovely little things to fly and, at 80 knots, were quite fast for those days.'[119] Flight Sub-Lieutenant G.F. Hyams, who was completing his training at Hornsea Mere in 1917, recalled his first encounter with the Sopwith Baby, 'They were absolutely perfect! They had Clerget engines ... They would all handle most beautifully and were really a delight to fly; you could turn them on a sixpence.'[120]

The Sopwith floatplane was never designed to be a fighting machine, being developed from a racing machines and their manoeuvrability was limited by the inertia of the floats. It was a willing horse, however, and could often be found with a Lewis gun or two pointing out at odd angles, or overloaded with bombs. Several of *Ben-my-Chree*'s aircraft were fitted to carry two Lewis guns over the wing centre-section. However armed, the little floatplane was out performed by almost any of the landplanes the Turco-German forces were able to bring against it.

Later versions of the Schneider were built with a 110 or 130 hp Clerget rotary engine in a redesigned cowling, they were known as the Sopwith Baby. Many of these were built by the Blackburn company and, officially, referred to as the Blackburn Baby. Whilst the Baby was an improvement on the original Schneider, later 'improved' variations of the design, known as the Fairey, or Parnall, Hamble Baby after their manufacturers, performed no better, even worse, than the original.

In total nearly 600 Schneider and assorted Babies were built in Britain, with approximately 100 more under licence in Italy. Many were still in service at the end of the war. The Norwegian *Marinens Flyvåpen* was the last user of the type, finally retiring their last remaining machines in 1930.

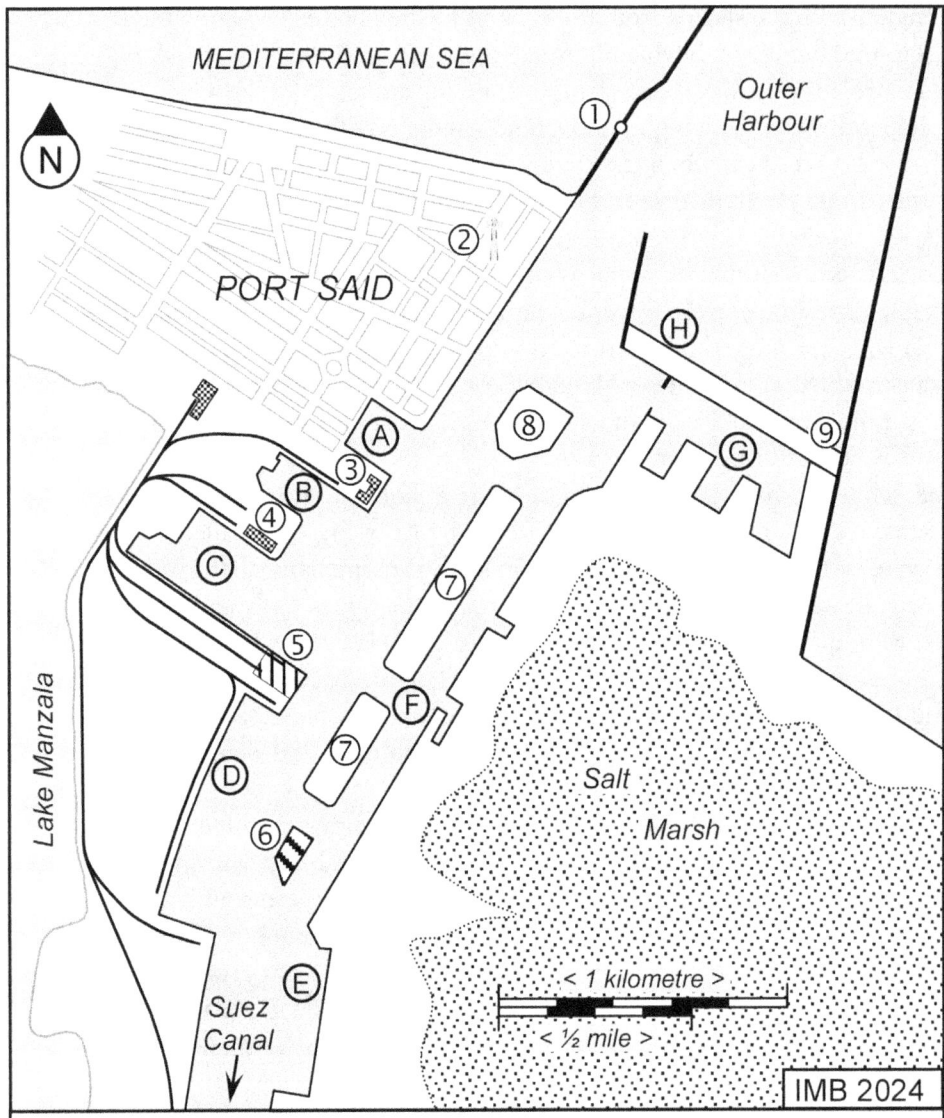

Port Said showing French and British Seaplane Bases.

An Island Base, 1916–1918

However, before returning to operational matters, there remained some essential problems to be resolved. The EIESS needed a land base. *L'escadrille de Port-Saïd* occupied the only available space within the port itself, land that was increasingly required for warehousing as the Egyptian Expeditionary Force grew in strength. It was suggested by the Administration Office, Port Said, that the EIESS should take over part of an island in the harbour as a Seaplane Depot.

> This station will be used for erecting and testing new seaplanes as they arrive, and for stowing spare seaplanes.
>
> By this means the fast ships can be kept going as continuously as necessary, damaged seaplanes or seaplanes requiring overhaul will be returned to Depot on ship's arrival in harbour, and new seaplanes drawn in lieu.
>
> The Depot will also be used:–
>
> a) For training Observers.
>
> b) Depot for spare Observers.
>
> c) For training pilots — if necessary.
>
> d) Depot for spare Pilots.
>
> e) Air Intelligence distribution centre.
>
> f) Miscellaneous duties at Port Said, such as supervising unloading of Air Service Stores on arrival from Home, a Lieutenant R.N.V.R. being specially detailed for this work, etc.
>
> This depot, which is situated on the "Turkish" Island, will be borne on the books of the *Ben-my-Chree* as a tender.[121]

Within the harbour at Port Said were a number of coaling islands, the southern most of these, Number 3 Island, was smaller than the rest and had, at one time, been used by the Turkish Authorities as a shipyard. Currently part of the island was being used as a shipyard for the coaling barges and tugs, operated by the Egyptian Coastguard. Known locally as Turkish, or Turk's, Island, opinions varied as to its habitability or otherwise. The island rapidly gained a number of nicknames; Treasure Island, Devil's Island and even, prosaically, *Ben-my-Chree II*. However it was named, and temporary and makeshift it may have been, the shore base served the RNAS and RAF up until the end of the war.

One of *Ben-my-Chree*'s pilots, Flt Lt G.B. Dacre, was appointed to the base and recorded the early days in his diary.

Feb 21 — I busied myself at the new air base where we are getting things into order. Rigging masts for wireless, 2 aeroplane tents, one for men's mess, one for officers' mess, and 5 bell tents were erected. A dark room is being erected, galley, and a motor ambulance and a motor travelling workshop have been carted to the Island.

Feb 22 — Got one new Short machine and 2 new Schneiders in several cases ashore at the base with the help of Arab labour. We also rigged up a fine field bakery by utilising old tipping trucks with one end biffed out. The existing Turkish Shed is having the entrance made bigger to house 4 big Shorts. A fine workshop & forge belonging to the Coast Guards is there for our use.

The island base of the East Indies and Egypt Seaplane Squadron 1916–1918. Seen here early in 1916 with the single, locally built, large white canvas hangar. The eastern side of the island remained in use as a shipyard for coaling barges and tugs throughout the war, the rest was slowly taken over by the seaplane base. *(EMK)*

Feb 24 —Busy at the shore station getting enormous crates full of seaplanes ashore with a small crane, removing parts of seaplanes & putting the crates aside for our living quarters. A large party of Arabs do quite a lot of heavy haulage.

Feb 25 —Busy on Station. Crocker & England arrived by P&O Caledonia. The former is going to be CO at the station.

Feb 29 —We commissioned the Shore Station. Flt Commander Crocker in command. At the time we have to be in tents, so we have had decks made of bits of packing cases & we have bought camp kit. Of course my camp bed collapsed absolutely directly I sat on it & sand was everywhere. Camping out is horrible after a comfortable existence aboard. Our mess tent, an aeroplane tent, is now fine & we have made ourselves as comfortable as possible. In spite of a small galley & necessary gear we had a fine 5 course dinner with desert [sic] & coffee. We have also got electric light in the messes & our tents run off the engine of the motor workshop.

Mar 3 —Two new machines in 6 cases arrived very badly damaged, about £10,000 damage, more work for us. Large gangs of Arabs lugged them ashore by force of numbers. Later two more Schneiders arrived, so now we have 9 machines here. The new shed is rapidly being erected, and my hut (half a seaplane case) is rapidly becoming a palatial residence.

Flight Commander William Reginald Crocker who had just arrived in Egypt, was killed overnight on 5/6 March as he was attempting to crawl underneath a train in the sidings at Port Said. Dacre recorded that he was taking 'a short cut to the jetty through the station. Here he apparently got caught between shunting trains & got squeezed. He was picked up by some sentries & taken to the hospital where he was operated on but died at 11AM this morning.' The investigation said that the Egyptian State Railway train was blocking a level crossing, suggesting that Crocker was too impatient to wait. In his stead, Dacre was appointed to command the base, making the first flight from the island on 8 March.

The Island Staff was also established at this time. It remained fairly stable throughout the life of the EIESS. One Flight Commander in overall command of the base also to act as test pilot, assisted by a Lieutenant, RNVR, two Military Officers —for the military guard provided for the base, one Military Officer —Adjutant, and clerical staff. The RNAS maintenance contingent was provided from the seaplane carriers and based at the island, being seconded to the carriers as required for operations. Pilots and observers, including a Senior Observer as discussed in a following chapter, were similarly based at the island.

Early days at the island 1. Two Sopwith Schneiders, 3721 and probably 3722, with another Schneider behind them. Schneider 3721 has a light bomb carrier installed beneath the fuselage. The variation in location of underwing markings is of interest. The bell tents suggest a very early date on the island possible February 1916. Schneiders 3721 and 3722 had been based on *Ben-my-Chree* at Gallipoli from August 1915, their only use with the EIESS was with *Raven* at Aden March/April 1916. *(Stone Family Collection)*

Early days at the island 2. Two Short 184s, 8004 and possibly 8075. Short 8004 was recorded on *Anne* 21/25 April, but 8075 was not operational until late July. The new hangar looms in the background, Flt Lt Dacre recorded it as being erected in early March.

The clerical staff provide a rare opportunity to recognise the work of the ratings and non-commissioned officers. Chief Writer P.R. Ridley, Writer Glass and CPO Bagge were responsible for all the office and clerical work aboard *Ben-my-Chree* (Ridley) and at the base.[122] PO McKenna was Draughtsman, mainly preparing maps and plans for the Intelligence Office, he was assisted by AM Penaluna. PO Frederick Cusden and Sgt Williamson had care of the Photographic Department, a vital adjunct to the Intelligence Office. CPO Harris was appointed for W/T duties. At a later date, Assistant Paymaster Bule was appointed coding duties; Writer McDonald was added to the clerical team; Writer Hicks took care of *Ben-my-Chree*'s ship's office.

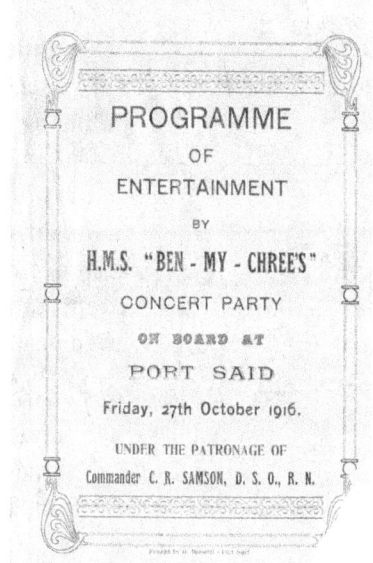

HMS "Ben-my-Chree's" Concert Party programme for 27 October 1917. *(Australian War Memorial)*

After the unavoidable asceticism of the Dardanelles, Port Said offered seemingly unlimited avenues for relaxation and recreation. Everything that could be desired, from the sublime to the salacious, could be enjoyed for a price. Several officers even maintained a small stable of horses. For the impecunious there remained the home grown pleasures of *Ben-my Chree*'s concert party, and the Hangar Theatre. For the concert party a basically string ensemble, led by a steward who had played in a London theatre orchestra, had sufficed. This was later enlarged to include winds, brass and a drummer. Their limited repertoire, played with enthusiasm rather than polish for the additions to the group lacked the talent of the earlier group, was called upon every time *Ben-my-Chree* left or entered harbour.

A Word about Photography

As Childers mentioned 'photography was now a necessary part of our work', an increasingly important part of the work of the EIESS. A dark room was a priority at the Port Said island, adjacent to the Intelligence Office, for the *Ben-my-Chree*'s Photographic Department, PO Frederick Cusden and Sgt Williamson. Cusden had been with *Ben-my-Chree* since 1915, rated as a PO Mechanic he was the ship's principal photographer, he was later promoted CPO and awarded the DSM for his work.[123]

Prior to the arrival of *Ben-my-Chree*, *l'escadrille de Port-Saïd* had made little use of photography. The Nieuport was not an ideal machine for the observer, his cockpit was located between the spars of the wings. However, some cameras were provided late in its stay at Port Said. Childers commented on it. 'Their camera is a German – Goerz one with film-pack & focal plane shutter the latter being regulated by a screw from outside. Three times as simple and rapid to use as ours, & much larger pictures.'[124]

In comparison, *Ben-my-Chree* had three Thornton-Pickard Type A aero-cameras, developed in early 1915. These were serial numbers 3A, 11A and 15A, all early production cameras. Unlike the Goerz they were large, heavy (approximately 10 lb) brass-bound tapered wooden boxes, with a fixed 8 inch lens set at infinity focus. The 5 × 4 inch plates were carried in a detachable Mackenzie Wishart slide specially strengthened for use with the camera. Although straps were fitted to the sides of the camera for the observer to grip, it was cumbersome and awkward to use, reportedly requiring ten operations to change and expose each 5 × 4 plate.

The Type A was designed as a hand-held camera, but using it leaning out of the open observer's cockpit, buffeted by the slipstream, with the machine being in any sort of position, was a hazardous operation to say the least. The observer quickly learnt to make sure his feet were firmly braced before indulging in such exercises.

The photographic department had its own set of problems. Whilst maintaining a darkroom at the island, where ice could be procured to maintain chemicals at a reasonable temperature, a photographer usually accompanied a ship on operations where no such luxury was available. In a report, written shortly before *Ben-my-Chree* sailed for Port Said, Malone listed just a few of the shipboard problems.

> Owing to circumstances, the work has usually been carried out under adverse conditions; many reconnaissances had to be made at, or immediately after, sunrise, resulting in frequent under-exposure of plates.

The Camera House on Imbros in 1915 with Petty Officers Atkins, William Garner, and Frederick Cusden, the latter from *Ben-my-Chree*. They are holding three Thornton-Pickard Type A aerial cameras. *(Bill Pollard – CCI Archive)*

The excessive heat experienced during the summer had also a most inimical effect on plates and paper. These and other minor difficulties such as impure water and dust have been responsible for the failure of eight dozen plates, owing either to deterioration prior to or blistering during development.

All in all, some shakes and blemishes are not to be wondered at, and the quality of images produced is all the more remarkable.

Island Life

Over time, life at the island base settled down to a routine.

6.00 a.m.	Hands fall in.
7.30 a.m.)	
to)	Breakfast.
8.30 a.m.)	
8.30 a.m.	Fall in — Turn to.
Noon.	Dinner.
1.20 p.m.	Fall in — Turn to.
4.00 p.m.	Watch on board clear up deck.
	Liberty men as per signal.
	Tea.
	Work in dog watches as requisite.

In hot weather a longer dinner hour should be given and the work commence earlier in the morning.

Tents were soon replaced by more permanent structures, especially modified aircraft packing cases, as these quotes from C.E. Hughes' *Above and Beyond Palestine* illustrate.[126]

'RNAS architecture' on the island, a sketch by C.E. Hughes.

As a place of residence, the island, like Port Said itself, could boast of few architectural attractions. Three or four brick and plaster buildings, used chiefly as hangars and workshops, were supplemented by many more structures of wood. Some of them were full-sized huts of a standard pattern; others were huts not of a standard pattern, though they were made of standard parts which the ingenuity of naval carpenters, with Arab assistance, had put together like a species of jig-saw puzzle that looked more or less right in spite of there being some pieces missing. Besides these, there were other wooden structures made out of that very useful perquisite of most air stations, the packing cases in which flying machines travel overseas.

The stranger threading the maze of passages among them was reminded of nothing within ordinary knowledge, and imagined, perhaps, a resemblance to some desert island peopled by shipwrecked mariners.

Yet it was not without a scheme and a design, though there was a good deal of desert about it. In a vague sort of way, the buildings had a focal centre in the "quarter deck", a patch of sand which had been at some distant date roughly cemented over. Perhaps originally the cement presented a continuous, square, level surface. It was not so in our time, but a few sandy depressions made it no less the quarter deck where the ratings would be mustered for various duties or for liberty, just as if there were a proper mast or a flag-staff, instead of an untidy pole with a more untidy yard, and a most untidy gaff. People at Navy House, with its spick and span masts and signal halyards, used to stop the war every now and then and gaze over towards the island to laugh at our crazy flag-staff. But it flew the white ensign with the best of them.

But although the island was barren of vegetation, it did not really resemble the desert island of romance. One might possibly mistake the R.N.A.S. personnel for shipwrecked mariners, but one could hardly so account for the Arab working party, which was ubiquitous during the hours of daylight. They were a picturesque crowd, these Arabs, from Big Ali who could do, and sometimes did, as much hard work as any ordinary navvy, down to the elderly couple of whom one spent his day watching the incinerator and feeding it with carefully selected morsels of rubbish, and the other walked round and round the water's edge gathering jetsam in a basket, and feeding himself with any particularly appetising tit-bits. Between these two extremes of native physique lay others of many sizes and hues, all of them answering to one or more of the three names, Ali, Abdul, or Mohammed, and all of them convinced that no job, however light, could be accomplished single-handed or without the accompaniment of a monotonous, droning incantation which contained, I am told, a reverent reference to Noah getting his charges into the Ark. These gentlemen arrived by boats early in the morning, and we got up, or listened to others getting up (as the call of duty demanded), to receive them. To say that we listened to others getting up is to say no more than the truth.

These cabins were formed by partitions made from the invaluable packing cases, hastily fixed into position in the first place, but subjected according to the taste of the occupant to many and various subsequent processes of decorative or utilitarian treatment. The show pieces were, I think, K.E.'s and Guy's,[127] each of whom had turned his quarters into a small museum of relics of his travels, set off with a finely conceived background of white and grey service paint, through which appeared irrepressible patches of creosote. This gave a mottled look to the white, but much of it was hidden under a careful arrangement of trophies celebrating the successful blandishments of local dealers in "antique" Soudanese weapons. There were also hangings of tent work embroidered with quaint figure-subjects taken from Egyptian tombs. These were the cabins de luxe. Another type was that which depended for artistic effect entirely on pictures of ladies cut from *La Vie Parisienne*, ladies who, if they wore clothes at all, wore garments eminently suited to a spot devoted to the intimacies of the toilet. Others, again, scorning silken dalliance, would contain the barest necessities for sleep and ablution. W.'s was the most austere of them all.[128] Whether from a natural distaste of all comfort or from a desire to enjoy the greatest possible pleasure in leaving his couch at a very early hour in the morning, he would lie down and sleep most unmistakably on a hard shelf of bare planks, and such furniture as he possessed was stamped with the same ascetic simplicity. A box served the dual purpose of a chair and a chest of drawers, and half a dozen nails in the wall were an adequate wardrobe. If one suggested that there was not much protection from the sandy dust which at certain times of the year permeated every fold and crack and crevice, he would ask in reply why one supposed he paid Private Bedford.[129] He objected on principle to what he described as "boudoirs." It was partly in defiance of this objection, and partly from the liking for a compact orderliness, that one of the military observers divided his cabin into two compartments. He called one his "bedoir" and the other his "bootoir."

'Bedoir and Bootoir,' a sketch by C.E. Hughes.

Port Said knew all varieties of weather. I never saw snow there, though some people profess to have seen it, but I saw almost everything else in the calendar, from the damp, clammy heat of the summer days to the thick overcoat weather of the winter nights. The island itself, was not the most peaceful spot in the world when the clerk of the weather had the wind up. One could look across the canal on a day apparently fine except for a feeling of oppression in the air and see what seemed to be a mist. Ships moored on the opposite side would fade away, and with only a few minutes' warning a [Khamsin] would be upon us driving its stinging sand into our eyes and faces, and playing havoc with our property. One of them lifted the roof clean off the Mess; another took an enormous Bessonneau hangar, a heavy timber structure covered with canvas, turned it completely over and piled it in wrecked confusion on some adjoining buildings.

One rarely hears from the other ranks, so the following is a refreshing look at life from their point of view. It was written by one AC J. Jones to his mates at Great Yarmouth and was published in *The Bat — The Organ of RNAS Great Yarmouth*, No.5 (probably May 1917).

> We have arrived! and I guess it was about time we did after four solid weeks travelling! It wasn't half bad sport tho' as we travelled overland for at least half of the journey. Unfortunately we travelled with the Army and under Army orders on the train journey; consequently we were unable to leave the platforms of any of the large stations, and had to break our journeys at their wretched rest camps. Ye gods! I didn't find much rest about them! Life in a rest camp is one continual round of bugle and bully beef with thousands of fatigues chucked in.
>
> The actual travelling wasn't bad, as the French trains are as slow as mud with stops at every hundred yards, and thousands of fruit trees en route. By the end of three days the sight of fruit made me feel sick! We got an enthusiastic welcome at every town we passed but nothing more substantial. The French evidently combine patriotism and parsimony. The Italians are much more open-handed, too open-handed! they chucked one tomato in the carriage and I caught it in my eye!! Italian girls are beautiful wenches, but unfortunately, Italian trains go very fast and stop seldom. We passed there some most magnificent mountain scenery, snow-capped mountains, water falls, and all the rest of the scenic paraphernalia.

Identifiable photographs of ratings on the island are rare, however some portraits and groups of non-commissioned officers do exist. The original caption identifies L-R: CPO (Mech) Charles Henry Stone, CPO (Mech) Charlie Attrill (torpedo specialist), Abdullah Arab Serang, CPO (Mech) S.J. Smith. *(Attrill Family Collection)*

CPO Stone again in his 'white' working uniform supervising general maintenance on a Short 184. The propeller has a canvas cover, the observer's cockpit fairing has been removed, and the mechanics appear to be filling the radiator. The serial is not known, but the location of the fuselage roundel suggests it may be one of the Shorts Ben-my-Chree brought from Gallipoli, probably 850 the longest surviving of the three. *(Stone Family Collection)*

This isn't half a bad show really, considering the God-forsaken part of the world, it is remarkably civilized. There are two seaplane ships attached to the Air Station which are responsible for messing arrangements, etc. when they go for a little expedition in more exciting regions, well, we go too. The Station is run on quite home Station lines, liberty every other night, and usual make and mends. And let me tell you that inspection for liberty is a real inspection. There are a few decent Army Clubs ashore which run whist drives, tennis club, and refreshments. Also surf bathing rather O.K. I have clicked with Natkiel for "sweeper",[130] an awfully soft number really, much the same as the sweeper's job in the "Met". Bill is the cook! He tried his hand on a rice pudding and shoved nine eggs in it! It wasn't a bad pudding, but not worth nine eggs!! Food is quite good, and a good supper every night. Sleep in hammocks of course aboard ship; I sling mine on the upper deck in the hangar, quite like the old duty night stunts in C Shed, isn't it? The female population is not particularly beautiful, and quite unapproachable. The status of an "airman" out East is quite different from his position in an English town; any service man out here (officers excepted) is considered a very inferior being and quite beneath notice. The chaps out here are an awfully decent crowd, the poor devils have been out here over two years already: what hopes for us!

Well, Charlie, time to sling 'ammicks so I suppose I'd better pack up. Kindest regards to all the Yarmouth irks yet remaining.

Yours, J. JONES.

P.S. We pack up at 4 p.m. daily, so there are no watches to squabble about, the Lord be thanked!

Another new arrival in 1917 was Flt Lt Arthur E. Popham, who was not so impressed.

> This is in fact the 22nd of May and the end of my second day at Port Said. I am now able to remember people's names a bit more clearly. The place itself, Naval Seaplane Base, is a miserable little island at the entrance to the Suez Canal. It was once, I am told, an Egyptian coastguard station and still is in part, though the seaplane station is encroaching more and more: there are two Bessonneau tents and one shed containing altogether 3 Shorts, one fairly new 240 and two aged 225's one from Mesopotamia. The sleeping quarters and the mess are made either of seaplane cases or boards there from and are draughty (which doesn't matter now) and insubstantial. One is of course meant to have camp equipment; there is nothing whatever of furniture. However they have knocked me up a bed out of boards, pinched a minute tin basin to wash in and a mattress and borrowed a seedy looking blanket. I must get a pillow, some curtains and some less lousy mats. There are also a few shelves at one of which I am writing. I should feel more interest in my cabin if I were certain of staying more than a week or two, but I suppose I shall come back eventually. There is no routine at the station. There seem only to be about a dozen air mechanics, the handling is done by Arabs. There is a military guard some of which I think act a mess waiters though as the air service ratings are mostly dressed in army kit it is difficult to tell which is which.
>
> [Two days later] A rumour reached me through the partition as I was dozing in the afternoon that we were to shift. This was confirmed and we were to pack up and shift at once. Thank God for that. But it appears we are merely not to live on the island, we are to repair thither daily for our "work." Still that is something. We were apportioned to the *Empress*, the *Manica* an observation balloon ship just returned from E Africa and the shore. I was for the *Empress* and came aboard about 7. Very comfortable deck cabin with a bunk, a sofa (and cushion) and drawers. I am very thankful to have somewhere where I can write and read in comfort.[131]

He was to remain on *Empress* until posted to Colombo on 2 June 1917 *(see Chapter 14)*.

This concentration on domestic matters has taken us well ahead of events. We must now return to record the first operations of the East Indies and Egypt Seaplane Squadron.

The Sunbeam engine of the Short 184 required constant maintenance and care. In use it was constantly overheating, so that most cowling panels were removed in service. The extensive use of a stencilled serial number was common on the Sunbeam engines. The dark coloured box on the fuselage was an external oil reservoir/cooler. *(EMK)*

Above: The 'office' of a Short 184, in this instance 8076 (May 1916), was an ergonomic nightmare. The layout was improved, slightly, in later production models.

Left: This Phoenix Dynamo built Short N1653 (October 1917) is somewhat tidier, the homely looking magneto switches are a nice touch, but far from perfect. *(via Colin Owers)*

A hazard operating in the congested waters of Port Said was the potential for collision. Sopwith Schneider 3773 was 'Damaged Beyond Repair' running into an Egyptian felucca on 16 March 1917. The pilot is unknown, but the caption in another album containing this photograph suggests it was Flt Cdr J.C. Brooke. *(EMK)*

The Short could also fall prey to local conditions. The identity and date are unknown, other than it is not 8031, and the original caption says 'Flt Lt Maskell's crash, Port Said.' *(Attrill Family Collection)*

Running into native craft was not the only problem. The temperature and humidity made maintenance a never ending struggle. As seen here, another unidentified Short has lost the bottom of its floats then crashed into the dockside.

A Sopwith Schneider being run up by FSL Bankes-Price. Evidently health and safety concerns were a low priority at the time. This photograph sends shivers down my spine, there is so much that could go wrong…
(Attrill Family Collection)

The RNAS ratings all too often go unrecorded except, of course, in photographs for the albums. None of the names, or the date, of this group are known, although the presence of several RFC style caps, and at least one maternity jacket, suggests this may date from the RAF era post 1 April 1918. No matter, as they represent the over 200 mechanics, riggers, carpenters, artificers, general hands, and at least five dogs, based on the EIESS island between 1916 and 1918. *(YEORN 2014-80-84. Courtesy of The National Museum of the Royal Navy)*

Raven and *Ben-my-Chree* together at Port Said. The difference between the two types of ship are quite marked. *Raven*'s heavy mercantile hull with cargo derricks in the well decks and bridge, wheelhouse, accommodations and engine room gathered in a central island, contrasts markedly with *Ben-my-Chree*'s sleeker hull and expanded accommodations spoilt only by the big slab-sided hangar. The ship's funnels speak to their engines as well. A single narrow upright funnel for *Raven*'s single four cylinder, triple expansion steam engine, whilst *Ben-my-Chree*'s two large funnels dominate her silhouette and were required for her three steam turbines.

Short 846 being fended off from *Ben-my-Chree*, the long padded poles were essential equipment on the seaplane carriers. The use of the Union Jack as a national marking dates this to *Ben-my-Chree*'s period at Gallipoli. When the aircraft were repaired or re-built the more recognisable roundel was added alongside the cockpit and the flag painted out. Also of interest are the additional struts under the fuselage supporting an extension for the torpedo carrier between the floats.

CHAPTER 6

Early Operations of the EIESS

Ben-my-Chree's first outing as flagship of the EIESS was a single day, 20 January 1916, sailing from Port Said at 00.30, to arrive off El Arish by 08.00. At 08.24 Flt Cdr C.H.K. Edmonds and Lt Childers took off on Short 849,[132] visiting Kossaima and El Auja. (Short 849 was a Short Admiralty Type 184. From this point forward any use of Short should be understood to refer to the Short 184. Any other model will be identified as such.) On their return they had a taste of the unwelcoming seas off the Palestine coast. During their absence the wind had risen, creating a very choppy sea. Edmonds bounced on landing, breaking a float strut and damaging a float. Quick work by the crew aboard *Ben-my-Chree* brought them aboard safely. An afternoon flight was made by Flt Lt G.B. Dacre and Childers, on Short 846, with a reconnaissance over Lake Bardawil. Dacre recorded his impressions of his flight with Childers.

> I went up with Childers & a machine gun & getting off a bad lop made for shore at 1000 ft there we saw several camels sitting down. So I came down to 50 ft & Childers eased off the machine gun. The bullets hit the sand all round but the camels got up suddenly & two men with rifles came running up from a place several yards away where they had been lying to avoid detection, & started to run off at the double with the camels inland to some grass patches. I circled round again low & the men left the camels & ran as hard as they could on their own. We eased off the machine gun again & as the bullets spattered in the sand one man dropped but no camels appeared to be hit. Both men had rifles & were dressed in a Greenish Khaki so were probably Turks. Jolly fine sport.
>
> We then flew westward to recon a large inland water [Lake Bardawil] where we took several photos & again came across 7 men on camels fording a part of the lake. We came down again & attacked these. The bullets splashed in the water but no one appeared hit & as they did not look as if they were armed we left them. I then got into a squall & had a hard fight with the machine flying down to the agreed rendezvous but no ship was there or anywhere in sight, & we were alarmed

to know where she was as a strong west wind brought us further along the coast & I only had ½ hours petrol left. Any how just as I was pretty fed up, & on the point of going over to the lake to land, I saw the ship 15 miles or so E of the agreed spot.

Dacre is a little vague about the landing, so here is Childers' version.

[The sea] was rougher when we returned after two hours to the ship & we damaged a float on landing and in coming alongside the propeller, still moving, hit the ship &, it is feared, damaged the engine.

Today therefore was hard on sea-planes but otherwise successful.

The ship proceeded back to Port Said arriving at 22.00.

Two flights and two Shorts in need of repair. *Ben-my-Chree* was learning that operations in the Mediterranean would be different from those they were used to in the Aegean. As with *l'escadrille de Port-Saïd* the EIESS had more than rough seas to contend with. They faced the same risks of flying a floatplane with an unreliable engine many miles over land. The Shorts always carried water bottles, some food and a reliable pocket compass on the off chance the crew survived the landing, immediate capture by local Arabs, and could march towards the sea. The smaller, shorter ranged Sopwiths appear not to have been similarly equipped. Time would prove that whilst landings were survivable, the Arabs were ever present.

On the same day *Anne* also visited Jaffa and El Arish, without seeing *Ben-my-Chree*. She had left Port Said on 10 January, two days before *Ben-my-Chree* arrived. The section comprised Nieuports N.19 and N.22, with pilots Grall and Bourgeois, observers Capt E.L. Chute and 2Lt K.L. Williams. Following a fruitless search for a submarine and mines at Beirut and El Mina on 12 January, the weather turned bad and *Anne* had to shelter in Famagusta, Cyprus, between 15–19 January. When the weather finally cleared she sailed towards Jaffa and El Arish. On January 20, when *Ben-my-Chree* was launching a Short at El Arish, Weldon wrote about the weather at Jaffa (about 150 kilometres up the coast), 'Heavy seas, low cloud and rain, flight out of the question.' Later they were off El Arish where flying remained impossible. Given *Ben-my-Chree*'s experiences that day, perhaps *l'escadrille*'s experienced, wiser heads prevailed. The sea was not calm enough to launch until the afternoon of 22 January, when a brief flight over El Arish was made. The following morning, having returned to Jaffa, conditions had greatly improved, Bourgeois and Chute making a good two hour flight on N.22 over the railway from Ramleh north to Tul Keram. *Anne* returned to Port Said on 24 January.

After a number of postponed and cancelled operations, on 30 January *Ben-my-Chree*'s airmen had another taste of what *l'escadrille* had been struggling against when they visited Beersheba. *Ben-my-Chree* arrived off Gaza at around

07.00 on the morning of 30 January. A morning attempt to reach Beersheba by Flight Lieutenant Maurice E.A. Wright, an experienced pilot who had been with the ship since May 1915, and Childers on Short 850, had to turn back due to low clouds en route after getting within a few miles of the town. Setting out again in the afternoon, low clouds remained a problem although this time they did reach the vicinity of Beersheba. In his diary, Childers recorded the afternoon flight.

Short 849 hoisting out from *Ben-my-Chree*, possibly on a test flight whilst at Milos in early December 1915, with pilot Flt Lt M.E.A. Wright and Lt Childers in rear cockpit. Whilst serving at Gallipoli the Union Jacks on the fuselage of *Ben-my-Chree*'s Shorts and Sopwiths were painted out and the serial numbers added on the resulting white panels. Under the fuselage just aft of the floats can be seen a 'V' strut used to support an extension for the torpedo carrier, this is also clearly seen in the photograph of 846. *(Attrill Family Collection)*

> We started again at 12.18 pm and this time got to Beersheba. Low clouds again made it necessary to fly low and outside Beersheba (which is 900 ft above sea-level) we were only 600 ft above ground level. This was most unfortunate as Wright was reluctant to go directly over Beersheba owing to danger from rifle-fire under these circumstances. All we did was to make one circular sweep south of the town and then home again. However we did obtain some important results: (1) Verifying the fact that the railway does not go beyond Beersheba, though the railway embankment and a light construction railway alongside do. (2) That the bridge across Wady al Saba, a splendid viaduct, is uncompleted. (3) That 200 or more loaded camels were leaving the town for Auja by the direct road. (4) That there is a large number of troops at Beersheba on both sides of the wady, and big barracks, shown in half a dozen photos. (5) That the approaches to the town from Gaza are entrenched for many miles out. (6) That three hitherto unreported bridges exist, two of them near on the Gaza road, one of them being broken (or uncompleted) as well as one of deviation bridges on a loop at this point, so that traffic is blocked (two trains standing at this point).

The following morning the ship was off the mouth of the Nahr Iskanderun. Although delayed by engine problems and a choppy sea Dacre was able to take-off at noon with Childers, probably on 850 again, to cover a Tul Keram–Ludd–Jaffa triangle and rejoin the ship at Jaffa. Shortly after realising they were too far north of their route, Dacre had to turn back with engine trouble, landing safely on a choppy sea. Childers then set out again with Edmonds as pilot at 13.47.

> It was a thoroughly pleasant & successful flight, our distance from the sea varying from ten to eighteen miles as we followed the railway south, the air calm with a slight northerly wind only, the light perfect. We found Tul Keram (ten miles in) to be heavily entrenched & the station packed with stores & rolling stock. The next most important feature was a mass of rolling stock at the junction for Jerusalem;

We next saw some fair-sized camps at Ramleh. A negative point of value was that only one train was seen moving on the section we covered — moving south: Perhaps the most interesting discovery as to the construction of the line was a seven arch viaduct 3m NE of the Wady el Sharia. I saw this yesterday with Wright, in the distance, or rather the big buildings & camp near it & wondered what it was.

The country is a paradise of fertility from Tul Keram to Ramleh, vast orange gardens around Ludd & distant Jaffa, & the only cultivated land everywhere. South of Ramleh the scene gradually changes until the railway is winding through a vast wilderness of barren hills & plains seamed by deep river beds now all full of water and in colour for the most part a deep chocolate. There is a big bend inland at the southernmost point, 18 miles from the sea and here the scene was infinitely desolate; & would have been more so but for the thin ribbon of metalled railway below us with its great arched bridges fretted a dazzling white against the gloomy background. As an engineering work the railway filled me with admiration. What the traffic & rolling stock indicates it is hard to say, The concentration at the Junction was altogether abnormal. Only one train on the move in an hour and a half, but hard to infer what that means. Other trains were ready to move & the break in the line seen yesterday must have caused delays.

At the Wady El Sharia viaduct we returned to the coast, making for a point north of Gaza & a little south of Ashkelon, where we found the ship steaming south at full speed & ready to receive. A rough landing again, and another fore strut broken but of no consequence. We alighted at 4.15, folded, & were hoisted in. The ship started at once for home and reached Port Said at midnight.

At the end of January *Anne* made what would be her final visit for the *escadrille* to the Gulf of Alexandretta. The section appears to have comprised a single Nieuport, N.23, with LV Destrem and Major Fletcher as observer. They made two flights one on 31 January and the second on 1 February, providing extensive coverage of the railway from Tarsus across the Chicaldere bridge and on to Toprakkale Junction. There were several engines with trucks in the rail yards but nothing moving on the lines, roads were almost empty but there were indications that the rivers were being used to transport material. *Anne* returned to Port Said on 6 February.

Ben-my-Chree set off on 9 February, sailing west to Alexandria where she was joined by General Maxwell, C-in-C Egypt, with his aides Lt Cdr Bolton Monsell, RN, and Captain Mervyn Sorley MacDonnell, Egyptian Civil Service and expert on the area to be reconnoitred. *Ben-my-Chree* then continued to the west in order to be off the coast near Sollum at daybreak on 11 February.

The plan called for Edmonds and Childers in one Short and Wright with 2Lt K.L. Williams in a second Short to examine the desert inland for a distance of 20 miles between Sollum and Bardia. They were to photograph the two ports, try to locate some mines reported to have been laid off both harbour entrances and to look for the Senussi army. A religious brotherhood, who before the war had co-existed amicably with the governments in both Egypt and Constantinople, the Senussi harboured no love for either the French or Italians. Early in 1915 the Ottoman government supplied them by submarine with arms and 'advisors', who had induced them to rise in holy war against the infidels in Egypt. This was one of the few successes the Ottoman call for a *Jihad* had achieved. Even then, being on friendly terms with the British, it had taken the entry of Italy into the war against Turkey to finally stir them into action.

Childers left an account of the following events.

> *Feb 10* — The General wants to land a force E of Sollum & march it to take Sollum which was evacuated by us when the Senussi took up arms against us owing to the intervention of Italy in the war. Hitherto they had been very friendly. It is here therefore as in earlier Egypt: we are evacuating most areas & leaving them to the enemy. Our furthest outpost west is Dabaa [Bir ed Daba, 50 kilometres west of Matruh]. Now we want to get back, but great caution is to be observed. It is thought that Sollum harbour is not sheltered enough to obtain constant supplies so that a protected coast-line is to be gained, to supplement sea-supply by land supply. Details I do not know. Much depends on position & intentions of Senussi. At present they are reported far away at Siwa Oasis. 130 miles S of Sollum, but they may have moved.

> *Feb 11* — A fine day, but a strong wind SSE making considerable sea. We stood to within 5 miles of the coast before hoisting out. No one knew exactly where we were. For the first time in these desert flights we did not take wireless, Edmonds wishing to save weight. I represented the necessity of taking it both to Malone & Edmonds. M was willing but E held out & went without it.

> We started in 849 at 9.32 & soon after rising saw Sollum clearly. We first made for Port Bardia [Mersa Burdi] a harbour 12 miles north of Sollum examined that & then came to Sollum. A pretty bit of coast from Bardia to Sollum, laced with little rocky inlets each with a white sandy beach and a gully with grass & steep rise to the plateau. Above & behind the vast desert plateau – brown & featureless. Sandstorms were sweeping over the desert but luckily not so as to prevent visibility at important points. The map of Sollum proved to be good:

some slight changes in the buildings. Impossible to see mines at either Bardia or Sollum. There was a small Arab camp near the beach and another 2 miles WNW. Both arranged in neat & regular rows of the large oblong camel hair tents used here-abouts. But our great find was a big scattered encampment four miles SW of Sollum — 300 or 350 tents — probably the main body of the Senussi. I was photographing steadily all this time & had a complete set of Sollum, Bardia & the camps.

We now bore off to find [Sawani] Augerin following the abrupt escarpment of the plateau which strikes away inland from Sollum to the SE. Augerin, if it was it, was a surprising little place a square walled enclosure with unroofed compartments: containing I suppose the wells; much vegetation around, and 25 tents. Away then for El Sabal which we could not identify but its district is a pleasant one: scores of little pools of water glistening in the sun some long cultivated strips of land and a good deal of grazing. Nothing else to report. Then north across the desert about 18 miles to Baq-baq: a tiny little settlement of a few houses, with a good motor-road running to it from the east.

Where was the ship? Nothing in sight. No wireless to signal with.

It was now 11.40 only twenty minutes more petrol and a strong off-shore wind. To make matters more difficult there was a haze over the sea especially below 2000 ft. (We had done the land flight at 5000–6000.) Knowing the ship would not come close inshore we went out for about five miles, an unlucky gamble, as it turned out as we had to fight out way back against the wind, trying to close the coast at an angle & heading about SW & lastly south. Edmonds had

Short 849 taxiing past part of the assembled fleet at Mudros. The ships, left to right, are the six funnelled French armoured cruiser *Jeanne D'Arc*, then two British pre-dreadnoughts, the first possibly *Agamemnon* the second a Canopus-class, and finally a French pre-dreadnought *Gaulois*, which had returned to Mudros in June following repairs after being severely damaged during the final naval attack on the Narrows in March. The Short arrived at Port Said on board *Ben-my-Chree* and was lost at sea on 11 February 1916, as related in the text.

realised of course that we were unlikely to find the ship and that our one chance was to land near the shore to allow for the rapid drift to sea which would begin directly we alighted & take us out of reach of any help. The petrol now gave out and at noon exactly we came down, making a good landing in a choppy sea about 2 miles offshore & 3 NW of Baqbaq. We had no sea anchor & no possibility of making one. We could do nothing in fact but wait for the ship.

The machine lay nearly head to sea & seemed fairly comfortable but the elevators broke almost at once and then the tail crumpled. However we appeared to be all right & I smoked a cigarette with great enjoyment. What we feared was that the ship, not finding us in what was evidently her mistaken position, would search inshore to the west thinking we had come down in the desert, while we were ever drifting 4 or 5 knots an hour to seaward further to the east. (This is exactly what happened.) It was not a bright lookout as the sea was growing steadily worse as we drifted away from the land — which at last became invisible. Altogether we were delighted after about three quarters of an hour to see a trawler about 2 miles away to seaward through the haze & still more to see her see us & alter course. She was just in time.

We shouted as she came up "Send a boat" and at that moment the seaplane seemed to suddenly give out. The port wing rose high in the air & Edmunds shouted "run out on the wing" which we both did & she righted, but once there we could not leave the wing though I badly wanted to save my negatives which I had left in my compartment together with everything else. In another minute the tail began to sink & the floats to rear up vertically so that the whole of the observer's seat was submerged. Edmunds stayed where he was on the end of the wing, knee deep in water. I nipped back to the floats & stood on the cross-bar between them dry except for my feet. The trawler's boat was not yet lowered, &, as we thought the plane was about to sink we got our boots off & I took off my coat which I could not button up over my inflated Gieve's waistcoat & which was therefore a danger. Nevertheless the machine remained afloat, the boat rowed up, & took us off in succession. I really think that now came our narrowest shave. A man on the trawler threw a rope as we got back and began to haul us in on the weather side, with a heavy sea running. We were within an ace of being cracked like a nut against her iron side before we succeeded in getting away & approaching from the lee side.

Our first step was to ask the Skipper (an ordinary fisherman) for his position which was 16m from Sollum on a course to Alexandria, proving definitely that we had been in the right & the ship in the wrong position. Next we asked him to alter course and pick up BMC which he did. They were very kind to us, giving us tea & Edmunds clothes (he was wet to the waist). After half an hour, BMC hove in sight & we were soon abreast. The Skipper threw me a couple of immaculate unused signalling flags & said "Here you are Sir, we don't know how to use them"! I signalled to send a boat which soon came alongside (1.50 pm) & took us off.

We reported to the Captain & General. I was bitterly disappointed at the loss of my negatives: all good, I am sure, as conditions were perfect; but from a military standpoint they were not, I think, of much importance as, according to my invariable custom, I kept a pencil log & sketched camps etc so that everything was in my memory. The main point was the presence of the big Senussi camp.

I was anxious immediately to turn back & have another flight to obtain photos of Sollum which the General had been very anxious to get, but it was decided to go back to Alexandria at once giving up Flight No2 to Thalatha & Mella. This latter decision was probably wise owing to the sandstorms on the plateau & the doubt as to the position of the two wells in question.

It was sheer good luck that we were sighted & picked up at all & we are deeply grateful to Providence & the trawler. £5 to the latter pleased them very much. It was a captured German boat "Charasin" by name,[133] manned by fishermen.

Returning eastward to Alexandria, *Ben-my-Chree* had one more encounter with a patrol trawler. Approaching the port around 18.00 she failed to see a challenge and was soon awakened by a shot across her bows. After hurriedly establishing their identity They proceeded into Alexandria. With the day's incidents behind them and their passengers landed, perhaps with more experiences than they had anticipated, *Ben-my-Chree* pressed on for Port Said, arriving early on 13 February.

Ben-my-Chree's log book provides an interesting list of equipment lost with the Short: 2 Webley pistols, 2 water bottles, 1 Webley pistol case, 2 waist belts, 2 30s braces, 2 WB carriages, 2 Couriers case pistols, 100 cartridges SA Ball pistols, 12 Very's cartridges, 2 haversacks, 1 Seaplane [sic]. Also lost was a toolkit, including 12 KLG spark plugs, and Thornton Pickard Type A Camera (No.11A) with a Mackensie [sic] Wishart slide back and 30 plates. In addition Childers noted that they had 'lost everything; machine, instruments, camera, binoculars, my own little compass, invaluable, & there is nothing to replace it; besides a lot of kit, revolver, etc.'

The Senussi, meanwhile, continued to be a threat and nuisance. Not until

April 1917 after a series of defeats, in which the RFC, Rolls-Royce armoured cars, and Ford light cars played prominent roles, were they sufficiently worn down to negotiate peace terms.[134]

Off the coast of Palestine on the morning of 11 February, *Anne* was preparing for what would be the *escadrille*'s final visit to Beersheba. *SM* Grall and Capt Chute set off on N.22 at 09.40 returning at 11.59. Their purpose was check on the state of construction of the railway extension from Beersheba south to El Auja. Crossing the coast south of Gaza, they followed Wadi Gaza then turned east at the junction with Wadi Imleih, following it in turn until reaching the railway and turning south towards Beersheba. From there they followed the railway until reaching Asluj, halfway to El Auja, there turning to the north west. Passing El Khalasa, they followed Wadi Gaza towards the coast. Halfway along the wadi they turned due west crossing the coast just north of Khan Yunis, midway between Rafa and Gaza, and back to the ship. Chute reported that camps at Beersheba had been removed but there were large work camps at Asluj and El Khalasa. Work was proceeding on the line, but there remained many gaps and uncompleted bridges and embankments. Even Childers rated it as 'a fine flight.'

A Squadron Operation

Early in March, *Anne* set out once more this time on a joint operation with *Empress* and *Ben-my-Chree*. The intent was to cover the railway from north of Jaffa to south of Beersheba in a series of flights on a single day, 7 March. The three seaplane carriers were to operate off different points on the coast; *Empress* was to visit Jaffa, *Ben-my-Chree* Gaza and *Anne* El Arish. With the loss of *Amiral Charner* to a torpedo from *U-21* on 8 February enemy submarines were now a very real threat, accordingly each ship would be escorted by a French destroyer, torpedo boat or armed trawler.

Starting in the north. In the morning *Empress* was about 15 kilometres off the mouth of the Nahr Iskanderun to launch the first of two flights during the day. This was flown by Flight Commander W.G. Sitwell with Lt E. *'le petit'* Williams as observer.[135] Flying Short 8088 they left the water at 09.05 bound for Tul Keram and Nablus before returning to the ship at 10.34. Sitwell noted in his log book, '10 miles off to Tul Keram, Samaria... and return to ship. Very interesting. Shot at a little. Observer took several photos but 30 miles inland is not at all amusing.'[136] *Empress* then proceeded south to a similar distance off Jaffa. The second flight, Flt Lt R.M. Field with 2Lt C.A. Bourne on Short 8381, took off at noon to fly over Jaffa and on to Ludd, Ramleh and, following the railway south, to El Falujeh before turning west to cross the coast near Ashkelon. *Empress* meanwhile had been steaming at full speed (18 knots) down the coast to meet them. The Short returned to the ship at 13.45. Both flights had been uneventful and brought back a wealth of intelligence.

Photograph of a loaded train heading south down the railway toward Beersheba. Taken from one of *Ben-my-Chree's* Shorts on 7 March 1916. (Attrill Family Collection)

Further down the coast *Ben-my-Chree* started the day off Gaza, intending to send two Shorts to Beersheba. The day commenced foggy but, as soon as it cleared overland, the Shorts were prepared and launched. Both were veterans from the Dardanelles operations as were three of their aircrew. First off at 09.01 were Flt Lt M.e.a. Wright and 2Lt K.L. Williams on 850, they climbed quickly and passed over the land just 15 minutes later. Next were Flt Cdr Edmonds and Lt Childers on 846, leaving the water at 09.18 they struggled to gain height before they could cross the coast.

Wright and Williams essentially followed the route used by Grall and Chute almost a month previously, but extending down to El Auja before turning for the coast. Williams' observations were very similar to those of Chute, although one bridge (at Tel el Sheria, north of Beersheba) was reported to have been completed. South of the town, work continued on the line without any major advances, although there were now more work camps. Wright brought 850 down alongside *Ben-my-Chree* at 11.18.

Edmonds and Childers had a far from trouble free flight 'staggering round the course at a maximum height of 3000 ft, dropping to 2500 whenever we turned round. It was 35 minutes before we even crossed the coast & then were carried too far north by the wind or an error in E's compass. I warned him several times but he persisted that we were all right.' They arrived over the railway a little south of where Wright and Williams reach it, so they cannot have been too far to the north of their intended route. However, with their on going engine problems, the flight was shortened and after circling Beersheba, attracting shrapnel and rifle fire, they headed south down the railway but soon turned towards El Khalasa. Here they reported much the same activity as had Grall and Chute. Edmonds and Childers reached the coast at 11.35 and landed alongside the ship a few minutes later, 'Coming alongside the engine, as so often before, refused to stop, the propeller touched the side & bang goes another engine & propeller!'

Both *Empress* and *Ben-my-Chree* returned to Port Said that same evening.

Anne had less far to travel, but was also much slower, reaching El Arish in time to send off an afternoon flight on 7 March. She had a section comprising two Nieuports, NB.1 and NB.2, and pilots, *SM* Jeanblanc and *QM* Roussillon. Their observers were Capt Chute and 2Lt A.D. Finney, the latter one of Childers' trainees. Accompanying her was *Torpilleur de Défense Mobile 250*, refitted for anti-submarine work with depth charges, and a *soixante-quinze* replacing her aft torpedo tubes.[137]

At 14.20 Roussillon and Finney set out on NB.2, crossing the coast at the Wadi el Arish into the Sinai as far as Gebel Libni and El Sirr, some 50 kilometres, then return. They were to search for any camps in the areas over flown. Gebel Libni was empty, the camp at El Murra was much reduced since the last visit eleven months previously, and one comprising 50 tents at El Arish. They returned to *Anne* at 16.18.

The following day, *Anne* was still off El Arish and sent Jeanblanc and Chute on NB.1 for an afternoon flight well inland to El Kossaima and El Auja. They left the ship at 14.38 and returned at 17.21. Jeanblanc had flown directly to El Kossaima, passing over El Magdhaba and Abu Aweigila, then to El Auja before returning to El Arish. El Kossaima is some 90 kilometres inland from El Arish, add another 120 kilometres for the return flight via El Auja, and they must have been very low on petrol when they reached *Anne*.

Nieuports NB.1 and NB.2 on *Anne* 7/8 March 1916. The port-side wooden hangar bulkhead on the aft well deck was installed by *l'escadrille de Port Saïd* on *Anne* during 1915. It was removed in Malta a few months later. *(EMK)*

Nieuport NB.1 came to an early end following this operation. On 13 March, with SM Jeanblanc as pilot and *docteur* Loyer, of *l'aviation Maritime*, as observer, set out from Port Said to take some photographs. The photographs were destined not to be taken as NB.1 suffered an engine failure, coming down at sea some five miles NW of Port Said. Damaged during the landing the Nieuport sank 45 minutes later. Jeanblanc and Loyer were lucky to be found by a pilot boat from the Suez Canal Company later that evening. The Nieuport was lost at sea.

On 10 March, Erskine Childers boarded the P&O steamer *Khyber* bound for the UK. His diary entries had been increasingly irritable over the past few weeks and he was probably over tired and over worked, he had been *Ben-my-Chree*'s sole observer for the past year. 'Is it a dream? No—I am homeward bound on the *Khyber*, P&O. No work: no worry: an ecstatic relaxation of mind & body.' Returning to Britain, Childers spent time at Harwich and Dunkirk working with high speed Coastal Motor Boats and later at Felixstowe as intelligence officer to the flying boat squadron. Towards the end of the war he began to be increasingly involved in Irish politics and in 1919 moved to Dublin to work as a propagandist for the Republican movement. He was never fully trusted by some fellow republicans, being seen as an outsider and too English, although both Eamon de Valera and Michael Collins thought well of him. He was well enough regarded to be appointed secretary to the London Peace Conference delegation in 1921. The resultant Treaty led to a split between the Republicans and Irish Free Staters and the ensuing civil war. Childers remained a Republican until his death, in front of an Irish Free State firing squad, on 24 November 1922.

Torpilleur de défense mobile 250, a regular escort for *Anne* and the ships of the EIESS. Built in 1903 she was of the same class as the Turkish *Demirhisar*. Refitted for local escort work, a navalised *modèle* 1897 *Soixante-Quinze* replaced the aft torpedo tubes, and up to eight depth charges could be carried. *(EMK)*

It was also time for *l'escadrille de Port-Saïd* to move on.

Final Flights of *l'escadrille de Port-Saïd*

By November 1915 the French naval *Commandant en chef*, vice-amiral Boué de Lapeyrère, had seen the utility and versatility of a simple cargo ship conversion to carry and deploy floatplanes. Further, he stated that they were necessary at any operational centre. So it was that *Campinas*, then serving as an underused hospital ship at Salonika, was ordered to Port Said under the orders of the *3ème Escadre* for conversion to a seaplane carrier, arriving on 29 January 1916. *Campinas* (3300 tons gross; length 102.4 metres [336 feet], 11 knots) was a cargo-liner owned by *la Compagnie des Chargeurs Réunis* plying mostly between France and Brazil from 1896. A little smaller than either *Anne* or *Raven* she was given a thorough conversion by *Compagnie Universelle du Canal de Suez* at Port Said, in accordance with a plan furnished by de l'Escaille in December. He had been informed that future plans for the *escadrille* included replacing the Nieuports with new FBA biplane flying boats, so the conversion was designed to accommodate the FBA. When completed in March she could house three machines on deck, one forward and two aft, with storage for several more in her holds. Machines on deck were housed in canvas hangars with stout wooden bulkheads 5 to 6 metres high, along the port side of the well decks, similar bulkheads were briefly installed on *Anne*. She was armed with a single 47 mm gun on the forecastle and a 100 mm gun aft.

Campinas anchored at Milos in the western Aegean in 1918, with the armoured cruiser *Victor Hugo* in the background. Her merchant origins are scarcely disguised and the port-side wooden bulkheads for her floatplane hangars can be seen. *(Robert Feuilloy)*

Work was completed by 9 March, and *Campinas* attached to *l'escadrille de Port-Saïd* for a brief work up. Following which, after taking aboard a *section d'aviation* comprising Nieuports N.19 and N.23 with pilots *LV* Destrem and *SM* Gramont and observers 2Lt J.M. Burd, RFA, and Lt J.H.B. Wedderspoon, RFA, *Campinas* sailed from Port Said arriving at the island of Castellorizo on 23 March.[138] Making several flights over the next three days, the Nieuports scouted the coast from Castellorizo west to Volos Island and east to Kalamaki Island and Fineka. They were to look for submarines and mines and perform a general reconnaissance of the area. Although no submarines or mines were seen, an old Byzantine castle on the mainland across from Kekova Island was bombed twice, one hit was claimed.

Having departed Castellorizo in the afternoon of 26 March, *Campinas* was a little north of Ruad on the morning of the 29th. preparing to launch a Nieuport to look for mines and submarines and scout the coast adjacent to the island. Gramont and Burd set off on N.19 at 09.55, but had to return with engine trouble 45 minutes later. In the afternoon Destrem and Burd tried to complete the reconnaissance on N.23, they left the water at 14.31 and returned an hour later after a trouble free flight. Together the flights had covered about

Campinas at Castellorizo 26 March 1916. She occupies the only anchorage available for large ships in the island's harbour, the first seaplane carrier to occupy the space, other EIESS ships would follow. Several buoys for the anti-torpedo net can just be made out across the harbour entrance. *(Nicholas Pappas)*

Nieuport N.23 being hoisted on *Campinas* in March 1916. *LV* Destrem is standing on the float strut by the engine, the observer either 2Lt J.M. Burd or Lt J.H.B. Wedderspoon is standing on the fuselage, most of the Nieuports had a plywood reinforcing panel installed behind the cockpit. *(ARDHAN)*

7 kilometres of coast above and below Ruad with little to report. There were no changes to trenches and most gun emplacements appeared empty, very few troops were observed. There were no mines or submarines.

Two days later *Campinas* was off Beirut. The last flight of the trip was to examine the harbour and local coast and to take photographs. Destrem and Wedderspoon set of on N.23 at 08.05, with the carbine and two magazines, two bombs, and a camera. In the harbour they reported a sunken tug and 'a large vessel sunk, grey funnel, 1 mast, and fore deck visible.' This was the wreck of the *Avnillah*, sunk in January 1912 during the Italo-Ottoman War.[139] There were a number of lighters, a crane barge next to the sunken tug, and a considerable number of local trading dhows and schooners in the harbour. Although no mines or submarines were seen, a large oily patch of water was visible but Wedderspoon thought this was due to a discharge from the town's drainage system. The railway station and sidings contained just ten trucks and no engines were visible on the line. Along the coast, at Nahr Beirut west of the city, a walled enclosure with two sheds was noted together with a small pier and a second sunken ship. Wedderspoon took some photographs; one of which has survived and is reproduced here.

The best of the few surviving photographs taken by *l'escadrille de Port Saïd*, is this one of Beirut on 1 April 1916. Taken by Lt J.H.B. Wedderspoon from the forward cockpit of Nieuport N.23 piloted by *LV* Destrem, it shows the unsuitability of the Nieuport for photography. The light coloured lower left corner of the image is the Nieuport's wing. (TNA, AIR 1/665/17/122/722)

For the duration of the flight a bomb was hanging, primed and ready, in the tube. Not having been dropped, it had to be delicately removed, disarmed and replaced on the wing before landing. Which was just as well, as Destrem made a heavy landing, damaging the machine, when returning to *Campinas* at 09.22.

By April it was becoming obvious that *l'escadrille de Port-Saïd* was coming to the end of its time in Egypt. Pressure was being applied on de l'Escaille to release a significant portion of his base, it was prime real estate, for use by the ever expanding requirements of General Murray and the Egyptian Expeditionary Force. Also, at the beginning of April, *vice-amiral* Boué de Lapeyrère began to look for additional floatplanes to open a base at Argostoli, on the Greek island of Cephalonia, in the Ionian Sea just west of the Gulf of Corinth. *L'Aéronautique maritime*, however, was still re-equipping and building its strength and had no spare capacity. So, eyes turned towards *l'escadrille* in Port Said. The French naval attaché in London was instructed to request the release of *l'escadrille* which was now required elsewhere.

Lt Wedderspoon's sketch map of Beirut from his report. The photograph of Beirut on the facing page is most likely taken from the position marked number 11 on this map. (TNA, AIR 1/665/17/122/722)

The situation in Egypt was now very different from that in November 1914, there was the growing RNAS presence and the RFC was also increasing in strength. So, as neither Admiral Wemyss or General Murray objected to the removal of the French floatplanes, *amiral* Boué de Lapeyrère ordered the *escadrille*, machines and personnel, to be transferred to Malta. Here they would be serviced and prepared to establish the new base at Argostoli. *Campinas* left Port Said for Malta on 7 April with N.18, N.19, N.20, N.21, N.23 and NB.3, on deck or disassembled in her holds, together with most of the pilots and maintenance personnel. She was to spend the remainder of the war as an aircraft transport or mobile seaplane base.

RNAS and *l'escadrille de Port-Saïd* maintenance personnel outside the hangar on *Ben-my-Chree*. In the short time they had together the French and British airmen appear to have got on well. As usual, names for non-aircrew group personnel are hard to find and most are unidentified. Standing left: WO F.H. Whitmore (standing next to two *Second-maître mécaniciens*). Sitting: PO Charlie Attrill (with rabbit); PO Smith. Standing, behind Atrill and Smith, PO Callender and *Mot* Mourdon. (Attrill Family Collection)

Before their departure there were multiple dinners and celebrations, as in the short time they had operated together the aviators and the rank and file of the two navies had quickly grown to appreciate the work each other was doing. Weldon recalled at the end of one dinner:

> One of the young British pilots — without at all meaning to be rude — turned to the Frenchman, Destrem, and said, "How is it that all you French pilots are such old men?" "Ah!" said Destrem, as quick as lightning, looking up and down the table at all the British boys, "we do not rob the nurseries." As a matter of fact, in spite of his beard and stoutness, Destrem was only thirty-three years old; but our men were boys.

De l'Escaille remained at Port Said with three Nieuports (N.15, N.22, and NB.2), a small maintenance section and pilots *SM* Grall and *QM* Roussillon. These would maintain local anti-submarine patrols and, one last time, provide a *section d'aviation* for *Anne*.

The morning of 16 April *Anne* was once more off the city of Gaza, aboard was a *section d'aviation* with Nieuports N.22 and NB.2, pilots Grall and Roussillon, and observers Bourne and Finney. At 08.20 N.22, with Grall and Bourne, attempted a reconnaissance of Beersheba, El Shellal and Gaza. The Nieuport hit a wave, capsized and sank on take-off. Although both crew were saved, the aircraft complete with instruments, camera and Bourne's carbine was lost.

At 13.15 Roussillon and Finney on NB.2, with a spare camera, set out to El Shellal. Crossing the coast 20 minutes later at 950 metres they reached El Shellal, 20 kilometres down Wadi Gaza, at 13.50. They reported several camps in the vicinity, but very few troops were seen. NB.2 returned to *Anne* at 14.20. The ship turned down the coast towards El Arish.

> We just had this 'plane hoisted on board when we heard a loud explosion in the sea between us and our escort [*Torpilleur de Défense Mobile 250*]. At first we all thought a submarine must have been spotted by our escort and fired at. But suddenly my little Sudanese orderly—Bakir Ahmed—called out, "Look, sir!" and pointed skyward. I looked up, and there, right over us, were two enemy aeroplanes. The explosion was caused by a bomb they had dropped on us. They then dropped a couple more, which went wide of us, and continued by opening fire on us with their machine guns. We replied with rifle fire, but as they were at a height of 2000 feet, did no good. Neither we nor our escort had any anti-aircraft guns. After about half an hour they retired, and when they were distant some considerable way we elevated our 12-pounder as much as we could and fired a common shell after them. The only damage done to us was that our wireless aerial was shot away. These were the first enemy aircraft seen by us in these parts so we at once reported their presence by wireless and proceeded to El Arish.[140]

The attacking machines were two Rumpler C.Is of *Flieger Abteilung 300* at Beersheba. They were still in the process of establishing a base at El Arish, so when summoned by a telephone call from Gaza, the Rumplers had to fly from Beersheba. One was flown by *Leutnant* Hans Henkel who recorded the attack in his memoir.

> When we arrived on site with our two biplanes, there was of course no sign of the enemy plane. On the other hand, very close to the coast, we saw a seaplane mother ship floating next to a small torpedo boat, with the outline of a biplane clearly visible on its black deck. The attack on our columns had obviously come from a seaplane that had already landed on the water and had been taken onboard. So we had no choice but to make our unfriendly intentions clear to the enemy below by dropping heavy ten-kilo bombs and machine-gunning from low altitude.
>
> The ships did not seem to have any guns on board for defence. In order to evade our bullets, they turned and wriggled through the sea below like blindworms in the sand.[141]

Fortunately there were no Nieuports in the air, they would have stood no chance against the Rumplers.

Anne's final flight for *l'escadrille de Port-Saïd* set out at 17.48, with *SM* Grall at the controls of NB.2 and Finney as his observer. The flight itself was unexceptional, a simple reconnaissance over El Arish, and arrived back at the ship at 18.20. Here is their report:

Hoisting Nieuport NB.2 aboard *Anne*. It would be nice to think that this shows SM Grall and 2Lt A.D. Finney on the final flight of *l'escadrille de Port Saïd* on the evening of 16 April 1916. Sadly the shadows suggest an earlier time of day, and it may have been the flight by QM Roussillon and observer 2Lt A.D. Finney. The purpose of the spike on each upper wing tip is unknown, they can only be seen on NB.1 and NB.2. Possibly they were associated with the Rouzet TSF wireless sets for which these Nieuports were fitted. *(CCI Archive)*

The tents reported in previous reconnaissance have increased to 80–100. These are not concentrated in one camp but are scattered down the Wady for a distance of about 2000 yards in groups of 10–20 tents.

Machine was under fire from anti-aircraft guns over El Arish.
There are apparently two guns as shots were fired in groups of two. Smoke was seen about 20 yards due west of "A" in the wady, indicating the presence of one or both guns.

Owing to the lateness of the flight, there was not sufficient light to produce the photographs.

It is appropriate that the stalwart Grall should have flown the final operational flight of *l'escadrille de Port-Saïd*. He had been there since the very beginning, and had probably spent more time over the desert than any other pilot. Probably, knowing this was the last cruise of the *escadrille*, Grall had wheedled permission to make this final flight.

Returning to Port Said, *Anne* off loaded the Nieuports and took aboard a section from the EIESS; two Short floatplanes, 8004 and 8054, with Flt Lt Wright as pilot and two new EIESS observers Lt T.V. Hughes, RFA, and 2Lt A.K. Smith, HLI. *Anne* anchored in the harbour at Castellorizo on 21 April, returning to Port Said on 26 April.[142]

At Port Said the final packing up of *l'escadrille de Port-Saïd* was proceeding, punctuated by further rounds of celebrations. Once everything was packed up it was loaded on to *Anne* for transfer to Malta, where she would receive a long overdue visit to the dockyard. Weldon recalled the departure.

> Late in the evening of the 3rd May, when all the personnel of the French squadron had come on board, and after Malone had said farewell, we sailed. I had been specially asked by Malone to go with them, and I was only too pleased to do so. We made for Malta, and had instructions on arrival there to transfer the Frenchmen to a French seaplane-carrier that would meet us. We had with us as escort the French torpedo-boat destroyer *Hache*.
>
> We sighted Malta on the 9th May, and arrived off Valletta Harbour about 9 a.m. We then had to wait till the gunboat HMS *Hazard* came to show us the way through the minefield. These mines were supposed to have been laid [by *U-73* on 27 April] just before we arrived. We began to transfer our Frenchmen to their new ship, the *Campinas*. They didn't appreciate the change a bit, and said she was not nearly as comfortable as the *Anne*. They had received orders to sail for Cephalonia as soon as they were ready.
>
> We presented de l'Escaille with a large silver cigarette-box with the French and British flying badges engraved on it, and he was most awfully pleased. On the 12th May we took an affectionate farewell of our good friends the French. As their ship passed the *Anne*, all our crew turned out and gave them three cheers, to which they responded. We were very sorry to part with them, and I never wish to meet a nicer or more gallant lot of men. Whenever one of our French flying men were asked if he could make such and such a flight, sometimes almost impossible, and at any rate most dangerous, his reply, without a moment's hesitation, was always the same, *Je peux essayer*.[143]

Anne returned to Port Said on 21 May 1916.

General Maxwell, always a supporter of *l'escadrille*, had the following to say in a despatch published in the *London Gazette* on 20 June 1916.

> The French Naval Seaplane detachment, with Headquarters at Port Said, under the command of Capitaine de Vaisseau de-l'Escaille, whose services were placed at my disposal for Intelligence purposes, was continually employed in reconnoitring the Syrian, and Anatolian Coast from the requisitioned vessels "Raven" and "Anne". The results of their work were invaluable. The "Anne" was torpedoed near Smyrna while employed by the Royal Navy, but was fortunately able to reach Mudros, where she was patched up and returned to Port Said. I cannot speak too highly of the work of the seaplane detachment.

Lengthy land flights are extremely dangerous, yet nothing ever stopped these gallant French aviators from any enterprise. I regret the loss of two of these planes whilst making dangerous land flights over Southern Syria.

Lieutenant de vaisseau de l'Escaille in his own final report, 21 April 1916, highlighted some the *l'escadrille de Port-Saïd*'s problems and achievements.

Arrived at Port-Saïd on December 1, 1914, the squadron ceased its services on April 17, 1916.

With an average of 6 aircraft and 5 pilots, it provided air reconnaissance services for countries between Alexandria and Makri [Fethiye, on S coast of Turkey] and in the Red Sea.

We participated in the Dardanelles operations from March 18 to May 30, 1915.

With a single type of aircraft we carried out strategic and topographical reconnaissance, carried out anti-submarine patrols, served as artillery spotters and bombers.

During this period 1072 flights were made, including more than 500 hours over land in enemy countries.

Difficulties of all kinds were encountered:

a) From the flying point of view the conditions were exceptionally bad. Extremely violent eddies, downdrafts, almost constantly suspended sand.

b) Given the weakness of our engines (80-hp) each reconnaissance was carried out at low altitude rarely above 1000m sometimes at 300m, often under fire.

c) The slightest weakness of the engine, constituted a breakdown, having no reserve of power.

Landing on land with a floatplane is always an accident.

In the desert, the slightest injury on landing meant a terrible death.

And if the Turks showed themselves to be known enemies, the Bedouins were untrustworthy.

d) Given the remoteness of the centres to be visited or the presence of mountains, the flights were generally of long duration over the land, although no reconnaissance has ever been more than 3 hours.

e) Our seaplane carriers were freighters with makeshift arrangements, offering few resources. Embarkation of the floatplanes on warships caused general wear and tear of the structure.

Despite the unfavourable conditions, we were able to respond to all requests from the English authorities, even when sometimes they brought two seaplane carriers into service at Port Said.

The British arrived at the beginning of January 1916 with much better equipment, in particular the Short 225-hp floatplanes, but they have never done better reconnaissances and have never even provided as many flights.

Whilst the Nieuport floatplanes showed exceptional qualities of flight and robustness, our observations would have been improved by having a more powerful engine. The Clerget engine was the best rotary engine we had. Its biggest fault is the requirement for a major inspection after 15 hours of flight.

The results obtained are due to the professional skill of the pilots, to their great qualities of courage and daring, and to their burning desire to do well.

The intense use of the machines was possible thanks to the professional skills of the repair teams and their hard work.

Campinas in Malta having brought most of the men and machines of the *l'escadrille de Port-Saïd* to the island from Port Said in early April. The Nieuport is N.23. N.22, which also lacked a fuselage serial, remained at Port Said and was written off on 16 April. *(Gallica)*

> As the squadron leaves Port Said, hoping to be able to continue to render service elsewhere, I believe I must bear witness to all the squadron personnel, officers, *Second-maîtres*, *Quartier-maîtres* & *matelots* whom I cannot praise too greatly.

Henry de l'Escaille commanded the base at Corfu for a year from August 1916. He then became Head of the Military Service for Aeronautics and Air Patrols at the Directorate General for Submarine Warfare until after the war. He was promoted *Capitaine de corvette* in June 1918. In October 1919, he became Secretary General of the Inter-Allied Commission of the Danube. He commanded the base at Saint-Raphaël (1920–1922) supervising flying experiments from the battleship *Paris*, and guiding the conversion of the battleship *Bearn* into an aircraft carrier. As professor of applied tactics at the *Ecole de guerre navale* (1922), he was appointed *Capitaine de frégate* in August 1923. Leaving the navy in October 1923, he founded the aeronautical branch of Bureau Veritas. He was promoted *capitaine de vaisseau de réserve* in July 1928. In 1934 he became general manager of Loire-Nieuport. During his service at Port Said he was appointed *Officier de la Légion d'honneur*, awarded the *Croix de Guerre* and the British DSC. He was appointed *Commandeur de la Légion d'honneur* in December 1933. Henry de l'Escaille died at Nice on 30 August 1954.

The East Indies and Egypt Seaplane Squadron Restructures

De l'Escaille's final report contains the somewhat self-serving comment, '…but they have never done better reconnaissances and have never even provided as many flights.' On the face of it the RNAS had not been very busy since *Ben-my-Chree*'s arrival in January. However, she arrived from active service at Gallipoli with a handful of, tired and well worn, floatplanes.[144] Also, as discussed above, there was only a single observer. *Ben-my-Chree* herself was also worn down and in need of a dockyard refit, Malta being the most frequently mentioned and hoped for destination. A visit to the dockyard dry dock became a necessity after the steamship *Uganda* struck *Ben-my Chree*'s bow a glancing blow on 19 February. Following temporary repairs, divers sent down to inspect the damage reported that she remained seaworthy but would require dry docking in the near future. Malta's dry docks could not accommodate her for some time, so the Khedivial Mail Steamship and Graving Dock Company Limited's dry dock at Suez was made available,[145] but not until the middle of March when *Raven* returned from a refit. On 13 March *Ben-my-Chree* entered the dry dock, returning to Port Said, after a thorough refit on 25 April.

As previously noted, *Empress* had arrived from Britain on 23 January, and now both *Anne* and *Raven* were brought into the fold. Towards the end of February Sqn Cdr Malone prepared a memorandum for Admiral Wemyss with suggested arrangements for *Anne* and *Raven*. Their current manning and management was not along strict RN lines.

> The ships were placed under the orders of the G.O.C. H.M. Forces, Egypt, at whose request they were fitted out and manned by the Ports & Lights Administration. All matters connected with their upkeep, coaling, watering, victualling, etc., were dealt with by the Naval Transport Office, Port Said. Sailing orders for the ships were prepared by the Administrative Commandant, Port Said, and issued by the Naval Commander-in-Chief. The ships were then flying the Red Ensign.
>
> The work in which the ships were engaged, and the fact of providing them with armament necessitated the addition of Naval Ratings and commissioning the ships under the White Ensign. The Naval Ratings are under the Naval Discipline Act and paid entirely by the Navy.
>
> In order to place these ships on a service footing it is proposed to treat these vessels, as regards personnel on the lines of the regulations in force in Commissioned merchant vessels. This would necessitate signing all ratings (not belonging to H.M. Service) on Form T.124.
>
> For obvious reasons it would be desirable for all these to be British or Allied subjects, except in special case such as stewards, carpenter or fitters which are difficult to obtain otherwise.

The actual complements of *Anne* and *Raven* did not change much, and *Anne*'s Syrian Boatmen were retained (albeit listed as prisoners of war). But all British, or Allied, civilian personnel would now serve under the T.124 agreement, this made them subject to naval discipline, at agreed pay rates, with clothing allowance and provisioning included. Existing arrangements covering Egyptian and non-allied personnel were retained.

Meanwhile, on 3 April, *Empress* was despatched to the Aegean. It may be recalled that before the evacuation of Cape Helles the requirement for three seaplane carriers at Gallipoli had been clearly stated. Following the evacuation, and transfer of *Ben-my-Chree* to Egypt, only *Ark Royal* remained in the Aegean, and she was too slow, and too valuable as a floating base, to be risked on operations around the Turkish and Bulgarian coasts. In consequence, *Empress* was ordered to join the Eastern Mediterranean Station at Mudros. Details of her service with the Aegean Squadron will be found in a subsequent chapter.

On 14 May 1916, Wing Commander Charles Rumney Samson assumed command of *Ben-my-Chree* and the East Indies and Egypt Seaplane Squadron.

CHAPTER 7

Aviation in the Middle East, 1916–1918

Fethi — A False Start

The Ottoman Air Force suffered from being at the end of a long, and interrupted, supply line. It commenced the war with a handful of obsolescent mostly French machines. Its most modern machine was a civilian Rumpler.

On 14 July 1914 Gustav Basser, with passenger Dr Hermann Elias, took off from Johannisthal on a Rumpler 4A biplane (100hp Mercedes) bound for Constantinople. Reaching the Turkish city four days later after an uneventful flight over 18 hours 12 minutes.[146] On the outbreak of war Dr Elias sold the Rumpler to the Turkish government. The Rumpler, now named *Fethi* after the first Turkish pilot to fly in Palestine, was chosen to provide aerial reconnaissance for the proposed Turkish raid on the Suez Canal. *Mülazım* Şakir Fevzi was chosen to pilot the aircraft. The importance of the aircraft unit was such that it was transported on the train reserved for the Command Staff of the Ottoman 4th Army. After a troublesome journey through the Taurus mountains in winter it arrived at Alexandretta in the middle of December. Here the aircraft was assembled to fly via Aleppo to Damascus.

Rumpler 4A *Fethi* probably at Aleppo 28 December 1914. It crashed when its engine failed on take-off for Damascus. The pilot *Mulazim-i-evvel* (Lieutenant) Şakir Fevzi survived but three spectators were killed and others wounded, and the Rumpler damaged beyond repair. *(Ole Nikolajsen)*

Also at Alexandretta was the British cruiser *Doris*.¹⁴⁷ One fanciful tale says that the cruiser shelled a railway tunnel hoping to destroy the Rumpler. As *Doris* did not actually shell the town, or immediate locality, and the nearest tunnels were well beyond the range of her guns, the story must be discounted. The Rumpler may well have been removed from the town to a place of safety but it was not deliberately targeted.

After the cruiser had left the area Sakir flew off to Aleppo on 28 December. He was welcomed by a large crowd which waited to watch the take off for Damascus. Unfortunately, an engine failure caused the aircraft to crash into the crowd killing three and wounding four. The pilot luckily was unharmed, but the propeller was broken and the rest of the aircraft almost wrecked. No replacement aircraft was available and it would be 1916 before Turkish or German aircraft began arriving in Palestine. The army lost its eyes in the sky, but the attack on the Suez Canal took place in February 1915.

Yüzbaşı (Capt) Şakir Fevzi later in the war.

Growth of the RFC

As two small units, the Egypt Detachment, RFC, and *L'Escadrille de Port-Saïd* had co-operated during the Turkish attacks on the Suez Canal and afterwards. Later in 1915 the RFC began a slow but steady build up that would eventually limit the usefulness of the ship borne floatplanes.

At the end of July Major Massy, commanding the EDRFC, was informed by the office of the Deputy Director of Military Aeronautics that 'the detachment of the Royal Flying Corps under your command became No. 30 Squadron on 24th March 1915.' Massy had been receiving signals, correspondence and stores addressed to 30 Squadron since early July, leading to confusion and delay at Alexandria and Port Said. Some weeks later he was informed that his unit was just a part of 30 Squadron and, pending the agreement of the Government of India, other units currently in Mesopotamia would also become part of the squadron. The inherent difficulty of a divided squadron would finally be resolved in November with the arrival in Egypt of 14 Squadron, shortly followed by 17 Squadron. The two, together with X Aircraft Park, comprising the 5th (Army) Wing, RFC. At this time the EDRFC/30 Squadron was transferred to Mesopotamia, although most of the observers, with their hard earned local knowledge, some senior other ranks and all the aircraft were transferred to 14 Squadron.¹⁴⁸ In June 1916, 1 Squadron, Australian Flying Corps, (67 Squadron, RFC, from September 1916) became operational in Egypt, taking over the aircraft of 17 Squadron which was transferred to Salonika. It too became part of 5 Wing. By early 1917, 5 Wing thus comprised 14 and 67 (Australian) Squadrons, mostly equipped with the BE2c and BE2e, BE12, and Martinsyde G.100/102 single-seaters, and a handful of other types not seen in France, including the DH1A. The Wing actively supported the Egyptian Expeditionary Force through the First and Second Battles of Gaza.

Vickers-built BE2c, 1757. An early production machine it retains the 70-hp Renault engine and the undercarriage of the BE2a. The BE2c served with the EDRFC, and 14 Squadron, from April 1915 until crashing into the harbour at Mersa Matruh on 20 Jan 1916. *(Ray Vann, via CCI Archive)*

The role of aviation during this period was mainly that of observation and reconnaissance, to watch the movement and build up of Turkish forces, bombing was secondary. The RFC strength was building but its aeroplanes were range limited. They kept watch over the Sinai, and the approaches to the Suez Canal, but were unable to reach much beyond El Arish in the north, Bir el Hassana in the middle, and Nekhl in the south. To keep watch on the coast beyond El Arish, and inland to Beersheba and the railway, still required the floatplanes of the EIESS. Their flights often requiring long periods overland on machines totally unsuited to such operations.

Prior to the Third Battle of Gaza (November 1917) the RFC in Egypt was reformed, becoming the Palestine Brigade, comprising 5th (Corps) and 40th (Army) Wings, and X Aircraft Park. The 5th (Corps) Wing consisted of 14, and 113 Squadrons, RFC, with BE2e and RE8 machines. 14 Squadron, RFC,

BE2c, 4306, joined 14 Squadron in January 1916 and was employed in the Western Desert during the Senussi Campaign. It is seen here, being refuelled, at El Dabaa (al-Dab'a) the end of the Khedival Railway from Alexandria.

was also active in the Hejaz area from November 1916 to late 1917, as detailed below. 40th (Army) Wing was formed from 67 Squadron (1 Squadron, Australian Flying Corps, from January 1918) and 111 Squadron, RFC, with a mix of machines including BE2e/f, BE12, Bristol F.2B Fighters, and small numbers of Bristol M.1C Monoplanes, DH.2, and Vickers FB.16 Bullets.

In April 1918, with the formation of the RAF, the 5th (Corps) Wing comprised 14, 113, and 142 Squadrons, with RE8s, Armstrong Whitworth FK8s, and a few Nieuport scouts. The Wing was disbanded on 1 April 1920. The 40th (Army) Wing was made up of 1 Squadron, Australian Flying Corps, 111, 144, and 145 Squadrons, RAF. At this time the squadrons were equipped with Bristol F.2B, DH.9, SE.5a, and a single Handley Page 0/400. The Wing was disbanded on 1 April 1920.

One of the Airco DH2 pushers used by 14 and 111 Squadrons, RFC, in Palestine during 1917. The pilot is installing a 97 round magazine on the Lewis gun. *(Air Force Museum of New Zealand, 2021-085.6)*

Bristol M1B, A5141, 111 Squadron, RFC, Palestine in 1917. The Vickers gun is mounted on the port side of the fuselage, easily accessible from the cockpit. *(Air Force Museum of New Zealand, 2021-085.4)*

Flieger-Abteilung 300 in Palestine 1916–1917

Flieger-Abteilung 300 was the first tranche of German aviation despatched to the Middle East. It was also the one most commonly in contact with the East Indies and Egypt Seaplane Squadron. Later reinforcements experienced similar difficulties in just getting to the Middle East which was at the end of a long and fragile supply line.[149] As discussed earlier, supply and reinforcement of the Turkish and German forces in Palestine was dependent on the Baghdad Railway. However, the two uncompleted sections through the Taurus and Amanus mountains required transfer of goods on to road transport at each incomplete tunnel.

> The primitive nature of the Turkish campaigns is nowhere more dramatically emphasized than in the crossing of the Taurus mountains by [the Pascha units]. Not only did they use the same ancient roads that Alexander the Great had used, but they crossed them carrying their supplies in ox-carts exactly as the Persians and Macedonians had done 2248 years before [Battle of Issus, 333 BCE]. There was a German motor truck unit of 150 trucks here but it was overwhelmed by the magnitude of the transport problem it faced.[150]

Flieger-Abteilung (FA)300 'Pascha' had been established on 22 January 1915 at Döberitz near Berlin, as an expeditionary unit. *Hauptmann* (*Hptm*) Hellmuth Felmy, *Abteilungsführer* from 23 August 1916 to 1 November 1917, recalled its equipment as follows.

> Their preparations were commenced in January 1916 with true German thoroughness. It was clearly going to be an experiment, and the technical equipment of the squadron had to be adapted to tropical conditions.
>
> The fourteen C.I Rumpler machines, with a 160 hp Mercedes engine and enlarged radiators, turned out to be excellent, as also the technical

equipment and the spare parts which were supplied. I can now think with pleasure of the handy crates in which Daimler delivered their spare engines, and which could be transported on camels, in contrast to the methods of other firms later on in the campaign, who packed our spares in enormous crates which could hardly be dealt with by cranes. The squadrons were provided with mechanical transport [three Maybach trucks] for carrying their supplies, but were denied motor-cars for themselves, because the 'experts' declared that they could not be used in the desert, a logic which even today I utterly fail to understand. Another expert declared that in the desert it was only possible to fly at night, or at very most at early dawn! [151]

There were also two Pfalz E.II single-seaters, whose 100-hp Oberursel rotaries did not take kindly to the Palestine climate. He also failed to mention the field weather station, a cartography unit, and the ten pilots and six observers. The initial *Abteilungsführer* was *Hauptmann* Hans-Eduard von Heemskerke, an observer.

The *FA300 Vortrupp* (advance party), with six Rumpler C.Is and the two Pfalz monoplanes, also including 41 NCOs and enlisted men, entrained for Constantinople on 20 February 1915, where they arrived on 5 March. Their troubles now began. It required six days to ferry the machines, supplies and

Pozantı, the northern terminus of the trans-Taurus route. The railway station on the right with the old town, and its mosque, beside it. Large storage sheds, stacks of barrels, and accommodations for the transport troops higher up the mountain side. The road toward Hacıkırı leads out of frame to the lower right, the photographer having just caught a Rumpler for *Flieger-Abteilung 300* preparing to set out through the mountains. The wings are stacked on a following truck, whilst the Rumpler will be towed all the way to Hacıkırı. *(Gunter Hartnagel Collection)*

A Pfalz E.II of *Flieger-Abteilung 300*. Leutnant Hans Henkel stands on the left of the group with his mechanics.

personnel across the Bosporus to Haidar Pasha, and to entrain once more. The new train arrived at Pozantı on 15 March. Here everything, and everybody, had to be transferred to road transport, ancient and modern, for the crossing of the Taurus mountains, thence again by train to the Amanus Mountains, and once more to carts, animals and trucks, to pass the next missing link. Then by train once more to Rayak to transfer to the narrow gauge railway, to finally arrive at Beersheba in the early evening of 1 April 1916. The base at Beersheba was ready waiting for *FA300*. In late February *Offiziersstellvertreter* (*OStv*) Edgar Dittmar, with a party of Palestinian workmen, had begun the assembly of six wooden aircraft sheds and several tents for accommodation.[152] This work had been observed by two floatplanes from *Ben-my-Chree* on 7 March *(see previous chapter)*, and was viewed as an ill portent.

FA300 had experienced a relatively quick passage. Even so, the *Vortrupp* had transferred its equipment seven times during the 4800 kilometre journey from Germany. Typically, stores consigned at Constantinople for the front line at Gaza took eight to ten weeks to arrive. If indeed they arrived at all. Inefficient as the system undoubtedly was, the level of traffic using it was a clear indicator of any upcoming activity by the Turco-German forces. In consequence, various locations were was subject to frequent visits by floatplanes from the EIESS.

Eventually, a main base was established at Ramleh, with advanced bases at Beersheba and El Arish. On 23 April 1917 *FA300* was reported to have available 12 Rumpler C-type aircraft, including three with wireless, 4 Fokker single seater monoplanes (this may have been a mix of Pfalz and Fokker monoplanes). In addition, 2 Albatros scouts were expected to arrive in the near future.[153]

It was *FA300* that most frequently interacted with the East Indies and Egypt Seaplane Squadron, especially the detachment at El Arish. *Flieger-Abteiling 300* despatched an advance party on horses to locate a suitable landing site near El Arish on 10 April,. Two Rumpler C.Is arrived at the new airfield on 14 April crewed by *Oberleutnant* (*Obltn*) Karl Stalter (O) and *Leutnant* (*Ltn*) Hans Henkel (P), *Obltn* Fritz Berthold (O) and *Obltn* Richard Euringer (P). The supply caravan, eight mechanics on horses leading twenty camels, bringing tents, fuel, bombs, ammunition, and supplies, arrived shortly after the Rumplers. A few days later, a forward landing site was established near Bir el Abd, some 60 kilometres west of El Arish. Their first attack, on 16 April, was against the seaplane carrier *Anne* and her escort *Torpilleur de Défense Mobile 250*, as recounted in Chapter 6.

On 20 April, the *FA300* detachment at El Arish made its first appearances over the Suez Canal. A Rumpler C.I, crewed by *Obltn* Stalter (O) and *Ltn* Henkel (P), flew over the Sinai and across the Suez Canal, to photograph the camps at Kantara. On the following morning the second Rumpler set out from El Arish, crewed by *Obltn* Berthold (O) and *Obltn* Euringer (P). They reached Port Said, dropping several bombs, before returning to base. The War Diary of the Administrative Commandant, Port Said, recorded the visit.[154] 'At 8–0 am one hostile aeroplane appeared over Port Said. The plane was flying at a very great height and difficult indeed to perceive with the naked eye. It dropped two bombs, both in the vicinity of the Basin Sherif.' The bombs were possibly intended for *l'escadrille de Port-Saïd*'s base, but de l'Escaille merely mentioned 'A Turkish plane came in the morning to fly over P.S. dropping 3 bombs.'

The Rumpler C.I was particularly successful in desert conditions, with a reliable 160 hp Mercedes engine fitted with specially enlarged radiators, they were the greatest danger the RNAS pilots had to face. The fuselage was a braced box-girder structure with a rounded top decking, with wood four

Rumpler C.I, C.1837/15, of *Flieger-Abteilung 300*. Despite the water cooled engine the rugged Rumpler was much better suited to desert operations than the comparatively fragile Pfalz.

longerons, aft of the rear cockpit the vertical struts were of ash, but forward of that point steel tube was employed. There was a three-ply sheet panel each side of the nose, extending as far aft as the front centre-section strut; the remainder of the fuselage was covered with fabric. The 160-hp Mercedes D.III six cylinder engine was installed on the upper longerons with curved metal panels partially enclosing the upper engine. A large, streamlined collector manifold was fitted on the starboard side, with a curved and raked back exhaust pipe ejecting over the top wing. Pilot was in the front cockpit with the observer behind, the latter having a single 7.92 mm Parabellum, belt fed, machine gun. Standard tail and control surfaces were fitted. The wings were two bay, parallel chord, with 5° sweep-back. The streamlined centre-section struts met to form an inverted V to which the upper wings were joined. Performance, whilst not spritely, was better than any floatplane it might meet. Capable of 152 km/h (94 mph) at sea level, it had an endurance of 4 hours, providing a range of just under 600 kilometres. In addition to the observer's Parabellum it could carry up to 100kg (220 lb) of bombs. Later production machines were also fitted with a fixed 7.92 mm Spandau on the port side for the pilot. *FA300*'s first Rumplers were not fitted with the forward gun. It was just as well that the Rumpler C.I was so useful as *FA300* had to soldier on with it until early 1918 when a small number of AEG C.VI became available.

Leutnant Hans Henkel and his observer *Oberleutnant* Karl Stalter in front of their Rumpler C.I. Henkel's improvised forward firing Spandau machine gun can be seen mounted on the side of the engine. On later production Rumpler's the gun was lower and partially enclosed in the cowling.

Leutnant Hans Henkel also flew a Pfalz E.II monoplane with a fixed forward firing gun, fitted with an interrupter gear. In his memoir, he wrote:

> If only I had had such a rigidly installed machine gun in my Rumpler biplane, which, thanks to its stationary engine, was particularly suitable for longer patrol and pursuit flights! At that time in Germany they had already started installing fixed machine guns in aeroplanes with stationary engines, but I didn't know anything about it in the desert. Since the desire to similarly enhance my machine had awakened in me, I quickly made a determined attempt.
>
> I soon became sufficiently clear about the main features of the execution that I was able to put the first sketches on paper. Then, not without first seeking the advice of my excellent first monteur [fitter], *Unteroffizier* Wiedemann, I set about producing an exact scale drawing. Of course, the design details gave me a lot of trouble. I walked with my draftsman from the drafting table to the plane a hundred times a day in order to repeatedly measure on the spot and look for new solutions. I breathed a

sigh of relief when, after endless back and forth, a workable solution was found and practical implementation could begin.

We tirelessly searched for the necessary material among the available scraps of iron and steel. Numerous things we could not find, others had to be laboriously heated and worked into a somewhat usable form. Then it was hammered, filed, machined, and riveted, into the required shapes.

After the supports and brackets that were to hold the machine gun in its new position, it was the turn of the pushrod that had to translate the movement between the engine and the gun. Then came the ammunition box, the cartridge feed, the trigger by the pilot's seat with connecting pieces and, finally, small modifications to the machine gun itself, which had originally been intended for use in the observer's seat.[155]

The great day of the eagerly awaited first test finally arrived. After a final check and oiling of even the smallest parts of the assembly, the engine was started – idling at first. When it was at full power, I cast an anxious glance around – because I wasn't feeling too good about the whole thing – and pressed the fateful trigger button. "Tack, tack, tack, tack."

In rapid, even succession the bullets passed between the blades of the whirling propeller without injuring any of them. We were lucky![156]

They were not lucky, but rewarded for a brilliant example of in field engineering. It is not clear exactly how the gear worked or when the modification was installed, but some time in May or June appears likely. A few weeks later a second Rumpler was similarly modified. When replacement Rumpler C.Is began arriving at Beersheba in September, they had factory fitted forward guns, which were installed lower than Henkel's improvisation, partially cowled, and had an interrupter gear controlled by a cam on the propeller shaft.

The EIESS held the German machines in healthy regard. Direct contact with them would time and again prove the Shorts and Sopwiths to be markedly inferior. However, General Sir Arthur Lynden-Bell, Murray's Chief of Staff, came close to calling the naval airmen cowards. In a letter to the War Office at the beginning of May 1916, he complained:

> ... the seaplanes have simply got cold feet and can do nothing in face of the big German machines which have now put in appearance. The consequence is that we really know very little of what is going on in the Beersheba area, which is beyond our radius at the present moment. The RNAS have now definitely informed us that they cannot take the risks of the Beersheba reconnaissance except in most urgent and exceptional circumstances.[157]

Given the strength of the EIESS at this time, five or six Shorts and a similar number of Sopwiths (some old and tired), the caution displayed was wise. However, when requested, they rose to the task, and sometimes paid the price.

In mid-June 1917 *FA300* received a couple of Albatros D.III scouts which were based at Huj behind Gaza. In addition a reinforcement of additional *Flieger-Abteilung*, *FA301* to *FA305*, began arriving in theatre during the latter half of 1917, as part of *Pascha II*. The first unit, *Flieger-Abteilung 301*, arriving at Ramleh at the beginning of October. Machines of *FA300* and *FA301* were operating from the landing ground at Huj during the Third Battle of Gaza. The *Pascha II* units came with AEG C.IV, and Rumpler C.IV, two-seaters and Albatros D.III (OAW) and D.V single-seaters. The Albatros single-seaters were eventually formed into a small *Jagdstaffel* unit. At the end of a long, and interrupted, supply line all the German units fought hard throughout the duration of the campaign.[158]

Turkish *3ncü Tayyare Bölük* and 14 Squadron, RFC, in the Hejaz

The German *Flieger-Abteilung* bore the brunt of the fighting over Palestine and Syria. In comparison, the Ottoman Air Force (*Osmanli Hava Kuvvetleri*) units were a mere token force. Whereas the German units in Palestine normally comprised 12 aircraft (including replacement machines), an Ottoman *Tayyare Bölük* (Aircraft Company, *Ty Bol*), could only be formed with a maximum of six aircraft, often less, due to lack of equipment and personnel. In December 1916 the total strength of the Ottoman Air Force was just 72 aircraft, of which 26 were trainers. The frontline equipment was less

Aircraft for the Turkish units also had to cross the Taurus mountains. Seen here at Hacıkırı, still on 600mm gauge feldbahn trucks, is a shipment of Albatros C.III in Turkish markings. The serial of the nearest machine is AK.67, painted on the upper white outline of the national marking, it was delivered to *13ncü Ty Bol* at Kifri, Iraq, in May 1917, later in 1918 it was converted into a fuel transport by installing captured British fuel tanks in the observer's cockpit. *(Gunter Hartnagel Collection)*

Personnel and aircraft of *3ncü Tayyare Bölük* at the airfield of Medina. To the right is *Yüzbaşı* Fazil commander of the company. Behind are the two types of aircraft operated by the unit: Albatros C.III and Pfalz A.II. Large crescent and star have been painted under the starboard wing of the Pfalz. Rubber tyres were short lived and irreplaceable, hence the bare rims of the Pfalz. *(Ole Nikolajsen)*

The Albatros C.III was the most widely used aircraft on Ottoman service; at least 73 were delivered from Germany. This example AK.58 served with *1oncü Ty Bol* in the South Caucasus from May 1917 until written off in October.

than 30 Albatros and Rumpler B and C types, plus a handful of Pfalz and Fokker monoplanes. Although during the war Germany delivered 295 aircraft and 44 floatplanes for Turkish use, few were available for service in Palestine.[159]

By spring 1916 increasing unrest among the Arab tribes on the Arabian peninsula forced the Turkish Hejaz Command to ask for reinforcements to protect the holy areas. As British aircraft had been reported being seen in the area an aircraft company was expressly requested.[160] Due to religious considerations, none of the German personnel already in Palestine could be used. Instead *3ncü Ty Bol*, which was originally to be sent to the south Caucasus, then forming up at Damascus was redirected to Medina in the Hejaz, and was hurriedly issued with 5 Pfalz A.II two-seater parasols (serials P.6 to P.10). An advance party, bringing 3 of the aircraft, 2 portable hangars, 50 bombs and 20,000 rounds of ammunition, with two officers, 2 NCOs, 2 mechanics and 97 other personnel, reached Damascus on 1 July 1916. They commenced to assemble and test fly the Pfalz, struggling to overcome engine problems. It was not until *Yüzbaşı* (Captain) Fazil, a veteran Turkish flyer trained at the Bristol School,[161] was brought in as commanding officer that the Pfalz aircraft became reliable enough to fly. The *Bölük* eventually arrived, by train, in Medina on 3 October.

The Pfalz were employed harassing Arab tribes thought to have accepted British aid. In December, with the three serviceable Pfalz and four crews, a total of 14 sorties were made, some may have been in support of the Turkish attack on Yenbo in December 1916 *(see Chapter 10)*. The final operational flight by the Pfalz A.IIs was on 7 March 1917 (by *Yüzbaşı* Fazil on P.7), they had flown in total 150 hours and were worn out. The first Albatros C.III (AK.28) had arrived from Damascus on 26 November 1916. It was immediately assembled but was lost next day when it disintegrated in the air due to very rough turbulence. By year's end, three more Albatros C.III (AK.30, AK.40 and AK.72) had arrived and were conducting patrols. The new aircraft struggled in the harsh desert environment and did not increase the company's serviceability. Still, two further Albatros, one C.I (AK.4) and one C.III (AK.31) were received in March 1917.

By mid-summer 1917 the Arab interdiction of the Hejaz Railway was increasingly effective, and vitally needed supplies for the aircraft unit could not get through. As a consequence *3ncü Ty Bol* was moved to Ma'an on 1 August 1917, although a single Albatros was left at Medina. Two ex-*FA300* Rumpler C.Is were waiting for the unit at Ma'an, four more being delivered in November. The next months saw daily patrols over and in the vicinity of the railway, helping to keep the Arab bands away. In this period occasional flights over and attacks on Akaba were made.

In November 1917 four rebuilt Rumpler C.I were received by the company.[162] With these reinforcements it was decided to split the unit into four separately located detachments— Ma'an, with one Albatros C.III (AK.31) and three Rumpler C.I (R.1150, 1837, 1847); Dera'a, with one Albatros C.I (AK. 4) and one Rumpler C.I (R.2626); Damascus, with one Albatros C.III (AK.59) and two Rumpler C.I (R.2628, 2636); Medina, with three Albatros C.III (AK.30, 40, 72) and one Rumpler C.I (R2627).

Yüzbaşı Fazil and *Mülazım* Orhan, both pilots of *3ncü Ty Bol* at Medina.

Whilst the Turkish machines were at Medina, 14 Squadron, RFC, had a small detachment, C Flight, operating with T.E. Lawrence and the Arab tribes mainly against the railway. They too were initially prevented by religious considerations from landing at Rabegh on the coast midway between Medina and Mecca. Following further negotiations they were permitted to land there in November 1916. The Detachment operated within the Hejaz from different landing fields and temporary bases until the end of July 1917 when it was withdrawn. A number of reconnaissance flights were made over Medina, but there was no contact with any Turkish machines.

After a storied history with 60 Squadron and Captain W.A. Bishop in France, Nieuport 23 B1566 found its way to Egypt and is seen here at Akaba with X Flight, 14 Squadron, RAF, in 1918. *(via Cross and Cockade International Archive)*

A second detachment, X Flight, 14 Squadron, was landed at Akaba in October 1917. Initially with a mix of BE2e, BE12 and a single DH.2 machines, later some Martinsydes and a single Bristol F.2B were provided. Based at Akaba, but also using temporary landing grounds, the flight continued supporting Lawrence and the Arab tribes. They also made bombing raids on Ma'an and other points along the railway. In February 1918 one of *3ncü Ty Bol*'s machines had been brought down by Arab rifle fire, and its pilot *Çavuş* (Sgt) Ismail Zeki captured. Brought to Akaba he provided information about *3ncü Ty Bol* stating that there were three Rumplers, some other two-seaters, and some Halberstadt D.II were expected. This information does not appear to have made any difference to the operations of X Flight. Although, in April 1918, two Nieuport 23 scouts were added to the flight line. As the fighting drew further north the Flight became less useful, but it was not removed from Akaba until October 1918.[163]

Turkish and German Aviation at Adana and Mersina 1916–1918

4ncü Ty Bol was formed in December 1915 to guard the railway line between the Taurus and Amanus mountains. It was provided with three obsolete Rumpler B.Is and commanded by a Turkish pilot *Mülazım* (Lt) Midhat Nuri with observer *Mülazım* Salih Rifat (or Rafet), 2 mechanics and 15 supporting other ranks. Winter weather in the mountain passes of the Taurus mountains prevented the company reaching Adana. Instead, in order to start the patrol missions, a small strip was prepared near Pozantı to the north of the mountains. On the 19 January 1916 *Mülazım* Midhat departed from there on a mission over the Gulf of Mersina. Meanwhile rain had turned the field into a mud puddle and the aircraft was damaged during landing. No more flights were attempted from this field as the weather cleared and the unit was able to travel on to Adana.

Established on a better airfield at Adana, with an out station at Silifke about 150 kilometres further west, over the following months *4ncü Ty Bol* made a series of reconnaissance flights over Cyprus. They also kept watch over the Gulfs of Mersina and Alexandretta.

By January 1917, *4ncü Ty Bol* was reduced to a single Rumpler B.I and a single Albatros B.I. At least the Albatros was equipped with a W/T set. At this time *Yüzbaşı* Hüseyin Sedat, in his monthly report to headquarters at Yeşilköy, suggested that seaplanes be sent to Mersina instead of using army aircraft to fly over such large areas of water. He considered that the duties of his unit more suitable for naval personnel as these were able to identify ships. Instead an Albatros C.III arrived in April, immediately becoming the only operational machine with the unit. The Rumpler and Albatros B.I were relegated to secondary duties and soon returned to Yeşilköy. The reconnaissance flights over Cyprus continued using the Albatros C.III, but the reports were not believed by the Germans. On the 19 October two AEG C.IVs of the German *FA302* en route to Palestine were diverted to Silifke to reconnoitre over Cyprus. The two German aircraft flew a short reconnaissance over parts of the island confirming previous reports, but bringing no further information to light. At least *4ncü Ty Bol* was given full credit for the observations it made. A second Albatros C.III crashed en route to the unit, which had to continue operations with just one machine. The final flight over Cyprus was on 13 November 1917, following which *4ncü Ty Bol* was sent to Amman.

Friedrichshafen FF.33L 1257 at Mersina in 1918. The minaret of a mosque, with two balconies, and the square windowed building can be seen in old postcards. The modified tail surfaces of this scout/fighter version, with reduced wing span and shortened fuselage, were possibly an attempt to improve manoeuvrability by removing the stabilizing fin. How many were built in this configuration is not known. *(Australian War Memorial, B01852)*

Shortly before its departure the first floatplanes, two German *Wasserfliegerabteilung* Sablatnig SF.5s, arrived at Mersina to assume responsibility for the work being done by *4ncü Ty Bol*. In 1918, both Sablatnigs were destroyed in accidents and replaced, over time, by four Friedrichshafen FF.33L and two FF.49C floatplanes. The floatplanes operated from a stretch of beach on the west side of Mersina, with an out station close to Silifke. It is also possible that two DFW C.V landplanes were provided in September 1918. All machines were withdrawn with the rest of the German units in October/November 1918.

Turkish *Tayyare Bölük* in Palestine and Syria 1918

Although not strictly relevant to the work of the EIESS's floatplanes we will quickly wrap up the story of *3ncü Ty Bol* and *4ncü Ty Bol*. Whilst the RFC/RAF grew, and finally got new equipment (see above), the Turkish and German air forces struggled to survive.

The British forces had broken the existing front line at the Third Battle of Gaza in November and by year's end a new one was being established north of Jerusalem. During 1917 *14ncü Ty Bol* was being prepared for deployment to Palestine, in December, equipped with some AEG C.IV two-seaters, was sent to the 4th Ottoman Army at Dera'a to defend the vital railway junction, and the area east of the Jordan river. In addition, *4ncü Ty Bol* at Adana was re-equipped with five AEG C.IVs in December 1917 and transferred to Amman. As we have seen, in August *3ncü Ty Bol* was withdrawn from Medina to Ma'an, leaving a small detachment at Medina. Limited reinforcements, in the form of reconstructed ex-*FA300* Rumpler C.Is, were made available. In addition *Flieger-Abteilung FA305* arrived at Dera'a in February 1918 bringing twelve AEG C.IV.

In 1918 the Sherifian Regular Army of around 4000 men were confident that they could take the town of Ma'an in a frontal attack against a network of trenches and redoubts surrounding it. The attack on the main garrison,

Reportedly taken at Huj, this photo shows the two Albatros D.III fighters received by *FA300* in June 1917. Due to conditions on the Palestine front, D.636/17 (left) had a Rumpler C.I radiator installed above the centre-section and D.2174/16 (right) has an airfoil shaped radiator installed. Both machines had brief careers, D.2174/16 was apparently written off on 1 September 1917 and D.636/17 was captured by the British intact with pilot *Obltn* Dittmar on 8 October 1917. *(LoC)*

numbering over 4,000 men, stalled amid fierce fighting on 16 April. While some of the Arab troops had penetrated the Turkish wire and advanced as far as the station on 17 April, they could not consolidate their gains and had to fall back on the Jebel Simnah. Virtually surrounded by Arab forces at Ma'an, and despite flying 25 sorties against the latter, *3ncü Ty Bol* was eventually forced to withdraw from Ma'an and relocate to Amman, as they were by this stage reduced to just four serviceable aircraft. One of their Rumpler C.Is was left at Ma'an, but this was subsequently wrecked on 8 May. The Ma'an situation settled down to a siege that would last until late September. The Turkish garrison commenced a retreat on 22 September, marching alongside the railway line which had been cut several months previously. Constantly harried by Arab forces, the survivors were able to surrender to a British force on 29 September.

By beginning of May *3ncü Ty Bol* was left with a single Rumpler C.I and *4ncü Ty Bol* with only two aircraft. British aircraft made a surprise attack on both Amman and Dera'a on 24 June during which *14ncü Ty Bol* lost two AEG C.IV and two pilots. In addition most of the equipment belonging to *3ncü Ty Bol* and *4ncü Ty Bol* was burnt when a hangar was hit at Amman. With too many aircrew and no aircraft to fly, many of the pilots and observers stationed in Amman were transferred to squadrons elsewhere, where they could still be usefully employed.

Attrition continued, *FA305* losing two aircraft in June, and another aircraft in July, *14ncü Ty Bol* having only one aircraft remaining at the end of July. By 19 August the hard working German and Turkish mechanics had managed to make two aircraft flyable each for *FA305* and *14ncü Ty Bol*. Two days later a British bombing raid destroyed three of the aircraft. Replacement machines dribbled in, two machines from Rayak in early September and eight from Jenin on 15 September. But the Arab attacks continued and the remaining aircraft struggled to keep the Arab forces away from the station and the railway line to enable the retreat of troops who were being sent by train from Amman to Damascus.

On 19 September 1918 the final British offensive broke through the Turco-German defences and an unstoppable advance commenced. Damascus fell on 1 October, and Aleppo on 25 October. The Ottoman government signed an Armistice on 30 October 1918.

For the air force units at Amman the last few weeks were disastrous. The remnants of the German aviation units in the 4th Ottoman Army area had been taken under command of *Hptm* Hermann Elias as a single unit. Most of the personnel of the Turkish units, *3ncü Ty Bol* and *4ncü Ty Bol*, were killed or captured when their evacuation train from Amman was attacked at Mafraq on the 21 September. Still more personnel of both *14ncü Ty Bol* and the German units were captured in a train outside Damascus on the 30th. A quick re-organisation was affected at the Rayak Aircraft Park on 31 September and despite the fact that many aircraft were serviceable only 15 could be flown north to Homs due to shortage of pilots. Five flyable aircraft had to be destroyed in addition to a score of aircraft in more or less derelict condition. At least ten more aircraft were lost in combat or in accidents during the remaining month of the war.

When the German Asian Corps under Command of General Liman von Sanders surrendered at Adana on 2 November only 600 aviation personnel of the *Flieger-Abteilung*, including 20 officers, were present out of an original force of 190 pilots and observers and 1400 other personnel which had been sent to Palestine since September 1917. Of the 155 aircraft delivered to the *Flieger-Abteilung* only three survived to be handed over to the seven surviving Turkish pilots and observers and eight mechanics at Adana on 1 November. They managed to escape to Konya and later formed the nucleus of the new Turkish Air Force of the independence forces in 1919.[164]

A Rumpler C.I of *FA300* coming into the attack. Not really. A nicely set up shot of a Rumpler flying towards the photographer who was probably on the balcony of the minaret of the town mosque in Ramleh. *(Australian War Memorial, H01937A)*

163

Wing Commander Charles Rumney Samson on his second favourite mount — a horse. Pre-war he was an avid rider, a pastime shared with a surprising number of British naval officers.

The charger was a German officer's mount captured in France on one of Samson's armoured car raids, retained by him and later brought to Tenedos along with the squadron's aerial mounts. Coal black, save for a white blaze on its face, and a sock on its right rear leg, the horse was given a name by Samson's bluejackets that we are unable to use today due to political correctness.

Seen here outside a village café in Tenedos shortly after the Gallipoli landings.

CHAPTER 8

Samson in Command

Charles Rumney Samson, born at Manchester in July 1883, entered the Royal Navy as a Cadet on 15 September 1897. After passing through *Britannia* he was appointed as Midshipman on 15 January 1899, and Acting Sub Lt September 1902. On appointment to Sub Lieutenant on 15 May 1903 he set out on a typical career path, spending time in cruisers, battleships and, following appointment to Lieutenant, assuming command of torpedo boat *TB81*. In 1910, he was one of the first four Naval officers selected to undertake pilot training, gaining RAeC Certificate, No 71, on 25 April 1911. Amongst the numerous experimental flights he carried out were the first take-offs, in the UK, from both a ship at anchor from *Africa* on 10 January 1912, and a moving ship from *Hibernia* on 2 May 1912. Both flights were on the same Short S.38, Sommer Pusher Biplane, RNAS No.2/T2.[165] He was made Acting Commander in April 1912, on assuming command of the Naval Flying School at Eastchurch, and appointed Wing Commander on 1 July 1914.

At the outbreak of war he led the Eastchurch Squadron, RNAS, to the continent on 27 August 1914. His flying duties increasingly took second place to an activity ideally suited to his buccaneering spirit, the development of the armoured car. His cars had several brushes with German cavalry units. They annoyed the Germans to such an extent that a price was placed on the heads of any captured armoured car men. Awarded the *Croix de Guerre avec Palme* for assisting French troops escape encirclement at Douai, he also received the DSO for his later aviation services at Dunkirk. In March 1915, his squadron transferred to the Dardanelles. He was passed over for overall command of the aviation services in the Dardanelles. Instead, Colonel Sykes assumed that command having to break the news to Samson. 'This was an unpleasant task, and Samson was extremely hurt, but he worked loyally under me and gave me every possible support. I was very grateful to him.'[166] Invalided home at the end of the campaign with an attack of jaundice, Samson had no sooner recovered than he was ordered to Egypt to take over *Ben-my-Chree* and the EIESS, taking command on his arrival 14 May 1916.

Samson had the ability to pick the right men and to inspire them with his leadership. Stocky of build with a ready smile, he favoured a pointed goatee style beard. His favourite close fitting white canvas flying helmet, beard and ready smile could often give him a quite devilish appearance. Vice Admiral Richard Bell-Davies, VC, writing of his own association with Samson at the Dardanelles, said of him, '… that nobody who had served with him under six months ever had a good word to say for him, but that nobody who had served with him over six months had a bad word to say of him … His manners were brusque, he was quite frequently rude and had no tact, but as a friend he was absolutely loyal.'[167]

Samson was not an ideal subordinate. His restless, thrusting, aggressive spirit was always coming up with new schemes for the employment of *Ben-my-Chree* and the Squadron. In Admiral Wemyss he found the ideal superior who gave his mettlesome junior a degree of freedom, and had little cause to regret his trust. Their relationship did not start happily. At the time Samson arrived in Egypt Wemyss had left on a tour of inspection, and at the time was in Ceylon. Faced with too few floatplanes to carry out the envisioned tasks Samson side stepped the usual channels, browbeating the SNO at Port Said into ordering replacement machines from the Admiralty. Consequently, his initial interview with Wemyss was difficult, but Samson won his point regarding the replacements and received a mild reprimand for his methods. After that initial meeting, despite being constantly bombarded with plans and schemes, Wemyss supported Samson as much as possible.

Ben-my-Chree herself Samson grew to love. With her speed and manoeuvrability he handled her like a destroyer. The classic story of his ship handling concerns *Ben-my Chree*'s first visit to Famagusta in Cyprus. The sea that day was rough, and the English harbourmaster had come out to pilot *Ben-my-Chree* through the narrow entrance, barely twice the ship's beam, to the then small port. Refusing assistance, Samson took his ship in at high speed stern first. After enjoying the pilot's astonishment for a while, Samson explained about the bow rudder that made her almost as manoeuvrable proceeding astern as forward. If, that is, one were a consummate ship handler to begin with.

He was not as satisfied with her armament, consisting of the four 12-pdr guns and two 3-pdr anti-aircraft guns she received at Liverpool. Considering the 3-pdr to be useless, Samson talked the captain of *Jupiter*, a pre-dreadnought serving as Port Said guardship, into letting him have a Vickers QF 2-pdr 'pom-pom' autocannon which was installed on the port side of *Ben-my-Chree*'s hangar roof. Subsequently, *Jupiter* also provided a 12-pdr with a high angle mounting, it was installed and test fired in late May.[168] Around this time both *Anne* and *Raven* were provided with a 12-pdr HA gun mounted on their poop. Samson also liberated a familiar friend, 'a travelling mounting to carry a 3-pounder provided with a shield; this was towed by a Rolls Royce car, armoured all over. This mounting proved excellent. We took it to the Dardanelles with us, and funnily enough in 1916 I found the gun and mounting on board HMS *Hannibal* at Alexandria. The Captain kindly let me have it, and I placed it on board my ship. It went into action several times.'

For Malone the situation must have been heartbreaking although, as mentioned earlier, not unexpected. In little over a year he had taken the first fully converted seaplane carrier with an inexperienced crew and forged them into a useful, adaptable weapon. Learning from their mistakes, together they had carried out the first ever aerial torpedo attacks, performed spotting and reconnaissance missions and inland bombing raids. Malone's plans always included comprehensive arrangements to rescue downed aircrew, if possible. After all his hard work establishing the East Indies and Egypt Seaplane Squadron, and being promoted in acting command, he must have hoped his appointment would be confirmed by the Admiralty, however he was junior to Samson. Like Samson before him, he had no choice but to buckle down and make the best of it. He remained with the Squadron, as Samson's second-in-command, in charge of the island base and taking out *Raven* or *Anne* as circumstances demanded, often flying as an observer.

Captain William Wedgwood Benn, like Samson he was also a horseman and they often rode together at Port Said. Samson quickly realized that 'in Benn I had found gold. He had a very keen brain, and a distinct flair for organization.' Benn had only flown once before joining the EIESS, a quick flight around Port Said in one of the French Nieuports. Quickly learning his trade he became an invaluable support of Samson, frequently flying as his observer. Benn is seen here in early 1918 with pilot's wings and the ribbon of the DSO awarded in June 1917 for services with the RNAS in Egypt. He was awarded the DFC in September 1918 for services with the RAF in Italy. *(ACME Newspictures, NY, 1928, in Author's Collection)*

Samson's pressing need was for a Chief Observer and Intelligence Officer to replace Childers. The officer chosen would also fly as Samson's observer on operations, for Samson intended to lead from the front. In selecting 2Lt (later Captain) William Wedgwood Benn, Middlesex Yeomanry, he secured a valuable and efficient assistant. A noted Liberal MP before the war, Benn had thrown all his abundant energies into organizing the National Relief Fund on the commencement of war. In October 1914 having raised over £2 million (worth almost £300 million in 2023) he resigned and accepted a commission with the Middlesex Yeomanry. Whilst 'sweltering in the sun in the desert' he had his first experience of flying.

> Major Fletcher, one of our squadron leaders, had joined the French seaplane flight at Port Said as an observer, and it was during a visit to him that I made my first acquaintance with the use of aircraft in war and conceived the earnest desire to join the Air Force. Strictly against orders, I was permitted to make a flight around the environs of Port Said with the French pilot de Saizieu in his quaint, cranky little monoplane.[169]

On his return to Egypt from the Dardanelles, where his unit had been serving at Suvla in August and September, he quickly attempted to join the group of army observers flying with *l'escadrille de Port-Saïd*. Refused permission, he began a campaign to be transferred to flying duties. Finally succeeding in May 1916 he joined the ongoing group of trainee observers initiated by Childers. When he commenced flying with Samson, Benn had made no previous operational flights. Whatever his previous experience, Samson soon realized that 'in Benn I had found gold. He had a very keen brain, and a distinct flair for organization… He soon became a very fine observer, although he frequently told me he never knew what I wanted, as I used to shout inarticulate remarks to him whilst in the air.'[170]

Samson's description of the work of a pilot and observer is particularly apt at this point.

> Benn was actively engaged in taking photographs, and with his well-known sleight of hand rapidly changing the camera for a bomb on my urgent demands; and then discovering that I wanted him to fire the Lewis gun.
>
> I must inform my readers that we generally carried the 16-lb. bombs loose in the passenger's seat. I leave to the imagination the job the observer used to have. He was in a restricted space with a Lewis gun hitting him in the neck every time he moved, nursing a camera on his knees, with three or four 16-lb. bombs somewhere loose at his feet. Somewhere handy he had to have a pair of binoculars, writing-pad, map, and pencil. Added to this he had to attempt to understand what an excited and, in his view, imbecile pilot wanted him to do. Of course, he couldn't often hear what the pilot said amid the noise of the engine and general turmoil of fight.
>
> I may add as a finishing touch to complete this actual picture of real life, that the 16-lb. bombs had a safety device, consisting of a revolving fan retained by a pin. Once you removed the pin, the fan had a nasty habit of revolving. When it had completed about three revolutions the bomb was liable to explode on the slightest provocation. It will thus be seen that the observer's life was a hectic one.
>
> The pilot, on the other hand, on one of the old Shorts in hot climates had no joy-ride. He had generally a really hard time. First coaxing, or most probably forcing, the seaplane off the water, he then had a tough job trying to make the machine climb in the gradually increasing heat of the atmosphere with the water in the radiator on the verge of boiling. He had to keep the engine at practically full revolutions the whole time to have sufficient power to maintain his meager altitude, and to have some sort of control in the fierce remous that constantly were encountered. At the same time he had to seize every chance, when he gained a few hundred feet to throttle down. Then when he had reached the required area for work, he had to convey to the observer what he required done. I know I frequently lost my temper with my observer, as

he seemed always to miss an opportunity for a bomb or a photograph, through being engaged at his map or writing notes. Meanwhile, when trying to attract his attention I would get hit by some colossal bump, and the old Short used to twist her tail round and I had to fly like a professor for about a minute to keep her from coming down to earth.

Still, we used generally to understand each other, and mutual recriminations were soon forgotten in the feeling that we had done our job to the best of our limited abilities.[171]

Benn should also have his say…

One of my main troubles as a would-be dutiful observer was the frequency and, I am afraid I should add, the occasional incompatibility of the orders shouted at me in the air by my trusted CO. Those who flew in well-equipped bombers and photographic machines in France will smile to hear that with us the wretched observer was accustomed to sit with his camera, note-book, map, Verey pistol, etc., scattered on the floor or in his lap; in addition to which he would perhaps have half-a-dozen 16 lb. bombs tied with string to the longerons or clasped with the rest of his "mixed bag" on his knees. Samson, who had a real eagle-vision, was always pointing out this or that minute object to be noted and was subject to sudden spasms of desire for a bomb to be thrown (bomb sights were not heard of then) or a photograph to be taken. At first I would, in response to the multiple orders hurled at me, attempt to do all processes at once, but I soon learned that calm was necessary to prevent the bomb being launched complete with safety pin, so that it would never explode, the note being taken at the wrong place, or the photographic plate being exposed with the cap still left on the lens.[172]

Whilst Lt Childers had begun to organize an Intelligence Office, its completion was left to Benn. His major innovations concerned the presentation of intelligence data. In the main intelligence office on the island he kept up to date wall maps of Turkish positions and dispositions, with which all observers were expected to be familiar. Supporting reconnaissance photographs and reports were also available. All obvious procedures today, but revolutionary at Port Said in 1916. Before sailing, each ship was supplied with two well laid out boxes containing stationary and intelligence respectively for the upcoming operations. The stationary box contained ink, pens, pencils, writing paper, drawing paper, tracing paper and special sun-printing paper (to be used to duplicate documents in the days before electronic copying machines), and drawing instruments. The intelligence box changed with each trip. It contained files on the operational area with itemized index cards

Part of Benn's tidy and well organized Intelligence Office at the island. *(YEORN 2014-80-73. Courtesy of The National Museum of the Royal Navy)* L-R: Racks for maps; two roller map cases (?); shoulder pistol holsters (?); 47-round 'hopper' for Lewis Gun; Petrol Bomb, Mark 1; Thornton Pickard Type A Camera; Daylight printing box; Stirling W/T set. On the table is a morse key and battery with what appears to be a signal lamp. The walls are covered with drawings and photographs of aeroplane types, airships and bombs.

summarizing this information for quick reference, books, maps, plans and photographs. Each ship's chief observer had the responsibility of ensuring that these boxes were prepared and transferred aboard ship.

Returning one more time to the observer's workload. Samson instructed that:

> When returning to the seaplane carrier, he [the observer] should get on the float when the seaplane is taxying back to the ship and look out for a heaving line. The practice is for the seaplane to taxi towards the ship, with wings spread, at right angles to the ship fore and aft, and, when close alongside, the pilot stops his engine; a heaving line is then thrown to the pilot who hooks on. The observer should be ready to catch the second heaving line if the first misses and also to bear the seaplane off from the ship's side. As soon as you have caught a heaving line remember first of all to catch a turn with it. Always remember to keep well clear of the propeller until it finally stops; also never touch the propeller unless you have first asked the pilot if the switch is to OFF.

Coastal Work

Preparations for an operation usually commenced the day before. During the day, boats towed the floatplanes from the island over to *Ben-my-Chree*, or the other carriers. At the same time their mechanics came aboard, the pilots and observers with all their gear following later in the day. With loading completed, the ship usually sailed late in the evening to be on station early the next morning ready to launch her floatplanes.

During 17 May 1916, *Ben-my-Chree* took on board two Shorts, 8054 and 8082, two Sopwith Babys, 8188 and 8189, both armed with a Lewis gun, also in the hangar were two Schneiders, 3774 and 3789. Samson later recalled the subsequent events.

> We set out with the French destroyer *Voltigeur* as escort on May 17th, and at dawn on the 18th joined up with *Espiegle*, a sloop under Command of Commander Betts, flying the Senior Officer's pennant, and the monitors *M.15* and *M.23*.[173] The objective was a bombardment of the forts and aerodrome at El Arish. At first the *Ben-my-Chree* was ordered to keep out of range and to operate her seaplanes in controlling the squadron's fire.[174]

The first flight left *Ben-my-Chree* at 04.45, Short 8054, piloted by Flt Lt M.E.A. Wright with observer 2Lt F.O. Baxter. The radiator started boiling

Ben-my-Chree at Port Said during 1916. The newly acquired 2-pdr 'pom-pom' can be seen on the aft end of the hangar roof. The unusual square ports in the fore part of the hull were for the First Class Saloon, now converted into seamen's mess decks, with lockers, plain wooden tables, benches, and hammock rails, replacing the original luxurious fittings. There is a canvas wind catcher rigged in an attempt to bring some air into the mess deck. *(Stone Family Collection)*

On 18 May the sloop *Espiegle* headed a small squadron of bombarding ships, including *Ben-my-Chree*, working just off the coast to bombard the forts and aerodrome at El Arish. Built in 1903 and intended to patrol the distant reaches of the Empire, *Espiegle* and her five sisters were an anachronistic throw back as, whilst fitted with six modern 4-inch guns, they were also fitted with a full set of sails on their three masts and were the last Royal Navy ships to have figureheads. *(EMK)*

after 10 minutes and the water was exhausted by 05.25, the Short landing back at the ship ten minutes later. In the interim, the crew had spotted for the monitors' 9.2-inch guns, observing two hits on the aerodrome, possibly damaging a hangar and starting a petrol or oil fire.[175]

Next to fly was Flt Lt T.H. England, on Sopwith Baby 8189, armed with a Lewis gun. He took off at 05.37, to direct the fire of the monitors. At 05.45 England was over El Arish at 2000 feet, 'Observed fire of *M.15* and *M.23* and corrected both. Corrections were made orally by return to ship.' Presumably he landed alongside the monitors and shouted corrections. 'Machine was under fire from shrapnel near Wadi [el Arish]. Observed about 30 camels and a body of men coming from Wadi to town. They were hit by 9.2" shell. Observed aerodrome sheds hit by 9.2 shell and bomb.' He returned to the ship at 06.12.

Samson now made his flight on Sopwith Baby 8188, taking off at 06.19 also to observe the firing of the monitors. 'Observed fire of four rounds from monitors, including one direct hit on Camp S of [El Arish]. Had to discontinue the flight owing to engine trouble.' He returned to the ship just 12 minutes later. It was also his only log book entry for the day.

> After the monitors had finished I received permission to take the *Ben-my-Chree* within close range and bombard. We hoisted an ensign at each masthead, and getting in as close as the soundings permitted, we started in with out 12-pounders, much to the delight of the crew, who dearly loved a bit of fighting in place of seeing the seaplanes get all the excitement. The *Espiegle* came in close as well, and I suppose she was one of the few rigged ships that fought in the War.[176]

Ceasing fire at 07.30 Samson directed *Ben-my-Chree* up the coast towards Rafa. Where Flt Lt J.T. Bankes-Price on Sopwith Baby 8189, armed with a Lewis gun and four 20-lb bombs, set out at 09.12, returning at 09.48. He provided a detailed flight report.

Struck the shore midway between Rafa and Khan Yunis. Turned to N and flew inland about two miles. There were no signs of trenches or any military works between this point and Rafa. The encampment consisted of about eight huts, four large and four smaller ones; also an Artesian well, but no tents. I dropped four bombs (20 lb) on the Camp, all four falling within its area. Continued in the direction of Gaza, distributing pamphlets over one or two farms and scattered buildings and a small village. Turned Southwest and made for the Coast about midway between Gaza and Khan Yunis. Saw scattered figures on the beach and fired a few rounds from the Lewis gun at them.

A final flight was attempted by FSL W. Man on Baby 8188. Hoisted out at 10.16, 'The machine refused to rise owing to roughness of the sea and propeller touching the forward cross-strut of the chassis.' The machine had to be towed back to the ship by one of the motorboats and was hoisted in at 11.08. This was only the first problem Man was to experience on the Sopwith. *Ben-my-Chree* now returned to Port Said.

During the afternoon of 22 May *Ben-my-Chree* sailed once more, this time to carry out a reconnaissance in the Jaffa area. 2Lt Benn was also to receive his baptism of floatplane operations. After a restless night he was called at 03.30 the following morning and stumbling aft through the darkness past coal bunkers, reached the hangar and entered the dimly lit space. Soon even these low lights were extinguished and the roller doors screeched open. Despite a stiff wind and rising sea Samson decided the flight should continue. He climbed aboard the Short followed by Benn, who struggled aboard loaded down with revolver, camera, several Lewis gun ammunition drums and a plotting board. The Lewis gun was handed up as he got settled in the cockpit.

Hoisted over the side and released to drift astern, Short 8087 soon became prey to the seas. Expecting the worst Benn removed his boots, put aside his heavy revolver and concentrated on supporting the camera. Samson opened the compressed air bottles, turned over the engine and succeeded in getting it started. Attempting to taxi and increase speed for takeoff, the seas took hold and after a rough ride smashed both floats. The heavy engine pulled the nose of the machine below the waves causing the tail to rise vertically above it. The fabric covering the wings provided enough buoyancy to support the Short, at least temporarily. Climbing out of their cockpits onto opposite sides, and standing on the wings, both Samson and Benn burst out into relieved laughter. Soon the wings could no longer support the Short, which sank suddenly, leaving the two airmen floundering in the water. By this time *Ben-my-Chree*

French destroyer *Voltigeur* (1910) in the years before the war. She was the regular escort for *Ben-my-Chree* during 1916 operations. This photograph has been identified as her sister *Tirailleur*, but the V on the bows suggests otherwise.

had been manoeuvred close alongside and launched a motorboat. Samson took hold of a heaving line thrown from the ship and was quickly hauled aboard. Benn, his head and shoulders through a life buoy thrown from the ship, was pulled into the motorboat, which then proceeded to tow the wreck alongside the ship.[177] Having proven by experiment that the sea was too rough off Jaffa, *Ben-my-Chree* turned south and returned to Port Said.

Left: Short 184 S.229 (8087) at the island in Port Said, possibly with Flt Cdr C.H.K. Edmonds as pilot and CPO Stone along for a test flight. *(Stone Family Collection)*

Right: On 23 May 1916, 8087 ended ignominiously on its nose in the sea off Jaffa. Cdr Samson and Capt Benn received a wetting but the Short sank. This is thought to show the event, but the lack of national markings is unexplained.

The items lost with 8087 make an interesting comparison with those lost with 849 in February (Chapter 6): 1 Buoy life cork circular, 1 Guns Lewis complete No.1759, 1 Guns Lewis mounting yoke, 6 Guns Lewis magazine for, 1 Pistols Webley No.4773, 1 Pistols Very's cartridge, 282 Cart. SA Rifle ball .303", 24 Cart. SA Webley ball, and 1 Bomb Heavy 65lb. Oddly no mention of Very's cartridges. This time there was no listing of tools or mention of a camera, although Benn does indicate that he had one.

The Jaffa operation had to be repeated on 27 May, this time successfully. Samson and Benn flew in Short 850, accompanied by Bankes-Price on Schneider 3774. Flying some distance apart both pilots spotted activity around a military post south of the town. Samson let them have two 65-lb bombs, whilst Bankes-Price came down low before opening fire with a Lewis gun. Bankes-Price then headed inland toward Ramleh in search of a large camp previously reported in the vicinity. The camp had been moved but he found a smaller one, on which he dropped five small bombs. Both floatplanes were recovered and *Ben-my-Chree* began retracing her steps towards Gaza.

As the ship approached Gaza Flt Lt England and his observer 2Lt F.O. Baxter set off in Short 850 to reconnoitre the town and its environs. *Ben-my-Chree* meanwhile continued south along the coast. England made a good flight over Gaza and down the coast to Khan Yunis and Rafa. Two 65-lb bombs and an incendiary were dropped on some tents near Gaza, the observer also making use of his Lewis gun. A further incendiary was dropped on some trenches at Khan Yunis where the Short came under rifle fire. After pausing to recover England, Samson now took *Ben-my-Chree* further south intending to pay a return visit to El Arish. Refuelled and rearmed, with two 65-lb, one 16-lb HE, six petrol and a single thermite bomb, Short 850 was sent off again with a different crew, Flt Lt Wright with 2Lt A.K. Smith, Highland Light Infantry, as observer. The bombs were distributed on camps and gun emplacements around the town. Damage to hangars at the aerodrome caused by the earlier bombardment had not been repaired. All the while 850 came under steady but inaccurate anti-aircraft fire, some of the bursts being seen from the ship.

Whilst the Short was away, *Ben-my-Chree* came under attack from a German aeroplane. Identified in Samson's report as 'of the LVG type,' it was actually a Rumpler C.I, crewed by *Obltn* Richard Euringer (P) and *Obltn* Fritz Berthold (O), from El Arish. Four bombs were dropped from about 5000 feet, two falling close enough to straddle the ship but causing no damage. The attacker then descended to 4000 feet to give the observer an opportunity to open fire with his machine gun. Throughout the attack *Ben-my-Chree* used her 3-pdr's, pom-pom, and Lewis guns against the assailant to no avail. The machine remained in the vicinity for about half-an-hour but fortunately departed a few minutes before the Short returned. The latter being hopelessly outclassed by the landplane. *Ben-my-Chree* returned to Port Said after a very busy morning's work.

Ben-my-Chree was fitted with two 3-pdr guns on anti-aircraft mountings on the forward corners of the hangar. These were often in action, albeit ineffectually. The variety of head wear is interesting running from naval officers and ratings caps, a civilian panama and what appears to be a Turkish kabalak. *(Ministère de la Culture, APOR070191)*

Ben-my-Chree was also provided with four 12-pdr guns, two forward and two aft, which Samson liked to use as often as possible. Here the crew of the forward starboard gun exercise for the camera. *(Ministère de la Culture, APOR070188)*

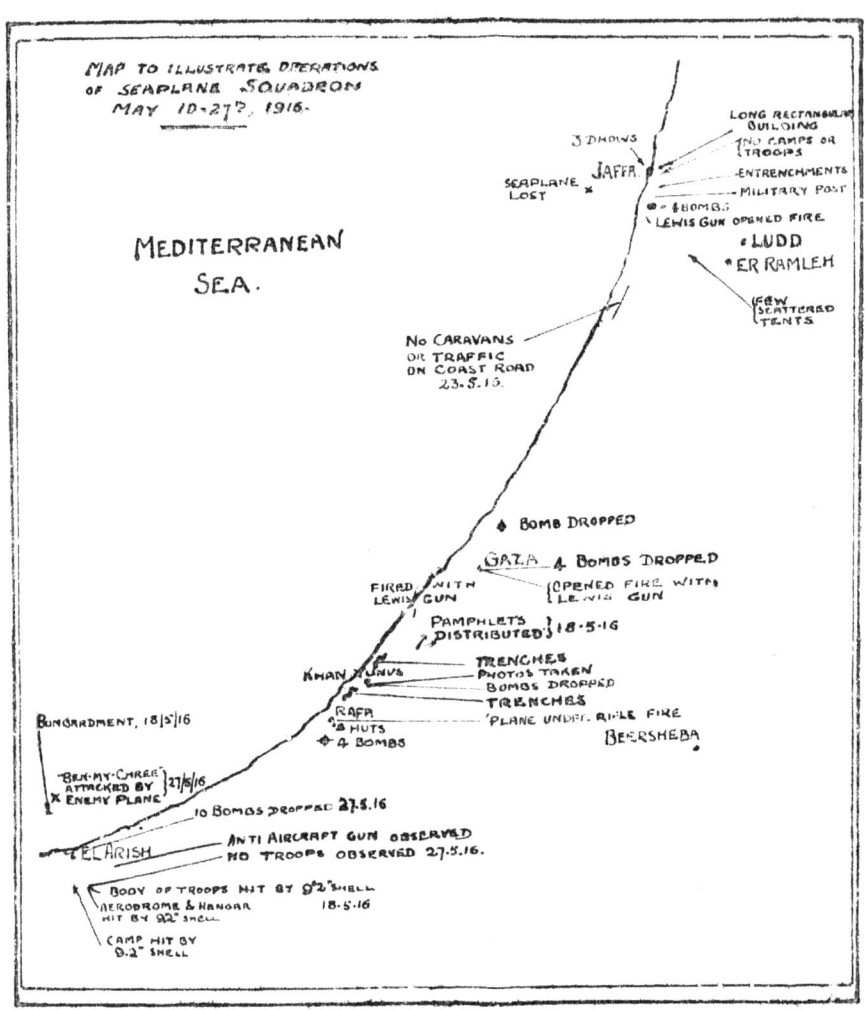

May 1916 was a busy month for the EIESS along the Palestine coast as shown by the map/table from one of Samson's reports. *(TNA, AIR 1/660/17/122/620.)*

Ben-my-Chree spent much of June at Aden and in the Red Sea, as detailed in a following chapter. During this period *Anne* had reverted to cargo ship and was delivering an assorted cargo of stores and gold, the life blood of the Arab Revolt, to Jeddah. On return from the Red Sea she was docked at Suez for refitting, returning to Port Said at the beginning of August. In the absence of the two ships *Raven* was able to handle the few requested observations along the Palestine coast on 7 and 22 June. Both were brief one day operations. The first was a reconnaissance of the El Arish area, and the second was to bomb camps near El Arish.

Raven left Port Said again on 1 July to reconnoitre El Arish and Haifa, and to bomb camps as opportunity presented. Throughout 29 June *Raven*'s crew were kept busy taking aboard stores, ammunition and fuel for the floatplanes. At the same time the ship's bunkers were being filled by the usual native work gangs. The next day two Shorts, 8090 and 8091, Schneider 3786 and Baby 8189 were taken aboard. The rank and file of the air detachment, 2 Petty Officers and 19 mechanics, armourers and photographers, joined ship during the day. The pilots and observers came aboard that evening. In addition to Flt Lt Dacre, who was in command of the RNAS contingent, the pilots were Flt Lt J.C. Brooke and FSL W. Man, observers Lts J.W. Brown and A.P. Ravenscroft, with 2Lt E. King.

Raven sailed from Port Said at 06.30 on 1 July, in company with the French armed tug *Laborieux* as escort. The tug was a more suitable companion for the ex-tramp steamer, 10 knots on a good day, than the destroyers that accompanied *Ben-my-Chree*. At 16.30 on the same day, the two ships were at sea 15 miles from El Arish. Dacre and Brown were hoisted out in Short 8091, loaded with bombs, Lewis gun, and camera. In a considerable swell they were both nearly sea sick by the time the floatplane left the water. The Short's Sunbeam engine did not fare very much better, beginning to boil after only three minutes of flight. Dacre throttled down to save his engine, loosing some of the 1000 feet altitude he had gained. Then, whilst still three miles from the coast, the oil began to overheat and they had to return to *Raven*, where they were safely recovered. Brooke on Schneider 3786 had more success. Over El Arish he dropped two small bombs on a suspected ammunition store, although without any resultant major eruption. Throughout his flight he was subjected to heavy archies and high explosives but managed to dodge them.' Following his safe return to *Raven*, the ship sailed on overnight to Haifa.

Dacre's diary provides details of the day's events.

> *July 2* — Arrived off Haifa in the middle of the Syrian coast at 6 am. I was hoisted out with Ravenscroft as observer, bombs & Lewis gun. [8091] We crashed into the atmosphere in company with Man on the Schneider [3786] & made a recon of Haifa district at 2000 ft. We dropped one small bomb on a road bridge &

just missed it. We then went north & flew around Acre, the last stronghold of the Christians in the Holy Land & the place from where my name is supposed to originate in the Crusaders days.

The town is surrounded by old fortifications in excellent preservation. Many boats were pulled up on the beach & several people looked excited. We saw in the distance an object just like an aeroplane hangar but on flying over it to bomb it we discovered it was a square store building with lean to ends all white washed. We dropped one small bomb on the railway bridge there & missed it.

Flying back over Haifa I had to descend to 1000 ft to avoid black clouds. Here I let go a 112-lb bomb at a large shed on the railway pier, used as a Customs House. This large bomb went off within 15 yards in the water alongside the shed with a terrific explosion chucking mud & water everywhere. A small bomb was also as near. We passed back close to the Convent on the high hill & waved in case there were any nice nuns there.

On being hoisted in I learned Man had not returned so after a short while waiting it was certain he must be down somewhere. The ship closed to 8 miles from the coast & the escort went in nearer. I ordered out the Short again, lightened her with less fuel, no bombs, taking Brown with me to use the Lewis gun if necessary & to render Man assistance if possible. We discovered the Schneider in a sinking condition under the wall of Acre & funnily enough no boat had gone out to him, nor were there many people on the beach. I landed alongside & Man who was standing on the planes that were out of the water dived in & swam to the Short. We then eased off 47 rounds at the Schneider to make certain it would not fall into the enemy's hands. As Man was wet I packed him on the floor of the passenger seat out of the wind. While Brown sat astride the petrol tank to preserve the balance of the machine. This position was most cramped & uncomfortable being just behind the hot air off the radiator & every time the machine bumped Brown's head got bumped too. However Brown seemed to like it & 3 up we flew back to the ship unmolested. Rather like a cinema stunt but quite nice, & Man lost his visit to Constantinople.

Man recorded that the Sopwith was forced down '½ mile W of Acre' after 20 minutes flight with 'Engine trouble caused by a tappet rod breaking in air; on landing two others were found to be broken.'
Raven now turned towards Cyprus, arriving at Famagusta at dawn on 3 July and mooring alongside the quay at 05.30. Almost immediately commencing the slow, filthy work of offloading 400 tons of coal, in 2680 bags, for the future

use of *Ben-my-Chree*. Unloading was finally completed at 6 pm the following day. Half an hour later the ship set out for the small island of Castellorizo, where her floatplanes were to carry out some reconnaissance flights for the French Governor *(see Chapter 11)*.

On their return from the Red Sea on 21 June, *Ben-my Chree*'s ship's company morale was at a zenith. But lying at anchor with the bright lights and attractions of Port Said just a boat's ride away soon tempted Jack astray. In any crew there were the incorrigibles, those for whom trouble always lay in waiting, as was retribution. The ship's cells were kept busy with those unfortunates caught outside the King's Regulations and Admiralty Instructions and put on Captain's defaulters, for Samson was a strict disciplinarian. It was to everyone's relief therefore that *Ben-my-Chree* sailed on 6 July bound, once more, along the coast of Palestine and Syria rich in Crusader history.

The Crusader's Coast

The Crusaders of earlier centuries had spent their blood and lives along the whole length of the coast from Ashkelon to Antioch. From the south to the north the coast was divided into three parts—the kingdom of Jerusalem, the county of Tripoli and the principality of Antioch—approximating to modern day Israel, Lebanon and Syria. Generally known as the 'Franks', for many were of French origin, these militant Christians held the coast and its hinterland for almost two centuries (1098–1291) against incessant attacks from the equally militant Moslem forces. In their wake they left great forbidding fortresses and romantic legends. Great War memoirs from this area are full of Biblical and Crusader references. It is clear that many of those involved felt deep down that they were part of a new Crusade. The French in particular still maintained their centuries long interest in the area.

Lieutenant de Vaisseau Picard, Intelligence Officer to *Contre-Amiral* de Spitz the French *Commandante Division Navale de Syrie*, brought maps and the latest intelligence when he joined the ship for *Ben-my Chree*'s next cruise. Aboard were Shorts 850 (S.2), 8054 (S.4) and Schneiders 3789 (B.1), 3790 (B.3).[178] Escort for this trip was the French destroyer *Dard*. By the afternoon of 6 July 1916 the two ships were off El Arish in a heavy sea. The importance of this town, which required the constant attention of passing ships from the East Indies and Egypt Seaplane Squadron, lay in the fact that on it rested the right flank of the Turkish front line. British intelligence considered that any future Turkish attacks would be indicated by troop concentrations in and around El Arish.

In the heavy seas a Short failed to get off, but Bankes-Price managed to wrestle Schneider 3790 off the water at 17.11 and headed off on a reconnaissance of the area. The Schneider had a Lewis gun and two 16-lb bombs.

Left water off the coast of El Arish and flew inland over Wadi El Arish which was honeycombed with trenches. Two bombs were dropped, one falling in a trench and the other near a gun pit. The machine was under very heavy shrapnel fire; also fire from rifles and machine guns. On proceeding inland towards the aerodrome I was headed off by two hostile aeroplanes; one was a biplane of the same type as previously attacked the ship [Rumpler C.I] and the other a small monoplane [Pfalz E.II?]. Both machines were considerably faster than the Schneider and they chased me about half way back to the ship, when they turned and flew towards the land.

An earlier raid by 14 Squadron, RFC, on 18 June, claimed to have destroyed at least two German aircraft, setting two hangars on fire and damaging others.[179] To gain surprise the RFC aircraft had approached from over the sea. So, *Ben-my Chree*'s Schneider coming from the same direction understandably set alarm bells ringing. The two machines from *FA300* quickly heading off the Schneider and chasing it back to the ship. Recovering the Schneider and its pilot undamaged at 17.31, Samson headed his command up the coast to Beirut, arriving there the following morning.

On 7 July, with the heavy sea of the previous day still running, the first Short attempting to take-off that morning only succeeded in smashing its propeller. Samson and Benn set out on a second machine, 8054, and finding some calmer water were able to get off. They made a reconnaissance of the town, harbour and defensive trenches. The harbour was busy with six schooners and innumerable Arab dhows. The pair dropped a single 65-lb and three 16-lb bombs on the ships but failed to cause any damage.[180] The flight over, Samson now took *Ben-my-Chree* further north to the mouth of the Nahr el Kebir beyond Tripoli. While her gunners were engaging and dispersing a camel caravan, Short 8054 (Flt Lt England and 2Lt Smith) set out to attempt to reach the railway station at Tel Kale, halfway to Homs. Due to engine trouble the Short was unable to reach the station, 20 miles inland, but did find two tugs sheltering in the Nahr el Kebir river. To end a busy day, *Ben-my-Chree* next proceeded to Ruad. Once the last stronghold of the Crusaders, *Lieutenant de Vaisseau* Trabaud and his garrison of French *matelots* now held the island for the Entente. The Governor requested information about trenches on the mainland opposite his small command. Samson obliged with flights, by Short 850 that evening (FSL Paine and Sub Lt Kerry) and the following morning 8 July (Flt Lt Maskell and 2Lt Smith). Before leaving for Famagusta he gave *Lt de V* Trabaud a set of maps and photographs detailing the Turkish entrenchments. While Short 850 conducted the morning reconnaissance, two Schneiders (Bankes-Price on 3790 and Samson flying 3789) flew back down the coast to bomb the two tugs in the Nahr el Kebir. The bombs missed, but *Ben-my-Chree* had not yet finished with the tugs.

Ben-my-Chree alongside the coaling wharf at Famagusta, 8/9 July 1916, with French destroyer *Dard* ahead of her. Modern Famagusta has spread far beyond the ancient city walls, although many of the buildings, including the Lala Mustafa Pasha Mosque (once the Cathedral of Saint Nicholas) seen here above *Ben-my-Chree*'s bow, can still be identified. *(EMK)*

On the way to Famagusta a submarine broke the surface about a mile astern. Turning away and increasing speed Samson took his valuable command out of danger. *Dard* wheeled about towards the submarine and opened fire. The U-boat quickly dived, enabling *Ben-my-Chree* to escape while *Dard* circled the area to keep it down. If there was a submarine, *Ben-my-Chree*'s log book only noted 'observed possible submarine.'

Arriving at Famagusta that evening, Samson put on his previously mentioned display of ship handling backing into the small harbour, then commenced coaling. The coal, delivered by *Raven* a few days earlier, had refilled *Ben-my Chree*'s bunkers by late afternoon on 9 July. At 19.00 she sailed and once clear of the harbour set course back to Nahr el Kebir, arriving there at 04.30, 10 July.

Determined to destroy the two tugs, Samson sent off two Schneiders, flown by Bankes-Price and himself, armed with bombs and Lewis guns. The bombs were dropped around the tugs, apparently causing little damage. During their attacks the floatplanes were fired upon by troops from a small post at the mouth of the river. Relieved of their bombs, the Schneiders dived on the troops, returning fire with their Lewis guns. Samson brought *Ben-my-Chree* inshore after sending Short 850 (England and Smith) to direct the fire of the ship's 12-pdr guns. After dropping a single bomb, they commenced spotting for *Ben-my Chree*'s forward starboard 12-pdr. Twenty-one shells were fired, at least two bursting alongside the tugs. Bankes-Price was sent out again on Schneider 3790 and set out again for the tugs, then returned for a third load of bombs. As a final gesture *Ben-my-Chree* turned her guns on the Turkish post, seventeen shells leaving it damaged and on fire.

Concluding that the tugs, if not destroyed, were at least severely damaged Samson turned his ship back down the coast towards Beirut. First off were Flt Lt Maskell with 2Lt Benn, on Short 850 at 11.05, to check for trenches along the coast north of Beirut and then to bomb any shipping in the harbour, for which they carried two petrol bombs. These fell into the water alongside a ship, and failed to ignite. Shortly before their return at 11.43, Flt Lt England on Schneider 3789, with two 16-lb bombs and two petrol bombs, set out to bomb a schooner moored in the harbour. 'Dropped petrol bomb amongst shipping; another near sailing ships lying alongside quay did not explode. Two 16-lb bombs exploded on the roof of the Customs House; a little smoke was observed.' Samson was very aware of the necessity of avoiding non-military damage and casualties and required that all bombs be dropped to avoid such targets.

Whilst they were away, Samson flew *Lieutenant de Vaisseau* Picard around the town (Short 8054) to give him first hand information for his chief to add to *Ben-my Chree*'s reports. Samson has two different versions of the flight, his memoir and log book tell a different story than his Report of Flight. In his log book,[181] he wrote, 'Short carrying Lt Picard, French Navy, 31^{min} 34'. 1–65lb bomb dropped at shipping in harbour (bad shot).' His memoir tells essentially the same story. However, according to his Report of Flight, the Short carried no bombs just a Lewis gun. They took off at 11.28 and landed at 11.59. The actual flight report, probably based on Picard's account, reads:

> Flew over BEYROUT Harbour. Observed four trucks on the line close to the railway station. Rifle fire was directed at the machine from the Harbour and from a point between the Harbour and the lighthouse. The following were observed in the Harbour:–
>
> 9 Schooners
> 12 Dhows
> 1 Crane Lighter
> 5 Lighters
> 1 Gun Boat sunk
>
> The water in the Harbour was too opaque for mines to be distinguished.
>
> Proceeded to JUNIE Bay. Rifles were fired at the Seaplane from RAS EL TIN. The water in JUNIE Bay is very clear and the bottom plainly visible. No signs of any mines.

Proceeding now further south, Maskell and Kerry on Short 8054 flew a final reconnaissance was over Haifa and Acre.

During the night a strong northerly wind blew up. Always a good sea boat *Ben-my-Chree* maintained a higher speed than her faithful escort *Dard*, and the two ships became separated. After searching back along the coast in daylight for the missing destroyer, Samson received a radio message that she was at El Arish. As *Ben-my-Chree* was now off Jaffa, operations were cancelled and both ships returned independently to Port Said.

More Coastal Work

Following *Ben-my Chree*'s brush with the submarine, Port Said had been placed on a full alert. Normally anti-submarine patrols were conducted by locally based trawlers, drifters and motor launches. When *Ben-my-Chree* returned her aircraft were called upon to conduct additional patrols. Between 12 and 20 July twenty-seven patrols, totalling only 17 flying hours, were flown.

Ben-my-Chree sailed again at 18.00 23 July, accompanied this time by *Arbèlete*, the two ships arriving off Gaza the following morning. Between 04.42 and 04.50 two Shorts crewed by Samson and Capt Benn,[182] and Flt Lt England and 2Lt Smith, were sent out to reconnoitre the Gaza to El Shellal roads and camps, the roads around Khan Unis and Rafa towards El Arish, respectively. Both had returned by 06.19. *Ben-my-Chree* now turned down the coast towards El Arish where, at 09.23, Short 8054 with Flt Lt Maskell and a new observer Lt N.W. Stewart, 7th Royal Scots, set off along the El Maadan to El Arish road and to search for mines off the latter. Inland a ground mist restricted visibility, but along the coastal strip some useful observations were made. A summary of the findings was immediately transmitted by W/T to headquarters, an innovation only recently adopted and henceforth employed whenever operations were within transmission range.

Shortly after Maskell and Stewart had returned one of the *FA300* Rumplers from El Arish approached the ships. After circling for a period, it came into attack dropping three small bombs. Two fell close alongside the port bow of *Ben-my-Chree*, the nearest within 20 yards (18 metres), the third even closer to *Arbalète*. *Ben-my Chree*'s collection of anti-aircraft guns kept up a brisk fire throughout the attack, somewhat diminished by problems with the pom-pom, which according to Samson would 'pom, but not pom-pom'. Ordered by W/T to return to port Samson cancelled the planned flying programme and the two ships retraced their course. They were allowed only a few hours at Port Said, just long enough to coal (from midnight to 05.15 for the weary crew of *Ben-my-Chree*), and receive new instructions, sailing again at 09.25 on 25 July.

It is probable that this recall was occasioned by increasing Turkish activity forward of their front lines. During an evening reconnaissance on 19 July a 14 Squadron machine had reported an enemy force advancing 'west from El Arish, and that it had established itself on the line Bir El Abd, Bir Jameil, Bir Bayud. The size of the force seen was estimated by the observer as between eight thousand and nine thousand men, of whom three thousand to four thousand men were in Bir El Abd, the remainder being distributed throughout the remaining areas'. These were the first moves leading to the Battle of Romani *(see below)*, and were countered by British reinforcements were hurried forward from the Canal Zone. Samson was briefed to carry out a series of flights between El Arish and Haifa to observe any supporting troop movements in the area.

A set of four images hoisting Short 8054 (S.4) on board *Ben-my-Chree* after a flight over Khan Yunis and Rafa on 24 July 1916. The crew were Flt Cdr T.H. England and Lt N.W. Stewart on his first trip as an observer. After landing off the port quarter of *Ben-my-Chree* England taxied around the stern of the ship to approach the starboard side by the hangar.

Ben-my-Chree arrived at a position close to El Arish at 13.30 on 25 July. Once more Short 8054, with Maskell and Stewart, set out to reconnoitre the area. They reported signs of recent movement on the roads, but no current activity. The camp at Bir El Mazar had greatly increased in size in just 24 hours. The two ships then headed north towards Gaza to continue the observations.

At 16.30 three schooners were sighted by *Ben-my-Chree* and *Arbalète*. Attempting to increase the flow of supplies to the front, the Turks made constant efforts to run supplies down the coast using small local sailing craft. French naval patrols constantly sought out these blockade runners, but unless they could be caught at sea and sunk the local boats were difficult to destroy. Designed to be run up on the gently shoaling beaches, on the approach of patrols their crews ran them ashore and took shelter in the sand dunes. Once ashore, they had to be burnt out to be sure of destroying them. If they were not, their crews would quickly repair them and continue their supply run. True to form these sailing craft headed into the shore. Before setting off in pursuit *Ben-my-Chree* stopped to launch Short 8054, Flt Lt England with 2Lt Smith as observer. The Short headed towards the two smaller schooners, whilst *Ben-my Chree*'s gunners engaged the leading vessel. This was a 250 ton vessel with a red hull, well known to the French as a frequent blockade runner. England with his Short drove one of the smaller schooners ashore, the other two being driven in by the two ships. After their crews had been given time to abandon the vessels, *Ben-my-Chree* and *Arbalète* opened fire on their prey. The red schooner erupted in explosion and fire as soon as *Ben-my Chree*'s first shell hit her, leaving no doubt of the nature of her cargo. *Arbalète* moved closer inshore, to finish off one of the smaller schooners, coming under rifle fire from a Turkish patrol. *Ben-my-Chree* closed to support her consort and scattered the patrol with a few well placed shrapnel shells. All three schooners were well alight when the two ships finally withdrew. Regrettably, *Ben-my-Chree* suffered her first loss since leaving the UK. Petty Officer J.A. Martin, gunlayer, collapsed and died of a cerebral haemorrhage whilst working his gun. At 10.22 on 26 July, *Ben-my-Chree* came to a stop whilst the body of Petty Officer Martin was committed to the deep.

Whilst the ships finished off the schooners, England had headed inland to spy out the roads between Gaza and Rafa. Again, no traffic was seen on the coast roads. Well satisfied with their day's work the little squadron headed further north, intending to be off Haifa at dawn on 26 July.

Shortly before 05.00 Samson and Benn set out on Short 8372 to follow the railway inland from Haifa to the junction at El Afule. During the flight numerous camps accommodating in total 5000 men were seen. At El Afule they found a train in the station. Benn, who had been nursing four of the tricky 16-lb bombs on his lap since leaving the ship, now 'threw' two of them at the train. To their surprise and delight one hit a coach setting it on fire. Returning along the same route they had flown outward, the Turks expected and awaited them. Approaching Tubaun they flew into a heavy shrapnel barrage. One of the port side interplane struts was smashed and holes appeared in the wings and elevators, but nothing vital was hit and they returned safely to the ship. Samson immediately set his air mechanics to work replacing the strut and patching the holes in the fabric, he required the Short to be ready to make another flight within a few hours.

Turning south back down the coast *Ben-my-Chree* and her companion encountered two more Turkish blockade runners. Given the previous day's losses, only urgent need could have persuaded the Turks to risk sending them out when the aggressors remained active along the coast. The effect of the blockade may be judged by the fact that even sacking to make sandbags was in short supply. Again, both were driven ashore and a 100 ton schooner and 50 ton dhow were left burning on the beach. Continuing down the coast Samson used *Ben-my-Chree*'s guns to disperse a camel caravan and to shell a road bridge.

Previous page, clockwise from top left.

1. With engine stopped 8054 drifts in toward the ship's side. England standing in the cockpit catches the hoist cable and hooks on. Stewart, following Samson's Instructions to Observers, has left his cockpit clambered through the wings and is standing on the float ready to catch hold of a heaving line if necessary. Men with long padded poles stand on the hangar and deck ready to fend off the Short to prevent damage. *(Ministère de la Culture, APOR070185)*

2. As hoisting continues the Short is carefully rotated through 90° to bring it over the deck. The two light struts aft of the main float struts to support the torpedo carrier are still retained. *(Ministère de la Culture, APOR070186)*

3 and 4. Now the tricky part. There is not much room for a Short 184 on the area of deck aft of the hangar and forward of the kingpost supporting the derrick hoist. Stewart remains on the float watching as the floats are lowered onto a handling dolly and the Short manoeuvred so that its nose is inside the hangar and the wing leading edges not too close to the steel structure. Once secure the wings can be folded and the Short stowed in the hangar. The Samson serial S.4 is clearly visible on the rudder. *(Ministère de la Culture, SPA 66 K 4030-4031)*

French destroyer *Arbalète* (1903) seen here at Monaco in 1905. She and her sister *Dard* were regular escorts to the seaplane carriers of the EIESS.

At 10.10 a fully repaired 8372 was hoisted out off the Nahr Iskanderun with England and Smith as crew, they were to attempt a reconnaissance inland as far as Samaria and Nablus. Overland the Short was buffeted about and unable to gain sufficient altitude to cross the Carmel mountains. Cutting the flight short England return to the ship and was hoisted in at 10.52. Cruising slowly south *Ben-my-Chree* was approaching Jaffa when, at 12.42, Schneider 3771 piloted by Bankes-Price set out, tasked with examining the railway between Ludd and Ramleh. Between Jaffa and Ludd, Bankes-Price saw no movement on the railway. But at Ludd he 'dropped 2 bombs on the station, one going through the roof of a shed causing a fire, which burned for a few minutes but went out.' Finding nothing at Ramleh, he returned to Ludd dropping his two remaining bombs, without result. Bankes-Price returned and was hoisted in at 13.24.

Approaching the Nahr Sukerier, between Jaffa and Ashkelon, at 14.03, Short 8054 with Maskell and Stewart aboard was hoisted out to carry out a reconnaissance to El Falujeh. They noted some activity on the railway and roads, and a large quantity of fodder close to the railway. The Short returned at 14.51, having come under rifle fire whilst crossing the coast, and was hoisted in. Later that afternoon, England and Stewart made a second attempt with 8372 to reach Samaria and Nablus. This time they succeeded, despite continuous buffeting and archieing at the low altitude they had to fly. The flight produced mainly negative information; empty or deserted camps, no trains or military activity. Samson was very pleased with the flight nonetheless, later commenting that it 'showed what a determined and skilful pilot can do.'

Following England and Stewart's recovery *Ben-my-Chree* and *Arbalète* returned to Port Said. It had been a busy and successful 48 hours. In his report to Wemyss, Samson commented that the support given by *Capitaine de Frigate* Monnaque and his *Arbalète* 'which went into very short range of the shore in face of rifle fire from concealed troops to shell the sailing vessels and displayed great vigilance in performing her duties as escort,' fully met his expectations.

Monnaque became a great friend of *Ben-my-Chree*'s officers, often serving them superb meals, cooked by his servant on a tiny stove on the open deck of the destroyer. Benn, not always generous with praise, noted:

> ... we left Port Said accompanied by our old companion the *Arbalète* under the command of Captain Monnaque. Monnaque was not only the best of good fellows and a most faithful watchdog — for it was not always comfortable for a small T.B.D. to follow a great ship like ours through bad weather — but he was also a warm friend, and

his poetical efforts in praise of the *Ben-my-Chree* were not the least attractive feature of the wonderful luncheons he used to provide in his little ship.[183]

It was about this time that Samson received a somewhat querulous signal from London requiring to know why *Ben-my-Chree* had expended so much ammunition in recent months. His reply was brief and pointed, saying 'that there was unfortunately a war on.' He heard no more on the matter.

Romani and the Advance to Gaza

Ben-my-Chree now entered the dockyard at Suez for a boiler clean and minor refit, remaining in their hands until the second week of August. During this period the military situation at Romani flared up. A skirmish on 28 July, five miles in advance of the British front line, proved to be the beginning of a Turkish offensive. The main attack developed soon after midnight on 4 August, with a holding attack on the front line and the main thrust coming against the British right flank. The ensuing series of engagements, ending on 9 August, became known as the Battle of Romani. It was a considerable, but qualified, success for the British inasmuch as the Turks although defeated were not routed. Despite losing 4000 prisoners, and maybe as many again as casualties, the Turkish army conducted a skilful withdrawal through prepared rearguard positions. Much material and equipment was captured but only a single battery of artillery. The battle also came to be considered the end of the Turkish campaign against Egypt and the Suez Canal. For the next moves the British forces were in the ascendance, Sir Arthur Murray leading a steady and methodical advance to El Arish, captured just before Christmas, and onto Gaza in March 1917.

At the time of the battle the RFC in Egypt comprised just two squadrons, 14 RFC and 1 AFC, equipped for the most part with BE2c two-seat biplanes. The BE's were able to patrol mainly over the battlefield area, but General Murray still relied upon the EIESS for long range reconnaissance. Whilst the RFC continued to grow in strength and capability, the RNAS was being starved of replacement machines. At present this presented no difficulties, but the future looked uncertain. For the moment, however, both air arms shared the tasks.

Anne, having completed her refit at Suez, took on board Short 8090 and Schneider 3777, with pilots Flt Lt Dacre, FSL Man, and observer Lt N.W. Stewart. She sailed from Port Said on 8 August, escorted by *Voltigeur*, arriving at a rendezvous off Mersina at 03.30 10 August. Here they joined a French squadron comprising the old (1893) armoured cruiser *Pothuau*, three destroyers, the armed tug *Labourieux* and five trawlers. The tug had swept a channel free of mines which the squadron followed to a position two miles off the town of Mersina, and there anchored.

Both the Short and Schneider made two flights, the two pilots taking turns on the machines. The Schneider pilots were instructed, 'To carry out a submarine patrol around the ships; to drop bombs [two 16-lb each] if a submarine sighted. At the end of the flight, bombs to be dropped on the shore.' No submarines were sighted and the bombs did no damage. Both Short flights were to spot *Pothuau* by W/T on to specific targets in Mersina. The shelling destroyed two factories, and a barracks, but the railway station was undamaged.

Whilst the floatplanes were over the target one of the Rumplers from *4ncü Ty Bol*, crewed by *Yüzbaşı* Hüseyin Sedat, a naval officer, observer *Mulazim* Hüseyin Bican sighted the squadron, reporting it as '1 cruiser, 2 destroyers, 2 large transports and 7 smaller vessels.'[184] After circling the ships for 1½ hours the Rumpler made a bombing attack. *Anne*'s report mentioned the attack. 'At 10.35 am while the Short was still spotting over the town, a large hostile biplane with Turkish markings approached the ship from the SW flying at about a height of 8500 feet and dropped four small bombs, none of which came closer to the ship than 400 yards.'[185] The Turkish machine departed to the south west shortly before the final Short flight returned at 10.45. The direction taken suggests that it was based at Silifke.

Anne received a signal from the French Admiral thanking the aviators for their co-operation. At 11.15 she sailed for Port Said again escorted by *Voltigeur*.

Raven returned to Port Said from ten days in the Red Sea during the morning of 8 August. Her crew's hopes of leave were dashed as, no other ship being available, in the evening of the following day she was dispatched to spot for monitor *M.21* harassing the retreating Turkish forces at Bir El Mazar, near the west end of Lake Bardawil. The spotting was not without problems.

The first machine, Short 8075 piloted by Flt Lt Clemson with 2Lt Williams as observer left *Raven* at 05.55 and at 06.24 was 3500 feet over the target camp, a little north east of Bir el Mazar. As they approached the camp two columns, each of 150 camels, were observed heading east. In the camp itself were an estimated 1500 camels and about 100 horses. The spotting did not go well. Wireless signals were not being received by the Monitor and very weakly by *Raven*.

> Only one shot was fired from the Monitor for which the correction 800 yds. S. and 100 yds. L. was given. 4 Bombs and 2 Very's lights were used as smoke balls. On passing over Monitor on way back she put out a strip signal to indicate that she was not receiving our signals. The Seaplane circled over the Monitor to obtain bearing of camp from directly overhead — 232 degs. which was signalled to Monitor on returning to *Raven*.[186]

The second Short 8091 (Flt Lt Brooke and 2Lt Smith) left *Raven* at 07.15, just over half an hour before the first returned. Arriving over the camp at 07.45, they noted in addition to the horses and camel a number of tents, three wooden huts, and 'many wheeled vehicles in rows.' Just as the signal to the monitor to open fire was made 'a hostile aeroplane (a two-seater) opened fire from underneath, hitting the petrol tanks. The radiator was hit just afterwards when a running fight ensued, the Seaplane landing beside the Monitor at 8.5. The seaplane was then towed out to *Raven II in* a sinking condition and hoisted in at 10 am.' The attacking machine was a Rumpler C.I from *FA300*, piloted by *Obltn* Richard Euringer with *Obltn* Fritz Berthold as observer, they were returning from a reconnaissance to Katia.[187] Brooke and Smith were very lucky to escape from this encounter uninjured.

Lt John Jenkins, RNR, commanding officer of *Raven* reported a further visit from a Rumpler of *FA300*.

> About 10.15 am hostile aircraft whose approach had not been observed, owing to great height about 6000 ft & suns rays, commenced dropping bombs on ship. Five in all were dropped the nearest being about a ships length off; they appeared to be heavy bombs, giving out black smoke on explosion & colouring the water black for a considerable area. The hostile aircraft then made towards the land & was not seen again. She was engaged by ship's gun but owing to difficulty in seeing her only two rounds were fired at her. She appeared to be a very large machine & the bombs might have been anywhere from 30 to 40 lbs. Ship was stopped at the time but was immediately got under weigh & zigzagged seaward. After hostile machine had disappeared ship was stopped & boat hoisted in, & we proceeded towards Port Said.

Once out of dockyard hands, *Ben-my-Chree* re-provisioned, re-ammunitioned and took aboard three Shorts (8054, 8080, 8372) and a Schneider (B.8). She sailed from Port Said, accompanied by a French destroyer, *Hache*, at 16.35 14 August. Once more they were to conduct a series of reconnaissance flights behind the Turkish front lines to examine lines of communication, to check up on the camps previously observed and to bomb targets of opportunity.

Commencing at 05.00 15 August all four floatplanes were hoisted out to conduct a series of flights. The area from Haifa, where the ship lay off the coast, to El Afule and south to Athlit was to be examined. The Schneider failed to get off due to a problem with the very basic electrical system. The first Short, 8372 with Samson and Benn aboard, flew inland to the railway junction at El Afule. There were six engines, some with steam up, 30 passenger coaches and 40 goods wagons in the sidings at the station. A 65-lb bomb hit a coach, starting a fire, two 16-lb bombs struck engines, which had been grouped closely together, a third 16-lb bomb falling amongst stores laid out in the station yard. Happy with their marksmanship Samson and Benn returned to

Samson's favourite, Short 8372, seen here over Port Said harbour with its fin painted red to be identifiable when he was leading a squadron attack. (Robert Cyril Vickers Photographs, John Oxley Library, State Library of Queensland)

the ship, as they neared the coast smoke from the fires they had started could still be seen billowing up into the sky behind them.

The second Short, 8080 piloted by England with 2Lt King as observer, flew down to Athlit before returning to bomb a tented camp at Jeida. The last machine, 8054, FSL M.G. Dover with Lt P.M. Woodland, RNVR, flew straight to the big camp at Tubaun, dropping bombs that demolished a big building. All machines came under ineffective rifle, machine gun and shrapnel fire, although at one time not less than six guns were observed in action. The locations of the guns were noted and their positions later marked on a map to be added to the increasingly voluminous intelligence files.

All the machines having returned and been hoisted aboard by 6.30 am *Ben-my-Chree* proceeded south. By 10.30 she was closing Jaffa where Bankes-Price was hoisted out on 8054 with Woodland as observer. They made a flight over Ramleh and Ludd finding the roads busy but no railway traffic. Dover and King on 8080 followed with a flight inland to El Falujeh, noticing no changes to the camps since the last visit a few weeks earlier.

The sixth flight of the day fell to Maskell and Sub Lt Kerry, RNVR, on 8054 again. Hoisted out at 13.00 they made for a large camp at Bureir, adjacent to the railway and inland from Ashkelon. The camp comprised 12 huts, 40 bell tents and 25 shelters, the whole area being protected by trenches. At one time this camp had been much larger. Returning to the coast by way of Shellal a camel caravan was scattered by fire from Kerry's Lewis gun. The final flight of the day was brief visit by Bankes-Price, on the now repaired Schneider, to the roads between El Arish and Bir El Mazar. He found the roads and camps quite deserted.

During a very busy day *Ben-my Chree*'s floatplanes had reconnoitred the coast from Haifa to Bir El Mazar, some 160 to 170 miles, up to 20 miles inland. A truly remarkable achievement and demonstration of the value of naval aviation. The results of the day had also given Samson much food for thought. His success at El Afule encouraged him to plan another visit, on a much grander scale. As described in the following chapter.

CHAPTER 9

Attacking the Railway

The Attack on El Afule Junction

Following the success of his 15 August solo raid on El Afule railway junction, Samson commenced planning a squadron attack. He flew down to visit Admiral Wemyss at Ismailia on 21 August. The admiral had moved his flagship, *Euryalus* there to be close to army headquarters. Samson proposed making 'as powerful an attack as I was able to make with my limited resources on the Turkish communications at El Afule.' The admiral quickly approved the plan and the French, who were very pleased with the recent work of the squadron, readily offered to provide the necessary escorts.

Samson's plan involved all the ships in the squadron, carrying ten floatplanes. *Ben-my-Chree* sailed with three Shorts and two Sopwiths, *Raven* two Shorts and a Sopwith and *Anne* carried one of each type. The two slower ships sailed during the morning of 24 August 1916, escorted by the trawler *Paris II* and destroyer *Hache*. *Ben-my-Chree* followed later in the afternoon with *Arbalète* as escort. The squadron assembled off Haifa before dawn on 25 August.

Ben-my-Chree nearly did not arrive. During the night she ran onto an uncharted sandbar off Athlit, uncomfortably close to a known coastal battery. It was 03.30 and all the aircrew were up and about gathering the equipment needed for the flight. As the ship came to a stop Dacre, who was back aboard for the raid and irrepressible as ever, recorded,

> I was rudely awakened before dawn by my camp bed giving way and discovered that the cause was the ship had properly run into Syria in the dark and we were hard aground. Much shouting of course, followed especially from the noisy 'Steen' [Lt C.H.C. Steen, RNR, one of the ship's officers]. Machine guns were mounted in the bows while the screws going full astern churned up plenty of sand and the syren called up our little TB escort. Some of us were rather pleased at the chance of something novel: the cause of this running aground I can hardly write as carelessness, but it was certainly rather comic. However, after an hour's going astern with the Chief Engineer's help — he always had steam up his sleeve for special occasions — we came clear, went on our way and arrived at Haifa Bay our rendezvous.[188]

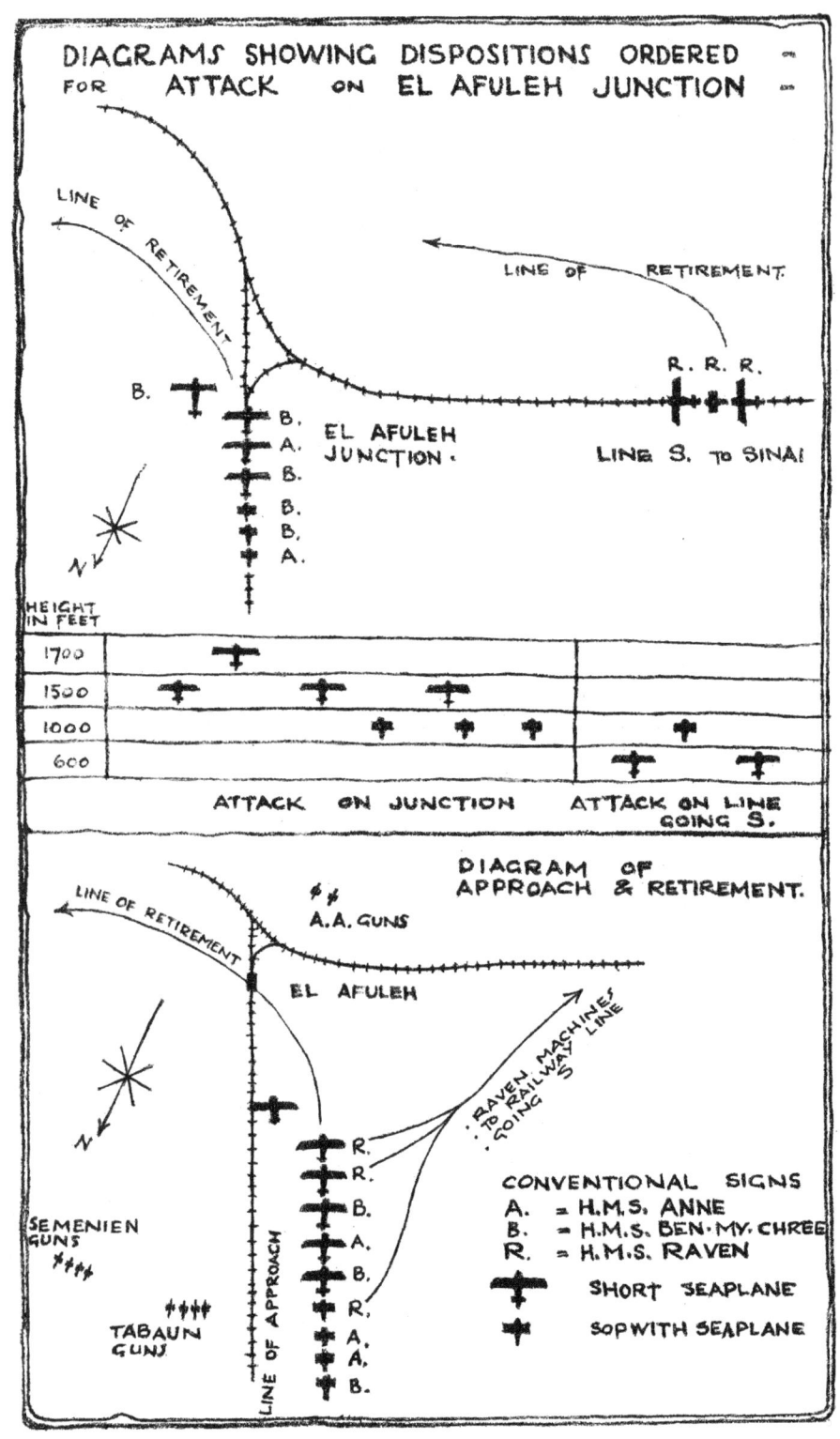

Samson's plan for the attack on El Afule. *(TNA, AIR 1/1707/204/123/69)*

Samson's plan called for all ten machines to participate in the raid on El Afule junction. Three separate objectives, split between the three ships, had been defined:

Ben-my-Chree	Rolling stock in the station, from a height of 1500 feet.
Anne	Buildings and stores at the station, from a height of 1700 feet.
Raven	Main line three miles south of the station, at a height of 6/700 feet.

It was hoped the 200 feet separation between the machines over El Afule would avoid collisions. The Shorts had also been assigned reconnaissances to verify observations made on previous flights.

Attack on El Afule, 25 August 1916

Ben-my-Chree

Wing Cdr C.R. Samson and Capt W. Wedgwood Benn	Short	8372	2 65-lb and 2 16-lb Bombs.
Flt Lt A.S. Maskell and 2Lt E. King	Short	8054	2 65-lb and 2 16-lb Bombs.
FSL M.G. Dover and Lt P. Woodland	Short	8080	2 65-lb and 2 16-lb Bombs.
Flt Lt Bankes-Price	Baby	8135	2 16-lb and 2 Petrol Bombs.
Flt Cdr T.H. England	Baby	8136	2 16-lb and 2 Petrol Bombs, a box of 50 Flechettes.

Anne

Flt Lt J.C. Brooke and 2Lt K. Williams	Short	8091	1 112-lb, 2 16-lb and 3 Incendiary Bombs.
FSL W. Man	Schneider	3777	3 16-lb and 1 Petrol Bombs.

Raven

FSL G.D. Smith and Lt V. Millard	Short	8045	2 65-lb and 2 16-lb Bombs.
Flt Lt A.W. Clemson and Sqn Cdr C.L. Malone	Short	8075	2 65-lb and 2 16-lb Bombs.
FSL L.P. Paine	Baby	8189	4 16-lb Bombs.

In the pre-dawn darkness the three carriers came to a stop a few miles from the coast, preparing to launch their machines, whilst the escorts kept watch to seaward. Samson was first off at 05.30, circling once to permit the remaining Shorts to join up. The fin of Samson's Short, 8372, had been painted red to aid identification. Using this, and his unmistakable white flying helmet, the machines formed up behind his leading machine in a starboard quarter-line (see accompanying plan). The faster Sopwiths followed, as much as 25 minutes later, joining the formation as it pressed inland. The intention being that, once they had dropped their bombs, they were to keep a look out for enemy aeroplanes.

Dacre, not due to fly until a follow up attack, watched the machines set off. 'Seaplanes were at once hoisted out and the Comdr with his machine having a red tail fin, got away, followed by 9 others. Away in the distance one could see a long line of machines disappearing on their errand of destruction.'

Leaving Haifa to starboard, Samson crossed the coast at 05.40 leading his squadron up the valley of Quishon, between the Carmel and Nazareth mountain ranges, on towards El Afule.

Taking advantage of the cooler morning air, all the machines were overloaded. In addition to the bombs listed in the table, the Shorts also carried the observer's Lewis gun with several 'hoppers',[189] and a camera. The Sopwiths were each fitted with a Lewis gun with three or four 'hoppers'. Heavily loaded as they were, the machines could not climb either quickly or very high. Part of the 20 minute flight could more properly be described as being through the valley, where they were tossed and bounced about in the winds flowing through the gorge. In places the floatplanes came under horizontal and plunging rifle and machine gun fire from camps around Tabaun on the north side of the valley above them. At one point England climbed out of the valley to drop his box of flechettes across one of the camps.

Ben-my-Chree at Port Said during 1916. The large square ports in the forward hull were originally for the first class saloon, and the three large round ports were for the first class dining room. Both now stripped of their fine finishes were seamen's and stokers mess decks.

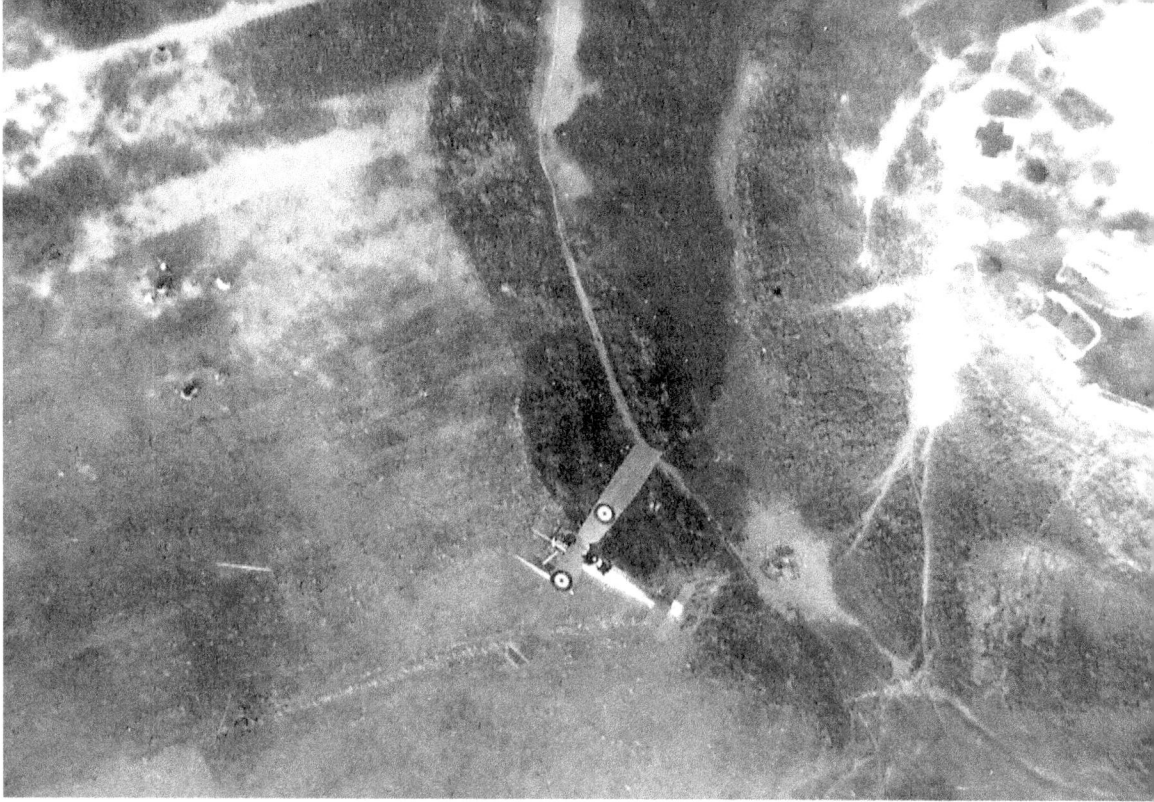

The original caption describes this as Samson leading the attack on El Afule 25 August 1916, and who are we to argue. The Short certainly has 8372's red fin, and there is a small white dot in the pilot's cockpit — Samson's white flying helmet. *(EMK)*

FSL G.D. Smith, a recent arrival on the squadron, recalled the flight in a memoir written for his family.

> Our reception was a hot one and was put under way without any loss of time. Anti-aircraft guns, the first we had yet encountered, opened shrapnel fire on us from about four miles inland. They were aided by fire from heavy machine guns and rifle fire from trenches by Tabaun. So intense was this attack upon us that I could hear the crackle of the guns above the roar of my motor.
>
> I conquered my desire to drop a bomb and tried to attract the attention of my observer, who could have made reply with the machine gun. He had his head tucked down in the fuselage and was making notes, paying no attention whatever to the firing.
>
> My observer, by the way, was a fellow without nerves of any sort. Battle might rage about him, but he never seemed conscious of it. If any part of the machinery needed adjusting, he thought nothing of leaning out over the edge of the [machine] and hanging on with one hand while he reached to the task with the other. The slightest slip would have meant a fall of thousands of feet for him, but in his attitude this was merely a part of the day's work. I used to tell him he should wear ten-pound weights on his ankles to keep him from losing his balance and falling out.
>
> I shouted to him several times, but could not make him hear, my voice

being drowned by the roar of the engine and the crash of the gun fire. So I picked up a biscuit I had been nibbling and threw it at him. His seat was behind mine, so that I could not take accurate aim, and the biscuit hit him on the side of the head.

He immediately developed a pronounced peeve, indicating that his sketch had been spoiled. At any rate, I attracted his attention to the situation, and he forgot his grievance and got busy with the machine gun. I don't know whether he accomplished anything with it, but I know that it was a big comfort to me. I settled back with the feeling that all was well once again and my nerves readjusted themselves.[190]

On the whole the fire was lighter than expected, although several machines were hit by rifle bullets without serious damage. Emerging from the defile the squadron was able to take up its attack formation.

Led by Samson, with the Shorts leading and the whole formation spread out between 1700 feet and 700 feet above ground level, the formation flew on to attack the railway and station. Looking back, Samson 'saw my little squadron of nine seaplanes keeping perfect station on me.' Shouting to Benn in sheer exhilaration and delight, he gave the signal to commence the attack.

As the squadron approached the junction a train steamed hurriedly out of the station, heading south. *Raven*'s flight, as planned, were first to break formation. Following the line south from El Afule they swooped on the train. Paine dropped his bombs in a single pass, the first three bombs marched along the line towards the train damaging the track, the final bomb hitting the last coach setting it on fire 'which burned for a few minutes then went out.' Smith takes up the story.

On the piece of line my observer and I were to attack stood a supply train. Throttling down my engine so I could make myself heard, I turned to him and said: "I'm going down very low and we'll see if we can bomb that train. I'll go around the train several times and then try to make a line over it. If I don't make a good line, don't drop any bombs." He nodded in agreement and we started gliding down towards our objective.

As we swooped down towards the train we could see soldiers jumping from it and scampering into the fields for safety. There came a few blasts from the locomotive whistle and then the engineer must have quit too, for there was no further sign of life.

The first time we made a line down the train we dropped one bomb, which demolished a car about in the middle of it. I circled around and we came back, hitting another fairly good line, our bomb tearing the roof off the rear end car. I missed my path the next time around,

> getting off the line, so the observer didn't drop a bomb. But on the fourth trip he let a couple of bombs go crashing down on the tracks behind the train, tearing two great holes in the earth.
>
> At this time we were flying at an altitude of only about 150 feet, and the concussion from these explosions was so great that it shook the airplane violently and seemed to lift me inches out of my seat.

Between Smith's second and third passes Clemson and Malone made two attacks. On the first at 500 feet, Malone dropped all their bombs on the line causing damage to the embankment and permanent way ahead of the train. They made a second attack at 300 feet permitting Malone to machine gun the train. Leaving the scene, Smith once more picks up the story.

> [W]ith one last look at the train, we turned the nose of our machine shipward, at the same time climbing as rapidly as possible, as we were subject to the anti-aircraft fire increasingly as we neared El Afule. Here a genuine battle was raging. Looking down on it, I think was the most wonderful and at the same time entertaining sight I have seen. The airplanes were darting in all directions, swooping down upon their objectives and dropping bombs that exploded with a terrific roar, kicking up great columns of smoke and filling the air with flying debris. Below the people looked like tiny ants, hurrying here and there. There was something tremendous about it all, and at the same time something unreal. I remember that I laughed — not at what I saw, but at the idea coming to me at that particular moment that a moving picture company might be staging the whole affair, and I'd see it later on with the hero doing a lot of daredevil stunts and at the finish rescuing the heroine from the villain's clutches.

The daredevil stunts were real. As the remaining machines came closer to El Afule the reason for the lighter fire in the valley became clear. The defences at the junction had been considerably strengthened since Samson's last visit, putting up a heavy barrage throughout the attack that holed most of the machines but again failed to cause serious damage. *Anne*'s machines opened the attack on the junction, Man dropping a line of bombs across the station buildings. He was followed three minutes later by Brooke and Williams, who released their bombs starting fires in the station and amongst the stores. *Ben-my Chree*'s flight concentrated on the rolling stock in the sidings, commencing their attacks as *Anne*'s flight climbed away. Samson dropped two 65-lb bombs, one of which burst between two trains damaging both. In turn the other floatplanes dropped their bombs across the station and sidings. Within 15 minutes the raid was over. One 112-lb, six 65-lb, fifteen 16-lb, and eight petrol/incendiary bombs had been dropped on El Afule.

A set of four bombing images taken during the El Afule raid on 25 August 1916. *(TNA, AIR 1/1707/204/123/69)*

The station buildings were left burning, a locomotive and 14 carriages and trucks were destroyed or damaged and an unknown quantity of stores burned or destroyed.

Anti-aircraft fire throughout the attack had been heavy, two guns in the station yard being particularly troublesome. Samson tells a story. 'I can still remember one old Turk standing in the open and firing at us each time we came over his head. He had apparently an old Snider, as when he discharged it I could see a puff of black smoke.' Both he and Benn were filled with admiration for this old soldier, Samson saying he flew low enough to be able to recognize the man if ever he met him. He doesn't say as much, but one feels that he would dearly have liked to have shaken the man's hand.

The floatplanes returned to their ships between 06.32 and 07.10. The Schneiders were first back followed by the Shorts. Last to return was Smith, who had done a bit of aerial sightseeing over Nazareth on his way back. Damage to the machines was mostly superficial and quickly repaired.

Paine's Baby had come 'under heavy rifle on return journey from low hills about two miles inland, floats, wing and main spar perforated.' This was the most serious damage reported by any of the attackers. The Baby was quickly made ready for further service, damaged parts either repaired or replaced.

Anne and *Raven*, now proceeded south for a rendezvous off Ashkelon. *Ben-my-Chree* loitered to send off two Shorts to make a final assessment of the damage at El Afule. She would use her superior speed to catch up with the two slower ships. Flt S/Lt M.G. Dover and Lt P. Woodland set out again on 8080 shortly after 08.00, reaching El Afule half an hour later finding fires still burning. After dropping the 65-lb bomb and two 16-lb bombs they had brought along on the rolling stock in the station, they flew south reporting that the 'permanent way south of the junction was completely wrecked. Rails uprooted, embankment destroyed.' They returned to the ship an hour after leaving. Samson, in his report, later concluded that, 'It may be assumed that the junction and south-going line will not be in operation for some time.'

Fifteen minutes after Dover, Dacre and Benn were hoisted out on 8054, loaded with a 112-lb bomb and two 16-lb bombs also destined for El Afule.

> A second flight was to be made to this junction [by] myself with Wedgwood Benn MP as observer. Benn had arrayed his tunic with many false medal ribbons thinking that if he were taken prisoner he would be treated as a General. I had the Comdr's machine [8372], but after hoisting out I found the engine was too bad to get off the water, so I was told to go away in another [8054]. Again luck was against me, the engine was bad and I could only get about 800 ft, dropping overland where there were bumps. From my point of view I was fed up as the other two machines returned and said the junction was then a sight worth seeing, while I had not been able to get there.

Steaming rapidly south *Ben-my-Chree*, with *Arbalète* in company, encountered two blockade running dhows. The larger dhow ran onto the beach where twenty soldiers were observed to scramble ashore and run for the sand dunes. *Arbalète* destroyed this dhow by gunfire. The smaller vessel decided to surrender. One of *Ben-my Chree*'s motor boats, with an armed party and mounted machine-gun, was sent to accept the surrender. Brought aboard *Ben-my-Chree*, the five crew members, reportedly wretched looking creatures, willingly admitted that they were engaged running supplies for the front line troops. In addition they said that Samson's previous raid on El Afule (15 August) had shut down rail traffic for five days. The small dhow was also taken aboard and, once in Port Said, the dhow was taken into RNAS service, becoming a valuable addition to the boats serving the island base. Stripped of her threadbare sails and fitted with an engine from a truck the dhow became a liberty boat for the island.

Previous page, clockwise from top left.

1. Quite the best known of the El Afule raid photographs, it appears in both Samson's and Benn's books. This copy has arrows drawn on it pointing to 'Bridging Material?'

2. This view is of a 'Bomb falling among stores', just to the left of the buildings. There is also a steam engine on the main line near the left hand arm of the reversing triangle.

3. The train attacked by *Raven*'s flight is shown here. The first and last coaches in the picture have been damaged by 16-lb bombs, whilst arrows point to two hits on the track from 65-lb bombs, possibly those dropped by Smith and Millard.

4. The arrow indicates a 'Bomb on store house', there is also a bomb exploding in the middle of a group of track switches/points, and one on an adjacent group of stores to the previous image.

The five 'wretched looking creatures' and their small dhow aboard *Ben-my-Chree* at Port Said, 31 August 1916. The dhow appears to be an ex-ship's boat rather than local-built, although rigged with two lateen sails. Similar in size to a 27ft RN whaler, but with a transom and not double ended, it fitted easily on her davits. It would have been able to carry between 1–1½ tons of bagged grain. *(Ministère de la Culture, APOR070178)*

The afternoon's flights covered various objectives in the Ashkelon and Bureir area. Co-ordination of the flights was not practical, so they were spread out over two hours, perhaps intending to cause maximum disturbance to the Turkish defenders. *Raven* started the afternoon's work at 15.20, sending Short 8045 (Flt Lt H.deV. Leigh and Sub Lt Kerry) to attack the camp at Bureir. Whilst well over land, the radiator was struck and holed by anti-aircraft fire. Losing water rapidly, the crew dumped their bombs and turned back towards the ship, reaching *Raven* just before the engine seized. Paine, on repaired Baby 8189, bombed the camp and returned unscathed half an hour later. Both machines had returned by 17.00.

Anne's two machines left the ship off Ashkelon, and flew down the coast to Wadi el Hesi. Following the wadi inland to the railway at Deir Sineid, where the track crossed the wadi by a bridge. First away were Brooke and Williams on 8091 at 15.55, returning at 17.38. Half an hour into the flight 'the engine began to vibrate about 8 miles inland, 2 bombs were thrown out to lighten the machine.' They were carrying three 16-lb bombs and a single 112-lb bomb. 'The machine was then turned towards the west to return to the ship, but the engine shortly picked up again and another start inland was made. The 112 lb and 16 lb bomb dropped on the bridge did not hit.' Man left on Schneider 3777 at 16.06, returning at 17.19. He dropped his four 16-lb bombs at the viaduct, damaging the embankment north of the bridge.

Meanwhile, *Ben-my-Chree* speeding down from Haifa arrived off Ashkelon to send off three more flights. First, England and 2Lt King, Short 8080, then Bankes-Price on Baby 8135, to the bridge at Deir Sineid. Both machines left within minutes of each other at 16.09 and 16.14 respectively. England's 112-lb bomb 'exploding 50–100yds East of wadi bed.' Bankes-Price, whose Baby was carrying four 16-lb bombs, dropping them 'on one of the 5 arches. 3 bombs missed the bridge and the line but the 4th fell in the middle of the permanent way.'[191] Both reported heavy rifle fire and, Bankes-Price, some shrapnel during their flights, but no damage.

The third flight was Dacre on Baby 8136 at around 16.30, he was to visit a camp near Bureir. When he failed to return the squadron, now re-assembled, stayed close to the shore until dusk, hoping that Dacre might make his way to the coast. As the sun set, *Ben-my-Chree* and *Raven* had to sail to conduct further operations leaving *Anne*, whose duties the following day required her to be only a few miles up the coast at Jaffa (a suburb of modern Tel Aviv), to keep a fruitless watch throughout the night. Samson wrote that the loss of Dacre 'one of the most brilliant pilots in the RNAS marred the success of the day's operations… He was a most experienced pilot and had made many magnificent flights during the war.' An intercepted Turkish Communique on 1 September indicated that Dacre had been captured and was now a Prisoner of War. By mid-October the American Embassy in Constantinople had confirmed that Dacre was a prisoner at Afion Kara Hissar.[192]

In his PoW Diary he wrote down the events following his departure from *Ben-my-Chree* on 25 August 1916.

George Bentley Dacre at the controls of a Bristol Boxkite whilst learning to fly at the Bristol School on Salisbury Plain. Dacre gained RAeC Certificate 162 on 28 November 1911, whilst a student at Bristol University. He joined the RNAS as a Flight Sub Lieutenant on 5 August 1914, joining *Ben-my-Chree* as a Flight Lieutenant on 23 March 1915. *(YEORN 1987-374-5. Courtesy of The National Museum of the Royal Navy)*

> I was allotted one of the new 110 HP Clerget Schneider machines which had only recently arrived at Port Said, the fastest thing in 'boots'.
>
> I left the BMC about 4.30 after some difficulty in starting up. Once away I climbed rapidly and made off rapidly inland to my objective a camp at XXX. My engine was going well and the weather was good. I watched the other seaplanes on their return journey with a careful look out for the Hun. I followed inland over a certain Wadi or dried up water course, over the sand dunes thinking what a nice machine this was, when at about 3000 ft up and 12 miles inland, my engine spluttered and died out. I waited a few seconds to see if it was going to continue, but no.
>
> Now I'm for it, I thought. I gazed at the petrol gauge as the nature of the stoppage was like a petrol failure. No petrol in the fore tank. Perhaps the fore tank had a bullet through it. I changed over the taps to the back tank and pumped up the pressure. No result. Taps were alright. Switch was on. The only explanation I could, on the spur of the moment, think of was that the tanks had never been filled. By this time I found myself 100 ft up, faced with landing this rapid machine on terra firma on its floats. I used this last height to turn head to wind and to square up, wondering if a few seconds later would see me alive. I got to within a few feet, pulling her back and back to do a pancake landing. I seemed to be actually travelling very slow just before I touched, but when I did hit Turkey she went over on to her nose like a flash.
>
> I obeyed Newton's laws and carried straight on landing mostly on my right shoulder and head and did two somersaults. I got up dazed with a pain in my shoulder; nothing broken thank God! I returned to the machine and having no matches or petrol to destroy the machine which was not badly smashed, I was looking inside for any possible cause of the failure and breaking the instruments with the butt of my revolver. I tore up the chart and tried to get the machine gun out of its

fittings without success. Arabs from all around were then running up and while I was fishing out my revolver and ammunition pouch, I was seized by the Arabs and held.

At first I shouted "Allemagne" hoping to bluff them that I was German and saying by signs I wanted to go to Gaza. This they took and my hopes of reaching the coast near the ships ran high, but alas! one blighter on a horse knew the Allied marks on my machine and shouted "Inglise" and then they fairly went for me, in spite of my arm which was hurting badly. They tore my clothes off me leaving me only in a vest and trousers. They drew their nasty looking knives threatening me and intimating they would cut my throat with a horrid grin on their faces. They took my watch, cigarette case, money, slitting my pocket down with a knife to get it, cap, water bottle and all.

By this time a whole Arab village was around me. The women being most anxious to have my blood, took the opportunity to hit me over the head with sticks and throw clods of mud in my face. They tried to get my shoes, but I kicked out and kept them. What a crowd of savages I had fallen into. Some were pitch black, others copper, dirty, evil looking brutes with hair matted like sheep's wool.

There were about 8 holding onto me squabbling like jackdaws for my possessions, and in the squabble they distracted their attention for a while and I broke away and ran like a hare followed by a howling mob of savages with swords waving. My arm pained fearfully to run, but I could have got out of their clutches and away had not a horseman ridden up and hit me over the head with his sword. The crowd rushed at me and I thought my finish had really come but they just pulled me to bits every way. One man made a dash at me with his dagger, a horrible glint in his eye, but the others kept him off. They sat me down and tied my arms behind my back. I knew then they were going to do me in by their manner and I sincerely hoped they would finish me off quickly. But no, they were about to cut my throat when a Turkish NCO rode up and drove them off with a whip saving me by a few minutes from the unpleasant operation of having my throat cut.

This treatment in the hands of the Arabs is very nice for a cinema show, but very different if you happen to be the unfortunate victim. The Turkish NCO, who could speak French a little, gave me water and a cigarette and attended to my arm. I complained to him about my belongings and he led me to understand they would mostly be recovered for me. I was grateful to him for his kindness and for saving my life.

A soldier was placed over me while the NCO went and looked over my machine. Later, more soldiers arrived bringing me black bread and dates and by the light of a lantern they requested me to take the bombs off my machine. This I couldn't do even with two hands, but I explained how the came off, requesting one man to each of the 4 bombs, while another pulled the lever. Volunteers were not forthcoming and with much strafing and fright, 4 were pressed into service and to the other's amusement the bombs came away. They got in another funk when I went near them to explain that the fan when unwound made the bomb dangerous, but I was kept away.

Later that night he was taken by donkey and horse to Gaza, where the NCO handed Dacre over to the local Commandant. It was now 04.00, but he was provided with a meal and amply supplied with cognac and cigarettes. Questioned through an interpreter 'to which I told as many lies, and in all seriousness, the Turk having little sense of humour, everything was put down in writing.' Then a doctor examined him and put his arm in a sling. With the dawn he was placed in a wagonette and taken on a rough eight hour journey to Beersheba. After a few days hospital care Dacre was transferred by train to Damascus, then on into Turkey proper.

Dacre remained a prisoner of war at Afion Kara Hissar until November 1918. Transferring to the RAF after the war, he served with distinction until his retirement in 1944 as an Air Commodore. George Bentley Dacre died on 4 January 1962.

Anne's business at Jaffa was a series of flights over Ramleh and Nablus. On their completion she and her escort returned to Port Said arriving later the same day. *Raven* went to the north west to search the southern Turkish coast near Antalya (Adalia), to the east of Castellorizo, for mines and submarine bases. Finding nothing she returned to Port Said, arriving 28 August.

Ben-my Chree's course took her back to the Nahr El Kebir, arriving early in the morning of 26 August. Here it was intended to fly a reconnaissance of the Tripoli–Homs railway. Samson and Benn, Short S.7 (8372), set out shortly after 06.30 but had to return with low oil pressure. The dropped their 65-lb and 16-lb bombs on the tugs which were still sheltering in the Nahr El Kebir and returned to the ship after just half an hour.

England and 2Lt King, Short 8080, set out shortly after 06.30 heading inland to Homs on a flight covering 75 kilometres each way. Samson later rated their flight as 'quite one of the finest I have seen.' Struggling up the valley, following the Homs Gap between the Ansarieh and Lebanon mountains, bucking a strong head wind and being tossed around between mountains and the clouds, England was lucky to find a hole that enabled him to slip through to Homs. Having carried out their observations, and dropping their bombs on the station, they benefitted from a tailwind, returning rapidly to the ship, being hoisted in at 08.35.

Whilst the Shorts were away, Bankes-Price took Baby B.10 (8135) to bomb a camp at Tel Kale. Hoisted out at 06.36 he was off the water two minutes later.

> Flew inland to junction of [the railway with] Nahr el Kebir at Long 36° 15′ at height of 3500. Clouds very low and thick. I was unable to get lower than this to drop bombs. I saw the camp through a gap in the clouds, and dropped 4 bombs into it, all falling within area of the camp. (6 bell tents and 2 huts one behind the other due S of tents). No firing was observed.[193]

He then climbed through the clouds, emerging at 8000 feet, and returned towards the coast and *Ben-my-Chree*. Bankes-Price landed at 07.13 and was hoisted in.

After recovering England and 2Lt King, *Ben-my-Chree* proceeded to Ruad. The sea was too rough to permit a flight requested by the Governor to examine new entrenchments on the coast facing his island. So, after a brief visit the ship now made for Cyprus, still accompanied by the faithful *Arbalète*. Arriving at Famagusta during the afternoon of 27 August, *Ben-my-Chree* remained until the morning of 29 August to coal and overhaul her well used floatplanes. Early on the afternoon of 29 August she was off the coast near the mouth of the river Seihun. At 14.24 Samson and Benn set out on Short 8080 to visit Adana. A few minutes later Bankes-Price, at 14.41 on Baby 8135 with four 16-lb bombs, to bomb some lighters at Tarsus Chai a little north-west of the mouth of the Seihun. 'On leaving ship steered for Karatash Burnu to the village of Karatash and then proceeded in a north westerly direction along the coast over the salt lake. In the Seihun Irmak 6 lighters tied up in pairs separated by about 200 yards along the N bank. Dropped 4 bombs damaging one of the lighters. All were empty. No firing and no troops observed.' He returned and was hoisted aboard by 15.11.

Samson and Been, meanwhile, had been experiencing a very different flight. Benn recalled it as 'a most bumpy journey.' Samson had much more to say about it.

> I took Benn with me in a Short, and we had rather a difficult flight, as there was a considerable downward trend of air in the cup formed by the Taurus Mountains on one side and the Amanus Mountains on the other. I could never get above 1100 feet; in fact, I arrived over Adana at only 700 feet, probably only being about 500 feet above the land. I had all my work cut out to keep the Short in the air at all. We let go one 65-lb and one 16-lb bomb at the railway station, where we actually saw a troop train, and one 65-lb and one 16-lb bomb at the railway bridge, which we missed; but this was easy to miss, as the bumps were terrific, and the old Short was wallowing about all the time. In

the middle of one bad spasm, when the saying goes, "all hands man the pump," Benn lent over me and shoved a bit of paper into my mouth. Thinking he had seen something highly important, I let go of the wheel with one hand, and after some struggle managed to spread out the paper and decipher what he had written. It read as follows, "Aren't the shadows on the mountains lovely?" I never felt nearer killing anyone in my life, and when we got back to the water I hadn't cooled down; but Benn took it all in good part. There was little that he missed as an observer, and I was delighted to read the voluminous report he had written. Personally, on that trip I had seen little, being otherwise occupied. I still look back upon this Adana trip as one of the toughest bits of piloting it has been my lot to undergo.[194]

Proceeding further along the coast towards the Ceyhan (Jeihan), at 16.50 Maskell and 2Lt King set out on Short 8080, 'to bomb any dhows or lighters in the Jeihan Irmak.' They had a 65-lb and three 16-lb bombs for the task. The 65-lb they dropped on a road bridge over the river at a place he identified as Kesme Burnu (Kesmeburun), as there were no dhows or lighters to be found. Returning to *Ben-my-Chree* they were hoisted in at 06.00. Samson then turned the ship towards Port Said, arriving on the afternoon of 30 August.

It will thus be seen that we kept the Turks busily engaged from August 25th until 29th, dealing blows at widely separated points on their lines of communication . There is little doubt that we not only did a good deal of actual damage, but gained important information, and made them doubtful of the security of their communications.[195]

The Price of Admiralty

The spate of attacks by *Ben-my-Chree* and her squadron were undoubtedly very damaging to the Turks. The loss of seven sailing vessels in little over a month, the attacks on the tugs at Nahr El Kebir and on the railway junction at El Afule, plus the ability of the squadron's floatplanes to appear anywhere at anytime were but part of the problem. The Turkish forces in Palestine and Syria increasingly became an army of occupation as the Arab Revolt spread. The pin-pricks of the East Indies and Egypt Seaplane Squadron were slowly eroding the support of the local tribes and the morale of the army. Samson meanwhile had much cause to be satisfied with his command, 'The *Ben-my-Chree* was of course the best, as her high speed enabled her to launch seaplanes at widely spaced objectives within the same day.' However, he could not be so sanguine with his ageing, obsolescent floatplanes.

Concerns about the increasing activity of enemy aeroplanes began to affect Samson's future planning. The superior performance of the enemy aeroplanes operating from their base at El Arish had already been

A deservedly cheerful Samson is recovered to *Ben-my-Chree* aboard Sopwith Schneider 3789. The engine had dropped down immediately after landing, as detailed in the text. Of particular interest are the two Lewis guns clamped to the cabane struts, one either side of the fuselage, as detailed in an accompanying photograph. *(EMK)*

demonstrated several times. Convinced that they would increasingly attempt to intercept his floatplanes, Samson instituted a series of Defence and Target Practices in early September.

> The defence practice orders were for the Short to operate along the coast with 2 Sopwiths, one keeping station on the port and the other on the starboard side about 800/1000 feet respectively higher than the Short. The object was to practice forming a protective screen over a Short on reconnaissance.
>
> Target practice was carried out in addition to the defence formations. The target consisted of a float carrying a Sopwith wing and tailplane, towed by a steamboat. Lewis gun with one round tracer and four Mark VI ball ammunition was used for the practice.[196]

He realised that individually, the Sopwiths were no match for a well handled landplane, but acting in tandem they might be able to distract the aeroplane long enough for the Short to escape. Samson remained pragmatic about the chances of the Sopwiths themselves, considering them 'absolutely useless for combat; but as they were the only seaplanes available I had to use them.'

The first opportunity to try out Samson's defensive theories came on

14 September at Gaza. As usual, *Ben-my-Chree* had sailed the previous evening from Port Said, escorted on this occasion by *Dard*. Starting soon after 05.00, Short 8372 and four Sopwiths were hoisted out within 12 minutes. Two of the Sopwiths suffered engine problems, so England and Benn set off on the Short escorted by Samson, on Schneider 3789, and Bankes-Price, flying Baby 8135. Some while later a third Sopwith, Schneider 3779, piloted by Flt Lt H.deV. Leigh was able to get away. Being so far behind the formation Leigh elected to stay close to the ship, climbing to 8000 feet from where he maintained watch until the other floatplanes returned. Intentionally or not, he seems to have flown one of the first recorded naval combat air patrols.

The reconnaissance flight ran into heavy anti-aircraft fire approaching Beersheba. One of the Short's floats was damaged, but the flight continued uninterrupted. Throughout the mission the Short kept in wireless contact with the ship. Messages being received at up to 35 miles, which was a good performance for the available equipment, probably an 8-volt Type 52 Sterling set, requiring used a 300 foot long trailing aerial. Samson having instructed observers during training to remember to wind in the aerial before alighting. On this occasion no other aircraft were encountered, which turned out to be fortunate for Samson.

During the last half hour of the flight the engine of his Schneider began to vibrate with increasing severity. After landing safely he commenced taxiing towards the ship, at which point the engine fell forward out of its mounting, smashing the propeller on the floats. The machine was recovered and later examination revealed that all four fuselage longerons were rotten and had broken, a result of age and extreme climate. Samson was lucky he had not had to engage in combat manoeuvres. As he said, 'I was glad it was kind enough to wait until I landed.' With all her aircraft safely recovered, *Ben-my-Chree* returned to Port Said, arriving there at 14.00 that afternoon.

A detailed view of the damage to Schneider 3789 back in the hangar aboard *Ben-my-Chree*. Samson was lucky the propeller did not smash the floats or struts. The installation of the Lewis guns can be seen. The muzzle of the starboard gun is angled out using a tubular support attached to the forward cabane strut and is probably fixed. However, the muzzle of the port gun appears to be clamped to the cabane strut and may be free to rotate as the pistol grip appears to be suspended by a system of cables. There is also a light bomb carrier below the fuselage. *(EMK)*

Since the Battle of Romani the military on both sides had been recuperating and preparing for the next round, relying on patrols and raids to maintain contact and keep up the pressure. One such minor operation, involving the Anzac Light Horse and Imperial Camel Corps under the command of Major General Chauvel of the Light Horse, was planned against Bir El Mazar to commence at dawn on 17 September. *Ben-my Chree*'s next operations were intended to be in support of this raid. Wemyss summoned Samson to his flag ship to discuss a bombardment of El Arish intended isolate the camp at Bir El Mazar from reinforcements. To his dismay, Samson received orders to provide two simultaneous spotting flights to work with *Espiegle* and monitors *M.15* and *M.31* respectively. He argued that as the German machines at El Arish were certain to intervene he should be permitted to concentrate his floatplanes in a single flight. This time he was overruled and had to provide flights for the two targets, more than 15 kilometres apart.

Having sailed the previous evening *Ben-my-Chree*, with *Dard* as escort, arrived off El Arish around 05.00 on 17 September. The first machine left the sea at 05.24, fourteen minutes later the sixth and last had got away. This morning there were no engine problems. Once airborne the machines split into two flights:

A Flight — Objective: El Arish camp, spotting for *M.15* and *M.31*

Flt Lt Maskell and S/Lt Kerry	Short	8080	1 Lewis and 7 'hoppers'. W/T transmitter. No bombs.
Flt Lt Bankes-Price	Baby	8135	1 Lewis and 4 'hoppers'.
FSL Nightingale	Schneider	3778	1 Lewis and 4 'hoppers'.

B Flight — Objective: El Arish to Bir El Mazar road, spotting for *Espiegle*

Flt Cdr England and 2Lt King	Short	8372	2 Lewis and 10 'hoppers'.[197] W/T transmitter. No bombs.
Wing Cdr Samson	Schneider	3770	1 Lewis and 4 'hoppers'.
FSL Man	Schneider	3777	1 Lewis and 4 'hoppers'.

B flight found no targets and, as 2Lt King was unable to make W/T contact with *Espiegle*, turned back to the ship. England landed and was recovered by *Ben-my-Chree* at 06.20. Man's Schneider chose this moment to have engine troubles. Passing over the coast at 1000 feet, nursing his engine and hoping to return to the ship, a burst of machine gun fire raked the Schneider's wings and holed the fuel tank. Whilst making an emergency landing the floats and undercarriage collapsed and the machine flipped over. Man struggled out of the rapidly sinking floatplane, to be picked up by HM Trawler *Rononia*. Samson, worried about A Flight, flew alone up the coast towards El Arish. He found the two monitors but could not locate the machines of the flight. Before returning to *Ben-my-Chree* he flew over the airfield at El Arish, noting several aeroplanes on the ground.

After leaving the ship, A Flight had flown north east to El Arish, reaching 5000 feet with a good view of the aerodrome. Unable to contact either monitor by W/T, Kerry passed a visual signal to commence firing. *M.15* let go a single round, but its fall was not spotted. At the same time as the monitor fired, an aeroplane was seen rising from the airfield, climbing rapidly.

The German airfield had been raided by the RFC during the night of 15/16 September,[198] and an attack from the sea was expected. Even so, the early morning arrival of the floatplanes caught them by surprise, the pilots and observers rushing to their machines clad only in pyjamas. One of the Rumplers, piloted by *Ltn* Walter von Bülow-Bothkamp with *Ltn* Friedrich von Hesler as observer, engaged A Flight on the morning of 17 September. Easily overtaking the slower machines von Bülow flew between the Short and its two escorts, intending to attack the former.

> He was first engaged by Flight Lieutenant Bankes-Price in the Sopwith but after a few shots this gallant pilot was killed and his machine fell to the water in a sheet of flame.
>
> The enemy was then engaged by the Starboard escort (Flight Sub-Lieutenant Nightingale) but owing to superior speed and armament he was able to hit our seaplane repeatedly through the tail and finally pierced the petrol tank, compelling a descent. Flight Sub-Lieutenant Nightingale was picked up by a Monitor [M.15] and his seaplane by a trawler.[199]
>
> The enemy then pursued the Short which [was] travelling rapidly downwards at a speed of 75 knots. The Short however was promptly overtaken and passed over by the enemy, both machines firing continuously. It is thought that the enemy was hit for he suddenly turned about and proceeded inland. 6 Hoppers were fired from the Short. The Short then descended to 800 feet and observed one disabled Sopwith, circling around the same to denote the position to the ship.[200]

Surviving German records are mute on the matter of possible damage to the Rumpler. However, von Bülow is reported to have landed and, after reporting his success, taken off once more with a fresh load of bombs for the British ships.

Sopwith Baby 8135. Bankes-Price was shot down in flames by a Rumpler C.I of *FA300*, piloted by *Ltn* Walter von Bülow-Bothkamp with *Ltn* Friedrich von Hesler as observer, whilst flying this machine on 17 September 1916.

John Thearsby Bankes-Price. Born in Chicago to a British family he was educated at Radley College from 1909 to 1911. He started learning to fly at the British Deperdussin School at Hendon in July 1913, but by November had moved to the W.H. Ewen School flying Caudron machines. He completed his tests with the Ewen School and was awarded his RAeC certificate (No. 760) on 3 April 1914. He joined the RNAS as a Flight Sub Lieutenant on 16 December 1914, joining *Ben-my-Chree* on 3 May 1915.

Hoisting aboard all the surviving machines the ships now headed north out to sea in order to distance themselves from the persistent German flyers. Over the next hour or so three more Rumplers emerged from low cloud cover to drop a total of eight bombs on *Ben-my-Chree*, none of which hit the ship. Other unsuccessful attacks were made on the monitors and smaller ships. The German aviators later referred to the day's events as the *Die Seeschlacht Von El Arisch* (Sea Battle of El Arish).

Not until after his return to Port Said did Samson discover that the attacking force had found the Turkish garrison to be on the alert behind strong fortifications. Lacking artillery support, Major General Chauvel had wisely decided to break off the action. With the raid called off the bombardments became unnecessary, the withdrawal having commenced about the time *Ben-my-Chree* had been launching her floatplanes, but the land force lacked the means to inform the ships of their decision. Samson had been proved correct, as Wemyss acknowledged to him in a letter shortly after his return.

John Thearsby Bankes-Price has no known grave. His name is recorded on the Jerusalem Memorial, in the Jerusalem War Cemetery. Of him, Samson later wrote:

> I had lost quite the best Sopwith pilot I had in Bankes-Price. He was only a youngster; but he had perfect hands, and was as plucky as they are made. He had a brilliant future in front of him. I felt his loss immensely, as did we all. I returned to Port Said, feeling that two seaplanes had been lost, and one gallant pilot sacrificed for nothing.[201]

Supporting the French

Through the remainder of September and all October *Ben-my-Chree* lay inactive at Port Said. A climate harsh on wooden aircraft structures, coupled with recent losses had brought about a shortage of floatplanes, keeping the mechanics at the island base busy. At about this time attempts seem to have been made to disguise *Ben-my-Chree*'s unmistakable profile by applying a form of disruptive camouflage to the hangar sides. The inactivity, though,

Sopwith Schneider 3778 with local labour at the island. FSL Nightingale was forced down on this machine shortly after Bankes-Price was killed. Repaired it was used by Flt Lt Clemson during the attack on the Chicaldere railway bridge, 27 December 1916. The pilot in the cockpit, with one of the many dogs populating the island, appears to be Flt Lt H.deV. Leigh.

was mainly due to a lack of work for the squadron. The army, for whom most of the squadron's missions were flown, were busy preparing for the next stage of their advance, to El Arish and Gaza, commencing in late December. Most of their local requirements were now being met by the RFC squadrons, which were growing in strength and capability as the EIESS was withering on the vine. The squadron was thus reduced to flying occasional anti-submarine patrols from the island and the work of *Anne* and *Raven* in the Red Sea, as detailed in the following chapter.

However, during October French navy patrols penetrating deep into Adalia Bay (Gulf of Antalya), a large indentation on the Turkish coast east of Castellorizo, and cruising close inshore were being opposed by a number of active Turkish coastal batteries. Deciding to make a move against these batteries, *Amiral* Henri de Spitz, *Commandant la Division navale de Syrie*, assembled a small bombarding force and called on *Ben-my Chree*'s floatplanes to provide reconnaissance and spotting. The heterogeneous bombardment squadron, led by de Spitz from his armed yacht *Ariane II* (one 12-pdr, two 6-pdr), comprised *Ben-my-Chree*, *Dard* and three French trawlers. One of the latter, *Canada II*, had been fitted with a 138.4 mm (5.4 inch) gun removed from the cruiser *Pothuau* at Port Said. Quite what effect the discharges of the oversized weapon had on the small trawler we can only guess, but the muzzle blast alone must have shaken her severely. It was arranged that the spotting Short would work with *Canada* and *Ben-my Chree*'s 12-pdrs, leaving the other gunlayers to take advantage of targets of opportunity.

A few miles to the south of the combat with *FA300*, FSL Man's Schneider 3777 crashed into the sea following engine damage due to machine gun fire from the ground. Man was saved but the Schneider broke up, parts later being washed up on the beach near Sheik Zuwaiid closer to Rafa. These were brought to the German airfield at El Arish. *(Gunter Hartnagel Collection)*

Ben-my-Chree arrived at the rendezvous before noon on 3 November and quickly sent off two Shorts to locate the batteries. The first machine found some trenches and empty gun positions near Lara (a reference to an 800 metre peak near modern Anamur), whilst the second made an exploration westwards

towards Adalia. Shortly after 13.00 England and Benn were sent up in Short 8080 to spot the guns onto the trenches. Benn later compared the exercise to golf, commenting, 'It is putting in excelsis to watch every shot and see how by careful advice and adjustment the next stroke can be made to hole out on the target.'[202] After recovering the Short, the little armada began cruising down the bay towards Adalia.

Only empty gun emplacements had so far been located. The Turks were employing a small number of mobile field gun batteries, moving them where and when necessary. At 15.15 one of the batteries opened fire from a position a little east of Adalia, quickly straddling *Ben-my-Chree* at 2000 yards range. Samson smartly hauled out of line, stopped, hoisted out Short 8080, crewed by England and Woodland, then got underway again. Samson claimed to have accomplished this within 30 seconds. During the brief time the ship was stationary she was narrowly missed by five or six shells. England quickly located the battery, then bombed the guns hitting one with a 65-lb bomb. The ships led by *Amiral* de Spitz now engaged the battery with their guns. His yacht, closing the shore so that his little six pounders could be effective, received special attention from the Turkish guns. The yacht was hit at least twice, wounding a seaman and blowing a hole in the funnel. In the midst of all the excitement, one of *Ben-my Chree*'s lookouts started a submarine scare by reporting a submarine's periscope astern. This was quickly subdued by an outburst of Samson's colourful invective, conveying the suggestion that the poor man would be seeing Zeppelins next. Samson had immediately realized that what the lookout took for a periscope's wake was in fact only part of *Ben-my Chree*'s wake, as she was steaming fast and zigzagging at the time. Having silenced the battery, the squadron continued cruising the bay.

At sunset *Amiral* Spitz held a council of war with all commanding officers aboard his yacht, a gesture reminiscent of an earlier century. Deciding that little more could be achieved in the bay the squadron split up. *Ben-my-Chree* proceeded to Castellorizo, arriving on the morning of 4 November. Since *Raven*'s visit in July (Chapter 11) the Turks were thought to have brought up some guns to command the harbour, and Spitz had requested that Samson's floatplanes try to locate the guns. After two flights failed to locate them, *Ben-my-Chree* returned to Port Said.

Another Loss

During the early morning hours of 2 December *Ben-my-Chree* and her escort *Dard* were once more off the familiar coast of Palestine. They were to spend the day carrying out what should have been a routine inspection of the camps and stores depots from Haifa to Gaza. First to be hoisted out, at 06.38, were Clemson and Benn on Short 8080. Their flight took them over the Nazareth valley and towards El Afule, where they were greeted by heavy anti-aircraft fire, but not hit. The Short returned and was recovered at 08.05. Two hours after their return, FSL Dover set out on Schneider 3778 to reconnoitre the Tul Keram—Samaria road, returning at 11.07.

The ship now proceeded south to Jaffa where, at 11.48, FSL Nightingale and Lt Woodland, on Short 8372 (this time with just a single Lewis gun), were sent off to cover Ramleh and Bureir. Their flight route kept them within sight of the ship as far as Ramleh. Samson's Chief Writer, CPO Ridley, was keeping a special watch on the Short. Approaching Ramleh it was seen to suddenly come under shell fire. Ridley saw a direct hit on the floatplane, and his shout of horror drew the bridge staff's attention to the plummeting machine. The Short had been hit by a 7.7 cm shell from *Flakzug 136* under *Leutnant* Karl Bader. Just before disappearing behind a ridge of hills watchers thought they saw the machine begin to level out. Samson immediately sent off England and Benn, on 8080 again, to search for survivors. They too were severely archied over Ramleh, which damaged one of the floats, and were unable to locate the tan coloured wreck in a tan coloured landscape. England made a careful landing and quickly taxied over to the ship where the Short was smartly hooked on and hoisted in. Accepting that Nightingale and Woodland were probably dead, a sadder *Ben-my-Chree* continued down the coast to Gaza. Here a Short and a Schneider completed the day's reconnaissance programme. Whilst *Ben-my-Chree* was stopped recovering the Short, two Rumplers appeared overhead. The ship rapidly got under way, whilst the Short took off and also headed out to sea. The German aeroplanes dropped six bombs, the closest missing by twenty yards.

Later in the evening, as *Ben-my-Chree* was returning to Port Said, she intercepted a wireless message from the German airmen reporting both

Short 8372 following its crash on 2 December 1916. Both crew members, FSL Nightingale and Lt Woodland, survived and were taken prisoner. The Short appears to have been arrested by a sabar cactus (prickly pear) hedge after a very short run. The wing roundels have been covered in brushwood.

A second view clearly shows the serial 8372 and Phoenix Dynamo, Bradford, trade mark located under the elevator, and the hand painted fin. The group of civilians, Turkish officers and enlisted men, and German officers, a local villager peering over the fuselage, not to forget the dog, is so representative of the population in the area. The officer with binoculars around his neck is possibly *Ltn* Bader, commander of *Flakzug 136* with two 7.7 cm AA guns, credited with shooting down Short 8372.

missing fliers as alive, injured and prisoners. Samson relates a tale, similar to Dacre's, that the local Arabs had arrived at the crash first and commenced stripping the airmen of clothes and valuables. When the Germans arrived, they laid about the Arabs with cudgels, driving them away, then rescued the injured men. Both Nightingale and Woodland remained prisoners, at Yozgad Camp, until repatriated on 30 December, 1918.

During the periods at Port Said Samson kept his mechanics busy modifying Short 8090 in accordance with his ideas. 'I cut off a good deal of the lower plane, and poked about in other ways, converting one of the old ones into a better performer than a new one. I gained six knots, and about 15 per cent in climb.' What he ended up with was a Short 184 that looked a lot like a Short 827 on steroids. Samson prosaically named his creation the 'Experimental Short', although some sources suggest that it was also known as the 'Cut Short'. First flight of this modified machine was on 23 November, but its baptism of fire came on 22 December over Gaza. *Ben-my-Chree* had as usual sailed the previous afternoon, this time with the French destroyer *Pierrier* in company. During the night her original orders were cancelled by a W/T message and she was redirected to Gaza. Two Shorts, 8080 and 8090, were hoisted out at 07.05 to reconnoitre camps in the Gaza and El Falujeh area. Samson, who was flying 8090 with his brother as observer,[203] was very pleased with the performance of his creation, 'the speed which she rose from the water and climbed fully justified the alterations made in her. The unconverted Short which was trying to keep station on her was left far behind.'[204]

The first appearance of Samson's 'Experimental Short' or, more popularly, the Cut Short, 8090. Angled water foils were later installed under the lower wing tips instead of floats. Another photograph taken within a minute or so of this image shows the Samson Serial S.5 on the rudder. *(EMK)*

The advance on El Arish having just commenced, army headquarters were waiting anxiously for the information garnered during these flights. As soon as the floatplanes returned, a summary of the observers reports was sent to Ismailia by W/T. Only small numbers of troops had been seen, so there must have been some relieved smiles amongst the staff.

Ben-my Chree's operations had been observed by a German aeroplane flying just over the coastline, but it did not approach the ship and soon disappeared. Shortly thereafter, as *Ben-my-Chree* and *Pierrier* were proceeding up the coast towards Jaffa, a second machine clearly alerted by the first came over to attack. The two ships' anti-aircraft guns kept the Rumpler from descending, and six bombs dropped from 7000 feet missed the ships by a good margin. Two more flights were made over Jaffa and, in the afternoon, over Haifa. No unusual military activity was noted, just the day to day business of conducting war. The afternoon flight dropped a few bombs on guns between Tabaun and El Afule, then returned to the ship. *Ben-my-Chree* now retraced her course to Port Said, arriving at dawn on 23 December 1916.

The Attack on the Chicaldere Railway Bridge

Samson was able to make one further squadron attack in late December, as he explained in his memoir. 'The next cruise was for the attack on the railway bridge at Chicaldere. This was about 30 miles [50 kilometres] inland from the Gulf of Alexandretta, and I had been endeavouring to obtain permission to bomb this important link in the Turkish lines of communication for some time. At last the powers that be gave permission, so I set off with the *Ben-my-Chree* and *Raven*.'[205]

This was, of course, not the first aerial attack on the bridge. As previously described *Rabenfels* had made at earlier attempt on 7 June 1915 with Nieuports from *l'escadrille de Port-Saïd*. On 20 August Nieuports from *Anne* and *Raven* bombed the bridge and, finally, *Anne* made a reconnaissance over the bridge in January 1916.[206] The converted artillery shell bombs used by *l'escadrille* had little chance of causing serious damage to the bridge even with a direct hit. Nor could the EIESS's 16-lb bombs do much damage, but the 65-lb and 112-lb bombs the Shorts and Sopwiths could carry might be more successful. However, any delay to Turkish supplies would be of help to the army in its ongoing offensive.

Hopes of a relaxed Christmas in port were dashed, and celebrations curtailed or rushed as Samson's plan came to fruition. *Raven*, having returned from the Red Sea only two days previously, sailed on the afternoon of Christmas Day after a delay whilst the channel was swept for mines. The speedier *Ben-my-Chree* following on the morning of Boxing Day. They arranged a rendezvous with their French trawler escorts *Nord Caper* and *Maroc* (for some reason no destroyers seem to have been available over the Christmas period!) at a rendezvous in the Gulf of Alexandretta during the forenoon of 27 December.

On entering the Gulf *Ben-my Chree*'s lookouts reported a ship at anchor in the Bay of Ayas. Thinking that she might be a submarine supply ship Samson sent out England and Lt W.C.A. Meade on Short 8091, with a single 65-lb and eight 16-lb bombs, to investigate. They dropped four bombs, probably 16-lb, although no signs of activity were seen aboard. Meade also reporting, 'The Tramp appeared to have been hit by shell fire in many places on deck'. It was later discovered that the ship had been an earlier casualty of war, having been driven ashore a year previously by the French Navy.

The delay leaving Port Said caused *Raven*, no ocean greyhound, to be late at the rendezvous. So Samson, champing at the bit as always, decided to stagger the attacks, launching the first at 11.00 comprised of a Short and three Sopwiths, all from *Ben-my-Chree*:

FSL G.D. Smith and Captain Benn	Short	8080	1 65-lb and 8 16-lb Bombs
Wing Cdr Samson	Schneider	3770	2 65-lb Bombs
Flt Lt Clemson	Schneider	3778	2 65-lb Bombs
Flt Lt Brooke	Baby	8188	2 65-lb Bombs

Guy Smith provides us with a colourful account of his flight. Leaving the ship at 11.05 and crossing the coast a few minutes later near Port Ayas (Yumurtalık), steering NNE. This would take him alongside the Jebel Missis with peaks reaching almost 1000 metres. They probably crossed the jebel near Kurt Kulak (Kurtkulağı) where there is a pass at around 300 metres. From there, a course to the NW would lead across the fertile plain of Artik Ova towards the railway bridge, just beyond the northern peaks of the Jebel Nur.

During the attack on the Chicaldere railway bridge on 27 December 1916 Short 8080 was flown by FSL Guy Duncan Smith with Capt Benn as observer. The fin has been painted over, suggesting Samson may have intended this Short as a replacement for 8372. The observer's Lewis gun is on Short designed ring mounting not a Scarff mounting. *(EMK)*

Below us spread rolling hill country, with wide, fertile valleys between the ridges, planted to well kept farms. The natives were at work in the fields as though no such thing as war was raging about them. The day was beautiful and I recall that I was in a particularly happy mood as we climbed upward towards the mountains. Just after we passed a range of hills we were Archied a bit. The gunners couldn't get anywhere near our range, however, so there was nothing to worry about. I recall that I was singing, and that my observer, [Captain Benn], was smiling in sheer joy of the beautiful day and kept drawing my attention to points of interest.

Sopwith Baby 8188 was flown during the Chicaldere railway bridge attacks by Flt Lt J.C. Brooke. (EMK)

We had been ordered to bomb a certain bridge. A huge mountain hid the bridge from view. Coming over the mountain's shoulder, we were almost above the bridge before we realized it. To our delight, there was a train on the other side of the bridge coming towards us. It would have been considerable feather in our caps to wreck bridge and train simultaneously, so I stuffed the nose of the machine down in order to increase speed, and there followed a race between the train and our airplane.

The train beat us to it, crossing the bridge before we could reach it. So I throttled down and turned around, chasing the train down the track towards Hamidieh [Ceyhan?]. The engineer must have had his throttle wide open, and as he didn't slow down for any of the curves I expected any instant to see the train jump the track and go rolling off down the mountain side. We dropped a couple of bombs which failed to explode, and then [Capt Benn] tapped me on the shoulder, signalling that it was time for us to get back to the bridge, the real job we had on hand.

Returning along the railway to the bridge, a 65-lb bomb fell in the river close the southern end of the bridge, which has a NW to SE alignment. Benn now takes over the story.

When we arrived at the other side we found a guard house, made like so many huts in the East, of reeds, with a reed roof. The personnel was clearly much excited by what was happening, and turned out to fire at us with their rifles. Smith banked the machine so that I could get a bearing with the tracer ammunition, and then I learned more in one second of the moral effect of tracer than ever I knew before.

It was bad enough for those old Turks to have a machine-gun turned on them from a distance of a few hundred feet, but actually to see the bullets arrive was too much for their nerve and they turned and bolted, like rabbits, into the hut. This however, availed them nothing, for the roof of the hut was no more impervious to fire than a sheet of paper. We may assume, therefore, although we did not actually hit the bridge, that we carried out the main part of our job, which was to keep the guard in check.

With the guard thus discomfited Samson, Clemson and Brooke were able to bomb without distractions. Attacking in a shallow dive from 700 to 400 feet Samson's bomb passed between the bridge girders exploding in the river bed. Clemson's bomb also narrowly missed, whilst Brooke managed to damage the approach embankment. All the machines had returned to *Ben-my-Chree* by 12.40, most with bullet holes.

Raven had now arrived at the rendezvous, hoisting out her two Shorts at noon:

Flt Lt E.J. Burling and Lt K.L. Williams	Short	8075	1 65-lb and 8 16-lb Bombs
FSL E.M. King and Lt N.W. Stewart	Short	8004	2 65-lb and 8 16-lb Bombs

FSL King, on his first mission for the EIESS, left the water at 12.05, followed by Burling a minute later. King claimed no hits on the bridge; Burling one hit with a 16-lb bomb the remainder missing by an average of 15 yards. Both Shorts had returned by 13.20.

A set of three images of the bombing of the Chicaldere railway bridge on 27 December 1916 (this page and next). All were taken by FSL Guy Smith's observer Capt Benn. *(TNA, AIR 1/1707/204/123/69)*

The bridge and the meandering river, with smoke from 'Commander Samson's Bomb falling on Bridge.' (A photograph taken ten months later, from the same angle, shows a very different pattern of sandbanks. See Chapter 15.)

The bridge over the river Ceyhan at Chicaldere. Three guard huts can be seen just to the left of the bridge backing on to the railway embankment, and another hut just off the right hand end of the bridge.

Below: An amazing image captured by Benn, who must have been hanging out of his cockpit, of the explosion of a 65-lb bomb close to the left hand bridge pier island.

A nice air-to-air of an EIESS Short over the Gulf of Alexandretta, returning from the raid on the railway bridge at Chicaldere 27 December 1916. *(EMK)*

Refuelled and rearmed, *Ben-my Chree*'s floatplanes made a second attack starting from the ship at 13.30:

Flt Lt Maskell and Lt W.L. Samson	Short	8080	2 65-lb Bombs
FSL Henderson	Schneider	3770	2 65-lb Bombs
Flt Lt Clemson	Schneider	3778	2 65-lb Bombs

The Short's two bombs missed the bridge, the Schneiders had more success. One of Henderson's bombs hit the southern end of the bridge, making a large hole that would have to be filled before the bridge could be used again. Clemson reported, 'First, direct hit on Permanent Way smashing away sleepers and trestles, and damaging rails. Second, hit eastern end of bridge damaging Permanent Way.' Which really was the most that could be hoped for as the squadron's bombs were far too small to seriously damage, let alone destroy, the bridge.

So, satisfied with the day's results, but feeling another day would be required to complete the task, the squadron retired to sea for the night. Anti-aircraft fire had been directed at the machines during the day, most suffering damage to some degree. One Sopwith, probably 8188, had taken hits through the fuselage, control wires and propeller, and one of the Short's had been hit three times in the fuselage and once in the wings. Whilst work continued in the hangar and on deck to repair damage, a summary of the operations was sent by wireless to Admiral Wemyss. Considering that they had achieved enough, he cancelled the next day's operations and ordered the ships back to Port Said. Intelligence reports later showed that the damage caused during the attacks was sufficient to halt traffic for several days.

Lieutenant Cecil E. Hughes, who had only arrived in Port Said from England a few days earlier, was aboard *Raven* for the Chicaldere attack. As Intelligence Officer he later received much delayed information about the success of the attack.

> The bridge had certainly been hit, though not destroyed, but the damage to the single line of railway had been considerable. The damage, moreover, had been inflicted exactly at the time—was it just a strange coincidence?—when the line was to have been used to convey big guns to the Turks defending Baghdad, and the rumour said that the guns never got there.
>
> True or not, it throws a good deal of light on the value of such operations as those in which the East Indies and Egypt Seaplane Squadron was engaged during 1916 and 1917. These operations were always liable to upset the enemy's calculations, very often, they actually did upset them, but whether they did or not they succeeded in creating in the mind of the Turks a feeling of insecurity which made it necessary to keep troops and guns many hundreds of miles behind the front line.[207]

A second attempt on the bridge, during the first few days of January 1917, to increase the damage had to be abandoned in the face of harsh winter gales sweeping out of the mountains. Forced to take shelter in Famagusta, *Ben-my-Chree* was quickly recalled to Port Said. The French, once more, had need of her services.

Samson's red tail 8372 landing in the outer harbour, Port Said. *(State Library of Queensland)*

A Vest Pocket Kodak image of *Ben-my-Chree* passing through the Suez Canal in June 1916, either heading to or returning from the Red Sea and Aden. The hangar doors were left open whenever possible to help cool what would otherwise quickly become an oven. Note the Australian troops relaxing on the bank and swimming in the canal. *(AWM C00071)*

CHAPTER 10

Aden and the Red Sea, 1916

Having, in the previous chapters, reached the end of 1916, it is time to go back to earlier in the year to look at the work of the EIESS south of Suez.

Events along the eastern littoral of the Red Sea called for the presence of the EIESS, often requiring multi-week absences from Port Said. During 1916 and into 1918 the EIESS made eight cruises into the Red Sea, and beyond; *Raven* made four visits, *Anne* two, and *Ben-my-Chree* and *City of Oxford* one each. Of these visits, three also visited Aden, and one sailed beyond the equator in the Indian Ocean.

Aden in WW1

When a deal with the Sultan of Lahej to purchase Aden for use as a coaling station was reneged upon, the British East India Company attacked and captured Aden in January 1839. The new acquisition coming, in 1858, under the control of the government of India. With the opening of the Suez Canal in 1869, Aden developed as an important coaling station. The centre of the Settlement was the port of Aden itself. Initially an open roadstead at the old town of Aden, gradually an inner port protected by the peninsulas of Tawáhi (Steamer Point) and Little Aden was developed. Fanning out behind is the desert, rising gradually to the town of Lahej approximately 50 kilometres from Aden. Lahej lies at the edge of a relatively well watered area below the jebel leading up to the Yemen border.

Water is the key to the survival of Aden. Whilst food and fuel could be imported, local sources of potable water were essential. An ancient system of cisterns in the Jebel Shamshan above Crater, the Tawila Tanks, capable of storing over seven million gallons of rain water, were restored by 1856. Wells were sunk around Sheikh Othman; an aqueduct constructed in 1867 channelled this water some seven or eight miles into Aden. Authorities considered the water fit for locals and animals only, steamships and Europeans demanded a better quality of water. Consequently, a number of condensers were installed around the harbour. By 1914 these were producing over 50000 gallons of distilled water daily, in addition there were refrigeration plants producing some five tons of ice per day. The latter was particularly welcomed

The Sultan's Palace Lahej, 1893. Aerial photographs taken in 1916 show few changes over the intervening 20 years. *(Qatar Digital Library, T.11308/15)*

by the European inhabitants as the climate is noted for stifling heat and humidity. Its most bearable temperatures fall between December and mid-March, with daily highs between 28–30°C, unfortunately accompanied by highest average humidities. During the remainder of the year hot *Khamsin* or *Shamal* winds prevail and daily temperatures often rise over 40°C. Average humidity is close to 70%.

An Anglo-Turkish Convention of 9 March 1914 agreed a border between the Yemen Vilayet and the Aden Protectorate. On the outbreak of war the Ottoman Yemen Army Corps, comprised two under strength Divisions, the 39th based in Southern Yemen, and 40th mostly distributed along the southern Red Sea coast.[208]

Facing them, the Aden Brigade of the British Indian Army on the outbreak of war comprised one Indian and one British infantry battalion and a local mounted unit, the Aden Troop of around 100 cavalry and cameliers. The port was protected by a number of forts, whose guns could not be trained inland. Within a few months the Brigade was strengthened to two Indian infantry battalions, one unacclimatized British Territorial infantry battalion, some field guns and engineers. Most of these were based at Aden, with a detachment at Lahej training the Sultan's forces on British equipment. With these forces only a limited area of the Protectorate, within a few kilometres of Aden, could be defended.

Turkish forces occupied positions at Sheikh Said, across a narrow channel from Perim Island at the entrance to the Red Sea. On 10 November 1914 troops of the 29th Indian Infantry Brigade, en route for Suez, were diverted to Sheikh Said and effected a landing covered by the guns of *Edinburgh*. The Turks were driven off, the Brigade then re-embarked and sailed off to Egypt (and Gallipoli), the Turks quickly reoccupying Sheikh Said. On Perim Island a small British garrison were able to repulse an attempted landing in mid-June 1915, no further attempts being made.

At the end of June a large Turkish force advanced on Lahej. A column, made up from Territorials and Indian units, left Aden on 3 July and advanced to reinforce Lahej. The weather was exceptionally hot and the column ground to a halt about halfway to Lahej, unable to proceed. Most of the British were in an appalling state, suffering from heat exhaustion and dehydration. Many were saved by the timely arrival of a medical officer in a motor car loaded with 100 gallons of water and 100 lb of ice. The British and most of the Indian troops returned to Aden, many to spend time in hospital recovering.

The defence of Lahej now rested on the small force based there plus a handful of Indian reinforcements rushed forward in motor cars. The total

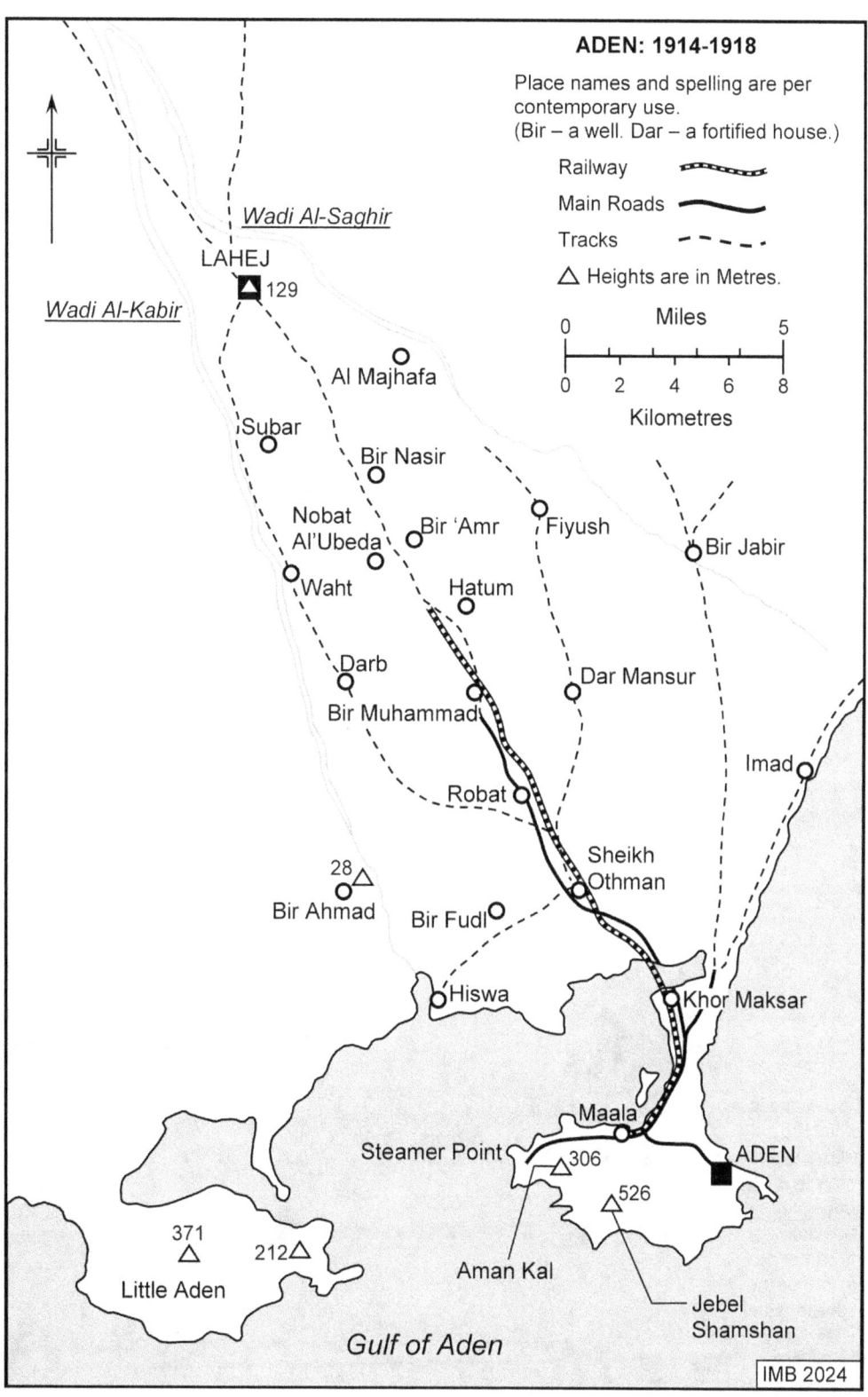

force available was around twenty officers and four hundred men, with a few machine guns, two 10-pdr mountain guns, and at least one of the Sultan's field guns. The Indian troops held gallantly until the following morning when they were forced back by overwhelming numbers. Once the retreat had started it continued past Sheikh Othman to the outskirts of Aden.

With reinforcements from Egypt and India, an offensive commencing 20 July 1915 quickly recaptured Sheikh Othman, and the vital water supply, then advanced to Robat, about a third of the distance to Lahej, establishing a defence line. The Turks established their headquarters at Lahej. Both sides created a series of strong points, relying on constant patrolling and minor operations for the remainder of the war. A metre gauge railway, and a metalled road, were constructed from Maala, through Khor Maksar, Sheikh Othman to Robat and beyond. By 1918 the British forces, renamed the Aden Field Force, had risen to four Indian and one British battalion, with field and heavy artillery, supporting and garrison troops, and a small RFC/RAF unit.[209]

A view of the western end of the Aden Peninsula from the Inner Harbour. Starting on the left with the height of Aman Kal (306 metres), below which lies Tawáhi and the coal yards, then the Hogg Clock Tower (locally known as *Little Ben*), the Signal Station and (on the extreme right) part of Steamer Point. Short 850 was aboard *Raven* and *Ben-my-Chree* on their visits to Aden. The photo was taken during *Ben-my-Chree*'s visit of June 1916, with Short 184, 850 (with Samson Serial S.2 on the rudder), taxying in the foreground.

Schneider 3790 was also on *Raven* and *Ben-my-Chree*, as seen here, also with a Samson Serial B.3 on the rudder. It has a single 65-lb bomb and a Lewis gun attached to the cabane struts on the port side ahead of the cockpit. *(YEORN 2014-80-201. Courtesy of The National Museum of the Royal Navy)*

Raven at Aden

Following the disaster at Lahej, the assistance of an aeroplane or a seaplane to aid the defence of Aden was requested. Aden would have to wait until November 1917 before a small RFC unit could be provided. Whilst the responsibilities of the East Indies and Egypt Seaplane Squadron included Aden and the Red Sea, they took second place to the Egypt, Palestine, and Syria. It would be the end of March 1916 before a seaplane carrier could be spared to visit Aden.

Raven sailed from Suez on 24 March, passed Perim Island during the forenoon of 30 March then headed out into the Gulf of Aden. Lieutenant J. Jenkins, RNR, commanded the seaplane carrier, whilst command of the RNAS detachment was vested in Flight Commander Charles H.K. Edmonds. The RNAS contingent included pilots Edmonds, Flt Lts T.H. England, Flt Lt J.T. Bankes-Price, M.E.A. Wright, and FSL R.M. Clifford, with a single observer Lt V. Millard, 1st Essex Regt. The aircraft were Short 850, and Schneiders, 3721, 3722, 3727, 3774, 3775, 3790.

> Well out of sight of land and of all shipping, we stopped to hoist up seaplanes from the hold, and erected and stowed them on deck. After dark we arrived at Aden, with five Schneider Cup Seaplanes and one Short Seaplane on deck, derricks plumbed, bombs in place—in fact all ready to hoist out the next morning; the Short for reconnaissance, or the Schneiders for bombing. The spare Schneider [3775] was kept below in the hold.[210]

Raven anchored in Aden harbour at 21.15 on 30 March. Edmonds immediately going onboard *Euryalus*, flagship East Indies, meeting with the C-in-C Admiral Sir Rosslyn Wemyss and the Aden Brigade Commander, Brigadier General William Crawford Walton.[211] It was decided that the Short would make an early morning reconnaissance to locate enemy camps, this to be followed as quickly as possible by bombing attacks by the Schneiders. Edmonds also requested the loan of an officer with a good knowledge of the country, Captain C.P. Paige, GSO, arriving onboard the following day.

At first light, Edmonds and Millard were hoisted out on the Short and airborne ten minutes later.

> I made a reconnaissance of the whole position on the Short. Low clouds hampered observation, but we found the enemy's principal camp near Subar, where we received with considerable rifle and machine gun fire. We returned to the ship at 7.50 am.
> I decided to attack the camp at once, explained its situation to the other pilots, and the first Schneider left the water at 8.37 am.

The Schneiders carried mixed loads of 16-lb and 20-lb bombs.[212] Edmonds led the way flying 3721, followed by Bankes-Price (3722), and Clifford (3774), all three reaching and bombing the camp at Subar. Bombs were dropped from between 100 ft and 600 ft. They were met by rifle and machine gun fire at the camp, each machine being hit several times. England (3727) experienced engine problems and returned after dropping his bombs on Waht. Wright (3790) failed to locate the camp, dropping his bombs on Sharaj and mule lines near Waht.[213] All machines had returned by 09.40.

The same pilots and machines flew in the afternoon, starting out at 15.00. The targets were probably Nobat Al'Ubeda and Bir 'Amr. Wright, Clifford and Bankes-Price reached the villages, but the latter's bombs would not release and he had to make a very gentle landing on his return. It was England and Edmonds' turn for engine problems, both dropping bombs on Waht before returning. Edmonds noted that 'The atmospheric conditions were very bad indeed.'

> On the following morning at daylight another bombing attack was made against the enemy's principal camp near Subar [Bankes-Price (3722), England (3727), Clifford (3774), and Wright (3790)] which I followed up by a combined bombing and reconnaissance flight on the Short, taking Capt Paige, GSO, as observer — his knowledge of the country proving very useful. It was evident that the attacks were proving effectual, for the camp appeared almost deserted and pilots met with much less opposition.

No flights were made in the afternoon, a strong wind with low clouds having developed. These had cleared by the following morning, Sunday 2 April. Four Schneiders set out to bomb the camp at Fiyush. Edmonds was flying 3775, with Bankes-Price (3722), England (3727), and Clifford (3774). Clifford failed to locate Fiyush but dropped his bombs on Waht, following up with 500 flechettes 'at some grass shelters near Abdullah-bin-Ahmed [probably Bir Ahmad] and observed men running away from these shelters.'[214]

General Walton had requested some photographs, but *Raven* was not properly equipped owing to a shortage of suitable equipment in the Squadron. However, Edmonds offered to do his best with the amateur equipment that was available. Wright and Millard set out on Short 850 at 07.47 to fly around Waht, Subar, Lahej and Fiyush, returning at 08.38. Although suffering with a failing engine they were able to complete the planned flight, taking some photographs and dropping propaganda leaflets. On examination the Short's engine proved to be damaged beyond immediate repair, no further flights could be made.

The coming and going of individual traders and caravans was little affected by the conflict, it was easy for them just to go around or between the various outposts and strong points. Consequently, the results of the bombing raids were known in Aden within 24 hours. The raw reports are a little difficult to follow, but it is apparent that some casualties were caused to both men and livestock, and damage to houses. One report, with suspicious precision, listing 'twenty Turks, six Arabs, twenty-nine Mullahs, seven guns, one camel and many military huts destroyed.' However, the main effect was on morale. The natives and camp followers are reported as having fled Lahej 'to the jungle' and to the desert. The Turkish troops are reported as being 'in terror [of] these seaplanes.'

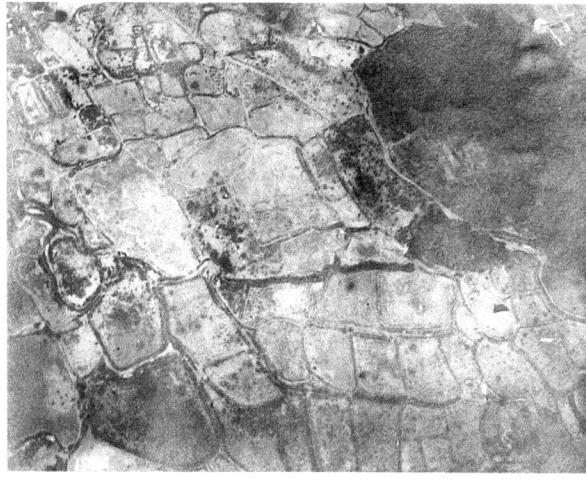

One of a small number of photographs taken during *Raven*'s visit to Aden. It shows a camp (the small white dots in the centre of the image?) 'S of Wood Situate N of Lahej.' *(TNA, AIR 1/1706/204/123/64)*

General Walton was very satisfied with what had been achieved, telegraphing the India Office in London, 'Four bomb attacks were made and considerable damage done to Fiyush and camp at Subar and to gun mules and bullocks. Damage to seaplanes immaterial. No casualties. Moral effect very great, damaging Turkish prestige with the Arabs.'[215]

There was actually one injury, Edmonds receiving a small scratch when a bullet grazed his cheek. The Short and Schneiders packed away on 3 April *Raven* steamed off for Egypt, arriving at Port Said 10 April. She would return the following year. But another EIESS ship arrived at Aden in June, becoming involved in the opening moves of the Arab Revolt when returning to Egypt.

Ben-my-Chree at Aden and Jeddah

Ben-my-Chree sailed from Port Said on 2 June to take passage through the Suez Canal and on to Aden. In her hangar were two Shorts, S.2 and S.3, two Schneiders, B.1 and B.3, and Baby B.5.[216] Besides Samson himself the pilots were Flt Lts Bankes-Price, England, Wright and FSL L.P. Paine. Observers were Lt J.H.B. Wedderspoon, 2Lt Benn (he was promoted Temporary Captain on 8 July, later confirmed and backdated to April 1916), 2Lt J.M. Burd, and 2Lt L. Clarke. *Ben-my-Chree* approached Aden at dawn 7 June, sending off Short S.2, with Wright and Burd to reconnoitre from Waht across the lines to Imad. The Short returned just as the seaplane carrier was coming to anchor.

Some quick work enabled Samson to present a full report of the flight, complete with photographs and maps, when he reported to General Walton. It was an impressive demonstration of the use of air power which delighted the General. With his full backing Samson embarked upon 'as intensive a bombing as my limited resources permitted.'

> I knew that our only chance of being able to fly with the Shorts was to try to get off very early in the morning or late in the afternoon, as the severe heat would inevitably not only boil all our cooling

water away, but probably affect our lift. As it was, we had a terrible time getting the Shorts off the water under existing conditions, and being unable to ascend beyond 1500 feet we soon began to lose our water whilst flying. On several trips it was touch-and-go whether we could get back before the engine seized up through this cause.[217]

As these were almost a carbon copy of the earlier work done by *Raven* only a few highlights will be mentioned. The bombs carried this time were bigger. The Shorts carrying bombs up to 112-lb, the Sopwiths up to 65-lb. The intent being to carry out as many flights as possible in the time available, bombing and reconnaissance flights were made morning and evening between 8 and 12 June. During the four days of attacks, 8, 9, 11 and 12 June, a total of fifteen individual sorties were flown, 10 June was a 'rest' day for maintenance and to hopefully catch the Turkish force unawares the following day. Three 112-lb, three 65-lb, seventeen 20-lb, four 16-lb, 14 petrol bombs, and one carcass (saltpetre and sulphur) were dropped, plus a single box of flechettes. All flights were met with accurate rifle and machine gun fire and some anti-aircraft shrapnel was reported but, although the floatplanes were hit several times, there was no major damage or injuries.

On the morning of 8 June Samson and Wedgewood Benn were the first off, at 05.20 on Short S.2, with a single 112-lb bomb, three 16-lb bombs, and four 'hoppers' for Benn's Lewis gun. Benn later wrote, 'We took a 112 lb. bomb on the rack in the undercarriage, and I carried a few 16 lb. bombs in the back seat.' The Short was also laden down with a W/T set and a camera, thus overladen the Short needed a two mile run to takeoff. Passing Waht they were unable to climb above 700 feet, at Lahej they were less than 400 feet above the roofs of the town.

> As this was too low to drop the 112 lb bomb at the Mosque owing to risk of damage to seaplane from the explosion of the bomb. The 112 lb bomb was dropped at and hit dug-out into which men were seen to go from camp in trees just N of Lahej. Two 16 lb bombs hit camp from height of 800 feet; emptied two hoppers from Lewis gun against camp. Lahej was full of troops and Arabs and firing was observed from 4 small guns, two of which were on house tops (probably 10 pdrs); one or two shells that burst were of larger size than 10 pdrs, one star shell was observed. Machine guns and heavy rifle fire were also encountered; two rifle bullets hit seaplane.[218]

They returned to the ship and landed alongside at 06.19. The Mosque, and similar buildings, were widely suspected of being used as ammunition or supply dumps and considered legitimate targets. Arab agents reported that the bomb dropped on the dug-out, actually a gun pit, killed seven men.

Samson and Benn's flight was typical of those carried out by the Shorts

View of Darb, lower centre, from the west on 9 June 1916, taken from Short S.3 (8082), Wing Cdr Samson and 2Lt Benn. They reported being 'Continuously fired at with rifles by soldiers in the town. From volume of rifle fire apparently about 200 men in town.' *(TNA AIR 1/1707/204/123/68)*

over the next few days, a mix of bombing and reconnaissance. The Sopwiths were mainly engaged in hit and run bombing raids. One by Flt Lt England on the first day is typical. Taking off at 05.49 he flew Baby B.5 (one 65-lb bomb, a box of flechettes and a Lewis gun with two 'Hoppers') inland to Waht reaching just 500 feet altitude. England was back by 06.14.

> Under fire from anti-aircraft gun in South corner of town. Dropped Flechettes on gun's crew of about 6 men. Dropped bomb which hit the South Corner of Central Mosque and Adjoining bldgs. Circled around village firing one hopper from gun. Observed few people.

Bankes-Price attempted the same flight the following morning, also flying B.5 but with four 20-lb bombs and the Lewis gun. His flight was quickly terminated.

> Flew inland to Waht at a height of about 1000 feet. About one mile south-east of Waht engine stopped dead. I dropped all my bombs at once and turned back towards Sheikh Othman, endeavouring to get into our lines before the machine dropped. The engine picked up slowly, and I managed to get back to the sea, landing in about 2 ft of water. I saw no signs of the enemy.

In an attempt to overcome the difficulties of operating overloaded machines in an unforgiving climate on the afternoon of 9 June one Short was flown solo. Short S.3 piloted by FSL Paine set out at 17.40 with a single 112-lb bomb. He was still only to reach 800 feet, but dropped his bomb into the middle of Waht, returning to the ship at 18.19. This is the last recorded flight for Short S.3. Two days later it is noted that it failed to take-off owing to engine problems.[219]

On 11 June 1916, Lt Wedderspoon, on Short S.2 (850) with Flt Lt England as pilot, took this photograph of Lahej. The three lines drawn on the original are from top to bottom: a small plume of smoke resulting from the explosion of a petrol bomb; the Sultan's Palace buildings; the city mosque. *(EIESS Samson Reports)*

On the same flight Wedderspoon took this photograph of the country to the west of Wejh. The settlement just appears in the lower centre and Wadi Al-Kabir cuts across the centre of the image. *(TNA AIR 1/1707/204/123/68)*

Operations over Aden wrapped up on 12 June after a bombing raid on Subar by Bankes-Price on Schneider B.1. Taking off at 05.26, he flew straight to Subar carrying two 20-lb and six petrol bombs. The first three petrol bombs were not observed and assumed not to have ignited. Of the remaining three, two started a small fire at one end of the camp, but the third started a 'large blaze, which was still burning strongly when the coast was reached 15 minutes later. The two HE bombs fell on the north-western corner of the camp amongst some tents, apparently doing some damage.' Samson noted that the fire could still be seen from the sea when *Ben-my-Chree* sailed that evening. Bankes-Price returned and was hoisted in at 06.01. A useful morning's work.

Once again the results and effects of the raids were quickly known in Aden. Again, the morale effect far outweighed the material results. Turkish stock falling even lower with the local chiefs and tribes. The Turkish command attempted to stop caravans running into Aden, a ban widely ignored by traders. All in all, as Samson noted in his report, 'In these circumstances, the results achieved, sometimes far inland, must be regarded as satisfactory.'

But *Ben-my-Chree* was not yet finished with the Aden Protectorate. Although due back in Egypt Samson, having first obtained the backing and approval of General Walton, decided to take some time to descend upon Sheikh Said. The Turks here, estimated at 500 men with five guns, occupied a small fort, some redoubts, and several tented camps on the adjacent jebels. They were conducting a desultory bombardment of the small British garrison on Perim Island. Colonel Alexander, Walton's GSO at Aden, sailed with *Ben-my-Chree*. No doubt the promise of additional photographic coverage was the main draw, although his local knowledge would be useful to Samson.

Ben-my-Chree arrived off Perim Island at 04.00 on 13 June. Planning to conduct a bombardment with her 12-pdrs Samson launched W/T equipped Short S.2 soon after arrival. When engine trouble prevented the Short taking off, it was replaced by Schneider B.3, with five 20-lb and two petrol bombs, piloted by Bankes-Price. *Ben-my-Chree* closed to within 2000 yards of the shore, in a position permitting enfilade fire on camps at Jebel Malu and Jebel Akrabi. Instead of attempting to direct fire using flares, Bankes-Price dropped the two incendiary bombs some distance apart to give the line to the camps. He then dropped five HE bombs on the camps, starting fires that gave the ship's gunners an aiming point. He later recorded seeing four shells burst in the camp. A single Turkish gun returned fire, but its shells fell short. Seeing troops moving about the hillsides Samson next turned his guns on this target of opportunity.

England now reported the Short repaired and ready for flight, so Samson turned *Ben-my-Chree* away from the coast to recover the Schneider and send off the Short. Getting off successfully, England and his observer 2Lt Burd directed fire on to two guns on the slopes of Jebel Akrabi. A brief exchange

A post card of HMS *Fox* in UK waters in 1901 before she proceeded east of Suez, not returning until 1919. Although old, she was launched in 1893, and obsolete she served throughout the war mostly as the flagship of the Red Sea Patrol. Not fast at under 20 knots, but heavily armed with two 6-inch, eight 4.7-inch, and numerous smaller guns. *Fox* will appear in several following chapters.

of fire now ensued. The ship was straddled, hit by shrapnel and a shell passed through the forward funnel without exploding. In return her gunners silenced one of the guns. While *Ben-my-Chree* shifted position to bombard a different camp, the Short's crew took a series of photographs of the Turkish positions and dropped bombs on one of the camps. The spotting continued for a while until *Ben-my-Chree* withdrew once again to recover the floatplane. After refuelling, topping up water, oil and rearming, with a single 112-lb and 20-lb bombs, and two incendiaries, the Short was sent off again at 08.03 with a new crew, Flt Lt Wright and Lt Wedderspoon. Its flight to the south of Jebel Malu, encompassed several camps at Jebel Barika and Khor Ghorera. The latter received the 112-lb bomb, and several more photographs were taken. They returned at 08.32, the Short quickly refuelled, and rearmed as previously, and the crew changed, FSL Paine and 2Lt Clarke taking over. The third flight commenced at 08.48, returning shortly after 09.30. The camp at Jebel Akrabi was bombed again and the observations completed.

Later reports suggested that *Ben-my-Chree* had literally caught the enemy napping, asleep in their tents, although reports of 40 killed is so similar to the reported casualties at Aden as to be suspicious. The bombardment had raised the ship's morale to new heights. Throughout the action the upper deck remained crowded with stokers and other crew members, whose duties normally kept them below decks, taking brief respites from their duties to 'see the fun.' Samson then set course for Port Sudan,[220] intending to take on coal, 'transferring Colonel Alexander, who had thoroughly enjoyed his first Naval engagement, to the Senior Naval Officer, who had arrived on the scene and joined in with his 4-inch guns.'[221]

Approaching Port Sudan they were ordered to immediately proceed to Jeddah where urgent developments required their attendance. Arriving on the morning 15 June, and after negotiating a tricky passage through some reefs, *Ben-my-Chree* joined ships of the Red Sea Patrol, *Fox*, *Hardinge*, *Dufferin*, and *Perth*.[222] Samson immediately met with Captain W.H.D. Boyle, commanding the Red Sea Patrol, aboard *Fox*. The ships were attempting to provide gunfire support for the Arab Army attacking Jeddah. Having been denied permission to land observers by the still distrustful Arabs, the British ships had to fire blind with mixed results. It was decided that in the evening the floatplanes would reconnoitre and bomb the Turkish positions. Then in the morning they would spot for the guns of the squadron.

By this stage in her cruise *Ben-my-Chree* had only three serviceable floatplanes available, Samson decided to employ them all. With Benn as his observer, he piloted Short S.2 armed with a single 112-lb bomb, two incendiaries and a Lewis gun. Two Schneiders, B.1 and B.3, piloted by Bankes-Price and England respectively, completed the trio. The Schneiders each carried a single 65-lb bomb and a Lewis gun. England had been tasked with attempting to breach a gate in the eastern wall of the town. He decided that the gate was too close to the mosque to risk dropping the bomb, instead dropping it on a group of soldiers west of the wall. Attacking trenches to the south of the town, Bankes-Price dropped his bomb from only 100 feet, receiving a shaking from the resulting explosion. He then flew along the trenches, firing his machine gun into them until all his ammunition was exhausted.

Samson and Benn flew across the town and defences, taking several photographs, noting some bombardment damage on the southern walls and south-western tower. They dropped the 112-lb bomb on two guns to the south east of the town, missing by 50 yards. Passing along the seaward side of the defences the Short came under heavy shrapnel and rifle fire. Whilst examining a redoubt to the north the Short was hit several times, one bullet removing the heel from Samson's right shoe, as well as nearly shattering the supports of his seat. Another hit started a severe vibration. Thinking his engine had been damaged Samson headed out to sea and put the Short down as soon as possible, immediately stopping the engine. One of the ship's motorboats came up and towed the floatplane back alongside. When hoisted aboard and examined, the propeller was seen to be badly splintered by a bullet whilst another bullet had almost severed the elevator control wires. Thankful for their narrow escape from disaster Samson set his maintenance crew to work repairing damage for the morrow.

At dawn, as *Ben-my-Chree* was preparing to launch her spotters, a white flag went up over the town as Jeddah surrendered. In his signal cancelling the flight Captain Boyle added, 'probably the seaplanes decided the matter.' To the garrison, already battered by several days bombardment, the appearance of the aeroplanes may indeed have been the deciding factor, their final straw. Samson's claim that, 'three inefficient rather antique seaplanes took Jeddah,' is somewhat stretching a point.

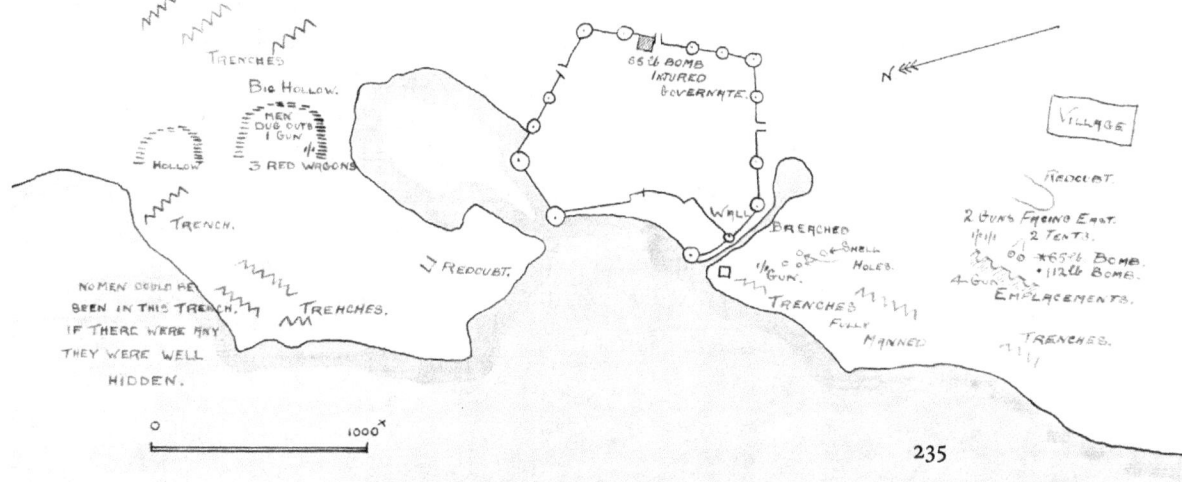

Map detail from a report on the 'Reconnaissance and Bombing Attack on Jiddah, 15.6.16'. (EIESS Samson Reports)

For now, *Ben-my-Chree* proceeded to Port Sudan, coaled and departed for Egypt, arriving at Suez on 21 June. 'I must here pay a tribute to my stokers, practically all Liverpool men. They had a tough time in great heat, steaming at high speed continuously in a ship constructed for English waters without any tropical facilities.'[223]

With the port of Jeddah in Arab, and British, hands the road to Mecca was open. The EIESS had still more contributions to make to the Arab Revolt, some more direct than others. On 28 June *Anne*, which had reverted to cargo ship, delivered an assorted cargo of stores and £30000 in gold, the life blood of the Arab Revolt, to Jeddah. She also brought two Egyptian mountain gun batteries, later used to subdue the Turkish garrison at Taif.

With Regard to Coal

Following *Ben-my-Chree*'s return to Port Said, the EIESS's work in the Red Sea was left to the two converted cargo ships, *Raven* and *Anne*. Although slower, they were designed for longer voyages and could, if need be, stay out for several weeks with minimum need to coal and resupply. Which provides us with a good excuse to discuss coal.

Navies and merchant shipping during the Great War relied on coal, oil was slowly being introduced but King Coal ruled. Not just any coal. The Royal Navy preferred Welsh anthracite, 'Best Cardiff' or 'Admiralty', coal as it 'generated great heat, but left only a small residue of clinker on the fire grates, as well as very little ash to fall into the bottom of the furnace or rise to coat the boiler pipes.'[224] Throughout the war there ran a long, thin black line of colliers connecting the coal docks at Cardiff, Swansea and Barry to the outposts of the navy. For the Red Sea there were limited sources of re-supply for ship's bunkers; coal could be taken on only at Port Said, Suez, Port Sudan, and Aden. Over 1000 kilometres separated Suez from Port Sudan, on the west coast of the Red Sea, then another 1000 kilometres to Aden.

When *Raven* left Suez for Aden on 24 March 1916 she had in her bunkers 499 tons of coal, when she returned on 10 April there were 251 tons remaining. *Ben-my-Chree*, however, when she sailed for Aden from Suez on 3 June had 676 tons of coal in her bunkers and in bags filling all available deck space, she arrived at Aden five days later with only 195 tons remaining. She coaled, taking on board 480 tons, before leaving Aden on 11 June. After diverting to Jeddah, she arrived in Port Sudan on 17 June with 313 tons to take on another 285 tons overnight. Returning to Port Said on 22 June just 236 tons were left in her bunkers.

Raven used less than a quarter the amount of coal than *Ben-my-Chree* over a similar cruise. Little wonder then that *Anne* and *Raven* were preferred for all future visits to the Red Sea.

Raven and a Survey of Akaba

Since last visited by *l'escadrille de Port-Saïd* in December 1914, only a handful of British and French warships had ventured into the Gulf of Akaba. The available Admiralty chart of the gulf dated from 1873, with some corrections in 1911, it was also detailed in the Red Sea Pilot of similar vintage. Intelligence suggested that the town was held by two companies of Ottoman Gendarmerie, but nothing was known of guns or fortifications. Samson was asked to make a visit and prepare a photo-mosaic of the area around Akaba itself, and to report on the possibility of making a landing to capture Akaba. *Raven* was available and given the task.

Having taken on board two Shorts, 8075 and 8091, with pilots Flt Lt Clemson and FSL Smith, observers Lt Millard and 2Lt Williams, and the necessary RNAS support personnel, *Raven* left Port Said for Suez on 26 July 1916. Samson, and probably Benn, accompanied the expedition, as *Raven*'s log book noted that the Squadron Commander, and three observers had come aboard at Port Said. At Suez an army staff captain, probably from the Department of Military Intelligence, Cairo, came aboard. Sailing from Suez at 03.20 on 28 July *Raven* proceeded to Akaba, at a steady 7 knots, being joined by *Hardinge* during the afternoon of the following day. Both ships anchored off Akaba at 06.00 on 30 July.

Flying for the vertical photographs commenced at 06.45, 30 July, with

The Akaba Mosaic. The original measures approximately 140 × 86 cm (55 × 34 inches). *(TNA, WO 319/1)*

Short 8091, Smith and Millard, followed ten minutes later by 8075, Clemson and Williams. All photographs were taken from a height of 2500 feet, using a 'Standard RNAS Camera' fitted with an 8 inch focal length lens.[225] The photo coverage requested included, in addition to the vertical mosaic, a panorama from sea level. As the available map barely covered the area, prominent points were provided with names such as, Mt Raven, Samson Bluff and Wemyss Ridge. A total of three flights were made, two on 30 July and one on 1 August (8091, Smith and Millard), all were morning flights. The finished photo-mosaic comprises both the vertical mosaic and panorama, it measures 32 × 54 inches (81 cm × 137 cm).[226] The mosaic was later used by the Survey of Egypt to produce a map of Akaba.

The town and vicinity appeared deserted, although a ship's boat examining the beaches at the head of the Gulf was sniped at from the ruins of Umrashash. 'After dark the ships at anchor at the mouth of the Wadi were heavily sniped at from both sides, and one man on board the HMS *Hardinge* was wounded, and with the exception of desultory fire (rifle), nothing else was fired at either the Seaplanes or the ships.'[227]

It was evident that following the visits from *l'escadrille de Port-Saïd* and *Minerva* in 1914/15, the town and its fort had not been reoccupied. Extensive trenches seen in the mosaic and mentioned in the Report were directed at defending against any movement inland from a landing near Akaba itself. The Report continued with observations, and cautions, clearly composed by a military mind, on making a landing at Akaba.

> BEACHES: During the daytime a few men landed without any opposition on the beach immediately south of the town, and at the north-west corner of the Umrashash Post. Several rifle pits were seen near this post. These landings were carried out in order to make quite certain that there were no outlying obstacles on the beach, and that they were quite suitable in all respects. The beaches here are without exception very suitable for landing from boats. The prevailing wind is from the northward, and the water is, therefore, invariably calm.

> AERODROME: It would not be difficult to construct an aerodrome in the bed of the Wadi. For carrying out extended reconnaissances, good climbing aircraft are essential owing to the high mountains which would have to be traversed.

> The country about 200 yds inland from the beach south of Akaba is quite flat and open; it then rises eastward into Wadi Malone and Franklin's Gully; the southernmost palms of the town commence at the northern end of the steep. A landing at Umrashash Post would appear to be accompanied by grave danger of surprise owing to the close proximity of sand hills and scrub. The necessity of providing

an adequate force to clear the sand dunes and scrub, patrol the heights and occupy whatever position is required, and the need for abundant support from ships guns is clearly obvious, but cannot be estimated in detail without further knowledge of the scope of any proposed operations.

Ultimately, it was decided not to make a landing at Akaba. With the Gaza Line still being held, the diversion of British troops was not possible. However, following the capture of Akaba by Arab forces approaching from inland on 6 July 1917, British forces were landed at Akaba. An aerodrome was quickly constructed, located where the Report suggested, and occupied by X Flight, 14 Squadron, RFC.

Leaving Akaba *Raven* and *Hardinge* headed back down the Gulf the turned south to Mowila. At 06.10 on 4 August, Short 8075 with Clemson and Williams set out whilst still seven miles north of Mowila. They flew down the coast over Mowila, then on to Sherm Yahar and Sherm Jubba, a similar distance south of Mowila, before returning. Over Mowila a 65-lb bomb was dropped on a well, one 16-lb bomb causing some mortality amongst the residents of a sheep pen and another just missed the fort. The Short returned undamaged and was hoisted in at 07.15.

Approaching Mowila a steam launch was sent inshore to parley under a white flag. Guy Smith provides a similar but more entertaining account than that in the Official Report.

> When we were about ten yards from the beach we gave a couple of toots on the whistle and several wags of the flag. Immediately there was a stir and a lot of loud and excited talking. A black soldier jumped up on the parapet of the trench, waving his rifle frantically and yelling at the top of his voice. He was answered by our Somalis, who told him to come down to the beach, as we wished to speak to him.
>
> He came, calming his excitement and assuming an air of businesslike dignity. Solemnly he shook hands all around, then explained that he was a slave and also a sergeant in the army and would fetch an officer of sufficient rank to talk to us.
>
> In a few minutes he returned with a Turkish officer, who saluted and then gravely and politely shook hands. "I'm sorry to inform you," said our commanding officer without loss of time, "that I shall have to demand the surrender of the town."
>
> The Turkish officer saluted gravely again and departed to carry the word to the Turkish commander, who sent back the information that he would take the matter under consideration with one of us if we would let our representative be led blindfolded

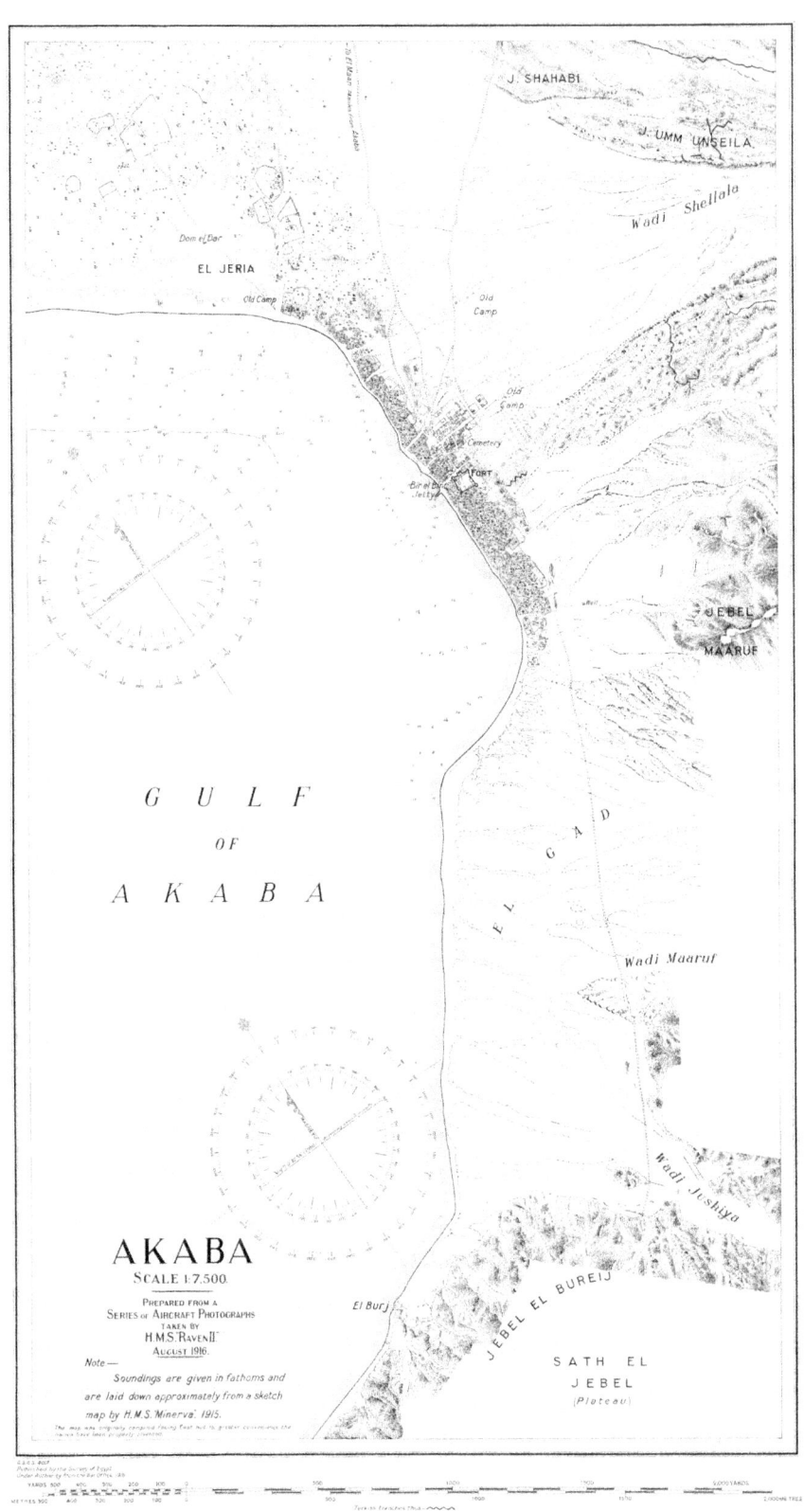

Akaba Map drawn from the mosaic made from *Raven*'s photographic survey. The original was coloured.

Akaba looking up 'Akaba Gully' (Wadi Shellala) with 'Mt Raven' (Jabal Um al Nusaylah, 538 metres / 1765 feet) indicated to left of centre, 30 July 1916. (TNA, AIR 1/2284/209/75/8)

through the defences. Our C.O. refused to permit this, requesting that the Turkish commander come down to the beach.

This brought from the Turk this reply: "I have 250 Turks and 500 Arabs, and I shall not surrender until every man has been killed."

We once again shook hands all around and said good-by, returning to the ship and proceeding down the coast. The Turkish commander probably flattered himself that he had frightened us away, as we made no effort to attack. He was left in peace for a few months, when Mowila was besieged and fell to Faisal's troops.

Arriving at Wejh (Sherm Wej in some reports, modern Al Wajh) the following morning, Smith was sent off with Millard, on Short 8091, for a reconnaissance. They also had a 65-lb and two 16-lb bombs. There was a heavy swell running and Smith had to drop the 65-lb bomb before he could get off.

I tried down-swell and upswell and across-swell, but every time I got the machine hydroplaning well a big wave would rush up and crash into my floats with a terrific bang, taking off all my flying speed. I was so anxiously trying to get the machine off that I hadn't realized how close I had got to shore until bullets from the snipers began to whistle about the plane, some of them dangerously near. It was a tight situation, but the old plane, apparently deciding it was no time to be balky,

Back at Port Said on 1 September 1916 *Raven's* preparations for a return to the Red Sea were brought to a halt by the airmen from *FA300*. Hit by a 10-kg bomb which caused sufficient damage for repairs to be required, and injured eight men. This is the hole made by the bomb in the foredeck. *(YEORN 2014-80-97. Courtesy of The National Museum of the Royal Navy)*

gave a few preliminary bumps and leaped into the air. I shot shoreward, so close by this time that I passed the town a quarter of a mile off and at a height of about four hundred feet.

[I flew] down the coast about ten miles, and then circling back over the town, where we flew round and round like a hawk for about twenty minutes. The enemy kept blazing away at us from the trenches, using black powder mostly, and the fellows on the ship, who could see everything quite plainly through field glasses, said the scene was exactly like a motion picture. All the lines of trenches were brought out clearly by the blue smoke of rifle and machine gun fire. Dropped two bombs [16-lb] on a camp, one falling within 30 yards.

Plenty of bullets whizzed by us, and our machine was hit in a number of places. However, the bullets merely pierced the fabric of the wings, so that no damage was done. The vital parts of the machine were not struck once – the vital parts including the engine, gasoline tanks, and I might also add the pilot and observer. Holes in the wing fabric are quickly repaired with a little patch applied by the airplane mechanics as soon as the machine comes down.

We finally flew back to the ship, landed alongside and were hoisted in. After I had washed up I came back to examine the machine, and was told by a mechanic that I had broken one of the floats a little. Considering the bumping the machine took in my effort to get away from the water, that was a small mishap indeed.

Raven now turned back north for Suez, arriving on the morning of 7 August. The army captain left here, to catch a train for Cairo probably with a set of photographic prints in his possession. In the afternoon *Raven* entered the canal, and reached Port Said the following morning.

Anne's Two Month Visit to the Red Sea

Early on the morning of 1 September, the enemy airmen of *FA300* came close to striking a major blow against the EIESS. Operating from Beersheba, with a refuelling stop at El Arish, three Rumplers, each with three or four 10-kg bombs, attacked Port Said. The intended target was the railway yards, but hits were made on local army camps,[228] and a ship of the EIESS. *Raven* and *Anne* were lying close together in Port Said. The former was taking on stores in preparation for an extended visit to the Red Sea, whilst *Anne* had her hatch covers off loading bombs. At around 06.30 the enemy aeroplanes flew across the harbour, dropping their bombs over the port area. One missed *Anne* by a mere twenty feet. Another struck *Raven* on her foredeck, making a hole in the deck, and wounding eight crew members and native labourers.[229] The damage to *Raven* was not extensive, '1 wooden derrick; steam pipe to No 1 port winch; cylinder of No 1 port winch; large hole in upper deck; 2 Sopwith seaplanes,' but sufficient to put her in dockyard hands for a few days.

Accordingly, all stores and two Shorts, 8004 and 8075, were transferred to *Anne*, which then set off for the Red Sea on a marathon voyage lasting from 1 September to 26 October. Aboard were the two Shorts, with pilots Flt Lt Clemson and FSL Smith with observers Sqn Cdr Malone, Lt Millard, and 2Lt Williams. It should be recalled that at this stage the Arab Revolt was in its infancy. Mecca and Jeddah were in Arab hands and Rabegh and Yenbo had only been captured at the end of July.

Leaving Suez on 2 September, the following morning *Anne* arrived off the settlement of Ras Abu Zenimeh (Abu Zenima), a fortified camp below the escarpment of the Jebel el Tih.[230] Located about halfway down the Gulf of Suez it was, in 1916, an open roadstead with a single pier running out from the camp. The uplands of the Jebel were regularly visited by small Turkish patrols based on Nekhl, and were sending scouts down the wadis towards the camp. As the sea was too rough for flying, in the evening a party landed and 'by the courtesy of Major Ottley, 23rd Sikh Pioneers we accompanied a camel patrol about 10 miles up country in order to inspect the nature of the surrounding country. During this trip the advance guard reported having seen two of the enemy's scouts, but they soon disappeared.'[231]

Anne stopped at Ras Abu Zenimeh between 2-3 September 1916, an isolated spot on the east coast about halfway down the Gulf of Suez. A single flight was made before *Anne* was called away on more important duties. A 'General view of Abu Zenima works & defences.' The pier and 'works' under construction, the railway yards and engine sheds, some accommodation and office buildings, and entrenchments are seen here. *(TNA, AIR 1/1708/204/123/73)*

A 'Typical View of Country' taken from 4000 feet looking to the north with Wadi Hamr, just inland from the coast, in the foreground. Definitely not floatplane country. (TNA, AIR 1/1708/204/123/73)

The following morning, Short 8004, with Flt Lt Clemson and Sqn Cdr Malone, set out at 05.35 to cover the wadis letting down from the Jebel and approaching the camp. They followed Wadi Hamr to its head, flying at just over 4000 feet the escarpment was at least another 500 feet above them, so turning northwest they skirted the Jebel to arrive at the head of Wadi Gharandal which they followed down to the coast. Crossing out to sea. the coast was followed back to Ras Abu Zenimeh, landing at 07.19. The country flown over was most unwelcoming to floatplanes; 'It consists of sharp and undulating sandstone intersected by deep wadis in gorges of one to two hundred feet sheer, and opening out in places to broad extents of sand.' Malone thought that, 'Two or three reliable land planes,' would be of more use in the future.

Any plans for further investigations were ended by an order from Captain Boyle to proceed immediately to Ras Malak, south of Rabegh, to make demonstrations to encourage newly formed Sharifian forces.

> *Anne* was [to be] constantly at work along the Hejaz coast. Spotting for naval guns, bombing and reconnaissance were all carried out as the occasion demanded. Frequent exhibitions of airmanship were given with the sole object of impressing the natives. Machines flew at low altitudes—it was difficult in any case to rise to higher ones—over the towns with their curious mud-huts and their massive though futile medieval-looking square forts surmounted by crenelated battlements.[232]

When considering this and future accounts of work in the Red Sea, the following admonition by Captain Boyle should be noted. Especially as the Admiralty Charts available dated from 1873.

In considering the Arabian Red Sea littoral as a sphere for naval operations, it should be borne in mind that the coast is of coral formation and studded with outlying reefs. The so-called harbours are mostly inlets between reefs and are only approachable at certain times of day. Navigation is difficult and dangerous. By the middle of 1917 a considerable amount of surveying had been done and some buoys and beacons placed, which made conditions easier, but they never became comfortable.[233]

Anne, unfortunately, experienced problems with the Red Sea charts on 10 September. Whilst approaching Sherm Yenbo harbour, ran on the coral. She was able to pull off, undamaged, when the tide (less than a metre, but enough) rose a few hours later. There is nothing marked on the Admiralty Chart, but the Red Sea Pilot mentions, 'Danger — There is a rocky patch 6 miles SW by S from the entrance of Sherm Yenbo, and SSE 3 miles from it is another patch, both of which must be avoided.' Possibly this is the coral *Anne* encountered.

Between 7 and 16 September, *Anne* operated up the coast moving from Rabegh, to Yenbo, Lejh, and finally Wejh. The flights were mostly routine reconnaissances until Wejh was reached on 13 September. Here she settled in, anchored in the lee of Marduna Island a few miles south of Wejh, for a series of reconnaissance flights and to spot for *Fox* and *Hardinge*. The intent appears to have been to land an Arab force of 800 in an attempt to capture Wejh. However, when *Hardinge* arrived at Wejh she only had three sheikhs and

The Red Sea was poorly charted, the then current Admiralty Charts dated from 1873, as *Anne* discovered on 10 September 1916 approaching Sherm Yenbo. A slight tidal rise helped her off the coral, undamaged, a few hours later. *(EMK)*

Wejh from one of *Anne*'s Shorts mid-September 1916. The tower on the headland no longer stands, and modern Wejh has grown behind the old town seen here. *(EMK)*

39 Arabs aboard, and the landing was abandoned. Nonetheless, over the course of five flights, *Anne*'s Shorts dropped 'twenty-six 16-lb bombs on military targets and 400 rounds fired from the Lewis gun at enemy troops and camels.' A final flight on 16 September added six more bombs to the total. They had also spotted an unknown number of 4.7-inch shells, including shrapnel, on to targets in and around Wejh. The main Turkish fort (Al Zareeb Castle), however, was some 10 kilometres inland and was only damaged by one or two 16-lb bombs.[234]

Anne was now ordered to proceed south to Hasani Island, mid-way between Wejh and Yenbo, to transfer a Short to RIMS *Northbrook*, then to proceed to Suez for coal and provisions. Meeting with *Northbrook* on 18 September, one of the Shorts with Flt Lt Clemson and Lt Millard, and ten ratings, were detached. Clemson and Millard made one flight from *Northbrook* over Yenbo 'to impress Arabs' on 21 September. They were then transferred to *Dufferin*, which carried them to Rabegh. A flight was made on 23 September 'to reconnoitre the country to the northward for a suitable aeroplane site, and to fly over the Sharifian Camps in order to impress Arabs. After a brief visit to Suez, *Anne* arrived at Rabegh on 26 September and retrieved the detachment.

Over the next month *Anne* only recorded two flights, on 30 September and 10 October. For the first flight *Anne* had sailed from Rabegh the day before to be off Ras Duleïdela, an inlet north of Rabegh, where the flight was commence. At 05.14 Short 8075, piloted by FSL Smith with 2Lt Williams observer, was hoisted out to 'find and reconnoitre the road which was reported to run from Medina to Rabegh.' They found the road eleven miles inland, where it was descending to the coastal plain, then running towards Rabegh. It was initially 8 miles inland but, further south, came within 4 miles – putting it within the range of naval guns. The tracks appeared to be 'clear of loose stones owing to the action of camel's feet, and are of hard surface. They are quite suitable for motor cars or cycles.' The Short returned and was hoisted in at 06.39. *Anne* now returned to Rabegh.

Before the next flight, *Anne* hosted a visit by General Murray, C-in-C Mediterranean Expeditionary Force, with 'a military commission including a representative of the RFC from Egypt.' The party had arrived in Rabegh on 26 September on *Hardinge* and were greeted by a 21 gun salute from *Fox*.

Malone reported that *Anne*, 'was sent with the representative of the RFC to show him the proposed landing grounds which had been selected by us at Rabegh, Duleidela, and Bureika. Bureika was considered most suitable for the advanced landing ground between Rabegh and Medina.' On 5 October, Malone accompanied Captain Boyle and Colonel Parker (who was in charge of the military commission) aboard *Dufferin* to reinspect Bureika and the sea approaches to the anchorage.

On 10 October, Flt Lt Clemson and Lt Millard set out on Short 8004 at 06.25, landing an hour later, to observe the environs of Rabegh and report on 'exactly what trenches, etc, had been made by the Sharifian troops.' An in person inspection was not permitted as Arab authorities did not permit any non-muslim on the holy land of Mecca. Which was the reason the RFC Detachment, which departed from Suez of 14 October, were not permitted to land at Rabegh, and had to return to Suez.[235]

A 16-lb bomb exploding adjacent to one the towers of Al-Zareeb Castle inland from Wejh. Dropped by Flt Lt Clemson and Lt V. Millard from Short 8004 on 14 September 1916. The castle has recently been restored as a tourist attraction. *(TNA, AIR 1/1708/204/123/73)*

The visiting party left Rabegh on 12 October. *Anne* sailed for Port Sudan two days later to coal and reprovision, also to bring 29 Egyptian gunners as reinforcements for the mountain gun batteries she had delivered to Jeddah in June. Initially ordered to Yenbo, General Murray insisted she return to Rabegh in case a Turkish attack developed. She was still there on 26 October when *Raven* arrived, with two pilots, two observers, but no machines. The two Shorts, with Sqn Cdr Malone, two PO's and 11 ratings, were transferred to *Raven* and *Anne* sailed 'for Suez and a long delayed refit. *Raven*'s stay at Rabegh was brief, she sailed for Suez on the afternoon of 30 October, and no flights were made. Which is perhaps as well, as Malone's report on *Anne*'s activities concludes, 'In any case the seaplanes now here are unfit for further service.' But *Raven* would return in December.

This was also Sqn Cdr Malone's final cruise with the EIESS. On return to Port Said he was shortly ordered to return to Britain to take up a post in the Air Ministry, with the rank of Wing Commander. For services the Government of Egypt awarded him the Order of the Nile. In May 1917 he was appointed to the battlecruiser *Lion*, apparently in his substantive RN rank of Lieutenant. Restored to his RNAS rank he was posted to the Plans Division of the Admiralty on 28 December 1917. He became Lieutenant Colonel, RAF, on 1 April 1918. In August he was appointed as the first Air Attaché to Paris, and Air Representative, Supreme War Council, Versailles. Elected as Liberal Member for East Leyton on 14 December 1918 he resigned his commission. Following a visit to Russia in 1919, he joined the British Communist Party, whilst remaining the Member for East Leyton until 1922. His outspoken support for the Soviet Union, and opposition to the Allied Intervention in the Russian Civil War, led to his prosecution and imprisonment, for six months, under the Defence of the Realm Act in 1920. His OBE, that had been awarded in 1919, was cancelled at the same time, and his name removed from the Admiralty's List of Officers. He returned to Parliament, this time as a Labour Member representing Northampton between 1928 and 1931, finishing his career as Parliamentary Private Secretary to the Minister of Pensions. During the Second World War, Malone was re-employed serving at the Admiralty Small Vessels Pool, 1943–45. Cecil John L'Estrange Malone died on 25 February 1965.

Raven at Yenbo and Wejh, December 1916

By the end of November both 8004 and 8075 had been prepared for further service and joined *Raven* at Port Said. With them were Flt Lt Burling and FSL's Man and Worrall, with observers Lt Stewart and 2Lt Williams, three PO's and 11 ratings. Passing through the canal, *Raven* left Suez for Yenbo on 1 December.

At about the same time, Fakhri Pasha mustered the maximum strength he could spare from Medina where he had two or three infantry battalions, a camel corp regiment plus some Arab cavalry, and artillery, and advanced on Yenbo.[236] They may have had limited support from *3ncü Ty Bol*, but this cannot be confirmed. Opposing his advance were two Arab forces, the first under Emir Ali was based at Rabegh with an advance guard under Emir Zeid closer to Yenbo. The second force was under Emir Faisal based outside Yenbo. In Yenbo itself were some Egyptian troops and, critically, several ships of the Royal Navy's Red Sea Patrol. Captain Boyle was then in command of *Suva* (*Fox* was being refitted at Bombay), *Dufferin*, the small monitor *M.31*, and *Raven*. The *Suva* and *Dufferin* were both armed with 4.7-inch guns and between them could present a 'broadside' of five guns, *M.31* was armed with two 6-inch guns. *Raven* had arrived on 3 December and anchored a few miles up the coast at Sherm

Aerial view of *Raven* at the anchorage of Sherm Yenbo, 4–16 December 1916. Showing one of the canvas shelters erected to protect the machines on the aft (and occasionally forward) hatch covers. *(YEORN 2014-80-144. Courtesy of The National Museum of the Royal Navy)*

Yenbo, a better anchorage than that offered by the town, the other ships were at Yenbo itself. The Admiralty chart shows the town surrounded by a wall and towers, however these were in ruinous condition and useless for defence.

Fakhri Pasha, advancing down the road from Medina to Rabegh, brushed aside Emir Zeid at the point where the road branched to Yenbo. He now controlled roads to both Yenbo and Rabegh. Emir Faisal's force, estimated at 4000 tribesmen, was based on the village of Nakhl Mubarak on another approach road north of the main road. At this point Fakhri Pasha appears to have divided his force, one part to continue down the road to Yenbo whilst holding off any attempt at relief from Emir Ali, the second moving across country to engage Faisal's force. On 9 December, the Arabs were driven in from Nakhl Mubarak. Both roads to Yenbo were open to the Turkish force. However, the defences at Yenbo, commanded by Lt Herbert Garland,[237] were being strengthened and had the support of the guns of the Royal Navy. Raven's machines were active daily, initially on reconnaissance, then bombing the Turkish force.

Short 8075 being hoisted on *Raven*, December 1916 probably at Sherm Yenbo. The Short was a regular on *Raven* sailing on six cruises. It was also had one of the longest operational careers of any Short with the EIESS, from July 1916 to June 1917. *(Kerry Family Collection, JLK-09)*

Between 8 and 16 December, *Raven*'s two Shorts had made nine flights, mostly just one each morning. On the 8th and 9th single reconnaissance flights were made. Each flight landing at Yenbo, to be briefed aboard *Suva* then debriefed after the flight, before returning to *Raven*. Flt Lt Burling and Lt Stewart made the first flight on 8004, leaving *Raven* at 06.20, returning at 08.45. They spent about 30 minutes aboard *Suva* before and after the reconnaissance, reporting, 'No signs of troops or men near Well or Dome Shaped Rock.' The location of the Dome Shaped Rock is given as Jebel Araur, some 15 miles due east of Yenbo.

The following day FSL Worrall and 2Lt Williams were sent to reconnoitre the road running NE from Yenbo. They had set out from *Raven* on 8075 at 06.05, but on leaving *Suva* a, 'Compression tap blew out on getting off, so wirelessed to ship that seaplane was returning.' Quick work by the mechanics at *Raven* replaced the tap and they were able to leave on their reconnaissance just five minutes after landing back at *Raven*. The NE road would take

Suva, in the foreground, and *Dufferin* probably at Yenbo in December 1916. *Suva* was an armed boarding vessel (three 4.7-inch guns) in the Royal Navy, previously used by the Australasian United Steam Navigation Company between Australia and Fiji. *Dufferin* was built for the Royal Indian Marine in 1904, an armed troopship she carried six 4.7-inch guns. *(EMK)*

them towards Nakhl Mubarak and Emir Faisal's force. They observed, 'About 100 Camels just outside Yenbo. Quite a large amount of traffic observed on road:—Civilians, Arab troops, Camels, and one gun, mostly proceeding to Yenbo.' These would have been troops retreating from the Turkish attack on Nakhl Mubarak. After reporting their findings onboard *Suva*, they returned to *Raven* at 07.50.

Over the following two days, just three bombing raids were flown, all to attack Turkish troops in the vicinity of Nakhl Mubarak. A total of 22 16-lb bombs were dropped, the morale effect of which was probably greater than any damage caused. On the 10th both Shorts were employed, in each case landing at Yenbo before and after the raid.

First to leave *Raven* was 8004, with FSL Man and Lt Stewart, they were ordered to carry out a reconnaissance of the foothills before bombing Nakhl. They flew out to Jebel Araur, then followed a wadi north, no troops were observed until they reached Nakhl. Here they discovered a camp NE of village and palm groves, comprising 12 large rectangular tents, one bell tent and several bivouacs. There were also some trenches. The bombs were dropped across the camp, demolishing one tent. They were fired on by two guns using HE & Shrapnel and by rifles and a single machine gun, being hit in five places by rifle fire. Returning to Yenbo to report, no further troops were observed. The Short had left *Raven* at 05.58 and returned at 09.15.

Five minutes after 8004 had returned, 8075, Flt Lt Burling and 2Lt Williams, set off for Yenbo then Nakhl. Heading straight for Nakhl they observed an advance guard of 80 camels and men five miles from the village. The main force was estimated to be 350 men on camels and horses, with two small mountain guns, at the camp reported by Man and Stewart. The Short was received with heavy rifle fire from men within the village and trenches. Six bombs were dropped, 'two being direct hits and the other four falling in close proximity.'

On 11 December 8075, FSL Worrall and with 2Lt Williams, left *Raven* at 05.49 landing at Yenbo twelve minutes later, then proceeding at 06.43. Yesterday's advance guard had apparently returned to Nakhl, and the two mountain guns had been emplaced together on a hill 2 miles SW of the village. The tents and bivouacs had been removed and the village occupied. Camels and horses being picketed under palm trees.

> Eight 16 lb bombs were dropped with what appeared to be good results, but bomb dropping was difficult owing to the scattered formation of the enemy. Eighteen to twenty rounds of shrapnel were fired at the machine, one making a large hole in the left float, and the remainder being very close to the machine. The seaplane was also under heavy rifle fire.

With the damaged float, Worrall returned directly to *Raven* and was quickly hoisted aboard by 11.31. Several days were required to repair the damage and 8075 was not flown again until 19 December.

The Turkish force was in position to attack Yenbo on the night of 11/12 December.

> ... old Dakhil Allah told me he had guided the Turks down to rush Yenbo in the dark that they might stamp out Faisal's army once for all; but their hearts had failed them at the silence and the blaze of lighted ships from end to end of the harbour, with the eerie beams of the searchlights revealing the bleakness of the glacis they would have to cross.[238]

Fakhri Pasha called off the advance on Yenbo, moving instead towards Rabegh. He advanced to within 48 kilometres of the town and stopped.

Fakhri Pasha inspecting troops at the railway station in Medina. *(LoC)*

Emir Faisal entering Yenbo in December 1916. Faisal is at the head of his mounted men, dressed in a dark robe and keffiyeh with white agal on the horse with white blaze.

Facing insurmountable supply problems, and with his rear being harassed by Arab attacks, on 18 January he began withdrawing to Medina. This was the final serious challenge to the Arab Revolt in the Hejaz.[239]

On 12 December, Flt Lt Burling and Lt Steward took 8004 on a reconnaissance around Yenbo to check on any movements by the Turkish force. After reporting the Boyle they again took off this time to test W/T communication with the monitor *M.31* and *Raven*, contact being established with both ships. The next flight was on the 14th when FSL Man and 2Lt Williams took the Short to Nakhl to 'verify Arab report of the presence of a large enemy force there.' The village and surrounding area were thoroughly searched. 'Force previously reported still at Mubarak, but no signs of tracks etc. to indicate any new arrivals. The 10,000 troops reported by the agent are certainly not there. Bombs dropped in village obtaining 2 hits on houses. Turned back—3 rounds of shrapnel fired at machine, from guns in the same position as previously reported.' Flights made on the two following days were similar, spotting practice with *M.31* and local reconnaissances. No Turkish forces were located within seven miles of Yenbo, Fakhri Pasha having commenced his withdrawal a few days earlier. The small force at Nakhl Mubarak being a rear guard.

Yenbo being considered to be firmly in Arab hands, *Raven* moved up the coast. With Lejh already being held by Sharifian forces, the next port Wejh was still lightly held by the Turks. *Anne*'s Shorts had last visited Wejh in September, and the purpose of this visit from *Raven* was to update information and to rattle the morale of the garrison.

Raven sailed from Sherm Yenbo during the afternoon of 17 December, nominally bound for Suez. Early on the morning of 19 December *Raven*, in company with *Dufferin*, was proceeding slowly past Riakha Island, a few miles off Wejh. Coming to a stop, between 06.30 and 11.00 she sent off six flights, three by each Short, to make reconnaissances and bomb Wejh and the local fort. Wejh has two forts, one within the village proper known as Al-Souk Fort, a customs fort built in the mid-19th Century, and a second some 10 kilometres outside up against the foothills, on a plateau in Al-Zareeb valley. Al-Zareeb Castle was built was built around 1620 on the pilgrim road to Mecca. It is built from local limestone, rectangular with a round tower at each corner. This second fort was the main target of the first three flights.

DETAILS OF BOMBING FLIGHTS OVER WEJH, 19 DECEMBER 1916

Short	Crew	Objective	Bombs Carried
8075	Flt Lt Burling and 2Lt Williams	Bomb inland fort 6 miles E of Wejh.	1 65-lb, 6 16-lb
8004	FSL Man and Lt Stewart	Bomb inland fort 6 miles E of Wejh.	2 65-lb, 8 16-lb
8075	FSL Worrall and 2Lt Williams	Bomb inland fort 6 miles E of Wejh.	1 65-lb, 10 16-lb
8004	Flt Lt Burling and Lt Stewart	Bomb entrenched camp 1 mile N of Wejh.	2 65-lb, 8 16-lb
8075	FSL Man and 2Lt Williams	Bomb entrenched camp 1 mile N of Wejh.	1 65-lb, 10 16-lb
8004	FSL Worrall and Lt Stewart	Bomb entrenched camp 1 mile N of Wejh.	2 65-lb, 8 16-lb

Raven's Report was very much to the point.

> 3–65 lb & 28–16 lb Bombs were dropped in all.
>
> 2–65 lb & 18–16 lb Bombs were dropped on the Fort.
>
> Hits were obtained with 1–65 lb & 2–16 lb Bombs, most of the others bursting just outside.
>
> 1–65 lb & 10–16 lb Bombs were dropped on the Camp, considerable damage being done.
>
> The seaplanes were heavily fired at over the Camp, but not when over the Fort.[240]

Only Man and Stewart's report contains any details of the attack on the Fort. Observations are included. '4–5 miles East of Wejh the wadi branches. Fort is on South Bank of North Branch. Around Fort are Tombs, Ruins, and Wells.' They observed a single Bell tent on the north side of the fort and 30

The original caption says it all. 'No.8004 seaplane HMS *Raven II* after a flight, during our stunt down the Red Sea, Arabia, Dec 1916.' (YEORN 2014-80-171. Courtesy of The National Museum of the Royal Navy)

Bedouin shelters to the north and west. From one of the shelters 'a white robed Arab waved a white flag to plane.' Very few troops and only 15 camels could be seen.

The entrenched camp was sheltered behind a bluff north of the village. It comprised about twelve huts or shelters and twenty tents, and was estimated to hold about 250 troops with another 150 in trenches adjacent to the camp. The trenches had increased since *Anne*'s visit in September. They were well aligned to resist a sea landing, with communication trenches leading to sheltered ground behind the bluff. No guns or emplacements were observed. Worrall and Stewart reported, 'The plane at a height of 1000 ft. was under severe rifle and M.G. fire from the Camp & Trenches—the planes, fuselage and floats were hit in many places. Some of the holes were made by bullets of large calibre (.455 approximately).'

Having stirred up the garrison *Raven* raised steam for full speed and sailed away for Suez at 11.38, arriving on 22 December.[241] At Port Said the following day all hopes for a Christmas in port were shattered, as described in the previous chapter.

Ships of the East Indies and Egypt Seaplane Squadron sailed the Red Sea, and beyond, on three further occasions, twice in 1917 and once in 1918. As will be related later.

Preparing bombs aboard *Raven* in December 1916, armourer's humour appears ageless. Flt Lt E.J.P. Burling (left) and FSL W. Man are steadying a HERL Light Case 16-lb bomb on a HERL 65-lb bomb. There are over thirty 16-lb and at least three 65-lb bombs to be seen, apparently intended for Wejh, if the writing on the bomb behind Burling is to be believed, dating this photograph to 18/19 December 1916. (EMK)

254

CHAPTER 11

Castellorizo

Let us first sort out the name of the island. Castellórizo or Kastellorízo, Château Rouge, Castelrosso, Meyísti or Megísti (Μεγίστη), Meis or Kızılhisar; take your pick as the island has had many names with infinitely variable spelling. Except when quoting directly, I will use Castellorizo throughout as, with one 'l' or two, it is the most common spelling in the British records.

A Brief History of Castellorizo

The island's story stretches back into pre-history. It has been fought over and occupied in turn by Hellenistic Greece, Rome, Byzantium, the Knights of St John from Rhodes, Egyptians, Venice and the Ottoman Empire. During the Greek War of Independence (1821–1832) the island was evacuated. Once restored to Ottoman control the population, of largely Greek extraction, quickly returned. Benefiting from a favourable tax situation the island's merchants created a profitable entrepôt trade in wine, olive oil and timber. Following the Young Turk Revolution of 1908, when the tax breaks were cancelled, the island's trade and population plummeted. Within four years Castellorizo's population had halved, largely due to emigration. During the Second Balkan War of 1913 the islanders sought union with Greece. However, Greece was able to provide little assistance merely sending to the island a series of unpopular and ineffective governors. By late 1915 trade was at a standstill,

A panorama of the harbour in 1905. The island's population at this time was around 10000 and, as witnessed by the number of ships in the harbour, it was at its peak of trading power. Within a few years trade and population markedly decreased, such that by 1914 there were less than half that number living on the island. *(Nicholas Pappas Collection)*

other than smuggling which the current governor attempted to suppress, and supplies running low. The islanders rose against the governor, hustled him off the island and established an emergency committee to determine their next move. Into this power vacuum the French moved. On 28 December 1915 they occupied the island.[242]

What made this small island so attractive to the French was its location. Positioned some 125 kilometres east of Rhodes and an equal distance south-west of Antalya, the major town on the Anatolian coast, the island was ideally located for intelligence gathering, it was also viewed as an ideal base for anti-submarine operations and a haven for merchant ships in an emergency. Before the war the islanders maintained farms and pastures on the mainland and adjacent islands, and they still had friends and business contacts there. Their local knowledge was invaluable and the basis for local intelligence network. The islanders also formed a local defence militia, which was later employed to raid the Turkish coast.

Known as *la pince du homard* to its occupiers (a glance at a map will show its similarity to a lobster claw) it is just 6 km by 3.5 km, largely infertile, arid and mountainous with a single town clustered around a small sheltered natural harbour in the open claw of the lobster. There is no source of water on the island, which relied on rain water catchment and water shipped in from the mainland. The harbour looks onto the Turkish coast, a headland named Gata Burnu, some 4000 metres to the north-east. Behind the headland lies the small town of Antiphilo (Kaş). The island is the largest of an archipelago of over twenty islands and islets, after Castellorizo itself the two largest are Ro to the west and Strongyli to the south-east. A number of islets fill the channel between Castellorizo and the mainland, so that the main steamer channel passes between Ro and the main island, around its northern tip and down into the small harbour.

Lieutenant de Vaisseau Henry Marie Joseph Octave de Bourdoncle de Saint-Salvy.

The new governor, *Lieutenant de Vaisseau* Henry Marie Joseph Octave de Bourdoncle de Saint-Salvy,[243] had barely 150 officers and men, including administrative and medical staff, to defend the island. Landed from the cruisers *Amiral Charner* and *Jeanne d'Arc* during the occupation, they were provided with two 65mm TR (*tir rapide* — quick firing) naval guns, two 65mm *canon de débarquemet*,[244] and two machine guns. Both types of 65mm guns fired shells approximately 4-kg in weight, and were capable of covering the adjacent Turkish coast. Anticipating a Turkish attempt to make a landing on the island, de Saint-Salvy installed one of the 65mm TR on the remains of the NW tower of the castle and the second at Point Niftis on the NE tip of the island. Together they covered the harbour and the probable approach routes of any attempted landing. One of the field guns and a machine gun were also posted at the castle, with the remaining pair at the *agora* in front of the Cathedral of *Agios Konstandinos kai Eleni* at the high point of the town.[245]

At the beginning of the occupation the main harbour contained approximately 45 trading brigs, schooners and smaller vessels belonging to the islanders. As time passed many of these sailed away to safer harbours. The French installed a torpedo net across the opening, behind which was a single mooring suitable for large ships which entered the harbour stern first. The main occupants of this mooring were French warships and British seaplane carriers. Naval command of the Eastern Mediterranean was vested in the French, at this time *Contre-Amiral* Henri de Spitz, *Commandante Division Navale de Syrie*. The majority of his ships were older French torpedo-boats and destroyers and numerous armed trawlers and tugs. Spitz himself maintained a hands on control of activities from his armed yacht, *Ariane II*, previously owned by the chocolate magnate Gaston Menier. These smaller vessels tied up with their sterns to the harbour wall.

Castellorizo and the Seaplane Carriers

The first seaplane carrier to visit the island was, appropriately, the French *Campinas* on 23 March 1916 *(see Chapter 6)*. Next to visit was *Anne*, a month later. Sailing from Port Said on 19 April, she remained at the island between 21–25 April 1916.[246] On the hatch covers of the aft well deck were two Shorts, 8004 and 8054. These were maintained by a small RNAS detachment with Flt Lt M.E.A. Wright in command and sole pilot, with Lt T.V. Hughes and 2Lt A.K. Smith as observers.

Anne in the harbour 21–25 April 1916. There was only space for a single large ship in the harbour, whilst the island's fleet of schooners and brigs occupies two sides of it, about sixteen are visible in this photograph. The torpedo net can be seen stretched across the opening of the harbour from the last white house on the left to a point beyond the minaret. The wooden bulkhead port side aft well deck from her days with *l'escadrille de Port-Saïd* is still in place. *(EMK)*

Somewhere at sea bound for Castellorizo, two Shorts aboard HMS *Anne*. The furthest aft is 8004, the one in the foreground 8054. There is a solid wooden bulkhead along the port side as part of a shelter for the floatplanes the remainder of which was removable canvas. A folded Short was too long to fit across the aft deck, the existing cut out in the well deck gunwales enabled it to be stowed with its floats protruding outboard. *(EMK)*

On 8 February 1916 *U-21* (*Kapitanleutnant* Otto Hersing) had sunk the old cruiser *Amiral Charner* off the Syrian coast near Beirut. There was only one survivor. This event made the allies overly conscious of the threat of submarines, which began being seen everywhere. In fact, at this time, only Hersing's submarine based at Constantinople was available for service in the Eastern Mediterranean. However, at Castellorizo, de Saint-Salvy received intelligence that the small harbours at Makri (Fethiye) and Marmaris were being used to refuel submarines. When Weldon went ashore he was requested by de Saint-Salvy to send flights to examine the coast to the north-west as far as Makri. Then, using the island of Tersana (Tersane Adası) as a temporary base, make further flights as far as Marmaris.

Under the command of Lt Hughes a small party of RNAS mechanics, with cans of petrol, a few spares and some bombs, joined the French armed trawler *Nord Caper* and set out for Tersana island some 96 km away. They were landed overnight (21/22 April) and settled down to await the Short from *Anne*. In some ways Tersana island is a little Castellorizo. It too has a small sheltered harbour facing the Turkish mainland; it too has a history dating back to ancient Greece; it too had a very small basically Greek population. Its chief was an old rascal, pirate and bandit named Marko, of whom more later. Suffice for now that he was on good terms with the French, and welcomed the visits of the RNAS floatplanes.

Piloted by Flt Lt Wright with Lt Smith as observer, Short 8054, left *Anne* at 06.32, 22 April. The Short was loaded with three 16-lb bombs, and Smith was provided with a 'magazine rifle' (probably the same type of carbine used when the French were aboard). Flying just off the coast they arrived over Makri about an hour later. After carrying out a brief reconnaissance of the area, including Levisi (Kayaköy), they headed across the bay to land at

Tersana at 08.12. The weather deteriorated and another flight was not possible until the following afternoon. Wright and Smith made an hour long flight around the bay, in poor conditions, checking for indications of submarines and dropping two bombs on some tents outside Makri. After both flights, they reported a 'blue oily film' on the water at Makri, but this could have been due to many causes.

Commencing at 13.09 on 24 April, in improved weather, Wright and Hughes set off further west along the coast to Marmaris. They searched the deeply indented coast line for some 50 km to the north-west, including Ekincik, Karagach and Marmaris for signs of submarines, and also looked for French mines in the passage to Marmaris. There was something at Karagach that was worthy of one of their bombs, the two remaining bombs were dropped on some boats beached near Marmaris. The Short returned to Tersana shortly after 15.00, refuelled and flew on to Castellorizo arriving at 19.00. *Nord Caper* and the RNAS detachment returned overnight.

Before sailing on the 25th Capt Weldon was requested by Saint-Salvy to transport forty-five Turkish prisoners to Port Said. Earlier in January the local militia had raided Antiphilo, bringing back the entire Turkish garrison of 50 men, another raid from Kekova island brought in more prisoners. Since most food and water for the island now had to be brought in from Rhodes, reducing the population by 45 was of great advantage. *Anne*, and her unexpected passengers, arrived back at Port Said on 26 April.

One of the photographs taken on 23 April 1916 during the expedition to Tersana island. A view to the south east of coastal Makri town showing the tortuous track up to the plateau above. The arrow on the original print points to a military camp. The town of Levisi, located on the plateau, can be made out through the haze at the upper left. The finger print is a comment on the makeshift photographic arrangements on the seaplane carriers. *(TNA, AIR 1/1708/204/123/73)*

At anchor in Castellorizo, Short 8054 is hoisted out from *Anne*. The gap in the gunwale is clearly seen in this image. The range of headwear from Victorian era, RN pattern, straw hats, through naval and civilian caps is interesting. The French delivered an RNAS party, with cans of petrol, a few spares and some bombs, to the island of Tersana in Makri Bay some 40 miles along the coast toward Rhodes. 8054 operated for several days (22–24 April) from the island. *(EMK)*

The next EIESS seaplane carrier to visit the island was *Raven* between 6–8 July 1916.[247] *Raven* had sailed from Port Said at 06.30 on 1 July, in company with the French armed tug *Laborieux* as escort. As previously related (Chapter 8) they first proceeded up the Palestine coast flights were made over El Arish then Haifa, before taking a load of coal to Cyprus for *Ben-my-Chree*. The ship then set out for Castellorizo, arriving at daybreak on 6 July. As had Weldon before him, Flt Lt Dacre went ashore to discuss the local situation with de Saint-Salvy. Once again he was asked to search the area around Makri, using Tersana as an advance base. This time *Laborieux* was detailed to carry the RNAS party to the island.

Whilst preparations for the Tersana expedition were proceeding, two local flights were made. At 16.15 Brooke and 2Lt E. King set out on Short 8090 to carry out an armed reconnaissance of the adjacent Turkish coast. They first headed east to the local headland of Tugh Burnu then reversed their course over Antiphilo and along the coast to Kalamaki and Volos Island, searching for signs of military activity. They came under rifle fire from two small patrols near Kalamaki, prior to which an overheating engine had forced Brooke to drop their single 112-lb bomb into the sea. No suitable targets were found for the two remaining 16-lb bombs. They landed just outside of the harbour of Castellorizo at 17.47 and were hoisted in twenty minutes later. In the evening, Dacre took out Baby 8189 armed with four 16-lb bombs. Flying east he 'saw very little but mountains, but dropped my bombs on an outpost which caused a scrub fire. This fire spread rapidly and by night lit up the places for miles. A fine sight indeed and very impressive for the natives of Castelorizo.'

Laborieux sailed at 23.30, to establish the base at Tersana Island. The tug carried two officers (Lt Fill (stores) and observer Lt Brown), seven mechanics, a supply of petrol and oil, a selection of engine spares and tools. Offensive ordnance included five 112-lb and 20 16-lb bombs, and a supply of fully loaded hoppers for a Lewis gun.

Short 8090, with Flt Lt Brooke and 2Lt King, remained aboard *Raven* to be available to de Saint-Salvy should need arise. There are no reports of any flights being made whilst the Tersana detachment was absent. The adventures of the Tersana detachment are told by Dacre from his report and diary.

Dacre with Lt Ravenscroft as observer on Short 8091 left Castellorizo at 05.15 on 7 July. The Short was loaded with five 16-lb bombs, a Lewis gun with four hoppers, and a camera. They were to fly along the coast and carry out a reconnaissance on the way to Tersana island. About twenty minutes after starting they saw FSL Man, flying Baby 8189, pass over them on a direct course to the island. Dacre's flight was not uneventful.

> The engine boiled very quickly & I could only get to 500 ft mountains 3000 ft high went right down to the shore so I dropped 3 of my bombs into the sea to lighten the machine & flying on got to 1800 ft over Makri district where important recon was to take place. We took photos dropped bombs & got a few rifles fired at us in return, then on & landed at Tersana [at 06.45] where I taxied into the creek & ran aground before reaching the shore. The astonished natives looked

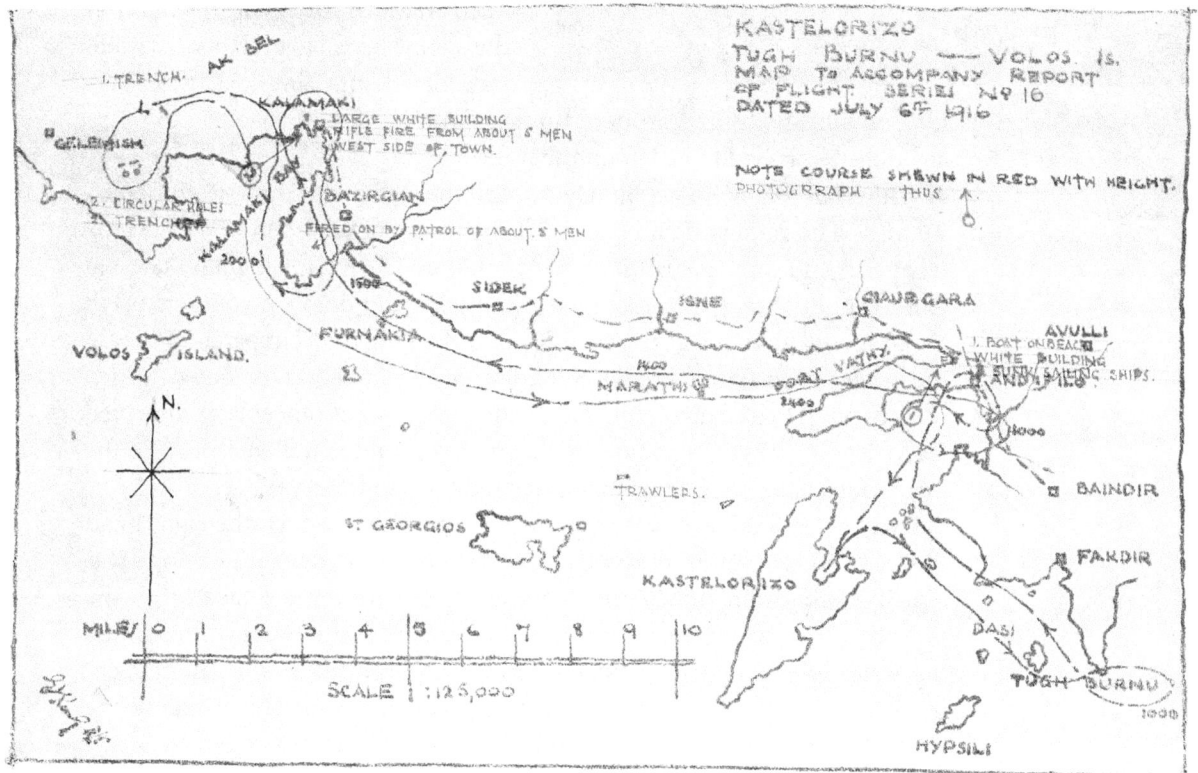

Sketch map of the route flown by Short 8090, Flt Lt J.C. Brooke and 2Lt E. King, on 6 July 1916. The map bears the signature of 2Lt E. King the observer. The photographs mentioned have not survived. *(TNA, AIR 1/1708/204/123/64)*

on in amazement & offered coffee & more got in the way than any thing in their efforts to assist. The Schneider had not turned up & I was beginning to wonder what had happened to Man again when he arrived having struck the wrong Is & found out his mistake. Soon after the French armed Tug *Laborieux* arrived & we enjoyed a breakfast.

We visited the 'King & Queen of Tersana', at his palace. He offered us liqueurs & Turkish Delight which was quite nice. The Queen was about his 15th wife, himself been a desperate pirate who delighted to tell us through out interpreter how he goes out before breakfast & stalks Turks, cutting their throats & heaving them over the cliffs.

Unfortunately for Marko his morning exercise would soon be cut short. He was ambushed and killed on 17 July.

When running aground, the right hand float of the Short was damaged, by the afternoon it had taken on water, requiring two hours of pumping to empty it out. At 17.10 Man taxied out with Lt J.W. Brown as observer, the Short was also carrying three 16-lb bombs, intending to carry out a reconnaissance around the coast of Makri Bay. Before leaving the harbour the engine began to boil then, as water was forced into the float, the Short took on a list with the starboard tip float under water. Man return to shore, again the Short had to be left on the water.

Raven visited the island 6–8 July 1916. Already many of the island's trading fleet have sailed to safer ports. The Turkish coast is seen in the background, *Raven* is pointed towards Antiphilo (Kaş), the slightly darker mass above the ship is the headland of Gata Burnu where Mustafa Ertuğrul will conceal his guns.

With Man unable to get off, Dacre took the Baby with three 16-lb bombs, 'To make a bombing raid on places of importance observed by 8091.' He took off at 17.18 and was back within twenty minutes. During the flight he saw 'No tents, camps, men, guns or submarines.' But he did bomb 'A concrete patch about 120 feet by 80 feet between Kanillo Kavoulli and Drepanaki Point at the water's edge. Three bombs were dropped from 2800 feet, one of which was observed to be a direct hit.' The French intelligence officer later informed Dacre it was 'being worked on by the Austrians and is well guarded.' So the assumption was that it was being prepared for a submarine's use.

Back at Tersana, as it was impossible to beach the Short, the right hand float was hoisted out of the water using the tug's boat davits. The hope was that it would drain sufficiently overnight to permit a rapid take-off in the morning. 'This was unsuccessful and the left hand planes were getting under the water with the right hand float out of the water.' Lacking an efficient pump, or a place to beach the Short, it was decided to partially dismantle the machine and bring it back to Castellorizo loaded on to *Laborieux*.

The Short was a nuisance as it was not possible to get it back by flying, there was only rocky beach, so it couldn't be hauled up & it was dangerous to get the *Raven* up. I either had to burn it, leave it or take it to bits on the water & get it somehow on to the tug. This was the most feasible. So, at daylight, this was done by strengthening the boat davits & hanging the machine as high as possible, then the right hand planes were removed, floated off and hauled across the *Laborieux*'s gunwale. Then removing the floats; got many men on to them to sink them clear of the body. The body & left planes were then trussed up to the tug's side with the tail hauled up out of the water & thus we rolled heavily back to Castelorizo. The wing tip dipped on every roll but beyond one wing tip float being broken no other damage was done.

Whilst the tug was being loaded, Man was sent off on the Baby to return to Castellorizo carrying a letter for de Saint-Salvy. Enroute he dropped three 16-lb bombs on a forested area close to Kalamaki, starting a small fire, then carried on to the island. Landing outside the harbour, at 05.50, Man taxied over the anti-submarine net damaging the tail float. The letter was delivered safely.

Laborieux returned to Castellorizo at 12.45, moored alongside *Raven* and the Short transferred aboard. Dacre presented de Saint-Salvy with a reconnaissance report and several photographs. The governor was pleased with the results, commiserating over the damage to the Short. *Raven* sailed for Port Said at 17.00, 8 July, arriving three days later.

Castellorizo -vs- Ottoman Empire

I mentioned the Castellorizo militia volunteers earlier, their leader was one Ioannis Yeorgiou Lakerdis. An islander, he was something of an *enfant terrible*, and an irredentist Greek nationalist. He previously led a force of volunteers to take Castellorizo from the Ottomans in March 1913. Then clashed with various Greek appointed governors and directly co-operated to bring the French to the island in 1915. Appointed *Chef de la Police de Castellorizo* by the French, he also organized the militia volunteers and led many of their raids. He worked with the French throughout the war in the defence of the island.[248]

The raids against vulnerable points on the mainland and islands were encouraged by the French, and transported by French warships. The first of these was on 27 January 1916, when 46 volunteers under Lakerdis were landed from two *patrouilleurs* (armed trawlers) *Cachalot* and *Surmulet* in a daylight raid on Antiphilo. The militia destroyed telegraph lines and also managed to 'recover' some of the movable property they owned on the mainland. The entire garrison, consisting of a captain and about fifty men, was brought back to Castellorizo. The prisoners were those evacuated by

The 65mm TR naval gun mounted on the site of the NW tower of the medieval Red Castle (château rouge). (Ministère de la Culture, SPA, K4057)

Anne in April. This operation, labelled by the Turks as the sack of Antiphilo (*Antiphilo'nun çuvalı*), was carried out without loss. Other raids, to destroy communications, disrupt supply lines (including some cattle rustling), and gather intelligence, continued for some time. The island of Kekova, which pre-war had been a source of timber and grazing land for the islanders, was occupied from 28 October 1916 to around 20 January 1917.[249] The raids were a constant irritation to the Turks, necessitating the retention of troops who could have been more usefully employed. It was, therefore, inevitable that a response would be mounted.

Towards the end of 1916, rumours began being reported from the mainland that the Turks were preparing an attack on the island. 'The Turks, embarrassed by the presence of our two observatories at Castelorizo and Ruad, undertook to drive us out. From the end of 1916, the emissaries who, at the risk of their lives, constantly commute between the islands and the enemy shore, announced the preparation of attacks which the proximity of the mainland makes particularly easy.'[250] *Contre-Amiral* Henri de Spitz requested another seaplane carrier visit the island to investigate the rumours. The opportunity came early in November.

Ben-my-Chree spent 3 November in Adalia Bay with a French squadron *(see Chapter 9)*. At the end of the day *Ben-my-Chree* and *Dard* proceeded to Castellorizo, in response to *Amiral* Spitz's request, arriving on the morning of 4 November. De Saint-Salvy had no positive information of the whereabouts of the batteries and, after two flights failed to locate them, *Ben-my-Chree* returned to Port Said.

They were too early. The guns did not arrive until the New Year.

In the autumn of 1916 local command was given to *Oberleutnant* Joseph Hesselberger. With headquarters at Antiphilo, he became the leader of the *Sonderkommando Hesselberger*, operating with a small group of Germans and Turkish troops from the 135th Regiment (*135 nci Alay*), commanded by *Binbaşı* Cemal Bey, along the Southern Turkish coast.[251] In a report to Liman von Sanders, Hesselberger wrote, 'Your Excellency gave me the order, to exercise coastal protection in the area of Meis and, specifically, to prevent the repetition of raids into Turkish territory. I believe the best way to execute this order is to force the enemy to evacuate his bases and outposts.'[252] Amongst other projects, including the re-conquest of Kekova, Hesselberger planned an attack upon, and possible capture, of Castellorizo. To attempt a landing, he would require artillery support.

Two batteries were allocated to the forces attempting to capture the island. *Major* Karl Schmidt-Kolbow was in overall command of the artillery and, whilst he outranked Hesselberger, operated as part of the *Sonderkommando*. A German battery (*Hauptmann* Max Ittmann) with Krupp 15 cm schwere *Feldhaubitze Typ 13* heavy field howitzers was detailed.[253] Each howitzer weighed 2135 kg, firing a 42 kg shell. The shell was separate from the cartridge containing the propellant. The cartridge could be opened to adjust the number of bagged propellant charges. These were highly explosive and required protecting from counter battery fire. The barrel could be elevated to 45°, so that the trajectory of a howitzer shell was high with a steep angle of descent. The second battery was Turkish, commanded by *Topçu Yüzbaşi* (Artillery Captain) Mustafa Ertuğrul, and equipped with four Erhardt (RheinMetall) *Gebirgskanone 7.7 cm M15* mountain guns. Weighing just 555 kg, the guns were designed to be taken apart into six loads to be carried on pack animals. They were intended to be used as field guns, firing a 6.85 kg shell with a relatively flat trajectory.

A posed photograph of Mustafa Ertuğrul and his battery somewhere on the Anatolian coast. Ertuğrul, in the light tunic, is standing behind a range finder. Each of the Erhardt *Gebirgskanone 7.7 cm M15* mountain guns has a crew of between nine and eleven, several sheltering behind armoured personnel shields. *(courtesy Mustafa Aydemir)*

Mustafa Ertuğrul in an undated photograph standing in front of a 7.5cm Feldkanone L30 Krupp M 03. *(courtesy Mustafa Aydemir)*

The difficulties of bringing the guns to the coast were manifold, especially for the heavy howitzers. The nearest rail head was at Burdur in the interior, over 160 kilometres in a direct line from the coast, the actual distance to be traversed being considerably greater. Schmidt-Kolbow reported that for the howitzer battery, 'the broad caravan road Burdur-Istanos-Elmali-Gömbe was chosen… Until Gömbe, the terrain was mostly flat and navigable for the batteries. Of course, due to heavy rain the road was very muddy.'[254] From Gömbe a narrow pack horse trail led on towards the coast. The first half of which, to Kasaba, was over a mountain pass 1500 meters above sea level. In less than three weeks, several hundred labourers made the 50 kilometres of pack horse trail into a three metre wide track sufficient for the howitzers to pass, albeit with great effort. In addition, over four hundred camels were required to transport the howitzer ammunition.

Mustafa Ertuğrul's guns had a much easier passage. Leaving the railway at Denizli (Approximately 100 kilometres west of Burdur.) they marched via Elmalı towards the coast. 'In good weather, after an enjoyable march, we came to Kaş on 23rd of December [5 January 1917].'[255]

Schmidt-Kolbow reported that:

> During the night of the 5th to 6th [January], the mountain battery reached Baindyr and was led at dawn to its position.[256] The battery was ready to fire at noon on the 6th. On the evening of the 6th, the howitzer battery reached Baindyr. During the night to the 7th, the last part of the track from Baindyr to the battery position was prepared, and on the evening of the January 8th, the battery was pulled over the section of the track which could be seen from the island, close to its position.

It was still being prepared for use on the morning of 9 January.

I know of no photographs of the German officers or the actual guns at Castellorizo. This image is of a *Krupp 15 cm schwere Feldhaubitze Typ 13* heavy field howitzer in an emplacement on winter watch in the Vosges 1917–18. *(National Archives and Records Administration, 17390244)*

Mustafa Ertuğrul made several sketch maps which provide the best indication of both battery and observation post (OP) locations. The guns were sited on a broad headland (Gata Burnu) facing the island's harbour. From the harbour this is seen as two low hills separated by a shallow valley. Having the lighter guns, Mustafa Ertuğrul was able to position them just below the crest of the western height (162 metres) with a good view into the harbour about 5000 metres away. The less mobile howitzers were located in the valley, probably in the shelter of the second height (100 metres), within 5000 to 6000 metre range, but with an obstructed view of the harbour. He shows an artillery fire control (*ateş topçu kumandaninin*) position almost two kilometres behind the guns on a 200 metre high ridge, this was probably Schmidt-Kolbow's station during the bombardment, and two battery commander OPs close to their guns. His was located a little to the west of his guns, giving an improved view into the harbour. *Hauptmann* Max Ittmann is shown having an OP a little in advance of his howitzers. During the forthcoming action both commanders directed their guns from these locations.

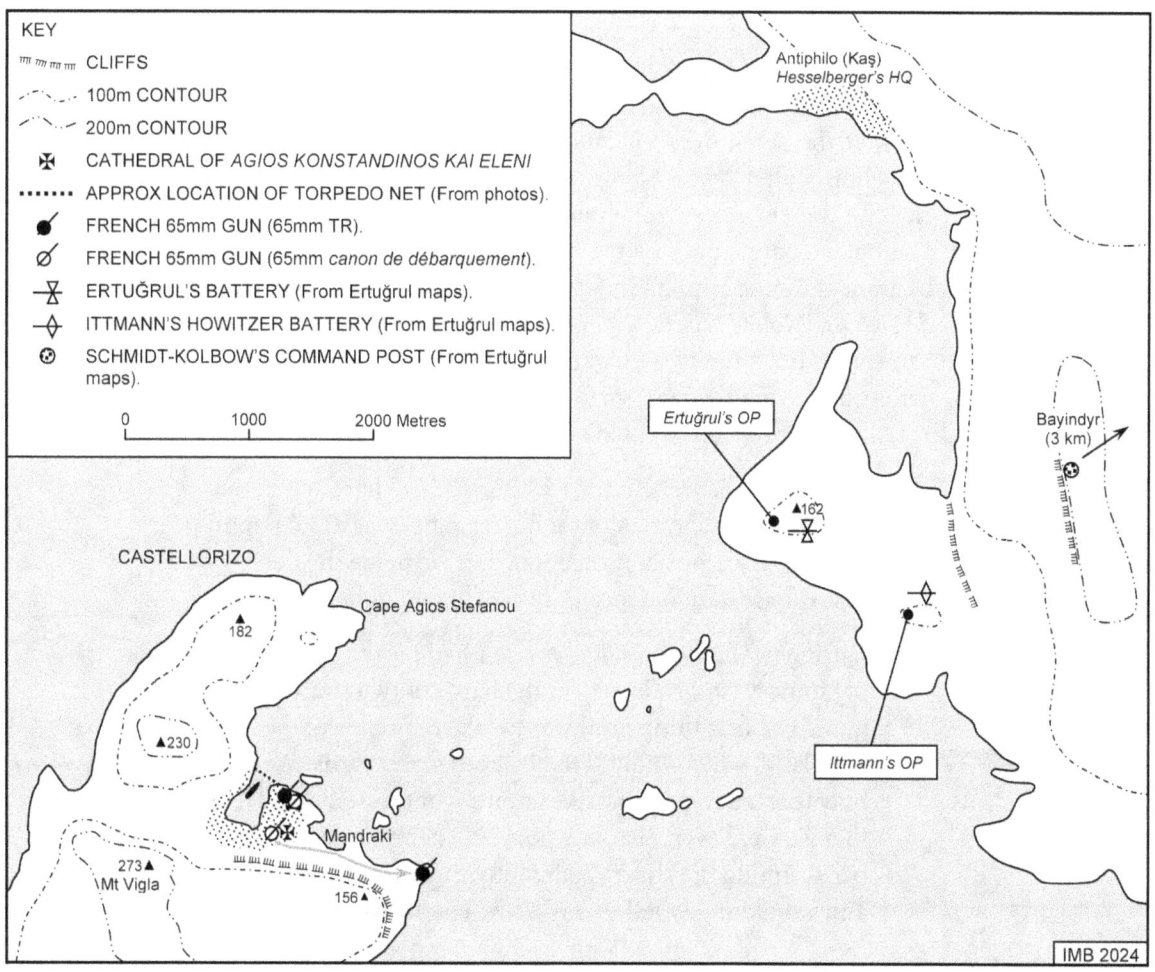

Battery Locations, Castellorizo, 9 January 1917.

Arrival of *Ben-my-Chree*

At Castellorizo Governor de Saint-Salvy continued to receive intelligence from the mainland. One of his most reliable informers was Kyrillos Romanos, abbot of a monastery which was attached to the church of *Ayios Nikolaos* at Myra (Demre). Myra was a small Greek Christian village about 35 kilometres east of Castellorizo, and was a mainland outpost of the island. Kyrillos had learnt that a battery was being built on the coast facing the harbour of Castellorizo, he was also informed that an invasion of the island was planned.[257] *Amiral* Spitz contacted Admiral Wemyss to request the loan of *Ben-my-Chree* so that her floatplanes could repeat their search of November last for the battery and to look for any invasion preparations.

Sailing from Port Said during the morning of 8 January 1917, with French destroyer *Pierrier* as escort. Aboard the seaplane carrier were a single Short, 8080, two Sopwith Schneiders, 3770 and 3778, and Sopwith Baby, 8188. Pilots aboard, in addition to Wing Cdr Samson, were Flt Cdr England, Flt Lt Clemson, Flt Lt Maskell, FSL Smith, and FSL Worrall, there were in addition four observers Lt Meade, Lt W.L. Samson, Sub Lt Kerry, and Capt Benn. When *Ben-my-Chree* arrived off the island at 07.15 following morning a strong gale was blowing, preventing floatplane operations. Samson decided to enter the small harbour. *Amiral* Spitz sent out an officer to act as pilot and shortly after 09.00 Samson backed *Ben-my-Chree* up to her moorings, close to the entrance to the harbour, and bows towards the Turkish coast. The net was drawn across the harbour entrance after she anchored. In the harbour she joined Spitz's yacht *Ariane*, the destroyer *Pierrier*, and *Torpilleur de Défense Mobile 250*, together with the island's much reduced trading fleet. Samson ordered the two forward 12-pdr guns to be manned and cleared for action, and to maintain steam for maximum speed at only five minutes notice.

FSL Guy Smith left his own version of events.

> We tied up securely in the Castilorizo [sic] harbour, about fifty feet from shore. A big submarine net was drawn up in front of us as special protection, and we made the decks ready for instant defence in case of an attempted attack.
>
> During my months in the service I had picked up a fair knowledge of gunnery, and so I was put in charge of two twelve-pounder guns. The first thing I did was to take the range of the Turkish mainland, which I found to be about 5,000 yards. We knew only that we were off hostile country, but were not at all certain that we were even near any point of enemy activities. I remember scanning the shoreline intently through the field glasses and wondering whether any of the enemy were there or not.

> I was on watch until 2 o'clock in the afternoon without discovering anything to get excited about, and had begun to feel grateful that the *Ben-My-Chree* was safely anchored in the harbour instead of being out in the open sea bucking that tremendous gale. Then I turned my watch over to another Flight Lieutenant, and after going below to change my clothes, started ashore.

Having prepared his ship for action, should it be required, Samson granted a make and mend,[258] and leave for half the ship's company, and gave approval for the ship's band to give a concert in the town square during the afternoon. Going aboard Spitz's yacht to be briefed on his mission, he was brought up to date with French intelligence reports 'that guns of large calibre were on the way from Kassaba to Antiphilo, and that a road was being constructed for the purpose.'[259] Spitz requested Samson to undertake a series of reconnaissance flights and bombing raids along the coast from Castellorizo and up to ten miles inland in search of the guns. In view of the prevailing weather conditions flights would be impossible that day but, if conditions moderated, Samson thought it might be possible to make flights from within the harbour itself on the following day.

Having avoided a time consuming French luncheon invitation, Samson returned to *Ben-my-Chree* at 12.15, had a look around and retired to his cabin to get some rest. Other officers resting in their cabins included Benn and Lt A.L. Braithwaite, RNVR, one of the deck officers. Those men not ashore or on duty were enjoying the make and mend, repairing clothes, reading, sleeping or just relaxing. The First Lieutenant, Lt H.C. Allingham, had the watch with Midshipman F.R. Smith but apart from the forward guns crews there was little activity excepting in the engine and boiler rooms. At 14.10 their reveries were shattered by an explosion and column of water less than 15 metres from the ship.

The arrival of the seaplane carrier had not gone unnoticed in the German and Turkish batteries on the opposite shore.

The Sinking of *Ben-my-Chree*, 9 January 1917

The arrival of *Ben-my-Chree* was unexpected. Both Hesselberger and Cemal had their headquarters some distance away from the guns, Schmidt-Kolbow, Ittmann and Ertuğrul were on the front line. Ertuğrul wrote about the tension which this caused among the artillery officers. Ertuğrul, Ittman and Schmidt-Kolbow had an exchange of opinions about what to do. They had seen many British sailors going ashore, and *Ben-my-Chree* could be fired at directly by the mountain guns and howitzers. This meant a great opportunity for a surprise attack. In Ertuğrul's words 'the three of us committed to each other with word of honour,' to the following new plan: The howitzer battery was to sink *Ben-my-Chree*, and the mountain gun battery was to shoot at the smaller French warships. Schmidt-Kolbow confirmed the plan in his report.

However, Schmidt-Kolbow had orders from Cemal Bey, confirmed by Hesselberger, limiting their freedom of action. 'I had received the order of the section commander Cemal Bey, to whom I was subordinated by corps commander Nureddin Bey, saying I had to wait for three more days before opening fire as the assault troops needed more time to prepare.' Cemal and Hesselberger's main concern was not to give advance warning of the attack, for which many people had worked very hard in the past weeks. Cemal, perhaps recognizing the value of sinking a major enemy warship, had added a final sentence to his orders: 'Only when the enemy warships move into the harbour, you may open fire without waiting for an order.'

Schmidt-Kolbow went to telephone Hesselberger to request permission to open fire. However, again according to Ertuğrul, Schmidt-Kolbow came over and said 'The telephone connection has broken I couldn't talk. Since we don't have any time to lose we should apply our plan.' The final sentence in Cemal's orders permitted this and, jointly, the artillery officers decided not to waste this unique chance for a surprise attack. Schmidt-Kolbow continued his report, 'While the howitzer battery prepared itself in its position, the cruiser launched boats into the water and started to set up the torpedo-protection-barrier which had disappeared a few days ago, making it more difficult for its own escape.' The artillery officers were waiting for these works to be finished in order to close the trap.

Times vary slightly in each report, a common enough problem. A 30 minute difference between Turko-German and Franco-British times that can be explained by local time settings. But, in general, the bombardment can be said to have commenced shortly after 14.00 island time, or 13.30 mainland time.

Schmidt-Kolbow. 'I was waiting for the work to be finished and opened fire at 1.36pm [14.06 island time] on the cruiser while the mountain gun battery started to zero-in on the gap in the barrier. After two range corrections, I had the first hit on the stern of the ship. It immediately caught fire, the spread of the fire was fostered by the strong wind.'

The two French 65mm at the castle were quick to respond. The 65mm TR at Point Niftis was slower, and was eventually joined by the second 65mm *canon de débarquement* which had been rushed over from the town. Schmidt-Kolbow, 'Shortly after opening the fire, two guns in front of the city started to fire on our batteries and achieved one hit in one cartridge-room of the howitzer battery.' Ertuğrul later wrote that after this hit the howitzers ceased firing and all subsequent shelling was done by his guns alone. However, there are no French or British reports of a serious fire or explosion on the mainland. Schmidt-Kolbow does not mention any casualties in his report. So, it can be assumed that any fire in the howitzer cartridge store was quickly extinguished without serious loss or injury. That the howitzers continued firing is clear from the French and British reports.

Ben-my-Chree's two forward guns were manned, but there is no evidence that they responded to the bombardment. The only surviving mention of them is from Mustafa Ertuğrul. After noting that the French guns responded, 'The two 21 cm cannons in the turret of cruiser began to turn towards us.' To Ertuğrul all allied ships, large or small, were *kruvazör* (cruisers) and all had their guns in *tarette* (turrets). However, this does suggest that the two 12-pdr (7.6 cm), guns were being trained towards the batteries. Events quickly overtook them and, per Samson's report, no shells were fired.

Midshipman Smith's testimony:

> When I heard the whistle of a bomb as I thought coming, I turned to the Quartermaster and the Bugler, and told them to sound off anti-aircraft stations. I immediately ran to the Captain's cabin to report to him, but he had gone out the other door.[260]

Samson, ran up to the bridge and, hearing another explosion, thought it was a submarine firing.

> Remaining on the bridge I heard a gun fired from the NE and heard a shell coming. I then knew it was a shell fired from the mainland. The fourth shell hit the hangar and the fifth shell hit the bridge as I left it to go to the hangar. I went to General Quarters and sent for the Chief Engineer to stand by the engines.

When he reached the engine room Engineer Lieutenant Robinson found that there was a lot of smoke coming from the petrol lockers located above the aft end of the engine room, and adjacent wood work was beginning to catch fire. Conditions rapidly worsened, by 14.30 the engine room was untenable and had to be evacuated. Before the last man left the boiler rooms the steam pressure safety valves were tripped. Now the sound of escaping high pressure steam added to the growing cacophony on deck.

Flt Lt Maskell had been in his cabin when the shelling commenced.[261]

> I ran on to the Quarter Deck thinking it was a bomb from a German machine. We went to anti-aircraft stations and fire appliances brought out. At the second crash, I thought it was a torpedo, so I ordered all the spare men down below out of it. That was when I knew they were shelling from the shore. I inspected the workshop to see that everything was correct, and I was just returning to the hangar when the first shell hit. I think it was the third shell that crashed into the hangar, and set it on fire. It was absolutely impossible to fight the flames in the hangar, as the machines burnt so quickly, and the decks being oily, it spread more rapidly. The drenches were turned on, but of absolutely no avail.

Samson, meanwhile, had sent away the ship's motor boats to stand by ready to slip the mooring wires, intending to take the ship out to sea away from the guns. These were now estimated to be 15 cm howitzers and 12 pounder field guns, firing at 5000 yards range and making excellent shooting.

> The seventh shell hit the hangar, and other shells very shortly afterwards came through the deck abaft the foremost guns and exploded between decks. Another shell fell close to the capstan on the Forecastle. It was not possible to reply to the enemy's fire as I could not see their guns and my guns were masked by the Town of Castellorizo. I therefore gave orders for the guns' crews to take shelter and both magazines to be flooded.

The ship's gunner Mr H.C.J. Evitt, carried out this order though not without some difficulty. The forward magazine and bomb store was flooded first. After hearing water begin to run into them Evitt headed for the after magazine, which lay under the workshop and fiercely burning hangar. 'When I arrived there, I could not see into the workshop. After that it was impossible to see at all. I believe I got the padlock off, and I hit the handle of the flooding arrangements with a piece of metal I got hold of, then came away and reported to the Captain.'

Seen from the town close to the cathedral *Ben-my-Chree* with the hangar on fire and steam escaping from the steam pressure safety valves. *(Kerry Family Collection)*

The shell that sealed *Ben-my Chree*'s fate struck and penetrated the petrol store before exploding, spreading burning fuel over the aft part of the ship. With fires now out of control, the engine room untenable and all steering out of action, the ship could not be saved. Samson ordered the crew to assemble port-side amidships, the most sheltered part of the ship, to prepare to abandon ship. Braithwaite was sent to lower the whaler, 'We had just got her hoisted clear of the chocks when she literally disappeared into thin air and we were left with the bit hooked onto the falls at each end. She had received a direct hit but fortunately no one was hurt.'[262] Benn and Samson were standing on the deck below the whaler when it disappeared, neither received so much as a scratch.

Representative of the four floatplanes lost on *Ben-my-Chree*, another view of Short 8080 on 27 December 1916. It was flown on this occasion by FSL Guy Duncan Smith with Capt Benn as observer and is seen here taxying up to the ship returning from the attack on the Chicaldere railway bridge. *(EMK)*

At 14.45 Samson gave the order to abandon ship. The sick and wounded having already been moved ashore to a place of safety, the ship's motor boats now returned for the remaining crew, each making several round trips. Some chose to swim the short distance to the shore. Malone and Samson had ensured that all non-swimmers learnt to swim in preparation for such an emergency. Benn was sent to check the engine and boiler rooms, other officers and CPO Ridley combed the mess decks, cabins, offices and sickbay to ensure no one had been left below. One seaman, viewing the sickbay as some sort of sanctuary made inviolable under the Hague Convention, almost had to be dragged to real safety. When he had seen everyone off the ship, and ensured that the confidential books had been destroyed, Samson boarded the last boat to leave the ship's side. *Ben-my-Chree* continued to burn throughout the night and into the following morning. Occasional explosions continued until 13 January. Slowly taking on water from numerous shell and splinter holes, she eventually settled on the bottom of the harbour with a list to starboard and most of the upper deck submerged.

Whilst Turkish attention was concentrated on *Ben-my-Chree*, Spitz had given orders for his ships to clear the harbour and head for safety at sea. *Pierrier* was quickly away, smashing through the anti-submarine net, followed by Spitz's yacht, *Ariane*. *TDM 250* took some time to raise steam and did not clear the harbour until much later. Guy Smith, who had just landed when the attack commenced, adds his own account.

> Another fellow and I got on the roof of a house where we could lie down and watch the bombardment. Our view was perfect and we could see the flash of the enemy guns and the burning ship as well.

The excitement got particularly intense when two French destroyers made a mad dash out of the harbour. Shells fell all about them, but they carried right on, and a great cheer rose from the shore when they eventually got out of range.

Schmidt-Kolbow's report continued:

> Subsequently, more direct hits were made [on *Ben-my-Chree*]. Afterwards, I directed the fire in the Southeastern corner of the harbour in which two torpedo-boats and an armed steamboat were still situated. One torpedo-boat [*Pierrier*] and the last-named vessel [*Ariane*] left the harbour quickly under full steam and passed through the net barrier without being hit by the firing batteries. The mountain gun battery initially shot at the torpedo-boat and several shots were in its proximity, also one hit. When this ship was out of reach, it directed its fire on the sailing steamboat and shot down its mast. The howitzer battery was not able to follow the fast-moving targets. I directed its fire on the last small torpedo-boat [*TB250*] in the harbour. Also this one was able to escape the barrier without damage, thereupon it received a hit by the mountain gun battery on the stern; one saw how, probably hit below the water-line, it jumped out of the water.

Ertuğrul was to claim, 'The torpedo boat took two hits out of eighteen fired and hardly passed the north foreland of the island.' The French ships actually escaped with minor splinter damage.

On The Island

Whilst *Ben-my-Chree* continued to burn the batteries turned their attention to the town and environs. As Schmidt-Kolbow reported.

> While the mountain battery focused on [the French] guns, the howitzer battery destroyed at first the radio station in the city and later a neighbourhood near the pier in which, according to testimonies of POWs, the depots of the garrison were located. Then it also turned its fire against the artillery of the island, which was silenced shortly afterwards. Meanwhile, it was evening and impossible to observe due to heavy rain.

Ertuğrul claimed that his guns silenced the French guns, then destroyed the radio station. After the escape of the ships, the French guns ceased fire, probably to save ammunition for the anticipated landing, and the radio station was undamaged. Following this, Ertuğrul ordered his men to shell the trading ships and smaller caiques in the harbour, expending fifty rounds on them. He claimed there were around 200 of them inside the harbour, actually there were probably less than fifty. There is photographic evidence suggesting that several of the larger sailing ships were sunk or damaged. All firing ceased around 17.00.

Schmidt-Kolbow concludes.

> I gave the batteries, of which the howitzer battery was left with only 14 rounds, the order to retreat to Baindyr as I did not wanted to expose them to expected enemy fire by naval artillery in the next days. In the harbour, the cruiser which we had set on fire and had not been able to reply our fire, exploded at approximately 11pm [23.30 island time].

Schmidt-Kolbow's mention of an explosion needs to be dismissed at this point. As already reported by Samson, and confirmed by Gunner, Mr H.C.T. Evitt, in a later deposition, the magazines were flooded before the ship was abandoned. Also, photographs taken during the salvage operations in 1920 show bombs and shells removed for dumping at sea. The most likely reason for the reported explosions was that the shell fire, coupled with the hangar and workshop fires, had caused a petrol vapour explosion in empty fuel tanks and cans.

Military casualties on both sides were surprisingly light. Ertuğrul mentions the loss of a single gunner to French counter battery fire, Schmidt-Kolbow no casualties at all. Samson's command suffered only seven men injured by shell fragments. The most seriously injured were Ldg Seaman William Billett who was hit through the lungs and doubly unfortunate AB Edward A. Bennison from *Raven* who, whilst serving time in *Ben-my-Chree*'s cells, 'had the misfortune to have half his hand blown off by shrapnel.' Guy Smith adds, 'But many civilians had been killed or wounded, and everywhere we

Below: The crew were evacuated to shelter behind the monastery of *Ayia Triadha* (Holy Trinity) where this photograph was snapped by Sub Lt J.L. Kerry. In the foreground is Flt Cdr Clemson in his Greek fisherman's finery. *(Kerry Family Collection)*

Left: another Kodak photo from Sub Lt Kerry. A group of the crew including Cdr Samson, the short figure in the white cap on the right of the image, sheltering behind the rear wall of *Ayia Triadha*. *(Kerry Family Collection)*

turned we encountered grief and desolation. Most of our time was spent in helping the wounded among the civilians.' He continues to describe the shelling of the town, the crew's rescue work of the islanders and the unending struggle by the French chief medical officer Charles Héderer, together with Temporary Surgeon L.S. Goss, RN, assisted by officer's steward S. Redington from *Ben-my-Chree* to care for the sick and wounded. Surprisingly, *Docteur* Charles Héderer, does not mention French or islander casualties in his book. He simply adds in a footnote 'all the civilian and military wounded received emergency care as required by their state.'

Samson had ordered his men to shelter behind the monastery of *Ayia Triadha* (Holy Trinity) which sits in a saddle above the town some distance inland, out of range of the guns.

> When the last boat had gone there was still one man missing, Captain Clemson. A few of us went back to search for him, but in vain. We knew he had been assisting with the fire and we feared he was caught in the burning hangar. Not finding any trace of him in our search we were soon compelled to return, for our boat was being shelled heavily. Great was our delight later to learn that he had been seen to jump from the ship and had swum ashore.[263]

Flt Cdr Clemson was actually the last to leave the ship, and inadvertently provided one of the few lighter moments of that day. Having swum ashore, people from a nearby cottage insisted on taking his soaked uniform to be dried, whilst the man of the house loaned Clemson his Sunday best. His arrival at the monastery in this attire, including some fancy trousers and Greek Fisherman's hat, garnered more than a few smiles and laughter.

Sub Lt Kerry recorded his impressions of the following hours.

> Tuesday, 9 January 1917:
>
> ... At 2 pm shore leave granted. With about 6 others wandered into the quaint town & up the hills. Returned to Cathedral Sq when suddenly a shell was fired. Much consternation. Children screaming running out of school, parents rushing to meet them. Awful panic. Shelling continues & some fell in town. Informed by French officer that BMC on fire. Returned to quay & found hangar well alight, ship been abandoned. Population started making for hills were shelled. Mainly women & children killed in town, houses brought down. Constant explosions from ship due to bombs & magazine. BMC people brought in & attended to. Eventually all made for church or hills in pouring rain. Many survivors no clothes or boots. Ship burning fiercely all night. At night all mustered in buildings at rear of church. Supper of bread & bully beef & were waited on by good old dame. 20 officers

including Cmdr [Samson] slept or tried to in a little room about 11 ft sq under a carpet. All in the best of spirits. Sleep almost impossible. Very heavy rain & frequent interruptions during the night.[264]

Many of the fit were co-opted by the island's Governor to keep an all night watch in case the Turks decided to attempt an invasion, and to assist with casualties. This was only a temporary measure, however, for 24 hours later most of *Ben-my-Chree*'s survivors were evacuated.

Wednesday, 10 January 1917:

Up at 4 in the morning. Stragglers began to come in. All quiet in town — evacuated by inhabitants. BMC still smouldering in hangar, down [by] bows and with a big list to starboard, a pathetic sight. No trace of my cabin and the majority of others. A few went on board and secured as much as possible. Returned to church for lunch of bread and sardines and drink of cordial. Ship's company split up into 5 parties and wounded brought up. Tea of ships biscuits and sardines and some cocoa. "Bed" at 6.30 but turned out at 7. All lined up in darkness and wounded on stretchers. Evacuation commenced. Trekked to a bay at rear of island a weird sight. 200 men at all stages of dress shuffling along in single file.

The evacuation was from Navlakas cove on the south east side of the island. Various accounts mention the overnight march, single file, across the island

In the days following the initial bombardment a number of RNAS personnel remained on the island, this is one of a series of photographs they took at the time. *Ben-my-Chree*, fires extinguished, lies on the bottom of the harbour. The hangar and stern are completely burnt out, yet the canvas sun screen over the bridge is apparently untouched. She would remain in this position for several years. *(EMK)*

for a distance of approximately two miles (3.25 kilometres). The rough path over stones and boulders destroyed their footwear, some having to complete the march barefoot. The final approach to the bay was down a steep decline, holding on to ropes to avoid falling. The French were waiting, taking them out to a waiting trawler in small boats. FSL Smith takes up the story.

> Weary we were indeed, but the Frenchmen did everything in their power to make the trip pleasant for us. That was difficult, because there were many of us and we were crowded like sardines into the little boat. The sailors actually gave up their bunks to us, and did it with such politeness that one might have thought they really enjoyed it.
>
> During this return trip, which took five days [by way of Rhodes], I never heard one word of complaint from anybody. The men were splendid. They had come singing off the *Ben-My-Chree* and they went singing back to the base, enduring cold, wind, rain and cramped quarters with a cheerful grin. A dirty, bedraggled lot we were when we went ashore at Port Said, but this was one time when Port Said really looked like heaven.

A small party, under Cdr Samson, remained behind to attempt to salvage what stores and equipment they could. This party included Braithwaite, Midshipman E.J. Nicolle, Assistant Paymaster J.M. Parker and several ratings. 'It was several days before it was possible to get aboard and the battery amused itself by taking pot shots at us. There was no cover, and we were driven from pillar to post.'[265] Despite the difficulties Braithwaite describes, the little party was able to dismantle the guns, eventually taking the breech blocks back to Port Said. Parker entered the ship's office by swimming through a submerged porthole, managing to retrieve papers and money from the safe. The surviving ship's boats were left for the use of the Governor and his small garrison of 200 sailors.

Initially the garrison was reinforced by a party of volunteers from *Ben-my-Chree*. Under Captain Wedgwood Benn they joined a contingent of French *matelots* guarding Mandraki Bay facing the Turkish mainland. Setting up his headquarters in a tiny whitewashed church, Benn thoroughly enjoyed the next week. His title *Commandant le Detachement Franco-Brittanique du Cimetière Turc, Castellorizo* was almost bigger than his *detachement* of seventeen *Ben-my-Chree* bluejackets and 46 Frenchmen. The force lacked arms and a common language, Benn having to institute basic language lessons, but more than made up for its shortages in spirit. When the order came for the British to withdraw, the departure of *Ben-my-Chree*'s men was more tearful than military. On 18 January Benn's group joined the salvage party at Navlakas cove to be evacuated to Port Said, arriving on 21 January aboard the trawler *Paris II*.

An Attempted Landing

The enemy artillery soon returned and further shelling of the island occurred on 13, 17 and 19 January. Followed by the attempted Turkish invasion of the island on 20 January.

The Turkish landing force had assembled at Faktora Bay, a small sheltered inlet about 7 kilometres east of Castellorizo. Boats had been assembled here from various harbours along the Antalya coast. Under overall command of Hesselberger, the landing force had a leavening of soldiers from the *135 nci Alay* but the majority were irregulars, followers of an Emin Aga.[266] They set out before dawn in twelve open boats to row to the island. It was daylight by the time they approached the island, urged forward by Hesselberger who appeared in a motor launch. Several of the boats were sunk by the French guns and the remainder took shelter behind some of the islets in the channel between Castellorizo and the mainland.

Throughout the day, whilst the French and Turko-German artillery duelled, the surviving attackers took such shelter as could be found on the islets. In the evening 86 men were rescued, 36 had been lost. A single wounded Turkish survivor, one of Emin Aga's irregulars, was taken prisoner by the French. He told his captors that the landing force thought the island had been abandoned as their guns had not replied to the previous bombardments.

One minor mystery remains to be mentioned, the appearance of a Turkish aeroplane. Both Mustafa Ertuğrul and Charles Héderer mention it bombing the French Trawler *Paris II* on 20 January 1917.

At the start of his account Ertuğrul mentions 'a German plane from *Binbaşı* Şule's command,' as being part of the force sent to Castellorizo, and that after the sinking of *Ben-my-Chree*, '*Binbaşı* Şule's aviators stayed behind.' Later, when writing of the attack on 20 January, he noted, 'Our plane at Demre [Myra] has flown. When it was over the island it encountered an intense machine gun fire. It dropped four bombs on the cruiser. The bombs missed the cruiser, and the plane returned to Demre without achieving anything.' The attack is confirmed by Héderer who, when writing about the attempted landing, wrote 'a plane attacked the trawler *Paris II* cruising to the SE of the island.'[267]

Ertuğrul's *Binbaşı* Şule was *Major* Frederik Schueler van Krieken, *Kommandeur der Flieger* of the Fifth Army's aviation units, based at Smyrna. He had previously served as an observer with *Feldflieger Abteilung 23* in France. Under his command at Smyrna was *5nci Tayyare Bölük*, equipped with the Albatros C.III. This Turkish squadron was commanded by a German observer, *Leutnant* Arthur Faller.[268] At Myra, a landing field could have been created on the flood plain, but supporting materials, fuel and armaments would have been difficult to bring in. This is perhaps why the Albatros only made a single attack.

After the attempted landing, the French strengthened the defences of the island with a battery of 120mm guns, several more 65mm guns and some 37mm canon, also an infantry platoon of Armenians from *La Légion d'orient*. However, apart from occasional bombardments, the island was left undisturbed for the remainder of the war.

Following the destruction wrought by the bombardments in the town, many islanders were living rough in ravines and tented camps on the heights of the island. Many chose to leave, sailing away to Cyprus, Egypt and Australia where an expatriate community had been established before the war. The population never recovered to its pre-war level.

Although Mustafa Ertuğrul's claim to have sunk *Ben-my-Chree* cannot be supported, there is no doubt that his guns played a significant role in the sinking, and his guns later sank the trawler *Paris II*. After Castellorizo he and his battery continued serving along the coast. They were on a peninsula near Kemer, some 35 km south of Antalya, on 13 December 1917 when his guns sank *Paris II*. The crew of seventeen were rescued and became prisoners of war. He was also involved in the sinking of another French trawler *Alexandra* on 8 March 1918 in the Gulf of Antalya. The exact circumstances of the sinking are not clear. However, it seems probable from Ertuğrul's writing, that the *Alexandra* was severely damaged by an explosion whilst investigating a drifting boat (possibly arranged to explode on contact), and then finished off by his battery. Eleven crew members, including the captain, were lost and the few survivors became prisoners of war.

Salvage of Ben-my-Chree

The burnt out shell that used to be *Ben-my-Chree* remained in Castellorizo harbour for the remainder of the war and over a year afterwards. The hull was essentially intact, save for a few shell holes, and was resting on the bottom of the harbour. During the war the French were able to salvage valuable supplies from the wreck using local island divers. They would have liked to remove the coal from the bunkers as well, 'The ship still has an important coal stock. There has been, since the first day, fires in the port side bunkers that still need to be extinguished. We might remove the coal gradually, but such weight displacement might affect the balance of the wreck.'[269]

Raised by the Ocean Salvage Company in 1920, *Ben-my-Chree* lies in the harbour. Alongside is tug *La Valetta* which will shortly take the salvaged ship to Piraeus harbour. The Greek salvage divers have installed a crucifix in the bows of the ship to ensure safety. Just aft of the anchor chain is a large wooden patch covering one of the shell holes, another shell hole can been seen in the forward funnel. *(Nicholas Pappas Collection)*

The hulk, however, occupied the only suitable mooring for larger vessels as well as being a general hazard to navigation within the harbour. In 1920 the Ocean Salvage Company arrived at the island to raise the wreck. They stopped the holes and pumped out the hull, but once afloat it was quickly realised that the remains were beyond economic repair. So, the burnt out shell was towed to Piraeus harbour and moored there, awaiting disposal. It was sold on to an Italian firm for scrapping at Venice in 1923.

CHAPTER 12

Empress with the Eastern Mediterranean Squadron

The Northern Aegean

In the aftermath of the Second Balkan War (29 June — 10 August 1913) Bulgaria lost most of the territories it had gained during the First Balkan War (8 October 1912 — 30 May 1913). It did, however, retain a short length of Northern Aegean coast between the rivers Maritsa (Evros) and Mesta (Nestos). The coast west of the Mesta was returned to Greece, becoming part of the region of Macedonia.

The Northern Aegean coast is divided by three main rivers. In the east is the Maritsa, which was the border between Turkey and Bulgaria. Next west is the Mesta which was the boundary between Bulgaria and Eastern Macedonia. Finally, there is the Struma (Strymonas), west of which lies the city of Salonika (Thessaloniki). The Aegean coast has been described as 'an inhospitable coast-line ... bounded on the North by the Rhodope range. The whole region is mountainous save for a somewhat extensive marshy plain on the coast formed on the mouth of the Mesta and of the three small rivers which enter the sea just to the E of it.'[270] Immediately east of the Struma, after it empties into the Gulf of Orfano (or Rendina), on the coast and rising out of the plain, are two outliers of the Rhodope range — Bunar Dagh (Mt Pangaion of antiquity), up to 2000 metres, and Simvolon Dagh, along the coast, rising to 600 metres. Past Kavala, flood plains then comprised the coast to a point just west of Dédé Agatch (Alexandroupolis) where mountains over 600 metres once again take over. East of Dédé Agatch lie the coastal marshes of the Maritsa delta.

Further east, beyond the Maritsa river, lies the Gulf of Xeros and Gallipoli. Beyond Gallipoli the Turkish coast turns south. This, the west coast of Anatolia or Asia Minor, is deeply indented where, over aeons, the lowlands had been swallowed by the sea leaving rocky peninsulas and offshore islands. Along this coast the city of Smyrna (Izmir) was blockaded by the Royal Navy; other cities, towns, and villages were periodically inspected from the sea and air.

Just off the delta of the Mesta river lies the Greek island Thasos. The most northern of the Aegean islands, it is separated from the mainland by the Thasos Strait, 5 kilometres wide at its narrowest point. It was a recent addition to Greece, having been captured by the Greek navy on 20 October 1912 during the First Balkan War, ending 350 years of Ottoman rule.

The Salonika Campaign, 1915 and 1916

Although allied to Great Britain, Serbia fought alone for over a year.[271] Attacks by the Austro-Hungarian Army in 1914 were repulsed. But when renewed in October 1915 the attack, with the support of Germany and Bulgaria, overwhelmed the Serbian Army. In a winter retreat through the mountains of Montenegro to Albania, over half of those who set out failed to cross the mountains. The survivors were evacuated to the island of Corfu to recover, regroup, and retrain. The rebuilt army would begin to deploy to Salonika in April 1916.

In a belated attempt to aid the Serbians, on 5 October 1915, French and British troops had landed at Salonika. Greece being unable, or unwilling, to honour its military convention to help Serbia in the event of an attack by Bulgaria, the Allies had been invited to intervene by Eleftherios Venizelos, the Greek premier. King Constantine and the Greek general staff opposed his policy, forcing Venizelos to resign, setting the scene for a potential confrontation with Greece.

The Allied armies moved up-country from Salonika, but proved too weak to open a front with Serbia, and they withdrew to Salonika to a series of entrenched positions which had been completed by the end of December 1915. The Allies, once reinforced by the Serbian army and by Russian and Italian detachments, made an attempt to break out of the Salonika entrenchment. On 2 August 1916, General Sarrail, the French Commander-in-Chief at Salonika, launched his Macedonian offensive. The Bulgarians at once replied by counter-offensives on the extreme flanks of General Sarrail's left and right wings. On the left the Serbian army held the Bulgarian attack successfully, on the right the position became more serious. Here the weak Greek IV Army Corps was spread out from Kavala along the Mesta river in the east and up to the northern Struma plains in the west. The Bulgarians marched into Greek territory at several points. However, the Greek forces had been ordered not to resist and retired on Kavala to await evacuation.

However, the ships would evacuate only those men who would support Mr Venizelos, and only one division was transferred via Thasos to Salonika and joined the allied forces. The rest of the IV Army Corps, nearly 6000 officers and men, were compelled to surrender on 11 September to the Germans, and were transferred to Goerlitz in Germany. By the last week in August practically all of Macedonia to the east of the Struma was in Bulgarian hands.

Eventually the Allies were able to establish a front line that ran from the Albanian coast through northern Greece to the Gulf of Orfano on the northern Aegean coast. The *Armée d'Orient* was a multi-national force under French command; French, Serbian, Russian and Italian forces held the western portions of the line. The British Salonika Force (or Army) held the area east of the Vardar River, the trenches west of Lake Doiran, and patrolled the Struma Valley to the east.[272]

RNAS in the Northern Aegean area after Gallipoli

Evacuated from Gallipoli, the Army was transported to Egypt, Salonika, and beyond. Remaining behind were the ships of the Royal Navy, Eastern Mediterranean Squadron, and aeroplanes of the RNAS. Vice-Admiral Sir John Michael de Robeck, VA Eastern Mediterranean Squadron,[273] was informed by Admiralty telegram in January 1916 that his role was 'to watch the Dardanelles, and safeguard the Greek islands in our occupation, to maintain the blockade and submarine patrols in the Aegean, and to support the army at Salonika.'[274] This essentially required a watch on the Greek, Bulgarian and Turkish coasts, and immediate interiors, from the river Struma to the Dardanelles and south down to Smyrna and beyond. For this duty the Squadron was divided into several Detached Squadrons, covering specific parts of the coast.[275]

> Admiral de Robeck stated that the air requirements arising out of the new naval policy in the Eastern Mediterranean would be met by one aeroplane wing at Imbros, one airship base at Mudros, two seaplane carriers, and two kite-balloon ships. Accordingly Wing Captain F. H. Sykes, together with No.3 Wing, returned to England.
>
> No.2 Wing, under Wing Commander E. L. Gerrard, which comprised three Flights, each of ten aeroplanes, remained at Imbros to keep watch on the Dardanelles, to spot for ships of the fleet during bombardments of Turkish gun positions, etc., and to bomb, as opportunity offered, enemy aerodromes, camps, and other suitable targets. The aircraft carrier *Ben-my-Chree* had been transferred from the Eastern Mediterranean to Egypt, while the *Ark Royal*, with five 200 horse-power Short seaplanes and two Sopwith seaplanes, was at Salonika, where she had been sent in November 1915. The kite balloon ship *Canning* was also at Salonika, and the *Hector* was lying at Mudros. The airship station had been moved from Imbros in October 1915 and re-erected at Mudros.[276]

In addition there was a French Air Force detachment, MF.98M, based at Mytilene from mid-January to 27 February 1916.[277]

When Wing Captain F.R. Scarlett arrived in February 1916, to take

command of the Royal Naval Air Service units in the Eastern Mediterranean, he quickly formulated proposals for an expansion of the air activities. Included in these were the return of *Ark Royal* to Mudros as a repair and depot ship, and a replacement for *Ben-my-Chree*. *Ark Royal* returned to Mudros Bay, Lemnos, on 15 March 1916 and *Empress* was ordered up from Egypt. In addition 2 Wing was to be maintained and expanded when additional machines became available. Over the next months a central base was established at Mudros with additional airfields on Imbros, Mitylene and at Stavros on the mainland, with a small Detached Flight based at each. Plans were also afoot for the Greek island of Thasos.

Empress at Mudros in 1916 with a Sopwith Schneider in the hangar. The two masts with a cross-bar supporting the W/T aerials, coupled with the rounded corners to the hangar, are a good recognition point for *Empress*. (Courtesy Maritime Photo Library)

For some time the island had been providing wood for the French, but covetous eyes were being cast upon it as the site for an airfield. A scheme for the redeployment of available aircraft approved by Admiral de Robeck included 'An Aerodrome on Thasos Island from which the entire lines of communication of the Bulgarian Army on the Grecian Front, are laid open to flank attack.' Despite interference from the Greek Civil Governor at Kavala, including an attempt to arrest the British representative on the island, by early May work had commenced to prepare a landing ground at Cape Kavamite (in British reports, alternatively Kazaviti, modern Prinos) to the north west of the island. Using local labour the landing ground, with an adjacent seaplane base, had been completed by 20 May and approved by Wing Capt F.R. Scarlett. A British military guard from Salonika disembarked at

Ark Royal, base ship for the RNAS floatplanes in the North Aegean area. The Short 166, 165, flying overhead was probably added to the original image. (Paschalis Palavouzis)

Kavamite the following day. A French military detachment landed on Thasos on 8 June to formally occupy the island.[278]

The first RNAS machine, a Sopwith Schneider, probably 3713, to land at Thasos was flown by Flt Lt Dunning from Stavros on 23 May 1916. The main RNAS unit, A Flight (later Squadron), 2 Wing, flew from Imbros to Thasos on 30 May, and the following day commenced flying operational missions to the Porto Lagos, Xanthi and Gumuljina areas. The detachment comprised three Henry Farman HF.27s, one Nieuport two-seater, and two Bristol Scouts. In addition, 'A flight composed of Nieuport and Maurice Farman Aeroplanes has been sent by the French Air Service to assist in operations from Thasos, and placed under the orders of the Officer Commanding A Flight, and the two flights acted in concert.' The *détachement* included three Nieuport 10 two-seaters and a Maurice Farman MF.11bis, from the *Groupe de Bombardement d'Orient*, based at Samli, north of Salonika, began arriving at Thasos on 26 May. It was briefly reinforced by a second Farman and a Caudron G.4. It was active over the same areas visited by the RNAS, but returned to Salonika on 28 June.[279]

Stavros Point is located at the south western end of the Gulf of Orfano, into which the Struma empties. The seaplane station located on the adjacent beach was operational from early February 1916, but was closed in July 1916 and the base, including the hangar, transferred to Cape Kazaviti, Thasos.

As far as available machines permitted, with these airfields the RNAS was able to cover the coast, and hinterland, from the Struma to Samos. Assisting the RNAS were several French units and RFC squadrons, mostly based in Salonika. However, until July 1916, when 17 Squadron, RFC, arrived from Egypt, the only British machines in the Salonika area were RNAS.

Bulgarian, German and Turkish Aviation in the Northern Aegean

The enemy machines the *Empress* crews would encounter were mostly German or Bulgarian along the northern coast line and Turkish along the Anatolian coast.

A small Bulgarian unit, *1-bo Aeroplanno Otdelenie* (*AO*, or Aeroplane Section), was based at Belitsa at the northern end of the Struma Front, with six LVG B.IIs and three Otto C.Is. On 19 January 1916, a detachment of three LVGs, under the command of *Kapitan* Dimitar Sakelarov, arrived at Xanthi to the service of the Bulgarian 10th Aegean Division. An airfield was established on some ground adjacent to the Rezi (or Régie, the Ottoman tobacco monopoly) tobacco factory. By July the Xanthi unit had become the *2-po AO*, commanded by *Kapitan* Nikifor Bogdanov with three LVGs and three Otto C.Is, remaining there until withdrawn to Hudova in August.[280]

Bulgarian LVG B.II of 1st *Aeroplanno Otdelenie*, with *Kapitan* Penyu Popkrastev at Xanthi 1916. The tobacco factory building in the background confirms the location. (Paschalis Palavouzis)

An unusual type operated by *Feldflieger-Abteilung 1* at Xanthi was this AEG G.II, G.23/15, *Sonnenvogel* (Sunbird) seen here in front of the tobacco factory. (Paschalis Palavouzis)

There were several German aviation units in the Salonika/Macedonia theatre during 1916.[281] *Feldflieger-Abteilung 1 (FFA1)*, had been based at Leskovac, on the railway some 35 kilometres south of Nisch (Niš), from late 1915. It was equipped with twin engined AEG G.II bombers and Albatros C.I/III two-seaters. In January 1916 a detachment from *FFA1* joined the Bulgarians at Xanthi. At the end of April *FFA1* was transferred to the Western Front, possibly leaving the AEG and Albatros behind, and the Xanthi detachment replaced with one from *FFA30*.[282] Increasing attacks on Xanthi by the British/French airmen from Thasos in May/June 1916, dictated a transfer of the units to a new location at Maswakli a few kilometres to the SE of Xanthi.[283]

Also operating, mainly reconnaissance flights over the Struma front, were *FFA57* based at Gumuljina (Komotini), 40 kilometres east of Xanthi, until late April when it was transferred with *FFA1*. Two more units, *FFA66* and *FFA69*, both based at Hudova, would fly occasional reconnaissance flights to Stavros, the Gulf of Orfano and the mouth of the Struma. All these reconnaissance flights were difficult for the various RNAS aircraft to intercept throughout 1916.

Early in 1916 two Friedrichshafen FF33 floatplanes from the Black Sea *Wasserfliegerabteilung* were based at Gereviz, on the western side of Lake Boru (Vistonida), south east of Xanthi near Porto Lagos. Administratively, they were attached to the Xanthi detachment. Over time the seaplane station became *Seeflugstation* Xanthi and continued in operation until the end of the war. The floatplanes were initially based there to help counter the RNAS

seaplanes from Thasos, making regular bombing and reconnaissance flights over the island and general area. The RNAS began reporting contacts with seaplanes from mid-April 1916.

In October 1916, following the Bulgarian advance into Eastern Macedonia, the *FFA30* detachment moved to a new airfield in the western outskirts of Drama, a major town located some 50 kilometres east of the Struma Front and 30 kilometres north of the coast at Kavala. An advance landing ground was maintained at Iralti (Eratino), on the flood plain about 20 kilometres east of Kavala and a similar distance north of Thasos.

The Turkish unit on the Anatolian coast was *5nci Tayyare Bölük* (*5 Ty Bol*), a small unit with a vast area, from north of Smyrna to Antalya, to cover. Initially equipped with four Gotha LD.2's (a Gotha WD.1 floatplane on wheels) and a Fokker E.I and based at Smyrna from March 1916. The inadequate Gothas were eventually reinforced by single examples of the Albatros C.I and C.III. With never more than a handful of machines *5 Ty Bol* struggled to maintain watch over the coast. Single machines were despatched as required to temporary landing grounds around the coast.

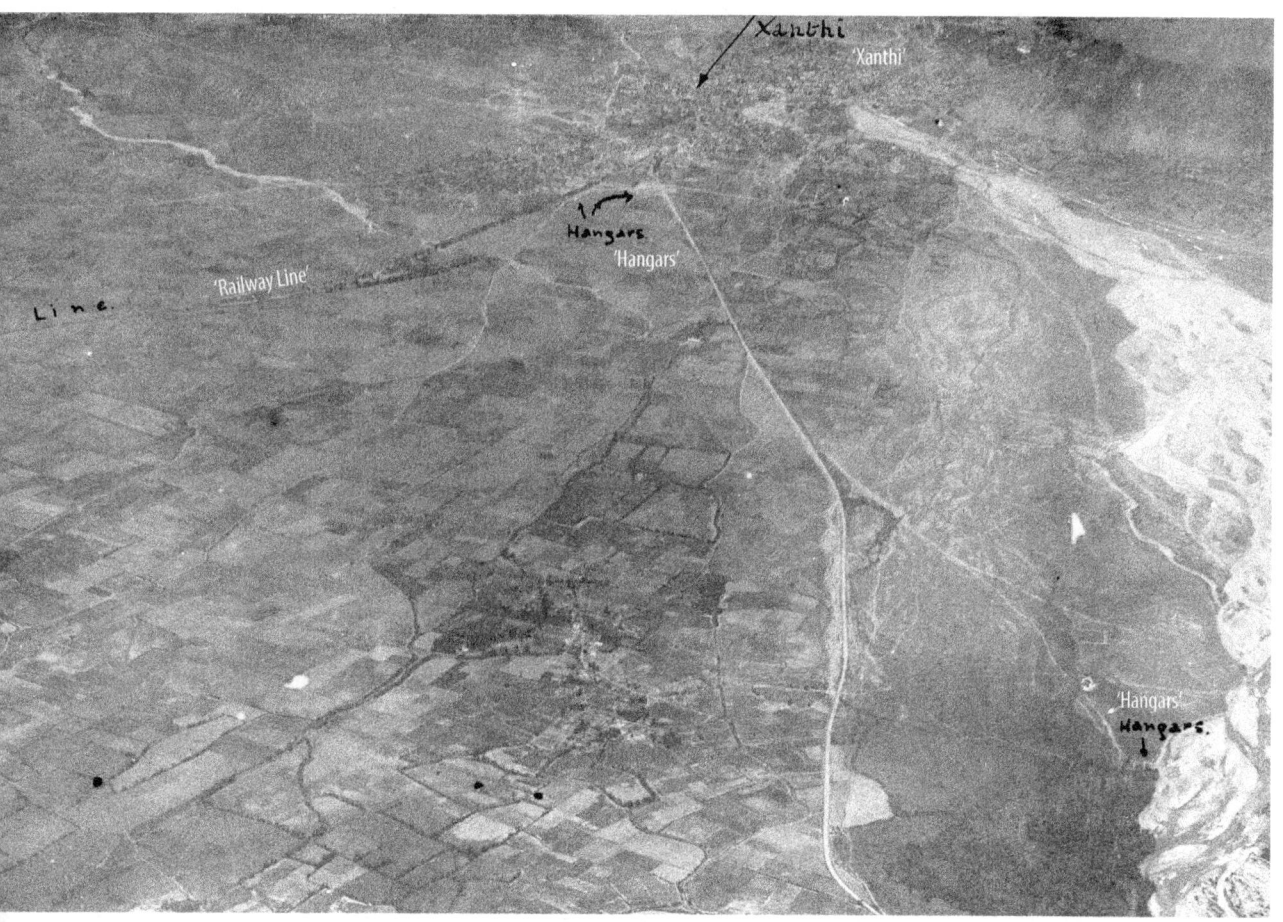

An aerial view of Xanthi (top) and Maswakli (bottom right) taken by A Squadron, 2 Wing, RNAS, Thasos, 8 August 1916. *(TNA, ADM 137/545)*

Friedrichshafen FF33L in a canvas shelter at *Seeflugstation* Xanthi 1916. *(Paschalis Palavouzis)*

Seeflugstation Xanthi, actually at Gereviz on the western side of Lake Boru near Porto Lagos. Universally regarded by those based there as 'hell on earth' due to the summer heat and humidity, endemic malaria, typhus, and other such diseases. The large hangars probably date this photograph to 1917/18. The two machines on the upper right are Friedrichshafen floatplanes probably a FF33L (dark) and an FF33E, the other two are possibly Gotha WD types.

Empress and the Eastern Mediterranean Squadron

Empress left Port Said on the evening of the 3 April 1916, and arrived at Mudros on the morning of the 6th. In her hangar were two assembled Shorts, 8088 and 8381, with a third 8095 still in transport crates. There were also unassembled Sopwith Schneiders 3772, 3787 and 3788. At Mudros the Short and Schneiders were transferred to *Ark Royal* for erection and testing. Her Flight was commanded by Flt Lt Roger Martin Field, with FSL Charles Vernon Arnold as the only other pilot, both had joined *Empress* in the UK in July 1915. The only observer aboard when she sailed from Port Said was Mid H.D. Thornton, additional observers were to be provided as required from 2 Wing.

On arrival at Mudros, *Empress* immediately had to spend a week cleaning her boilers. Meanwhile, Field and Arnold commenced practice bomb dropping, and made test flights with the Sopwith Schneider, with which they were unfamiliar. From 29 April a Sopwith from either *Ark Royal* or *Empress* was detailed for daily guard duties to attack enemy aircraft should they appear over Mudros.

An attack employing all available aircraft, British and French, had been planned for 3 May along the length of the Bulgarian Aegean coast. Poor weather prevented landplanes from the French unit at Salonika and 2 Wing aeroplanes from Imbros taking part. Only the floatplanes carried on board the ships and from the seaplane base at Stavros were able to participate.

On 2 May *Empress* proceeded from Mudros to Kephalo Bay, Imbros, arriving early in the afternoon. A Sopwith Schneider was transferred to the monitor *Abercrombie* together with a pilot and small maintenance party.[284] The monitor then sailed to be in position off the coast between Dédé Agatch and Porto Lagos the following morning, as part of a co-ordinated series of bombardments and bombing raids along the Bulgarian coast from Salonika to Gallipoli. *Abercrombie*'s report fails to mention the Schneider, other than it was aboard. However, from another report, we learn that 'the Schneider patrolled around [*Abercrombie*] for 4 hours without sighting any hostile aircraft.'[285]

Empress had been 'allotted the Eastern end of the line for the purpose of attacking Ferejik with her seaplanes.[286] A concentration of transport and troops had been reported at Ferejik and it had also been stated that there was a Zeppelin Shed, but this report was regarded as doubtful.' She sailed from Kephalo at 10.00 on the morning of 3 May, arriving off the mouth of the river Maritsa, just over an hour and a half later.

The Abercrombie-class 14-inch gun monitor *Raglan* leaving Harland & Wolff, Belfast shipyard shortly after completion, June 1915. The large area of open deck aft of the funnels, the boats are just temporary, was allocated for 'Stowage For Seaplane P&S', a kingpost and derrick boom can be seen port and starboard. *Raglan* was originally commissioned as *M.3*, as seen here, then briefly as *Robert E. Lee* and then as *Lord Raglan*, before finally becoming just *Raglan*. The typical big-gun monitor's broad beam, emphasised by the anti-mine and submarine bulges, is evident in this view.

Running up a floatplane on land always attracted an audience. Sopwith Schneider 3713 seen here at Stavros seaplane base around the time of its attachment to *Raglan*. Of interest is the extensive use of tree branches to camouflage the hangar. *(Paschalis Palavouzis)*

Short 8088 left the water at 11.45 piloted by Flt Lt Field with Midshipman H.D. Thornton as observer. It was carrying three 65-lb bombs and a Lewis gun for the observer. They reached the coast at 12.20 and, following the river, Ferejik fifteen minutes later. As expected there was no Zeppelin Shed anywhere in the vicinity.

> Dropped bombs at Barracks. About 50 men standing about Barrack Square in small groups. About 30 trucks in or near station. Gun and about 20–30 men seen 1 mile NE Ferejik. One AA Gun on railway truck on junction. Left Ferejik 1–5 pm. Observed gun emplacement containing two AA guns about 5 miles NE Dedeagatch. No rolling stock seen on Northern or Southern railways between Ferejik and Dedeagatch. About 20 trucks by East Station and 30 on sidings by West Station Dedeagatch. Guns in Dedeagatch not placed. Heavy firing from the guns mentioned was experienced and the machine was hit but not damaged.

The second Short was less successful. Piloted by FSL Vernon Arnold with CPO J.W. Bell as observer, 8381 was loaded with three 65-lb and eight 16-lb bombs, in addition to Bell's Lewis gun.

> Left ship at 12–4 pm, but owing to the engine not giving its full power and to the fact that there was no breeze, the machine refused to lift. Returned to the ship and removed the 8 16 lb bombs, making a fresh start at 1–24 pm. The machine lifted this time, but the engine continued to give trouble, and when at 2000 feet a compression cock blew off, making it impossible to gain further altitude. Returned to the ship at 1–56 pm.

After retrieving the Schneider from *Abercrombie*, *Empress* returned to Kephalo then, on the following day, proceeded to Mudros.

Other floatplanes had been supplied to the monitor *Raglan* and cruiser *Lowestoft*. The monitor had taken onboard Short 166, 166, and Schneider 3713, the cruiser being provided with two Schneiders, 1577 and 1579, all from *Ark Royal* detached to Stavros for the operation. *Raglan* was to be stationed off Porto Lagos to bombard any targets of opportunity, whilst *Lowestoft* was stationed to the south-east of Thasos.

Both of *Raglan*'s machines were soaked during overnight and morning rain storms, and struggled to take-off to attack Porto Lagos on the afternoon of 3 May. When they did take off the Short, Flt Lt J.S.F. Morrison and observer Sub Lt A.G. Errington, was unable to bomb as its bomb gear jambed 'probably due to the chassis of the Seaplane being strained in getting off the water. [Three] attempts were made to release the bombs under heavy fire from Anti Aircraft guns and hostile aeroplanes.' The Schneider, flown by Flt Lt E.H. Dunning,[287] was escorting the Short.

> [At] 1.10 pm. German seaplane 'FF' type seen flying low beneath Short. Attacked 'FF' Seaplane from below at 3500 feet. Schneider's gun fired most of one disc, then jambed. Enemy retired in a Northerly direction and was not seen again. On clearing jambs a German aeroplane, a 2 seater tractor biplane yellow in colour, and with vertical engine was seen following Short at about 5000 feet. [I] attacked from ahead and 200 feet below at 8000 feet. Gun behaved as before. After clearing jamb Schneider approached enemy machine which retired towards Xanthi.

The rotary engine was notorious for its liberal use of oil as well illustrated in this image of Schneider 3713. The mounting of a fixed Lewis gun through a cut away centre section is unusual. The pilot is Flt Lt E.H. Dunning, later killed attempting to land a Sopwith Pup on *Furious*. *(Paschalis Palavouzis)*

Lowestoft's Schneiders were tasked with escorting a Short 166, 165 (Sqn Cdr Kilner and observer Sub Lt Roberts), which had left Stavros at 11.30 to reconnoitre the Mesta River. Kilner was able to complete his reconnaissance returning to Stavros at 14.40 without interference. The Schneiders had departed *Lowestoft* and almost immediately ran into a severe rain storm. The first away, 1577 (Flt Lt Douglas) landed safely alongside the small monitor *M.28*. Sighting the Short he took off again but was unable to catch up then returned to Stavros. Landing out of fuel he had to come down between two caiques, hit a mooring buoy and flipped over. The second Schneider, 1579 (Flt Lt Pulford), crashed close to *M.28*. Both pilots were uninjured but the Schneiders were out of action.

Captain T.W.B. Kennedy (*Lowestoft*), commanding the operation, commented, 'The various flight officers concerned were undoubtedly greatly handicapped by the weather, and I respectfully submit that their perseverance was most praiseworthy.'

Budrum, the Aydin Coast, and the Gulf of Scalanuova

After a few days at Mudros *Empress* was ordered to proceed to Port Lakki, Leros, where she was to conduct a series of reconnaissance flights over Budrum (ancient Helikarnassos) and area. In the hangar were two Shorts, 8088 and 8361, and two Schneiders, 3787 and 3788. Pilots were Flt Cdr W.G. Sitwell, Flt Lt Field and FSL Arnold, with observers Lt Lord Torrington,[288] and CPO Bell.

At 04.35 on 10 May *Empress* sailed from Port Lakki, sailing south past Kalymnos island before heading east to reach the launching point off the western end of the mountainous peninsula guarding Budrum three hours later. Two Schneiders and a Short were sent off between 07.45 and 07.51, flying around the southern coast of the peninsula to reach Budrum. First away was Short 8381 with FSL Arnold and Lt Torrington, they were quickly followed by Flt Lt Field and Flt Cdr Sitwell on Schneiders 3787 and 3788 respectively. The Short was carrying three 65-lb and eight 16-lb bombs, Schneider 3787 had eight 16-lb bombs and 3788 just four 16-lb bombs. All three machines also had a Lewis gun. *Empress*, meanwhile, proceeded north around the peninsula to reach a rendezvous near Hagios Apostoli island (Apostol Kilisesi).

George Master Byng, 9th Viscount Torrington, was attached to *Ark Royal* but served aboard *Empress* during the May 1916 operations. (Paschalis Palavouzis)

Sopwith Schneider 3788 was brought to the Aegean aboard *Empress* and operated from the ship throughout May 1916. The Schneider was transferred to *Ark Royal* shortly afterwards. It is seen here in its original light coloured fabric possibly being hoisted onboard the cruiser *Endymion* in September 1916.

Flt Cdr Sitwell was first to reach Budrum, approaching at 4000 feet with the sun behind him. 'Dropped a bomb which fell 50 feet SE of castle, turned and repeated the evolution at 2800 feet, as there was no sign of hostile fire. Of the 3 remaining bombs, one missed north of, and the others exploded in the castle, but caused no obvious damage. By this time the two other seaplanes from "Empress" had commenced bombing and 6 of their bombs were observed at point indicated.'[289] Flt Lt Field's report was brief and to the point. 'In the harbour west of the promontory on which is the fort were a number of fishing craft. The bombs fell in the region of the castle or fort, and amongst the buildings at the root of the promontory. No movements were seen in the castle or town, and no shooting was experienced.'

Lt Torrington's report from the Short was more comprehensive.

> The objective of the first attack was the Fort on the small peninsular to the South of the town. This was attacked from a height of 3400 feet, and three hits were obtained with 16 lb bombs, but it was not possible to ascertain the exact damage done. At the base of this small peninsular, just North of the Fort was a large building, which was taken to be the house referred to, as the quarters of German Officers. A 65 lb bomb was dropped on this, and appeared to cause some damage.

The next point of attack was a new building on a small promontory to the West of the Fort, which appeared to be a Coastguard Station. One 65 lb bomb was dropped, but fell into the water about 50 feet to the north of object.

The last point of attack was a group of large new buildings in the Centre of the Turkish part of the town near the Mosque, but no damage was apparent.

Nowhere was there any sign of concentration of troops or war materiel, and at no place were we fired at, to our knowledge.

After photographing the town, we proceeded to cross the neck of the Peninsular to Durvandra Bay, and reconnoitred the Bay, but although the water was very clear, no signs of mines were seen.

All three machines safely crossed over the peninsula and found *Empress* 10 kilometres NW of Hagios Apostoli. The were hoisted aboard between 09.03 and 09.15. The ship then set course for the small islet of Kapota (Toprak Adasi), some 20 kilometres north of Hagios Apostoli, from whence an afternoon series of flights were to depart. These commenced a series of investigations of the Gulf of Mendelia with special attention for possible submarine bases, and to bomb any suitable targets. The same machines and crews set out from *Empress* between 12.04 and 12.21. All three returned to *Empress* between 13.08 and 13.41.

The results are best summarised in the one of the weekly Intelligence Reports prepared by Air Services HQ at Mudros. Whilst there was nothing to report elsewhere, Asin Kalesi Bay produced some interesting observations.[290]

On the north side of the bay, 'Geuren has a deep water harbour closed by 2 short breakwaters. At the northern end of the harbour there are 2 small

Bottom left: A photograph of Budrum looking to the south east taken by Lt Torrington from Short 8381 on 10 May 1916. Not a great image, but it shows the castle of St Peter, built in 15th century by the Knights of St John, on its penninsula, and the walls of the old arsenal on the lower headland. *(TNA, ADM 137/365)*

Below: Flight Commander Sitwell's sketch map of Budrum from the flight report, 10 May 1916. A is noted as 'a white building with a flagstaff alongside', Lt Torrington thought it might be a coast guard station, whilst B, C, and D are 'conspicuous buildings which might be stores.' *(TNA, ADM 137/365)*

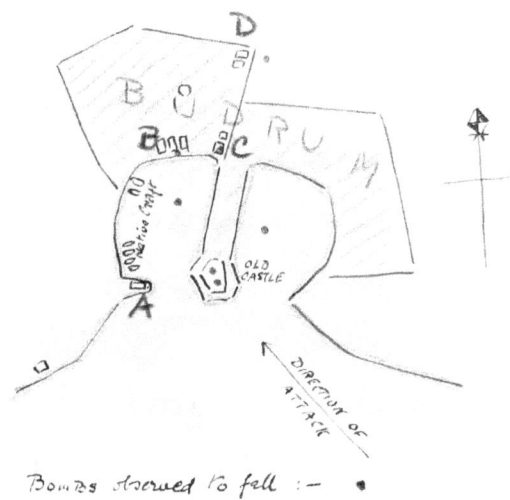

Again from Flt Cdr Sitwell's report, a sketch map showing the area investigated by the afternoon flights on 10 May 1916. (TNA, ADM 137/365)

piers and some buildings. This place has the appearance of a possible base for Submarines.' On the opposite shore, at Kuluk (Güllük), 'There are 5 jetties and on the coast there are several large, new red buildings which are probably store sheds. The road from Milas via Kuluk to Budrum seems in excellent condition near Kuluk.' Sitwell reported dropping three bombs on the store sheds, but did not see any hits.

The bombing at Dalian, north of Kuluk at the then mouth of the river Sari Chai, but no longer extant, was carried out by FSL Arnold with Lt Torrington on Short 8381.

There are two entrances to the Sari Chai Lagoon, the southern of which appears to be artificial. There are 4 large buildings, apparently new, which were bombed and damaged. As these appeared likely places for storing materiel for submarines, should the bay be used for that purpose, these were attacked, the remainder of the bombs being dropped here. Great damage appeared to be caused as, when we left, large clouds of dense smoke were seen rising from them.

Poor weather prevented a return until 13 May, when three morning flights were made over the same area. First off at 07.30 were Short 8381, FSL Arnold with Lt Torrington, escorted by Flt Cdr Sitwell on Schneider 3788. The Short had one 65-lb and four 16-lb bombs and the Schneider four 16-lb bombs. They were to reconnoitre Milas then join a second Short, 8088, Flt Lt Field with CPO Bell, to bomb a road bridge at Kujeli (Koruköy), south west of Milas. The second Short took off at 07.44 and flew directly to the bridge carrying two 100-lb bombs.

Short 8381 led the way to Milas, with Sitwell trailing 1000 feet above, following the road from Kuluk to Milas. Sitwell reported dropping four bombs on a supposed barracks, no results were observed. Arnold and Torrington also dropped two bombs 'at a large building to the SE of the town' (possibly the same building) again without effect. Both machines now turned back towards the Kujeli bridge.

Field and Bell dropped their two bombs from 1800 feet, 'the first falling about 20 yards to the South, and the second on the road about 10 yards Northward of the bridge, making a large hole in the road.' Sitwell was flying over the bridge at 7000 feet and 'observed one heavy bomb fall just west of Kujeli bridge, one in the centre of road about 30 yards NE of bridge, and a third in soft ground to South of bridge.' This was probably one of the bombs dropped by Arnold and Torrington, who 'failed to obtain a hit, the nearest falling about 20 feet short. A 100-lb bomb from the other Short seaplane was observed to fall on the road near the bridge.'

Whilst returning to the ship Field and Bell re-visited the targets of 10 May.

> Of the sheds at Dalian, two appear completely destroyed by the bombs dropped on 10th May. At Kuluk some damage was caused by the bombs of the 10th May [Sitwell's], the large square shed in the centre of the bay being in ruins. The number of jetties was seen to be eight, and not four as previously reported, and there were a number of sunken boats in the north side of the Harbour.

All the machines returned safely by 09.09, and *Empress* returned to Port Lakki, where she remained except for two short trips. An outing on 19 May by a Short with a Schneider escort quickly returned when the Short suffered engine troubles. A trip later in the month, on the 25th, was more successful. *Empress* steamed to a position SE of Samos Island, where at 05.00 a Short and two Schneiders flew off. The ship then steamed generally in a NE direction to recover the machines between 06.14 and 06.36.

The machines having left *Empress* proceeded to the coast and up the valley of the Maeandere river (Büyük Menderes, or Meander of antiquity), to some lignite (brown coal) mines located at Sokia (Söke). Although a poor quality coal, once made into briquettes it was exported to Smyrna by a branch line from the main Anatolia railway. The mines were dug into the hillsides, without the need for pithead lifts, consequently they were not located.

Port Lakki, Leros, in June 1916 with *Empress*, two destroyers probably including *Wolverine* lying on the landward side of the seaplane carrier, the sloop *Aster* to seaward, and three merchant ships. Of the latter, one was the collier SS *Wathfield* which twice suffered minor damage whilst alongside *Empress* coaling in the open roadstead. At the bottom of the photograph are the headlands guarding the entrance to the deep bay that is Port Lakki almost cutting through the lower part of the island. *(Flt Cdr W.G. Sitwell)*

'The Railway and Station were accordingly selected for attack and 14 16-lb bombs and three 65-lb bombs were dropped. One large bomb hit the corner of the Station and two other bombs dropped on the Railway line to the North of the Station.' *Empress* was back in Port Lakki by 09.00.

The following morning *Empress* sailed at 02.38 passing Pharmako (Farmakonisi) Island at 04.11. At 05.17 she stopped to launch one of her machines, then again at 05.59 a second machine. Both machines returned to the ship and were hoisted in by 06.35. One of the machines spotted for the destroyer *Wolverine*,[291] 'firing at a large two storied red roofed building near Ieronda (Yenihisar), to the South of the Biyuk Mendere River, and destroyed it, the seaplane also dropping bombs at a similar building but without scoring a direct hit.' The two ships returned to Port Lakki following this bit of demolition.

On 28 May *Empress* was ordered up to Port Iero, Mitylene, to act under the orders of the Senior Naval Officer, Port Iero, 'for the purpose of assisting in the further bombardment of Railway bridges in the Scalanuova Area.'

Leaving Port Iero in the early evening of 1 June, *Empress* proceeded south, passing through the Khios Strait between the island and mainland during the night. By 04.15, 2 June, she was in the Gulf of Scalanuova preparing her machines for the day's work, spotting for *Grafton* and the 12-inch monitor *Earl of Peterborough* on various targets in the area.[292] The *Earl of Peterborough*, assisted by a Short from *Empress*, obtained two hits on a railway bridge close to Ajasoluk (Selçuk). *Grafton*, meanwhile, was 'Firing on Skala Nuova barracks & Kusadasa fortifications. Set fire to magazine.'[293] Sitwell mentions the spotting in his log book, 'Spotting for "Grafton" supposed to be on guns N of Scalanuova but they suddenly opened up on the town!! They took no notice of my spotting by Vereys lights. Rather an unpleasant trip and a bad sea for the Schneider.'[294] Various observation and reconnaissance flights were made during the day. At one point *Empress* came to a stop and presented an irresistible target for local shore batteries. She was not hit and hurriedly got under way. The following day a final reconnaissance was flown over the area around Chesme, on the mainland at the southern entrance to the Khios Strait, finding nothing of military interest. *Empress* returned to Port Iero at 12.45 on 3 June, and was immediately ordered to Mudros, proceeding at 15.00 and arriving seven hours later.

Empress spent the next couple of weeks undergoing self maintenance and, with assistance from *Ark Royal*, 'stripping and truing up Machines and engines.' Short 8088 was condemned as beyond local repair, although no reason was given. On 12 June Short 8381 was transferred to *Ark Royal*, and returned four days later. When *Empress* sailed on 19 June, to join a gathering fleet at Milos island, she had on board Short 8077, 8078, and 8381, and Schneiders 3772 and 3787.

The Problem of Greece

The Allied armies in Salonika had more than the Austro-German and Bulgarian armies to be on guard against. Behind their backs was a politically volatile Greece, with a large army, and whose long term goals were at best uncertain. Salonika was effectively in a state of siege. A fact that constantly affected plans of General Sarrail's staff.

> The [French] naval authorities urged that action should be quickly and impressively taken. A partial blockade of the Greek coasts was instituted; and detailed preparations were made for a demonstration against the capital. By 20 June [1916] the men and ships were assembled in Milo Bay, and lay crowded in the land-locked, airless harbour in the sweltering heat of the Aegean summer.[295]

On 21 June an Allied Note was presented at Athens demanding primarily that the Greek army should be demobilised. The Greek government agreed, and so, the crisis passed. The Franco-British Fleet at Milos was dispersed, though arrangements were made to reassemble it at a moment's notice. The ships returned to their previous duties. Within a week *Empress* had returned to Port Lakki (see below).

As the Greek Crisis did not go away, a reinforced fleet was reassembled at Milos at the end of August.[296] Briefly, the powerful Bulgarian riposte to General Sarrail's August 1916 offensive seems to have unnerved the French government, already struggling with Verdun. In a move to prevent the Greek Army from taking advantage of the situation, *Amiral* Dartige du Fournet, had been instructed to, 'Immobilise the Greek fleet and the Greek General Staff; seize all enemy ships at Eleusis; seize all enemy agents and establish control over all telegraph post offices, wireless telegraph stations, coal and petrol depots and railways; and endeavour to cut off railway communication between Athens and the Peloponnese.' In response to requests from Paris, the British Government instructed the Vice-Admiral, Eastern Mediterranean,[297] to detach a force to Milo to co-operate with the French fleet.

Under his control *Amiral* Dartige du Fournet had the dreadnought *Provence* (Flag), and the *3ème Escadre de Ligne* comprising six pre-dreadnoughts, two armoured cruisers, 12 destroyers and many smaller vessels, also the seaplane carrier *Campinas*. The Royal Navy's contribution of the pre-dreadnought *Exmouth* (Flag of Rear Admiral Hayes-Sadler), two old light cruisers, monitors *Earl of Peterborough*, *M.29* and *M.33*, and four destroyers paled by comparison. However, the

Amiral Dartige du Fournet, commanding the assembled fleet at Milos.

His flagship, the modern dreadnought *Provence*. Commissioned in March 1916 *Provence* was armed with ten 340mm and 22 138mm guns. She was almost 25m shorter than the British battleships of the same period but of similar beam. Seen here possibly on trials when she reached 20 knots.

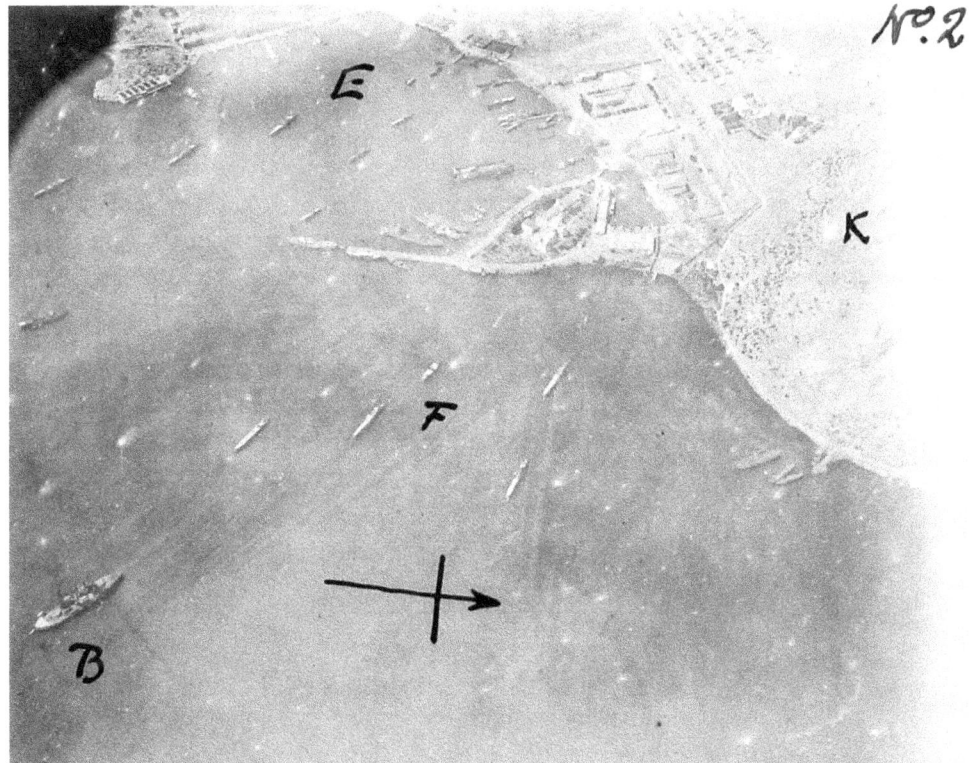

The Greek Fleet at Piraeus photographed on 1 September 1916 from 5000 feet by Short 8077, FSL Arnold with Sub Lt Holcombe as observer. B is the Greek pre-dreadnought *Lemnos*, ex-USS *Idaho*. E and F are torpedo craft with, in the harbour, a floating dock and a Hydra-class ironclad (1889) alongside one of the breakwaters. Not seen on this photograph are the second pre-dreadnought *Kilkis*, ex-USS *Mississippi*, and the modern cruiser *Averof*. (TNA, ADM 137/545)

Royal Navy did include *Empress*, whose Shorts made an important, and overlooked, contribution to succeeding events. The fleet commenced leaving Milos late on 31 August, the monitors, sloops, drifters, and the auxiliaries (including *Empress*), proceeding first, the heavy ships following on the morning of 1 September.

The two carriers had special orders. *Campinas* was to have two FBA flying boats available for spotting fire on to shore batteries should they open fire. Once in the vicinity of Phleva Island, *Empress* was ordered to 'put out two seaplanes which will reconnoitre approaches to Salamis Bay and Strait and the anchorage, reporting as soon as possible the presence of mines or submarines.' At 10.25 the ship stopped just long enough to send off Short 8078. Ten minutes later she stopped once more, this time to send off Short 8077, which returned to the ship at 11.40. 8078 was hoisted in at 12.54. There is no record of the crew or flight of Short 8078, however a complete Record of Flight for 8077 has survived.[298]

Short 8077 was piloted by FSL Arnold with Sub Lt A.C. Holcombe as observer. There were no bombs, just a Lewis gun for the observer. Their report includes a tracing from Admiralty Chart No.1513, and three photographs. Leaving the water at 10.45 they headed directly to the naval dockyard on Salamis Island at the northern end of Salamis Strait close to Eleusis Bay. Here they found the Greek fleet at anchor: 'Consisting of 24 vessels, including eight large ships, the *Kilkis, Georgios Averof, Lemnos, Spetsai, Hydra, Helle* and 16 torpedo craft.'[299] They also reported a floating dock and a number of small craft. In Eleusis Bay were 'about 20 merchant vessels to the north of the bay.' At Arabi Point on Salamis Island there were 'three large gun emplacements commanding the approach to the dockyards from the south.' Returning down the Salamis Straight, protective nets were seen running from Georgio Island to a shoal close to the mainland, effectively closing the shipping channel. There were also on Keromos Point, facing Lipso Island, 'four guns, about nine inch, in permanent open emplacements, commanding the entrance to the Straits.'

Early in the afternoon of 1 September the main units of the Allied fleet reached Phleva Island. From here the battleships were escorted up a channel, which had been swept and marked by drifters, to an anchorage within the Salamis Strait. As soon as the ships anchored, a torpedo net was laid behind them between Point Keromos and Lipso Island. The guns at Keromos Point observed by *Empress*' Short had apparently been disabled. The cruiser division under *contre-amiral* Biard seized all Greek ships lying off Piraeus and towed them to the fleet anchorage. The main Greek fleet lying at anchor before Salamis dockyard, and behind its own torpedo nets across the Giorgio Channel, was completely immobilised. To get to sea it would have to pass between the batteries of the entire battle squadron in Salamis Bay.

Flight Sub Lieutenant C.V Arnold. *(Dulwich College Roll Of Honour)*

Amiral Dartige du Fournet subsequently directed Admiral Hayes-Sadler to congratulate the Senior Flying Officer, *Empress*, on the quality of the report. With the completion of the flights *Empress* had no further role to play and, on the evening of 5 September, she was released and proceeded to Mudros, arriving the following morning. At this point the politicians take over and we quietly follow *Empress* to Mudros.

Suffice it to add that the Greek fleet was seized by the Allies in October. The larger ships were demilitarised; the cruiser *Helle* and the destroyers were incorporated into the French Navy. When Greece declared war on the Central powers in July 1917 the ships were returned as Greece found the crews to man them.

Back to Anatolia

When first released from Milos in June 1916, *Empress* proceeded to Mudros. After coaling, and exchanging some of her floatplanes, she sailed for Port Lakki on 26 June. She had onboard two Shorts, 8095 and 8381, two Sopwiths, including 3772, three pilots (probably Flt Cdr Sitwell, Flt Lt Field and FSL Arnold) and two observers. She remained at Port Lakki until the end of July. Her log book does record some flights, but the weekly Intelligence Reports contain little information. Strong winds, whipping up impossible seas for floatplanes, cancelled flying for much of the period, but successful flights were made on 12 July, and the last few days of the month.

In her previous visit to Port Lakki, flights had concentrated on Budrum and the coast around the Gulf of Mendelia. This time an area centred on the delta of the Maeandere. On 12 July, according to *Empress'* log book, whilst a little to the north of Pharmako Island: '5.19am: 3772 left water; 5.33am: 8095 left water; 5.56am: 8381 left; 6.06am: 8095 returned, engine trouble; 6.59am: In 8381 & 3772.' Flights covering the whole coastal area were made, 'from Port St Paul to Meander River, Kavella Bay and Ta Kokina. Photographs of the Bays were obtained and four 16-lb bombs were dropped on the breakwater at the entrance to Derin Geul. Visibility was not good and no damage was observed.' Derin Guel was a lagoon at a silted up old mouth of the river, a modern irrigation scheme has brought the present river mouth back to a point just north of Derin Geul. At the time of the flights the river's mouth was well to the south, it having curved around Lade Island, landlocked by the ever growing river delta. In antiquity Lade Island had been the site of two naval battles between the Greeks and the Persians in 494 BCE and Macedon and Rhodes in 201 BCE, by 1000 AD it was being surrounded by silt from the river.

Further flights were not possible until 26 July when, on three consecutive days, the areas around the mouth of the Maeandere were revisited. Once again from the vicinity of Pharmako Island, two Shorts set out, 8381 shortly after 06.15 and 8095 about 20 minutes later. Their target was a stone road bridge located crossing the Maeandere some 12 kilometres inland at Sari Kemer (almost due east of Derin Guel). Three 65-lb bombs were dropped from 1500 feet. Whilst all were close to the target, the nearest was within 10 yards of a pier, a direct hit would have been required to cause any damage.[300] Both Shorts had returned by 07.50.

The following day just 8095 set out, leaving *Empress* at 06.10 and returning at 07.20. The target for the day was Lade Island, where defences were to be reconnoitred and bombed. The Intelligence Report recorded that three emplacements were observed on the summit of the hill. One was reported as 'very conspicuous – probably a dummy. A direct hit was obtained on one of these with a 65-lb bomb, and caused a large explosion.' The attack was followed up on 28 July again by 8095 this time carrying eight 16-lb bombs. They were dropped over the guns emplacements 'obtaining a direct hit on one emplacement which contained a gun.'

Empress was now required once more at Port Iero, sailing at 01.52 on 30 July. Steering a course to past west about Samos Island, she passed Cape Katavasis at noon. Some time earlier Short 8095 had set out, returning an hour later. There is an indication that this flight investigated the Samos Straits, between the island and mainland, but no details have survived. The carrier dropped anchor in Port Iero at 19.31.

Her first flights here were on 6 August, spotting for the monitor *M.33*. *Empress* had left Port Iero at 01.00, escorted by the destroyer *Jed*, an older (1904) River-class ship. The target was a single gun position on Cape Bianco in the Khios Strait. Short 8095 left the ship at 06.03 and commenced spotting at 07.25. The shoot was completed by 08.35 and the Short recovered at 08.44. 'The gun did not open fire and the results are considered satisfactory.'

Bad weather prevented any further spotting trips and over night 17/18 August *Empress* proceeded to Mudros 'to replace worn Seaplanes [including 8095, and 8381] with new ones which had been erected & tested in readiness for her.' Remaining there until on 27 August, escorted by the destroyer *Scourge*, *Empress* sailed for Milos, as recorded above.

Stavros

Her work at Milos completed, *Empress* returned to Mudros, remaining there until 11 September when she was ordered to proceed to the anchorage at Stavros Point. Arriving on 12 September she became part of the 6th Detached Squadron, under command of Captain Edgar Grace, *Grafton*.[301] With a few brief absences *Empress* was based at Stavros until the end of December 1916. *Empress* rarely had more than two Shorts, and a Sopwith Schneider or Baby available, but she had the benefit of mobility. Many operations were carried out in conjunction with, or with the support of, the RNAS units based on Thasos and at Stavros. Much of the work of the 6th Detached Squadron was along the coast between the rivers Struma and Mesta. The Royal Navy ships forming the Squadron included *Grafton* (Flag), the large monitor *Sir Thomas Picton*, with several smaller M-class monitors. To support and protect the 'Firing Squadron' were several armed trawlers and Motor Launches (ML).

In early October 1916 *Empress* exchanged her Short 184s for Short 166s, including 166 seen here earlier in the year whilst operating from the beach at Stavros. On a much calmer day, with ample willing hands if required. *(Paschalis Palavouzis)*

Stavros seaplane base had closed down several months earlier. On 10 September, D Flight (later Squadron), 2 Wing, RNAS, was transferred from Mudros to Thasos, and an advance base was prepared at Stavros. The airfield was located further north than the old seaplane base, behind the present day Vrasna beach, close to where Lake Beschik (Bolbi or Volvi) empties into the Gulf. At this time the unit was mostly equipped with Nieuport 12 two-seaters. The Stavros detachment was small, rarely more than two Nieuports, and relieved regularly by fresh personnel and machines from Thasos.

Following the ill-fated August offensive, the Struma Front had largely stabilized into a form of trench warfare that endured well into 1918. Essentially the Bulgarians, with Turkish and German support, held the east bank and the British the west bank of the river from the coast at Cajagzi (Chai Aghizi in most reports), past Lake Tahinos and through the marshes beyond. Once past the marshes both lines were east of the river, until they turned west towards Lake Doiran. The lake and marsh areas were abandoned during the summer malarial season, both sides moving to higher land, pushing occasional patrols into the area. So, aerial observation was essential.

The area of flat land east of the Struma, between Bunar Dagh and Simvolon Dagh, was the main area kept under observation by the floatplanes from *Empress*. Orfano (Orfani) was centrally located on the Ak Dere stream and about six kilometres from the then coastline. Tuzla farm (or *chiftlik*) was located west of Orfano and close to the coast; it had been the site of a French airfield earlier in the year. The village of Dranli, about six kilometres north west of Orfano, is another frequently mentioned location.[302] The advantage of a regular beat is obvious; flight crews become familiar with the area and can quickly detect any changes. *Empress*, weather permitting, had machines over the area every morning and afternoon from her arrival until her departure. Flights usually included some bombing and reconnaissance combined. If a suitable target presented, they were in W/T contact with the duty monitor, usually one of the M-class small monitors. Each monitor had its own fixed position in which it anchored off the mouth of the river. The naval guns on the ships of the squadron replaced local field artillery support for the troops. Additional spotting for the monitors was provided by a Naval Observation Station, manned by two midshipmen from *Grafton*. The station was established on high ground to the west of the river mouth.[303]

In his report for events for the week prior to 17 September, Captain Grace noted that *Empress* had 'arrived at Stavros and is settling down to steady routine work.' Most of the local flights could be accomplished from the ship at anchor behind the torpedo nets protecting the anchorage, and very little information about them is available. Most taking the form of the following note for 20 September. '*Empress* and D flight carried out reconnaissances. *Empress* reported camps have been removed from Karjani and Orfano.'

On 25 September one of *Empress*' Shorts spotted for *Grafton* and *M.18*, on gun positions located close to Cajagzi. The position of the gun had been incorrectly reported, but the Short found the gun and successfully directed the two ships' fire on to the target. On 26 September, shortly after dawn, FSL Arnold and Sub Lt Hampton

A rare view of Short 166 in the air over Stavros. The oddly proportioned underwing roundels are interesting. The float spreader bars were optimistically arched to carry a torpedo. *(Paschalis Palavouzis)*

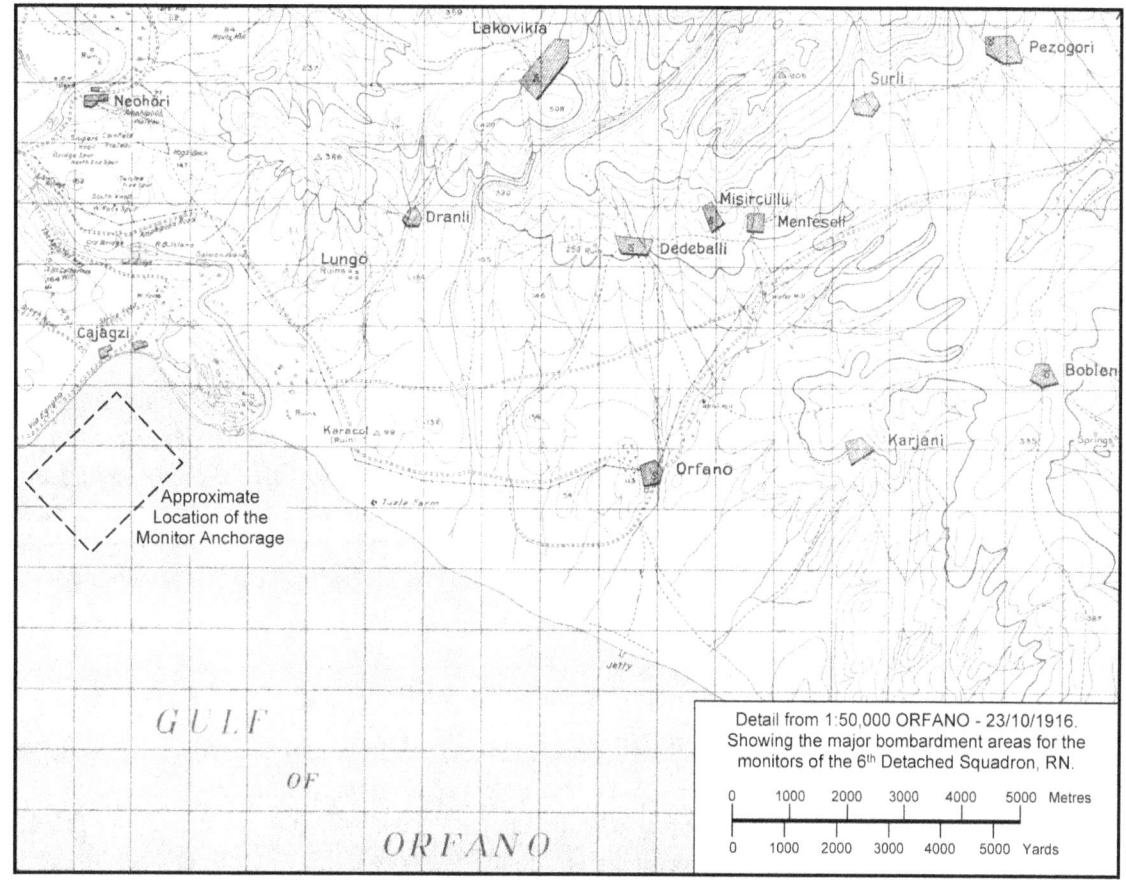

Detail from 1:50,000 ORFANO - 23/10/1916. Showing the major bombardment areas for the monitors of the 6th Detached Squadron, RN.

set out to reconnoitre the lower Struma area. 'They were heavily engaged by hostile anti-aircraft guns, seven of which were located, and much useful information about the defences of the Lower Struma area and on the coast line was obtained.'

On 29 September one of *Empress*' Shorts spotted for *Grafton* and D Flight spotted for *M.28*. Both ships onto the same target, the village of Dranli. A Nieuport from D Flight, Cdr Smyth-Pigott and 2Lt Barry, observed about 4000 troops drawn up in squares outside the village. On this date *M.28* was the duty monitor and, at 12.30, commenced firing her 9.2-inch gun on Dranli, directed by the Nieuport. *M.28* later reporting, 'From the spotting corrections received ... it was difficult to understand what object the aeroplane wished to hit & it was presumed he was chasing troops up the hill.' A shell had burst amongst a group of 50 men. Not surprisingly at this the troops scattered, many of then heading up a gully leading into the hills. The Nieuport was directing fire at this group 'with good effect.'

Later in the afternoon, a Short from *Empress*, Flt Lt Field with Sub Lt Hampton, was over the same location. They had been briefed before taking off, Dranli being identified as 'Target I', and the gully as 'Target II'.

Grafton was flagship of the 6th Detached Squadron commanded by Captain Edgar Grace. One of nine Edgar-class protected cruisers all launched between 1890 and 1892 they were armed with two 9.2-inch guns one forward and one aft, and ten 6-inch guns five on each beam. In late 1914 *Grafton*, and several of her sisters, was refitted for shore gunfire support and the 9.2-inch guns replaced with 6-inch guns. Anti-torpedo bulges were also fitted as seen here. The multitude of boxes appear to be providing protection for swan-neck vents leading from the bulges.

Having familiarised themselves with the targets, at 16.33 they commenced directing *Grafton*'s 6-inch guns on to the targets. The first four salvoes all burst on the village, causing an outbreak of fire. The following four were directed into the gully. Two rounds from *M.28* were also directed at Dranli, but 'It was found impossible to distinguish between fire of *M.28* and Grafton owing to *M.28* being invisible in the glare of the evening sun.' At 17.20 the Short's engine began to run rough, but it kept going for another hour before the machine returned to *Empress*.

Captain Grace reported for 29 September:

> Apparently the enemy considered this to be a prelude to a general attack as they were observed to open fire from three different positions. The firing appeared to be directed at British lines and not at the ships firing. Seaplane proceeded to spot *Grafton* on to these in turn and reported that four guns were destroyed. There were obvious explosions, and finally at 8.30 pm in the evening a heavy explosion was observed in the gun position to the north of the village. This gun position blazed up and flames were observed there for several minutes afterwards by the Monitor on duty. The explosion was also plainly visible from Stavros.

Then on 30 September:

> An operation of considerable importance arranged by the military for Saturday afternoon but further to the north of this position, our co-operation with 80th Brigade was asked for in an artillery demonstration. This was successfully carried out by *Grafton*, *Picton*, *M.32* and *M.28*, assisted by spotting from aeroplane, seaplane, and Naval Observation Station. The Military authorities were very satisfied with results of operation.[304]

Changing Shorts

At 02.10 on 6 October 1916, *Empress* slipped out of Stavros anchorage for a brief visit to Mudros. She returned at 23.52 to following day. Whilst at Mudros she was alongside *Ark Royal* exchanging her floatplanes. Nothing unusual in that as *Ark Royal* was the central maintenance base for the RNAS in the area, what was unusual was that her two Short Type 184s were exchanged for two Short Type 166s, 166 and 9755. At the same time, Flt Cdr W.G. Sitwell and Flt Lt R.M. Field[305] packed their bags and moved to *Ark Royal*, their cabins being taken by Flt Cdr R. Whitehead and FSL H.G.R. Malet. Flt Lt C.V. Arnold, who was promoted on 1 October, remained with ship until late January 1917. They were later joined by FSL G.S. Abbott from the Thasos seaplane base. Observers during this period included Sub Lts S. Hampton, A.C. Holcombe and D.P. Rowland.

According to the Mudros weekly report for 29 September:

> The 225 HP Short Seaplanes at present carried by *Empress* have a poor rate of climb and are sluggish on control, rendering them difficult and tiring to handle in the disturbed atmosphere experienced in the Gulf of Ruphani. It has been thought advisable therefore to exchange these 225 HP Shorts for 200 HP Canton-Unné Shorts, the exchange being accomplished by an exchange of Pilots, those who are experienced in the 200 HP Shorts being sent to *Empress*.[306]

Changing from 225 hp to just 200 hp does not, on the face of it, appear to make much sense. True, the Short 184 was not the most sprightly of machines, but then none of the Shorts had a reputation for speed or manoeuvrability. The Type 166 was smaller and lighter than the Type 184, although the power to weight ratio for the two was almost identical. The Type 166 was, if anything, slightly slower than the Type 184, and neither was a good climber. There is, however, an account to indicate the Short 166 may have been more pleasant to fly. FSL Norman Woodhead was completing his training in March 1917 with a conversion course to floatplanes at Calshot. He had first flown a Short 184, going solo on 13 March. He later had the opportunity to fly the Short 166 which he found to be 'lighter on the controls than the 184.'[307]

One advantage the Short 166 had was its engine; the Canton-Unné was considered more reliable than the Sunbeam of the Short 184. In addition the mechanics in *Ark Royal* had at least a year's experience with the engine. *Ark Royal*'s log book notes that '9 Air Service Ratings exchanged with EMPRESS' suggesting that some of that experience was shared with *Empress*.

Back to Stavros

Following her return from Mudros until mid-January 1917 *Empress* operated on a daily basis with just two Short 166 and a single Sopwith. A typical, and particularly effective, flight was made by FSL Malet and Sub Lt Holcombe on 18 October. Flying Short 166, 166, they left *Empress* at 14.50 and returned at 16.32. 'Two hundred enemy cavalry were seen under the trees between

Small monitor *M.18* worked regularly with *Empress'* Shorts whilst based at Stavros. The single 9.2" mounted forward came from one of the Edgar-class cruisers, the monitor was essentially designed around the gun.

Orfano wood and Orfano village. Incendiary bombs were dropped on the wood and the cavalry were attacked with Lewis gun fire. The cavalry took refuge in a ravine, accordingly the observer informed HMS *Grafton* and *M.18* thereupon shelled the ravine.' The two continued their harassment of ground forces on 5 and 6 November, this time whilst flying 9755.

During a reconnaissance of the Orfano area on 5th Nov, a large camp was observed on the Northern slopes of Hill 154.[308] It appeared to be occupied by over 1000 men and about 80 horses were tethered in front of the huts. On the following day the same pilot and observer dropped 6 bombs on the camp and attacked the horses with Lewis gun fire from 2000 feet. *M.32* was then spotted on to the camp, 60% of the shots being direct hits on the camp. Later in the day, *M.32* was again spotted on to the camp. Nine rounds of 6" HE were fired, 3 of which fell among the huts and horses, while the other 6 were all within a few yards of the camp and must have done considerable damage.

Co-operation with the RNAS detachment at Stavros airfield was common. A Nieuport 12 'Gunbus' (8736?) which was based at Stavros, although seen here at Kucos forward operating base, near the outlet of Lake Tahino, in late 1916. There is a Lewis gun on an Etévé mounting for the observer and another fixed forward firing Lewis on the starboard (and possibly port) side of the forward cockpit. (FSL N.H. Starbuck's album, via Paschalis Palavouzis)

On 31 October an operation was carried out along the length of the Struma Front. The major action took place north of lake Tahinos and the marshes, gaining the name of the Affair of Bairakli Jum'a (Iraklia) in the Official History.[309] For the 80th Brigade, at the mouth of the Struma, this involved a trench raid in strength, supported by a heavy naval bombardment of enemy positions. Weather was unfavourable for flying operations, although D Flight and *Empress* (Short 166, 166) were able to make spotting flights late in the afternoon. 'Continuous and most valuable work had, of course, also been done by the aircraft.' In particular, 'A great deal of valuable information has been obtained recently by *Empress* seaplanes who usually carry out reconnaissances daily at 2000–3000 feet over an area where AA guns are at times very active.'

A signal was subsequently received from 80th Brigade Headquarters. 'Turkish deserters report that naval fire was terrible and was splendidly directed and caused many casualties and consternation. Both prisoners' clothes are stained yellow by lyddite. Many more deserters will come. No Bulgar troops or artillery remain here, only 45th Regiment Turkish.'

The raiding was not always one way. On 18 November two attacks were made on the ships off the coast and at the anchorage. At 07.20 a seaplane dropped three bombs at *Sir Thomas Picton* lying off Cajagzi, missing by 800 yards, and six more close to *Grafton* and *Empress* at Stavros. At 10.00 three machines were approaching Cajagzi but turned back when the ships opened fire and they caught sight of a Nieuport from D Flight and Schneider 3772 from *Empress* coming to attack them.

At 11.35 on 30 November, Short 166, 9758,[310] left *Empress* to make a reconnaissance over the Karjani-Boblen (Akropotamos) area, east of Orfano on the north western edge of Simvolon Dagh. The clouds were low and the machine was damaged by machine gun fire. The pilot, FSL Abbott, was wounded by a bullet through his right leg. Abbott was able to bring the machine back to *Empress*, making a good landing at 12.36. His wound cannot have been too serious as it is not documented in his service records. The Short was repaired and back in the air the next day.

Sir Thomas Picton, the seventh of a class of eight *Lord Clive*-class monitors. They were almost identical in size and appearance to the preceding *Abercrombie*-class. The main difference was their main armament of a twin 12-inch gun turret from *Majestic*-class pre-dreadnoughts. The 12-inch turret was more compact than the American design and mounted on a higher barbette. Only two served in the Mediterranean, *Sir Thomas Picton* and *Earl of Peterborough*.

On 6 December, *Empress* was instructed to proceed to Salonika. Arriving on the following evening, she was to remain until 18 December before returning to Stavros. *Empress* and *Sir Thomas Picton* were being held available to bombard the Athens-Larissa railway near Lamia 'should the necessity arise.' The concern apparently being the ongoing political disturbance in Athens and concerns about the possible reactions of a large number of Greek Royalist troops in Northern Thessaly. A landing of 4500 British troops, with artillery, was made on 11–16 December at Skala Vromeri, near Katerini, to block any potential advance of the Greek troops. Within a few days of the landing it was reported that the Greek troops were retiring. No longer required at Salonika, *Empress* was released to return to Stavros, arriving on the morning of 19 December. Not before time, as Capt Grace had been complaining 'the absence of HMS *Empress* is felt.'

In the afternoon of 19 December, *Empress* was back at work flying a reconnaissance of the Dranli area. On the following day Dranli was revisited and bombs dropped on objectives in the vicinity of Orfano. Thereafter, returning to the daily round of flights.

On 22 December Captain Grace handed over the duties of Senior Officer, 6th Detached Squadron, to Captain Francis Clifton Brown, *Edgar*.[311] We shall encounter Captain Grace and *Grafton* again in a later chapter.

Christmas Day 1916 was just an ordinary day.

> Two submarine patrols were carried out in the gulf of Orfano and two reconnaissances of the Dranli, Karjani, Orfano area were also made.

During the first of these signalling was observed to be taking place from a trench near Tuzla farm. The trench was immediately attacked with four 16 lb bombs and no movement was afterwards observed. During the second reconnaissance two 16 lb bombs were dropped on the new redoubt on Hill 113, just to the W of Orfano village, which had been located by a seaplane on 20th December.

Two days later *Empress* was off to Mudros, and *Ark Royal,* to exchange her floatplanes. Returning to Stavros on 30 December in her hanger were three Short 166, 9758, 9763 and 9764, and Sopwith Baby 8202. Arriving at 08.05, she made her first flight at 11.00. On 1 January 1917, the report of an Albatros type circling over Cajagzi at 7000 feet apparently taking photographs, caused *Empress* to send off Baby 8202. Climbing up to the Albatros the pilot attempted to attack, but it turned away and quickly escaped.

Four days later Short 166, 9758, FSL Malet with Sub Lt Rowland as observer, spotted the monitor *M.18* on to a target in the village of Boblen. The monitor was at anchor off Cajagzi firing at a range of 19000 yards; *M.18*'s BL 9.2-inch Mark X gun had a maximum range of 25000 yards.[312] Lt Cdr R.G. Stone, *M.18*'s CO, reported after firing only seven rounds 'Spotting by aeroplane the signals of which were very faint.' Faint or not, three direct hits were obtained. The Short then made a reconnaissance of the Karjani area, locating several new gun emplacements, before returning to *Empress* at 12.47 after just over a two and a half hour flight.

Not all locations in *Empress*' area took their names from local features. For example, on the afternoon of 13 January, Short 166, 9758, made a reconnaissance of the Orfano-Dranli area at the end of which four 16-lb bombs were dropped on trenches and redoubts in Latrine Nullah, just over two kilometres southwest of Dranli. Then on the following day Short 166, 9763, escorted by Baby 8202, spotted *Edgar* on to targets in the Dranli area, including Latrine Nullah, destroying a camp and some sheds. Then again on the 15th, 9764 spotted monitor *M.32* on to the same targets. The escorting Baby 8202, dropped bombs in the area but the pilot spotted an 'enemy machine of the Nieuport Scout type' preparing to dive on him. 'He therefore dived and turned to get into a better position, and on flattening out could not locate the enemy machine which, after having been discovered made no attempt to engage either the Schneider or spotting machine. The enemy machine was much faster than the Schneider and flew off to the Eastward.' Later in the afternoon an Albatros C-type flew over Stavros at 12000 feet. A Schneider took off in a vain attempt to catch it, the German machine disappearing in the direction of Thasos.

On 16 January, Short 166, 9758, piloted by FSL Abbott with observer Sub Lt Holcombe, made a morning spotting flight for *Edgar*. They ranged on to several targets in the Dranli area where a camp and some sheds were destroyed.

Not the best photograph but images of *Empress'* machines in the Aegean are rare. Short 184, 8095, sometime in 1917 whilst operating from Tenedos. Beach operations in a choppy sea are not easy but always entertaining for casual observers.

Next the guns were registered on Debeballi, east of Dranli and north of Orfano, where a gun emplacement was hit. Then across the valley to the north east of Orfano, where some horse lines were hit killing several horses and stampeding the remainder. Three widely separated targets in one shoot, a good demonstration of the value of airborne spotting.

Empress' last day of operations in the Aegean was on 17 January 1917. Between 10.22 and 10.33 the Baby and all three Shorts set out together to bomb Boblen, which was thought to house the headquarters of a regiment of Turkish irregulars. Twenty five bombs, 65-lb and 16-lb, were dropped with unknown results. All machines were safely back by 12.33.

On 19 January, *Empress*, having been ordered to proceed for a refit, began transferring men and machines to Thasos. In the early afternoon Short 166s, 9758 and 9763, with Sopwith Baby 8202 flew off to Thasos, where they were attached to A Flight. In addition, 25 maintenance ratings, stores, petrol, and bombs were dispatched to Thasos aboard *Trawler 298*.[313] *Empress* sailed from Stavros at 03.19 on 20 January, arriving at Mudros at 12.30. She remained at Mudros for several days completing transfer of stores, spares and armaments to *Ark Royal*, including Short 166, 9764. On 23 January, *Empress* sailed once more, proceeding to Genoa for an extensive refit commencing 30 January.

From contemporary reports it is clear she was expected to return to the Eastern Mediterranean Squadron following her refit. However, the recent loss of *Ben-my-Chree* changed these plans as *Empress* was now required back at Port Said as the new flagship of the East Indies and Egypt Seaplane Squadron. Sailing from Genoa on 30 March 1917, she stopped at Suda Bay in Crete where on 7 April, 'All Royal Naval Air Service stores [were] discharged to air station.'[314] *Empress* arrived and moored up at Port Said at 17.00 on 11 April, where the next few days were spent receiving replacement stores and the first of her new machines, Short 8020.

CHAPTER 13

Anne Carries On

Return of the *Ben-my-Chree* survivors

In that mystical way they have, rumours about *Ben-my-Chree* began circulating at Port Said within hours of her loss. Eventually, the rumours became official, but it was not until her survivors returned to Port Said that the details became known.[315]

> Having lost the accommodation of the ship's cabins, they had to find temporary quarters ashore, and one came upon them unexpectedly in the town refreshing themselves at the expense of eager circles of listeners to whom they were apologising for the curious cut of their hastily purchased ready-made clothing. The sinking of the ship had already faded into insignificance in their minds in the shadow of the big fact that they possessed only the gear in which they had paraded in the market place of Castelorizo on the eventful afternoon. To some of them it was not the least trying of their experiences that they should have been compelled to face the world of Port Said in reach-me-downs. Luckily one of them still retained his eye-glass. An eye-glass gives an air of great distinction to a party whose uniforms are full of unexpected bulges and wrongly-placed creases.[316]

One the ship's deck officers Lt A.L. Braithwaite, who had remained behind with the salvage party, recalled his arrival at Port Said.

> I was instructed to go to a large hotel on the sea front temporarily, so I walked in looking like nothing on earth wearing an old jacket, disreputable trousers, sea boots which I had jumped into when the fire started, and a three weeks growth on my chin. Also I had had no wash worth mentioning in all that time. There was a Saturday night dance in full swing and all the girls gazed at me in horror and amazement, then burst into fits of laughter. But I didn't care: I went and enjoyed one of the best hot baths I have ever had and a real good dinner.[317]

Samson, of course, had to undergo the formality of a Court Martial for the loss of his ship. Assembling at the Royal Naval Depot, Port Said,

HMS *Egmont*, on the morning of 25 January 1917, under the President, Captain Philip Streatfield, MVO, RN, commanding HMS *Hannibal*, the Court briskly conducted the business at hand. The forenoon was devoted to hearing evidence from Samson and several of his officers concerning the events leading to the loss of *Ben-my-Chree*, adjourning shortly before noon.[318]

The Court delivered its Finding to the re-assembled Court later that afternoon, concluding that:

> We are of the opinion that no blame is attributable to Commander Charles Rumney Samson, DSO, the Officers or crew of HM Ship *Ben-my-Chree*.
>
> We are of the opinion that great credit is due to Commander Charles Rumney Samson, DSO, for the manner in which he extricated his Ship's Company from an untenable situation under a heavy fire, this operation being performed practically without casualties.
>
> The behaviour of all Officers and men is considered to have been highly satisfactory.[319]

Shortly after the Court Martial, Admiral Wemyss in a letter to the First Sea Lord, Admiral Sir John Jellicoe, summarised his view of the events.

> I do not think that under the circumstances Samson is in any way to blame. Had I had any information, I should, of course, have given it to him, and since the French Admiral was in harbour himself, and invited, if not actually ordered him in, I cannot think that he did wrong in going in there and concluding that he was safe.
>
> The fact of the matter is that the French Intelligence is extremely bad. The Military authorities are always complaining that they can get nothing out of them as regards Syria and Asia Minor, and that is the reason for their strong desire for a ship to land their own agents. The Syrian Coast being in French hands makes matters somewhat complicated. They are so awfully touchy, and every time I send a ship for Intelligence purposes into their sphere they are full of suspicions, not so much of me, but of the Military Authorities.[320]

The actual fact of the matter was that in this case the French intelligence proved to be all too accurate. Had it been made available earlier it is possible that Samson would have been able to avoid sailing into what became a trap.

More critical is the question raised of responsibility for *Ben-my-Chree*'s loss. It was a question that the court martial had attempted to answer, of the 50 questions put to Samson by the court ten directly refer to the decision to enter the harbour. The one fact emerging being that a pilot was sent out to *Ben-my-Chree* around 09.00 by *Amiral* Spitz without being requested by Samson.

When asked directly if it was his decision to enter the harbour, Samson replied, 'No.' He then muddied the waters by adding, 'I do not say I would not have gone, as I probably would have gone in, as I considered that outside was a very dangerous position. In waters where they are always sighting submarines, it is not only a dangerous position but a useless position.' *Amiral* Spitz had not been questioned by the court, but did provide a written statement, dated 5 February, which was forwarded to the Admiralty. His opinion was that, 'the officer in command [Samson] said of his own accord that the weather was too unfavourable for the seaplanes, and that he was going to anchor in the harbour.'

After reviewing the files the Admiralty forwarded a request to Wemyss that Samson, 'may be directed to report on the apparent discrepancy in the reports of the incident.' As a commanding officer must, Samson assumed responsibility replying, 'I take the whole responsibility of entering the inner harbour. I would submit that I stated at the Court Martial that I took the whole responsibility.'[321] At that point the matter was closed, no further action being taken by the Admiralty.

Now came a time of parting and farewells. Samson remained in command of the rump of the EIESS, most of the pilots, observers and mechanics also staying on. For the crew of *Ben-my-Chree* a lucky few got to return to 'Blighty' for leave and reassignment, the majority were distributed through the fleet at Port Said. For the ex-Isle of Man Steam Packet Company crew members, a few of whom had been with the ship since 1908, it was a terrible time. Samson recalled that, 'Poor old [Engineer Lt G.] Robinson was like a father bereft of child.' Probably he was not alone in his feelings, for *Ben-my-Chree* had been a happy ship.

Captain Wedgwood Benn returned to England where he learnt to fly. Benn himself admitting that, 'I fell into the class technically known as "the World's Worst Crashers."' After completing his training he was then sent to Italy to work on the Adriatic Barrage. He was later appointed as an observer with 51 Wing, RAF. While in Italy, Benn took part in the operation that resulted in the first successful parachute drop of a spy behind enemy lines. During the First World War Benn was awarded the Distinguished Flying Cross, the French *Croix de Guerre* and the Italian Military Cross. In December 1918, he was elected to Parliament for the Liberal Party in Leith, Scotland. He remained a Liberal MP until 1927 when he left the party. As a Labour MP he was Secretary of State for India in 1929–31. Although out of Parliament from 1931, he won a by-election in 1937. Benn rejoined the RAF in 1940 and was made an Air Commodore, he was also made a Viscount in 1942 to increase the number of Labour peers in the House of Lords. As Viscount Stansgate he worked on planning the reconstruction of Italy and after the 1945 he became Secretary of State for Air before being reshuffled out of that post the next year. He then sat in the House of Lords as a Labour peer until his death 17 November 1960.

Red Sea, January 1917

Returning to Port Said where, in truth, only *Anne* was immediately available. *Raven*, having commenced a refit at Suez in early February 1917, did not return to service until March 1917. *Empress*, although recalled to Port Said, was about to commence an overdue refit in Genoa. So, until both refits were completed, *Anne* was left to carry on alone.

The first call on *Anne*'s services was in the Red Sea, between 16 and 26 January. Initially she was tasked with making a reconnaissance from Mowila to Noman Island (Al-Numan) to seek any Turkish troops marching south to reinforce Wejh. Aboard were two familiar Shorts, 8004 and 8075, with Flt Lt Burling, FSL E.M. King, and FSL L.P. Paine, and observers Lt N.W. Stewart, and 2Lt Williams. Both machines made early morning flights on 18 January, finding no evidence of recent troop movements near Mowila. *Anne* then proceeded south to Hassani Island to rendezvous with *Fox* (SNO Captain Boyle), *Espiegle*, *Hardinge,* and *Suva*, arriving at 17.30 on the 19th. Later that evening Burling was briefed on the proposed attack on Wejh.

Hassani island is located just off the coast from Lejh, which had recently been occupied by Sherifian forces, some 280 miles north of Jeddah. At the time it was the northern point of advance of the Sherifian forces. According to Boyle, 'The Arabs remained halted at Umlejh for a month, during which time plans were made for a further advance to Wejh. Stagnation prevailed in the Medina district and it was known that the Turks had received reinforcements to allow of a march on Mecca. It was thought that the presence of an Arab force as far north as Wejh would effectually check any such attempt.'[322]

Early in January Lawrence rode out to Faisal's camp with Boyle and Major Charles Vickerey.

Anne, probably in the Red Sea January 1917. There is a single Short with wings folded on the after well deck, floats protruding through the cut outs in the starboard gunwale. There is a small steam powered lighter alongside, too small for an X-lighter, it is probably bringing supplies and water. *(EMK)*

Short 8004 being hoisted out from *Anne*. The observer is standing on the starboard wing, the pilot cannot be seen perhaps with his head down in the cockpit. The air mechanic on each float aided balance during the hoist and were available to help spread and lock the wings once the Short was on the water. The ship's gig, seen under the floats, was available in case of emergency and to retrieve the air mechanics once the wings were spread. *(EMK)*

We decided to break the army into sections: and that these should proceed independently to our concentration place of Abu Zereibat in Hamdh, after which there was no water before Wejh; but Boyle agreed that the *Hardinge* should take station for a single night in Sherm Habban – supposed to be a possible harbour—and land twenty tons of water for us on the beach. So that was settled.

For the attack on Wejh we offered Boyle an Arab landing party of several hundred Harb and Juheina peasantry and freed men, under Saleh ibn Shefia, a negroid boy of good courage (with the faculty of friendliness) who kept his men in reasonable order by conjurations and appeals, and never minded how much his own dignity was outraged by them or by us. Boyle accepted them and decided to put them on another deck of the many-stomached *Hardinge*. They, with the naval party, would land north of the town, where the Turks had no post to block a landing, and whence Wejh and its harbour were best turned.[323]

Before sailing for Wejh, a flight was made on 21 January to exercise the ships with spotting gunfire using wireless and smoke bombs to mark targets. *Hardinge* then took on board 400 Arabs, and the ships proceeded up the coast to Marduna Island, an anchorage opposite Sherm Habban some 20 kilometres south of Wejh. From there, on 22 January, Flt Lt E.J. Burling took Captain Boyle, on a flight over Wejh to update him on the existing defences. Some new trenches and blockhouses were seen, and mapped, outside the town.

> In accordance with the plans we had made, four hundred Arabs were embarked in the *Hardinge*, a ship always to the fore, and particularly useful for this work. In the *Fox*, with four other ships,[324] I went to an anchorage [Sherm Habban] to wait for the Arab army which was to march up the coast. As twenty-four hours after the appointed time there was no sign of it, and being anxious to prevent the escape of the garrison as had happened at Umlejh, I decided to capture the place with the four hundred Arabs we had on board, backed by a naval landing force. Major Vickerey and Captain Bray took charge of the Arabs, the majority of whom proved most unsatisfactory. The naval force was held back at first so that the Arabs should have the satisfaction of capturing the place, but by dusk they had failed to do so, being more intent upon looting than fighting.
>
> So the following morning [24 January] at dawn the naval party advanced and occupied Wejh, and some eighty prisoners, but the bulk of the enemy force had got away during the night and only slight resistance was offered. Twenty Arabs were killed, but we had only one officer and one seaman killed and three wounded. In this skirmish I again had the *Anne* with me, her seaplanes proving most useful.[325]

Clear of the ship the Short is rotated so that it can be lowered onto the water. The top of the pilot's flying helmet is now just visible behind the radiator. *(EMK)*

Wejh from the sea with *Fox* in the foreground. Al Souq fort is just visible behind the buildings. *(EMK)*

Lieutenant N.W. Stewart, 7th Royal Scottish, was killed by a single rifle shot from the ground on 24 January 1917 whilst observer on Short 8004 over Wejh. *(YEORN 2014-80-173. Courtesy of The National Museum of the Royal Navy)*

Anne sent out three flights on 23 January; two to search, fruitlessly, for the Arab army and one to bomb the trenches at Wejh and then to spot for the 6 inch guns of *Fox*. This last flight was piloted by FSL King with Lt Stewart as observer. They were flying Short 8004 with six smoke bombs and six 16-lb bombs. As they had full wireless transmitting and receiving equipment, Stewart did not have a Lewis gun. They took off at 08.17, and once airborne they saw the Turkish forces sheltering behind a coastal ridge, inaccessible to gunfire or their bombs, although two were dropped. Spotting for *Fox* was by W/T, all the time the Short being fired at from the ground, a single shot killing Stewart, who was the only officer casualty during the capture of Wejh. In a surviving fragment from a letter home King provides an account of the flight.

> It was a very pleasant trip but for one unfortunate episode which was this. I was flying with an observer over some Turkish lines when we were sniped. Though the machine was only hit by a very few bullets one unfortunately hit my observer and it entirely cut the main artery for he bled to death in a few seconds. It was only a freak shot as we were fairly high up at the time, about 1400 ft., and only one rifle was fired just then and I heard it, then the observer leant forward (he was in the back seat) and touched me on the shoulders to show he was hit and promptly fainted from loss of blood. I at once headed for the ship, about 2 miles away and landed as fast as possible. We got him aboard and attended to, but in my opinion he must have been dead before even we landed. His death naturally cast a gloom over the ship for a while, but being wartime and most of the chaps young, our spirits couldn't be kept … [and there the page ends.]

Lieutenant N.W. Stewart, 7th Royal Scottish, was buried at sea the following day by the Chaplain of *Fox*.

Returning to Boyle's account.

> The Arab army arrived the next morning [25 January], having been delayed by water difficulties which a ship had to be sent down the coast to rectify. Faisal made a triumphant entry into the town; the Arab forces streaming in across the desert made a most picturesque spectacle, and the many types of Arabs added interest to the scene.
>
> The occupation and use of Wejh marked a definite advance in the revolt. Not only did it paralyse the Turks by its threat to the railway, but it enabled Sherif Faisal to get into touch with the northern Arabs, who were only waiting for a leader to join the insurrection.
>
> Wejh gradually became the most important base; the majority of the Egyptian troops moved up there, a water-distilling ship, the *El Kahira*,[326] was moored in the harbour and a motor lighter was sent to help in the disembarkation of stores, etc.,[327] which was made more difficult by the fact that the harbour was too small to take any vessel larger than a sloop. One of the sloops—there were now three on the coastvwas always kept in the port in order to maintain W/T communication with Cairo, a constant problem in the Red Sea, where conditions, especially by day, were always difficult. Later a W/T station was established on shore, largely due to the indefatigable efforts of the captain of the sloop *Espiegle*, Commander R. Fitzmaurice, who was always to the fore when difficulties or dangers arose.

By the 25 January, *Anne* was proceeding north and returning to Suez. At 12.47 she stopped whilst off Sherm Dumeïgh (about 50 kilometres north of Wejh) to launch Short 8075, FSL King and 2Lt Williams, to conduct a reconnaissance of the coast, up to Noman Island, returning at 14.26. FSL King's log book contains a brief report of the flight. 'From *Anne*, flew up the Arabian coast from Uweindiya Is to Noman Is (50 miles) searching for party of Turks escaped from Wej. No luck!' After recovering the Short and its crew, *Anne* reached Suez in the evening of 26 January.

'Hafenstädtchen el Weğh von Osten' (Harbour town of el Wejh from the east), 1914. Al Souq fort, built in 1875 and recently restored, is clearly visible in the right centre, with a similar vintage lookout tower, now demolished, on the adjacent headland. *(Bernhard Moritz, Library of Congress)*

Anne with a Short 184 being towed back to the ship.
(Sub Lt H B Burd, RNR, album courtesy Australian National Maritime Museum)

The Damascus Flight, February 1917

Anne's next cruise commenced a month later, leaving Port Said with the French trawler *Nord Caper* as escort during the early morning of 26 March.[328] She had aboard two Short 184 floatplanes, 8021 and 8022 both on their first outing for the EIESS, pilots Flt Cdr A.W. Clemson, Flt Lt H.V. Leigh and FSL G.D. Smith with observers S/Lt J.L. Kerry and 2Lt G.H. Pakenham-Walsh. The two ships made their slow way, *Anne* could barely maintain 10 knots and *Nord Caper* about the same, up the Palestine coast to be off Haifa the following morning.[329]

The two flights inland from Haifa were typical for *Anne*. First off were FSL Smith with S/Lt Kerry on 8022, at 06.47 returning at 08.29. They followed the road down to Jenin, then north to El Afule and Nazareth before returning to the ship. The second flight, by Flt Lt Leigh and 2Lt Pakenham-Walsh on 8021, left at 07.10 returning at 08.30. They had flown inland down the coast past Athlit to Zimmarin, some 35 km, before turning inland to pick up the Jenin–Haifa road back to the sea. Both flights covered 90 to 95 km overland.

Patrouilleur auxiliare Nord Caper seen here at Sidi Abdellah, west of Algiers, 16 October 1915. Typical of the escorts to *Anne* and *Raven*, armament was a single 65mm forward and a 47mm aft. The wooden shed-like structure forward of the bridge was apparently additional accommodation. *(Corsaires du XXe Siècle, fp.62)*

The ships then headed further north towards Beirut, arriving just off the coast the following morning, 28 February. Flt Cdr Clemson and S/Lt Kerry were briefed to fly to the important railway junction at Rayak. It was the terminus of the standard gauge (4ft 8½in / 1.435m) line from the north. For onward transmission all materials, ammunition, supplies, etc had to be transferred to 1.05m gauge rolling stock. At Rayak there was also a fully equipped repair shop and foundry. There was no mention of Damascus.

EIESS Officers Football Eleven. Sub Lt Kerry commented 'After lunch the football match -v- Mechanics. Very good fun and splendid exercise but officers lost 5-2.' Left to right: Flt Cdr T.H. England, Flt Cdr A.W. Clemson, Flt Lt E.J.P. Burling, FSL E.M. King, Warrant Officer F.H. Whitmore (Senior Torpedo WO), 2/Lt G.H. Pakenham-Walsh (Cheshire Regt, observer), 2/Lt A.D. Ferguson (Highland Light Infantry, observer), Lt W.R. Kempson (RFA, observer), Sub Lt J.L. Kerry (observer), FSL H.V. Worrall, Warrant Officer A.T.E. Witt (Carpenter). Oddly, there appears to be a cricket ball on a mound just in front of the group. *(EMK)*

Short 8022 setting out on the morning of 28 February, Flt Cdr Clemson and S/Lt Kerry were briefed to fly to the important railway junction at Rayak. They went far beyond that, all the way to Damascus. In addition to the observer's Lewis gun there is a second on a pillar mounting beside the starboard side of the pilot's cockpit. *(Kerry Family Collection)*

Damascus Flight Map. *(TNA, AIR 1/271/15/226/119)*

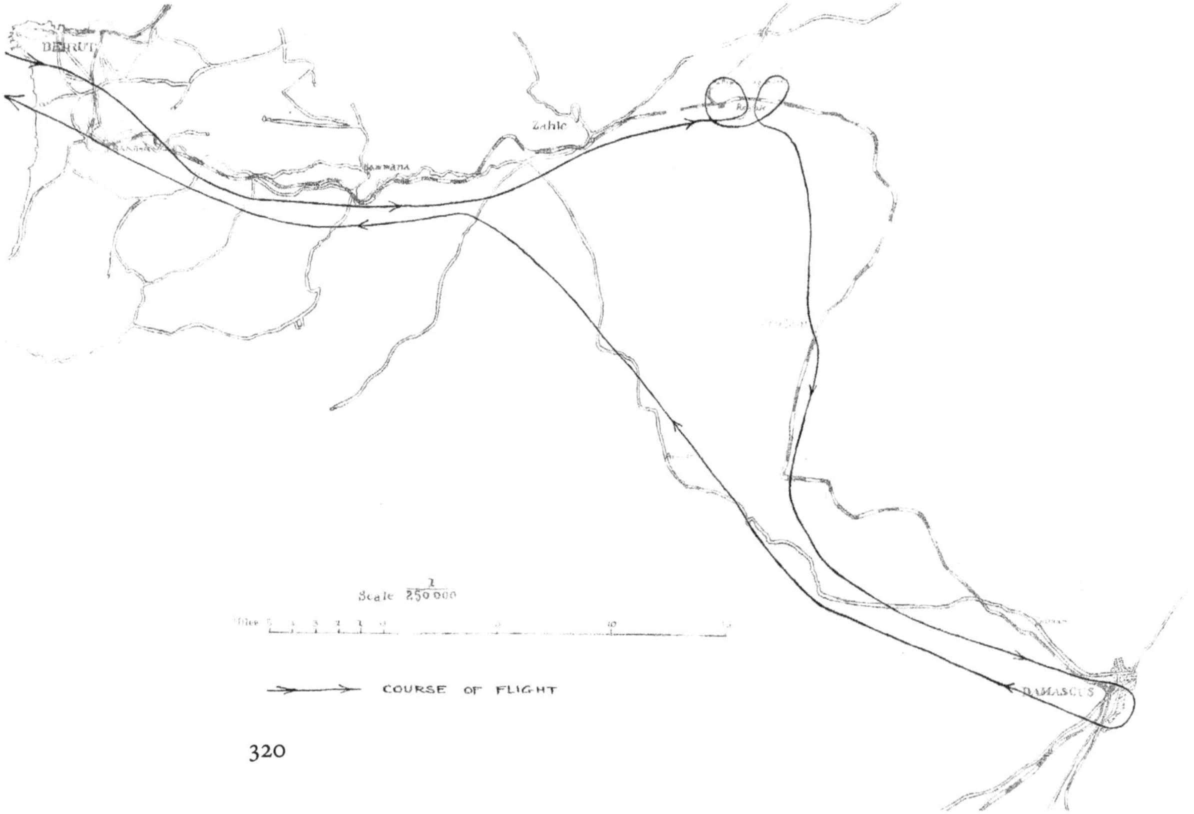

The two set off on Short 8022 with two Lewis guns,[330] nine 'hoppers', W/T transmitting and receiving sets and a camera, but no bombs. Getting off the water at 07.03 Clemson circled the Short, climbing slowly to 5000 feet, before crossing the coast half an hour later a little south of Beirut. To reach Rayak they had to cross the Lebanon Mountain range, which east of Beirut reaches 2695 metres (8842 ft) at Mount Sannine. The railway takes a somewhat lower route, crossing the mountains at a maximum elevation of 1487 metres (4880 ft), before descending into the Bekaa Valley at Rayak, 929 metres (3050 ft) above sea level. Between Rayak and Damascus lay the Anti-Lebanon range. In its southern reaches this inland range includes Mount Hermon at 2814 metres (9235 ft), often snow covered year round. However, the railway crosses the mountains at a maximum of 1372 metres (4500 ft).

Clemson and Kerry's report was dry document, albeit replete with the required observations. For example, at Rayak Junction there were '23 large sheds and a chimney from which smoke was issuing.' Kerry also counted '120 trucks and carriages. Two locomotives were seen.' Whilst at Rayak Station 'there was a locomotive and 6 coaches.' As his engine was running, well Clemson decided to press on to Damascus itself, much to the surprise of his observer. At Damascus 'No hangars or camps were observed. No hostile aircraft were sighted.' Which, of itself, was fortunate as most of the EIESS's encounters with hostile aircraft ended badly.

C.E. Hughes' account provides some authentic colour.[331] As the EIESS's Intelligence Officer he had first hand experience of shipboard operations and would have interviewed both pilot and observer. His account indicates that he was on board *Anne* for this cruise.

> A two-seater machine got off the water at 7 a.m., the ship's company went unconcernedly to breakfast and to their various jobs, expecting to see it back somewhere about halfpast eight. By nine they were beginning to get anxious, and when by nine-thirty there was still no sign of the seaplane, there was much speculation in that suppressed half-hesitating manner which people adopt when their feelings fight for expression in spite of themselves. It was ten minutes to ten when the machine was at length sighted high in the distance.
>
> The pilot—Noisy Dan, we used to call him, because he was the quietest person who ever gave a word of command—having taken the machine in as far as Rayak, forty-five miles or so from the ship, and found that it was running well, just went on to Damascus in spite of signals from his observer, who thought he had mistaken the direction of home. But he was not making any mistake. He had that instinct which some pilots have (though it sometimes plays them false) for the exact extent of the demands they can make on their engines. Theoretically it was a risky thing to do, and it would never have been ordered beforehand. But Dan happened to be in command of the operations. He had a feeling that what he wanted to do could be done, and he did it.

First in a series of photographs made on the Damascus flight by observer Sub Lt Kerry.

Looking south-east, Rayak Junction repair sheds and transfer station. The standard gauge line from Homs enters from the lower left and, passing the repair sheds, runs into the main yard. All goods were transferred on to 1.05m gauge rolling stock and moved out along a spur line, at the top right, to the main Beirut–Damascus line. Steam can be seen rising from an engine in the yards and smoke from an adjacent workshop. *(Kerry Family Collection)*

 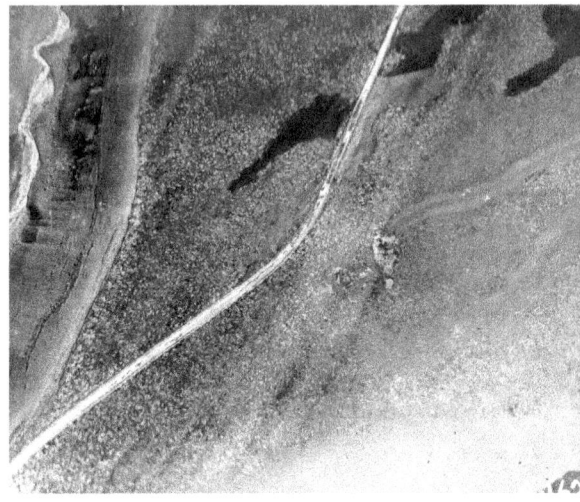

Above: Looking north-west, Rayak Station was a typical small way station on the Beirut-Damascus line. The 1.05m main line runs from lower left to upper right, and the spur line from Rayak Junction can just be seen curving in from the left above the main line. *(TNA, AIR 1/271/15/226/119)*

Above right: A body of troops and horses marching along the road to Damascus from Zahle. *(Kerry Family Collection)*

The flight to Damascus followed the railway the whole way out, and the machine reached its highest elevation at Zahle, where it flew at 7,600 feet, though Zahle is not the highest point on the railway. It was, however, near enough to the original objective, Rayak, to make a safe altitude advisable. Rayak was inspected from 7,000 feet, and several photographs were taken showing a most elaborate arrangement of railway workshops and sidings. Notes were made of the rolling stock, which was very plentiful, of tents and of such evidences of the practical use of the workshops as smoking chimneys.

Then the journey was continued in the direction of Damascus. About two miles from the town the white course of the highway revealed what to the RNAS observers in these parts was a good deal of a novelty. There was a battalion of troops on the march. One might suppose that there was nothing so very unusual in this in a land in which there were everywhere signs of war. Trenches, gun emplacements, camps, were common objects of the country-side, but these were fixed things which could not hurriedly take cover at an alarm. With moving bodies of troops it was different. Seaplanes seldom approached the Syrian coast without awakening into life certain smoke signals which conveyed the news to watchers on the lookout inland, and there was always time for mobile bodies to find some place of concealment. But neither Rayak nor the roads leading to Damascus had ever before been visited by hostile aircraft, and doubtless the warning signals which had produced the utmost alertness some miles in from Beirut had stopped far short of this distant spot where the highway crossed the bare open slopes of the low hills. At any rate, the compact little parties of soldiers, with their accoutrements gleaming in the morning sunlight, and the straggling tail of baggage animals at their rear, were evidently taken by surprise.

Possibly the machine was mistaken for a friendly one arriving for the training school which was about to be started in Damascus, a conceivable error, since at 7,000 feet the difference between a land machine with its wheeled under-carriage and a seaplane with its floats might not be readily distinguishable by inexpert eyes. But the mistake was quickly discovered, and the troops got five or six guns into action with remarkable speed. Their shells flew wide, however, and the machine continued its flight uninterrupted, and in another minute or two it was hovering above one of the oldest cities in the world.

Damascus has been called by the Arabs "The Pearl set in Emeralds", and to see it from the air with its pale mass of houses lying in the midst of its luxury of trees and fields, so I learned from the lucky observer, is to realise something of the truth and beauty of this description…[332]

But neither the pilot nor the observer had much leisure for such thoughts as these. They were there to make as close an inspection as time allowed, not of pearls or emeralds, but of trenches and gun-pits, camps and hangars. The machine circled over the town and

A view of Damascus looking toward the east. Taken from 7000 feet, Damascus is at the at the top of the photograph, and the settlement of Er Rabwe at the lower right. The white line running through the image and passing under the tailplane is the post road from Damascus to Beirut, which Clemson followed on the return flight. Running alongside the post road is the Nahr Barada which rises in the Anti-Lebanon, running alongside the railway for almost 30 miles before entering Damascus, then waters the agricultural area to the east of the city. The railway can just be made out curving towards Er Rabwe then up to Damascus. *(Kerry Family Collection)*

C.E. Hughes wrote, 'The saddle of Lebanon was crossed once more at Zahle, where a photograph was obtained of a prospect of snow-clad peaks with the white drift clinging in the hollows and filling the zig-zag grooves of the watercourses. This photograph, the last of the series taken, had no military value excepting possibly that it gave negative evidence. It showed no trenches, no fortified posts, no guns, no men ; but I think when the authorities saw it they forgave the waste of a plate.' *(Kerry Family Collection)*

the plain, and, the necessary notes being made, set off again on the return journey. This time the main road was followed in preference to the railway, which separates from it seven miles from the city, and the reconnaissance of this showed another battalion of troops and several smaller parties all marching in the same direction [towards Damascus]. Their faces could be seen as they gazed upwards in wonder and bewilderment, but few of them were ready with their weapons, and no damage was sustained from the very spasmodic fire which they put up. The saddle of Lebanon was crossed once more at Zahle, where a photograph was obtained of a prospect of snow-clad peaks with the white drift clinging in the hollows and filling the zig-zag grooves of the watercourses. This photograph, the last of the series taken, had no military value excepting possibly that it gave negative evidence. It showed no trenches, no fortified posts, no guns, no men; but I think when the authorities saw it they forgave the waste of a plate.

Crossing the coast, again south of Beirut, they landed alongside *Anne* at 09.48. Clemson, Kerry and Short 8022 were hoisted in a few minutes later. Sub Lt Kerry's diary for 28 February 1917 briefly states:

> Up at 5.45 am. Hoisted out at 6.45 with Clemson and in air at 7.3. Flew over Beirut and up between break in mountains at 4000 ft high. Beautiful country. Reached Rayak and photographed repair sheds. Then Clemson flew to Damascus. A fine looking city. Passed troops on the way there. Flew 130 miles over land a record for seaplane. In air 2 hours 45 minutes. Glad I have done it now that is safely over.

For a floatplane it was an exceptional flight. The straight line distance from the coast to Damascus and back was 170 km. The route actually flown was nearer to 220 km (135 miles), the Short spending more than two hours overland. They returned with photographs and up to the minute information about supplies and troop movements. Just part of the intelligence that the RNAS and RFC were gathering in advance of the First Battle of Gaza on 26 March 1917.

Both Clemson and Kerry were awarded the DSC. 'In recognition of their conspicuous gallantry on the 28th February, 1917, when they carried out a reconnaissance of Rayak and Damascus in a seaplane. During this flight they crossed two mountain ranges whose lowest ridges are 4,000 feet high, and brought back valuable information.'[333]

The flyers hoisted in, *Anne* and *Nord Caper* turned back to Port Said, arriving 11.00 on 1 March, where she remained for three weeks. *Raven* however, having returned from her refit at Suez on 3 March, spent several days preparing for a long cruise commencing a week later, not returning until the middle of June. This is described in the following Chapter.

First and Second Battle of Gaza, 26 March and 19 April 1917

Raven's departure left just *Anne* available to support the First Battle of Gaza. *Empress* returned to Port Said a few days before the Second Battle of Gaza. However, only *Anne* was required to make a reconnaissance before the battles. The reason is not hard to find.

As the Egyptian Expeditionary Force advanced across the Sinai, airfields were established close behind the front line. Throughout 1916 the RFC had slowly expanded *(see Chapter 7)*. By the beginning of 1917 both squadrons of the 5th (Army) Wing, RFC, were based behind the British lines. Between them 14 and 67 Squadrons had 25 aircraft at their disposal: 17 BE2s and 8 Martinsydes. They made frequent attacks on Turkish military installations including the aerodrome at El Arish, as well as continuing the artillery direction and reconnaissance roles. 14 Squadron also used w/t to direct artillery. Essentially taking over the many of the roles the EIESS had supplied up the that point. Thus the EIESS's participation was limited to more distant reconnaissance to observe any troop movements the Turkish forces may have been making.

Anne sailed from Port Said on 21 March, with a small air component of two pilots, one observer, and a single Short 8020 (a Model D bomber—see below). At 07.05, whilst 20 miles west of Haifa, the Short was hoisted out with Flt Lt Leigh and Lt Millard as crew. Their orders were to 'Reconnoitre Haifa, Nazareth and the Railway East and South of El Afule. Particularly to note concentration of troops, if any, also traffic on Railways and Roads.' Airborne ten minutes later, they crossed the coast near Haifa at 07.44, flying at 4800 feet. Over the next hour and a half they flew inland over Tabaun, Jennin, Zerin, Beisan, Nazareth and back to Haifa. No concentrations of troops, just a few small camps near Tabaun, were seen. There was no traffic on the railway, and no stores visible. The roads were equally bare of traffic. The machine was fired on only when approaching Haifa at the end of the reconnaissance. Landing safely, they were hoisted in at 09.20, ending *Anne*'s 54th, and final, flight for the EIESS.

Anne had supported almost 200 flights for *l'escadrille de Port Saïd* and 54 for the EIESS. However, she was now employed as a depot ship for the squadron until 8 August 1917, when she was decommissioned. She then served as a collier under the Red Ensign from 29 January 1918 until the end of the war, managed for the Shipping Controller by F.C. Strick and Co, a company with long experience for the Egyptian and Red Sea trade. *Anne* was a survivor, as her experiences at Long Island in 1915 showed. Between the wars she was owned by S.N. Vlassopoulos of Greece between 1922 and 1939, and renamed *Ithaki*. Bought by a Romanian company in 1939, renamed *Moldova* she was transferred to the Panamanian flag, and survived WW2. In 1949 she was sold to Hong Kong based Wallem & Co. Renamed *Jagharat* in 1954, she resumed the name of *Moldova* in 1955. *Aenne Rickmers*, or whatever name she still bore, was scrapped at Hong Kong in November 1958.

Empress' role in the Second Battle of Gaza was equally limited. Just returned a few days earlier from her refit at Genoa, she sailed from Port Said on the evening of 16 April with just a single Short, 8020, in her hangar. After dropping the pilot she, and her escort the French destroyer *Coutelas*, made 15 knots towards Deir El Belah, just south of Gaza. A protected anchorage had been created for bombarding ships off the coast, and *Empress* passed through the boom at 06.10 the following morning. Fifteen minutes later, proceeding slowly, she ran on to a sand bank. Using her own engines, and assisted by two trawlers, she was soon released and proceeded to anchor with no sprung plates. Whilst this was going on the Short, with pilot FSL H.V. Worrall and observer Lt A.D. Ferguson, made an anti-submarine patrol outside the anchorage returning at 08.44 after a flight of just under two hours. Sighting nothing, their excitement came on landing, as the port float carried away. The machine was salvaged and rebuilt at the base after returning to Port Said the following day.

C.E. Hughes was aboard *Empress*, later recording his observations.

> French and British battleships were conducting a coastal bombardment, and the seaplane carrier executed submarine patrols until the machine was put out of action through the breaking of a float on landing. Having nothing else to do, we watched the battle through our glasses from the deck or played cards in the wardroom.

There was only one thrilling interruption. High in the air a distant aeroplane was sighted, and as we stood on the bridge it seemed to be making a line straight for us. Was it an RFC machine, or were we to be bombed by a German? This question was not decided until the airman wheeled landwards as he reached a point almost above us, and we saw that his wings bore not black crosses, but coloured circles.

Not a particularly high point in EIESS history, but a shortage of machines limited possible participation. Many of the available Shorts had been with the squadron since early 1916. Both 8004 and 8075 were undergoing overhaul and repair following their return from the Red Sea, and 8091 appears to have been unserviceable since August 1916. After *Raven* had taken the best of the remaining machines on her extended cruise (see next Chapter), only 8020 and 8022 were available at Port Said. Both were recent arrivals at Port Said, although 8020 had been built as a single-seater.

Short 184 Model D Bomber

The Short 184 had not been designed for Middle East weather conditions, they were usually operated without engine cowlings to improve cooling, and was a poor load carrier. With pilot and observer, one or two Lewis guns, and full load of fuel it could lift either a W/T set and a few 16-lb bombs or, without the W/T set, a mix of 65-lb and 16-lb bombs. EIESS Shorts had a pair of bomb carriers for the bigger bombs installed directly under the fuselage between the float struts.[334] There must have been light bomb carriers for the 16-lb bombs, the days when Benn nursed 16-lb bombs in his cockpit were long gone, but photographic evidence is missing. Later production Shorts, with 240-hp or 260-hp engines, had a factory installed long bomb rail with several bomb carriers along its length, some of these began arriving in Port Said during September 1917. Back in the UK they often carried a pair of 230-lb bombs on anti-submarine patrols, anti-submarine patrols from Port Said and Alexandria were more limited as we will see later. An earlier attempt had been made to make the Short 184 into a bomber.

The Short 184 Model D was intended as a single-seat bomber. The pilot was relocated to the aft cockpit and the forward cockpit converted into an internal bomb bay 'to carry nine 65 lb bombs vertically tail downwards inside the fuselage, the bombs being released by the Woolwich pattern electrical releases operated by a single control switch in the pilot's seat so arranged that any number of bombs required may be dropped at once.'[335] The forward cockpit was covered by a fairing, curved to match the upper fuselage contour, which appears to have been hinged on the starboard side and clipped, or toggled, closed on the port side. The number built was quite small, less than twenty, most were used for trials or based at Dunkirk to bomb Ostend

Short 184 Model D single seat bomber 8016 was damaged by anti-aircraft fire during an attack on Zeebrugge 10 November 1916. Forced to land on the sea with a shattered propeller it was towed into the harbour and is seen here on the mole at Zeebrugge. The pilot Flt Lt G.G.G. Hodge was unwounded and taken prisoner, the Short was repaired and tested at Warnemünde in 1917. Before modification by the EIESS the three Model D's delivered to Port Said, 8018, 8019, and 8020, were identical in appearance to this machine. *(via Colin Owers)*

and Zeebrugge. Somehow, three were supplied to the EIESS—8018, 8019 and 8020. Two more Shorts, 8021 and 8022, arrived at the same time but were standard machines. All five were part of a 30 aircraft contract (8001–8030) built by S.E. Saunders Ltd, with a 240-hp Sunbeam Gurkha engine. They were completed in November 1916 and delivered to Port Said in early 1917.

It is doubtful that as single-seat bombers they were well received. Most of the EIESS's work involved reconnaissance flights with the requirement for an observer to operate a camera, W/T set, and a Lewis gun in self defence should the need arise. All three were mostly flown as two-seaters, quickly converted with the observer in the forward cockpit.

An indication of the modifications carried out on 8020 is provided by Lt C.F. Thomson Walker, RNVR, the EIESS's Armament Officer.[336] 'Short No.8020 is fitted with Handley Page bomb gears, and carries two 65-lb bombs suspended from the nose, inside the Observer's seat.' Short 8020 was flown as a single-seater just once, with an external bomb load, but it is known that the two cells were used when being flown as a two-seater. The modifications carried out on 8018 and 8019 are not known, only their bomb loads when flown as single seaters. The bomb loads varied from a maximum of 36 16-lb bombs (576 lb) to two or three 65-lb bombs plus 16 16-lb bombs (386 lb or 451 lb). Exactly how 36 16-lb bombs were carried is not known, although they appear from the Reports of Flights to have been loaded and dropped in multiples of four. Thus opening the possibility that four 16-lb bombs could have been suspended from each of the nine internal 65-lb bomb releases. The bombs were widest at the square formed by the tail fins. For a HERL 65-lb this was 12.75 inches, and for a 16/20-lb bomb the square formed was 5.1 inches max. So, four would fit comfortably in the space of a 65-lb bomb. The 16/20-lb bombs would probably have been suspended from the existing loop at the tail of the bomb.

Providing a Lewis gun for the observer also required some ingenuity. With *Raven* in the Maldives both 8018 and 8019 had a pillar mounting on the port and starboard side of the fuselage. Back at Port Said one of the machines, probably 8018, had the upper part of one of the forward cockpit fairings removed and replaced with a flat panel to provide seating for a Short designed ring mounting with a Lewis gun. Another photograph, probably of 8020, shows yet another arrangement with a rather flimsy looking mounting for an aft facing Lewis gun, or guns, on a cross bar supported by struts. Short 8020 was carrying two Lewis guns during the Gaza flights from *Anne* and *Empress*.

CHAPTER 14

Raven's East Indies Cruise

During the First World War, once the cruisers SMS *Emden* and SMS *Königsberg* had been neutralised, the Indian Ocean was a long way from any active naval front. Were it not for the German practice of using disguised merchant ships as commerce raiders it might have been completely undisturbed. SMS *Wolf*, formerly the DDG Hansa freighter *Wachtfels*,[337] commanded by *Fregattenkapitän* Karl August Nerger, sailed from Kiel at the end of November 1916. Her armament, well hidden but quickly exposed, included eight 150 mm guns of modern design and four 50 cm torpedo tubes. She also carried over 450 mines. A few months later she was operating in the Indian Ocean. Minefields were laid off the Cape of Good Hope in January 1917, and at Colombo and Bombay between 15–19 February. *Wolf* remained between Aden and Bombay between 26 February and 6 March, taking several prizes, before sailing towards Australia and New Zealand. On her return voyage, *Wolf* lingered south of the Maldive Islands, from 26 September until 8 October then set course for Germany.

During the height of the search for the raider in the Indian Ocean, February to April 1917, over 50 British, French and Japanese warships were employed. These included an old battleship, cruisers, destroyers and sloops and a single seaplane carrier, HMS *Raven*. *Wolf* eluded *Raven*, her seaplanes, and every other searcher to return safely to Germany at the end of February 1918. During a 451 day cruise *Wolf,* assisted by its own floatplane, Friedrichshafen FF.33E *Wölfchen*, had accounted for 14 ships totalling nearly 40000 tons, plus at least 13 more sunk by mines laid by the *Hilfskreuzer*.[338]

SS *Wachfels* (aka SMS *Wolf*) of DDG Hansa pre-war. Very similar to *Raven* (*Rabenfels*) except for the doubled kingpost ventilators aft.

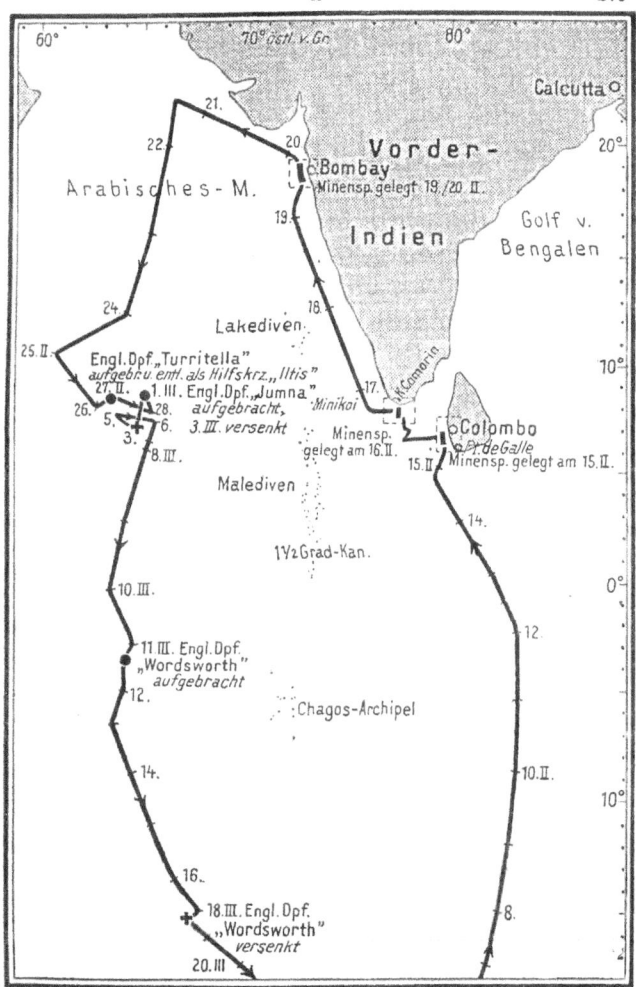

Above left: Observer *Leutnant* Alexander Stein and pilot *Oberflugmeister* Paul Fabeck on *Wolf*'s return to her home port of Kiel on 24 February 1918. *(US National Archives, NARA 533181)*

Above right: Friedrichshafen FF33E *'Wolfchen'* aboard *Wolf* with pilot *Oberflugmeister* Paul Fabeck with cigarette and observer *Leutnant* Alexander Stein. All markings were painted out but a *Kaiserliche Marine* ensign was flown from the aft starboard inboard wing strut. *(Australian War Memorial, P05338.110)*

Left: *Wolf*'s movements 2 February to 20 March 1917. Also a useful map showing the relative locations of the Laccadive (*Lakediven*), Maldive (*Malediven*) and Chagos archipelagos. *(Der Kreuzerkrieg, Band 3 – Die deutschen hilfskreuzer)*

Raven Heads South

Shortly after her return to Port Said on 3 March 1917, following an annual refit at Suez, *Raven* commenced preparations for an extended cruise that would take her into the Indian Ocean, Ceylon (Sri Lanka), and across the equator. Not that this was known at the Base, as C.E. Hughes explains.

> I think from first to last three days covered it. During that time material in stupendous quantities was loaded in the *Raven*. Machines, erected and still in their packing cases; spare engines, spare propellers, spare floats, spares of all sorts, made the passage from the island to the ship in a constant succession of tugs and lighters. The Stores and Engineer Officers had the time of their lives, and recalled in their brief moments of relief the distant days of mobilisation.
>
> And there was a further feature which stamped the affair as unique. One of the officers was entrusted with the task of providing a supplementary larder for the mess. This was distinctly unparalleled. For short voyages up the Syrian coast, and even for longer ones along the shores of Arabia, it had been usual to rely on the ship's steward. Evidently there was a chance now of the ship's steward making an under-estimate. The work of prevention of this possible catastrophe was done no less promptly than the other things which required doing in spite of the numerous gastronomic suggestions of other members of the party told off for the expedition.
>
> More intimate personal preparations were carried forward simultaneously. Neither officers nor men detailed for the trip were given particulars as to their destination (or they were given them in strict confidence), but the conviction gradually grew in intensity among those not detailed that they were bound for tropical waters in the Red Sea direction. There were signs which could be read by those versed in such matters. For trips up the Syrian coast officers' servants seldom had orders to pack camp bedsteads; for trips down the Red Sea always. Frequently a camp bed on deck offered the only chance, and that not of the best, of sleep at night. This was one of the signs. Another was the anxiety displayed in many quarters to exchange the cosmopolitan coinage of Port Said, which accepted almost any recognisable European money, for some currency more likely to be negotiable farther east. There was talk of rupees and there were hints of Mesopotamia.[339]

The work included overhauling seaplane handling gear, preparing a platform for a Short machine on the forward hatch covers, and receiving and

Raven at Port Said. The Japanese flag on the main mast is presumably being flown in honour of the Japanese destroyer flotilla in the Mediterranean, possibly to celebrate the Emperor's birthday (Emperor Taishō, b. 31 August 1879). If you look closely, there are a pair of Short 184 floats poking out of a cut out in the aft canvas hangar cover. The aft kingpost-ventilator is clearly seen, as is the 12-pdr anti-aircraft gun on the stern. *(Courtesy Maritime Photo Library)*

stowing bombs and seaplane stores. An additional steam launch was also taken onboard. On 9 March, '5 flight officers, 4 observers, 1 warrant officer, 5 CPO mechanics, 5 POs & 18 mechanics joined; 5 officers' servants joined.'³⁴⁰

The 'flight officers' and 'observers' noted above were Wing Cdr Samson, with Flt Cdr T.H. England, Flt Lt E.J.P. Burling, Flt Lt A.W. Clemson, FSL G.D. Smith, and FSL T.G.M. Stephens, and observers Capt R.E.C. Knight-Bruce, Lt W.R. Kempson, Lt W.C.A Meade, and Sub Lt J.L. Kerry. *Raven* left Port Said and proceeded down the canal commencing at 09.44 on 10 March. Onboard were four Shorts, 8018, 8019, 8021, 8090 (the Cut Short), and a single Sopwith Baby, N1014, probably with a second, N1016, crated in the hold. Shorts 8018 and 8019 were single-seat bombers versions of the Short 184 modified at Port Said into two-seaters with the observer in the forward cockpit. After leaving Suez they proceeded initially to Aden, arriving on 16 March. Whilst at Aden, between 17–19 March, *Raven* sent off eight flights; a mix of reconnaissance and bombing, mine searches, and Aden Staff flights. The reconnaissance flights were over the same areas covered in previous visits *(see Chapter 10)*, and provided the local military authorities with up to date information and photographs.

Two mine search flights were made 'At the request of the Senior Naval Officer, Aden … to look for mines which were believed to have been laid by an enemy raider off the Rasmarshy W/T Station. No mines were found.'³⁴¹

There the account might have closed, but for a curious coincidence. The raider in question was not *Wolf*, it was SMS *Iltis* the former's first victim, taken in the Indian Ocean on the early morning of 27 February 1917. The name of the captured ship was *Turritella*. When *Rabenfels / Raven* had been seized off Port Said on 16 October 1914 several other DDG Hansa ships were taken, including *Gutenfels*, which, renamed *Turritella*, was managed by the Anglo-Saxon Petroleum Co Ltd. If *Wolf* was *Raven*'s younger sister, then *Gutenfels/Turritella/Iltis* (built in 1905) was the middle sister.

The Cut Short, 8090, was one of the machines flown at Aden on 17 March 1917. Wing Commander Samson, with Lt Meade as observer, flew over Fiyush, Waht and Subar, dropped a few bombs and took a few photographs. Not much had changed since Samson's last visit in June 1916.

Fregattenkapitän Nerger quickly decided to convert *Turritella* into an auxiliary minelayer. Provided with a single 52mm gun and 25 mines SMS *Iltis* was commanded by *kapitänleutnant* Iwan Brandes, with 27 men from *Wolf* and the Chinese crew from *Turritella*. The two ships separated on the evening of 27 February. However, *Turritella*'s career as a warship was neither long nor fortunate. Just five days later, after having laid mines off Aden, the ship was sighted by *Odin* (Lt Cdr E.M. Palmer).[342] The ship was showing no lights and was ordered to stop engines, but failed to do so. Continuing the chase, *Odin* signalled the course and speed to all ships on patrol. Shortly after 04.00 the steamer slowed down and came to a stop. The moon having set, *Odin* steamed slowly around the mystery ship, its searchlight being kept on the steamer until dawn around 06.00. Shortly after daybreak the ship's boats were lowered and left the ship. A few minutes later two explosions were heard and *Turritella/Iltis* began sinking. The boats were rounded up, and *kapitänleutnant* Brandes, another officer, his crew and 46 Chinese taken prisoner. The mines were quickly swept, claiming just a single victim, SS *Danubian*, which was damaged.

A full description of *Wolf*, and her seaplane, was received from the Chinese crewmen and, a few hours later...

> [a] general warning was being sent to all shipping afloat to let them know that there was a raider in the Indian Ocean. Captain Nerger, lying in wait on the trade route a few hundreds of miles away, took in the Admiralty's warning, and after cursing the Chinaman's invariable fidelity to the interests of his de facto masters, acknowledged, regretfully, that the description of the *Wolf* and her seaplane was singularly accurate. He now decided to leave the Indian Ocean for a time, and made towards Australia, on a course which took him to the westward of the Chagos archipelago through the most deserted part of the Indian Ocean.[343]

This however remained unknown to the British for some weeks.

The old French armoured cruiser *Pothuau* (1895), *Raven's* companion on her Indian Ocean cruise.

The flights for the local Staff were at the request of Major-General J.M. Stewart, GOC Aden. The first flight was made on the afternoon of 18 March when Samson flew Lt Col E.C. Alexander, GSO (I), Aden Field Force, on a tour of the Turkish positions between Waht and Fiyush. He was up again the next morning with General Stewart to view positions at Darb and Subar. Following a quick change of pilots and passengers, ten minutes after Samson returned with General Stewart, Flt Cdr England took a Lt Benning from the General's Staff, for a quick trip around Aden harbour.

The flights ended at 07.53 on 19 March and *Raven* sailed from Aden at 15.30 bound for the Laccadive Islands. She sailed in company with the French armoured cruiser *Pothuau* to provide big gun support. Commissioned in 1896 *Pothuau* was long past her best, but her two 194 mm and ten 138.4 mm guns were capable of taking on *Wolf*. Slow by the standards of the period at 19 knots on trials, she still might be able to overhaul the German raider if necessary.

On leaving Aden, and as a general practice, during the day *Raven* proceeded 6 miles ahead of *Pothuau*. At night the gap reduced to one mile, *Raven* showing all steaming lights, whilst the cruiser darkened ship. The intention being that

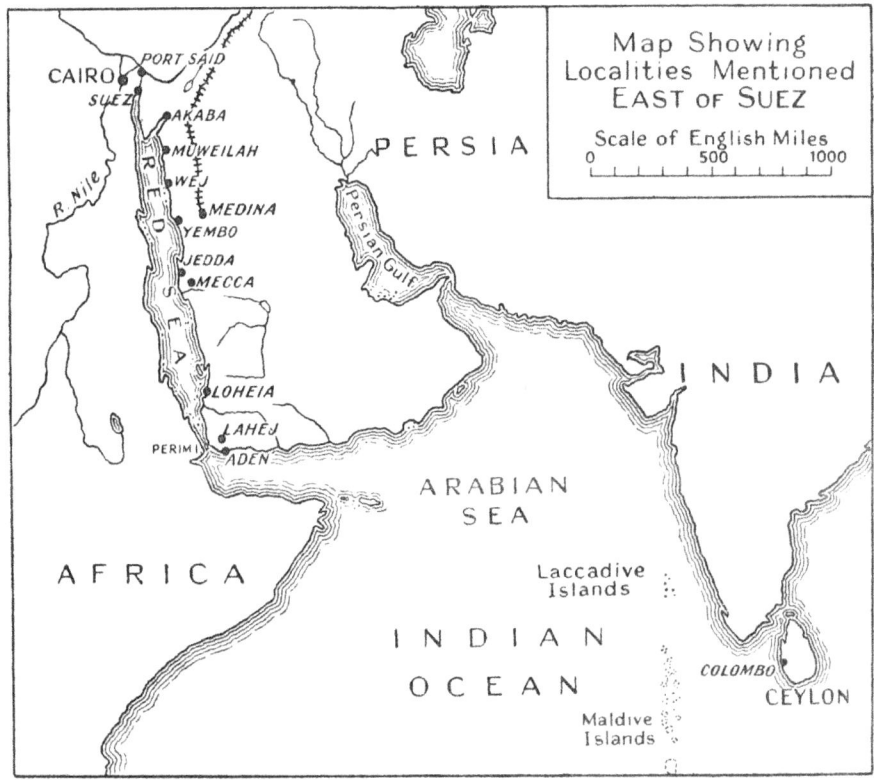

Red Sea and Indian Ocean map from C.E. Hughes' book, *Above and Beyond Palestine*.

Raven would be perceived as an ordinary merchant ship, leaving the cruiser to close in the darkness if she were attacked. During the day seaplane flights were made whenever sea conditions permitted. *Raven* stopping to hoist out the machine, then steaming ahead at 11 knots until stopping once again to recover the machine. 'The general routine was for a seaplane to fly 20 miles on the port bow, then cross over to a position 20 miles away on the starboard bow, then return to the ship.' Flying at 2000 feet the crews were able to spot vessels at up to 40 miles radius.

The Laccadive and Maldive Islands, and Chagos Archipelago

Three groups of coral atolls and islands are located on the Chagos-Laccadive Ridge, running between the Northern and the Central Indian Ocean, from 12° N to just south of the Equator for the Laccadive and Maldives Islands. The Chagos archipelago is centred on 6° S. From the northern most Laccadive island to Diego Garcia is over 2000 kilometres. The low lying islands cover a vast expanse of the Indian Ocean, and could have been employed as temporary hiding places for a German raider. Hence the requirement to search them for evidence of such a visit, *Raven*'s floatplanes enabling a wider area to be covered in a shorter time.

The islands are much changed over the past century, so the following brief descriptions have been edited from the most up to date source available to the two ships—*The West Coast of Hindustan Pilot* (1909) for the Laccadives and Maldives, and *The South Indian Ocean Pilot* (1911) for the Chagos group. The available charts dated back to 1904.

The Laccadive islands (now Lakshadweep, a union territory of India)...

> [c]onsist of fourteen islands, with a few detached reefs and banks. There are nine inhabited islands, two uninhabited, and three open reefs. The northern portion of the group is attached to the Collectorate of South Kanara,[344] the remainder belong to Ali Rájá of Kannanur, and form part of the administrative district of Malabar. Each of these islands is situated on an extensive coral shoal, with an area of 2 to 3 square miles; no part of these formations rises more than 10 or 15 feet above the level of the sea; the outer edges are higher than the body of these shoals, and generally enclose a regularly formed lagoon in which the water is still even in the worst weather. The receding tide leaves the outer edge of the reef nearly dry, and the tide runs out of the lagoon by breaches in the edges, which are sufficiently large to admit the light native craft into the natural harbour. The island of Minikoi is an acquisition from the Sultan of the Maldives, and may be considered as part of that group.

The Maldive islands (a British Protectorate from 1887 until 1965, when they became the present day Republic of Maldives) are:

> grouped in clusters termed atolls, occupy a space 470 miles in length, north and south, and 70 miles east and west. There are nineteen atolls or groups: in the centre of the chain they lie in double rows—east and west atolls, from 10 to 25 miles apart; at the north and south extremes of the chain the atolls lie singly. Barrier reefs encircle most of the atolls. The islands are in general not more than 5 or 6 feet above the level of the sea, so that the cocoa-nut trees on them appear on first approach to be growing out of the water. There is no bottom at a depth of 200 fathoms, close to the sea face of the islands and reefs.
>
> These islands are governed by a Sultan, who acknowledges the suzerainty of the British Government, as rulers of Ceylon, on which island the Maldivians have always been dependent, but the British Government in no way interferes with their internal affairs. The title and rank of the Sultan are hereditary; under him are six viziers or ministers of state, of whom the first in rank is styled Dorhimena, the chief or general of the army; but above these, and second only to the Sultan, is the Fadiyaru, the head priest and judge, civil and religious. The other chief officers are the Hadégiri, or custom-master and public treasurer; and the Míru Baharu, or master-attendant of the port. All these functionaries reside at Mále, or Sultan's island. The Sultan sends an annual embassy to the Governor of Ceylon, claiming the protection and favour of the British Government, and presenting a tribute of cowries, fish, and cakes. The Governor, in return, stipulates for succour to Europeans shipwrecked on the islands.
>
> These islanders have been kind in their hospitality to shipwrecked mariners; as exemplified in their humane and liberal conduct towards the crews of vessels wrecked on their atolls.

Finally, the Chagos Archipelago which during the Napoleonic War was claimed by Great Britain, and ceded to Britain by treaty after Napoleon's defeat in 1814. From 1814 to 1965 it was administered from the island of Mauritius.

> The extraordinary assemblage of islands and coral reefs included under the name of Chagos… from the North comprises Speakers Bank, Blenheim Reef, Peros Banhos Islands, Salomon Islands, Victory Bank, Great Chagos Bank (on which stand Nelsons Island, Three Brothers, Eagle Islands, and Danger Island), Egmont or Six Islands, Pitt Bank, Ganges Bank, Centurions Bank, Wight Bank, and Diego Garcia Island. Between the northern side of Speakers Bank and Addu Atoll (the southern group of the Maldive Islands), there is a clear channel 240 miles wide.

Poor quality view of Short 8021 being hoisted on *Raven* during the East Indies cruise. It was with this Short that FSL Smith and Sub Lt Kerry had their Laccadive Islands Adventure. The Short has a split side mounted radiator which was more common on 240-hp Renault powered Shorts, but can be seen on some, mainly Robey built, machines also powered by the 240-hp Sunbeam Gurkha engine fitted to 8021. Also of interest is the raised Short-designed ring mounting, similar to those fitted to other EIESS Shorts, and what appears to be a light bomb carrier under the forward fuselage. *(Kerry Family Collection)*

Laccadive Islands Adventures

On reaching the Laccadive Islands on 26 March the ships entered the archipelago from the north before proceeding through the centre of the islands to exit near Androth Island on 28 March. Several flights were made each day investigating and photographing the islands. Most of these were routine. But on 26 March, a busy day with a total of six flights, FSL Smith with Sub Lt Kerry as observer on Short 8021, experienced engine problems. They had to come down close to an island, later identified as Chetlat, and were able to taxy across the reef into the lagoon. There were warmly welcomed by the islanders, fed and entertained.[345]

When the Short failed to return *Raven* instigated a prearranged search procedure. *Pothuau* and *Raven*, 'steamed 10 miles apart towards Chetlat; in addition a seaplane was hoisted out to search. This seaplane located the missing seaplane at Chetlat and came back and reported the fact. Communication was established with the shore and the seaplane was got back to the ship. The trouble was found to be a broken piston, necessitating changing the engine.'[346]

On 28 March, the Cut Short crashed. Just after 06.00 FSL Stephens and Capt Knight-Bruce set out on Short 8090 to reconnoitre Androth island and Kalpeni atoll, comprising Cheriyam and Kalpeni islands. At each island they observed a few fishing boats but little else. Returning to *Raven* at 07.38, the Short 'was severely damaged in alighting.' The struts for the starboard main float collapsed, and the float itself was smashed into two parts, the port float

Returning from a flight over the Laccadives on 28 March 1917 the Cut Short 8090 crashed on landing. *(Kerry Family Collection)*

Although the crew FSL Stephens and Capt Knight-Bruce escaped with a wetting, the Short was a write off. *(EMK)*

remained attached to the fuselage by a single strut. The Short flipped nose first over on to its back. Stephens and Knight-Bruce escaped uninjured. The engine and any reusable parts were salvaged from the wreck.

As another Short, 8019, had to be erected to replace 8090, at 10.20 the ships proceeded southwards towards the Maldives 'as the direction of the swell allowed of a steady platform to erect a seaplane whilst under way.' Their route took them across the Nine Degree Channel, past Minicoy island at 02.00 on the 29th, then across the Eight Degree Channel to the northern atolls of the Maldive Islands. Commencing on the afternoon of 29 March they made a series of flights reconnoitring the northern Maldive atolls—Ihavandiffulu (Ihavandhippolhu), Tiladummati (Thiladhunmati) and Milandummadulu (Miladummadulu). A total of 14 flights were made, without incident, and the two ships sailed for Colombo at 17.00 on 31 March, arriving at midday on 2 April.

The following ten days were spent coaling, reprovisioning and servicing the seaplanes. Commander Samson also found the time to marry Miss Honor Oakden Patrickson Storey on 7 April 1917.[347] He flew his wife on a brief 10 minute flight over Colombo on 11 April. The following day *Raven*, with Samson onboard, and *Pothuau* sailed for the Chagos Archipelago, south of the equator, and the Maldives. Flights were not made during the passage owing to heavy seas with squalls of rain and high winds. The ships

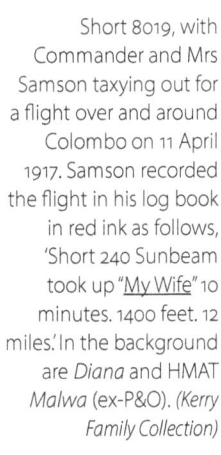

Short 8019, with Commander and Mrs Samson taxying out for a flight over and around Colombo on 11 April 1917. Samson recorded the flight in his log book in red ink as follows, 'Short 240 Sunbeam took up "My Wife" 10 minutes. 1400 feet. 12 miles.' In the background are *Diana* and HMAT *Malwa* (ex-P&O). *(Kerry Family Collection)*

had to reduce to half speed for part of the passage. Arriving on 16 April *Raven* commenced a series of reconnaissance flights through the Chagos and Maldive archipelagos, terminating on 23 April with their return to Colombo. There were three Shorts erected and available 8018, 8019, and 8021.

Raven's visit to Chagos was brief. On 16/17 April, a total of eight local flights were made, examining the northern Saloman Island and Peros Banhos Atoll, then the western islands of the main Chagos Atoll—Three Brothers, Eagle, Danger, and Egmont—before departing for the southern-most of the Maldive islands, Addu Atoll. The flights were not without incident.

Sub Lt Kerry recorded a flight with Samson on Short 8019 in his diary on 16 April 1917.

On his next flight, five days later, Samson was flying Observer Lt Kerry (he had been promoted 1 April 1917) over the Chagos archipelago. On landing the starboard float of Short 8019 broke in two but, as seen here, apparently without damaging the struts. Samson has left the aft cockpit and is on the port wing root, Kerry is hidden behind the box radiator. A pillar mounting for a Lewis gun is clearly visible on the starboard side of the forward observer's cockpit. *(YEORN 2014-80-119. Courtesy of The National Museum of the Royal Navy)*

> Much calmer in morning, things looking more settled. [Smith and Meade] made a flight in morning to test conditions. My turn at 2pm [Samson and Kerry], flew for 53 mins between Islands Peros Banhos [Atoll] and Salomon. Very cloudy but pleasant flight except on landing smashed and broke off completely starboard float. Machine was given a violent list but did not sink. Eventually hoisted on board and machine was saved.

Whilst his Report of Flight does not mention the landing, Samson's log book confirms the above, adding 'Very cloudy (rain storms and windy). Stbd float collapsed on landing machine salved practically intact.' The Short was repaired and flying again by 22 April. The following day Samson, piloting Capt Knight-Bruce on Short 8018, also recorded damaging the chassis whilst landing following a flight over the Eagle Islands of the Great Chagos Bank. Damage must have been minor as his log book entry is the only mention of the incident, and 8018 was flying over the southern Maldives two days later.

Making the passage north to the Maldives on 18 April, a single scouting flight was made, flying a box pattern ahead and both sides of the two ships. Adu Atoll was sighted in the early hours of 19 April.

Pothuau anchored off the King's Island, Malé, 23 April 1917, during the search for the missing Short and crew, Flt Lt G.D. Smith (he had been promoted on 1 April 1917) and observer Lt W.C.A. Meade. The anchorage is within the atoll and this view is looking to the south, toward South Malé atoll. *(TNA, AIR1/1707/204/123/67)*

Marooned in the Maldives

A planned series of flights up through the islands was commenced. As *Raven* and *Pothuau* proceeded at a leisurely rate up the channel between the western and eastern line of atolls, flights were made early in the morning and late in the day to take advantage of better flying conditions. Usually just one machine would be sent out, scouting the islands either side and ahead of the two ships. During the night, speed was reduced so that the two ships would reach the furthest point scouted by the seaplanes the previous day.

By the morning of 21 April they had progressed to the north end of the Mulaku Atoll, some 120 kilometres south of Malé the Maldives's capital. At 16.00, as they were approaching South Malé Atoll, *Raven* stopped to hoist out Flt Lt G.D. Smith (he had been promoted on 1 April 1917) and his observer Lt W.C.A. Meade on Short 8018 to conduct a reconnaissance flight of Ari Atoll, across the channel from their present position off the Felidu (Felidhu) Channel, between Felidu and South Malé atolls. They were to fly to the eastern edge of the atoll, then turn north along the atoll to rejoin the ship off South Malé Atoll.

When the Short had not returned by 17.45 the two ships separated by 5 miles and steamed slowly west towards Ari Atoll. At 20.00 *Raven* turned north and *Pothuau* south continuing a pre-arranged scheme of search. At 20.30 'a Green Very's light was seen about six to eight miles distant. A bearing was taken and HMS *Raven* steamed towards the place but no further signs could be seen.'

> As no further Very's lights or other signs were seen there is little doubt that the seaplane was capsized owing the floats having carried away. In this position the seaplane would be unable to support the Officers, but would float for some considerable time herself when freed from their weight. If the seaplane had been floating in a normal position the Officers would have fired more Very's lights (they had 12), and also fired their Lewis gun and Rifle which were both in the seaplane.

Although a thorough search was made by both ships during the night nothing more was seen.[348]

Raven's log book contains the following entry at 23.10, 21 April: 'Flight Lieutenant Smith age 22, Lieutenant Mead [sic] RNVR age 29, believed drowned.' Nonetheless, the search continued the following day.

A morning flight covered the lower line of islands of Ari Atoll, finding no evidence of the missing Short. A second flight, Short 8019 with FSL Stephens and Sub Lt Kerry, set out at 09.30.

> At 10.45 am this seaplane sighted a large patch of oil on the water [at the south of Ari Atoll, but closer to Filidu Atoll] within half a mile of HMS *Raven*. The ship approached the oil and something was observed to be floating on the water. A boat was lowered and picked up a ship's biscuit. As biscuit is always carried in the seaplanes it is considered that the lost machine is very likely sunk here; the current which was running very strongly would have drifted it to about this position, which is at the NW corner of Felidu Atoll.

However, as there was no further sign of the missing Short, the two ships proceeded to Malé island, anchoring at 17.25 on 22 April.

> The Sultan sent his secretary on board the *Raven* and he, speaking good English, surprised the ship's company with his knowledge

King's Island, Malé, taken from Short 8019 by Obs Lt Kerry piloted by Flt Lt Burling. This natural island forms the core of the modern island city of Malé, increased many times in size by land infilling. Little of this photograph survives in the modern city island, but part of the large clear compound just right of centre is the modern Sultan Park which also contains the National Museum. *(TNA, AIR1/1707/204/123/67)*

of the progress of the war, gathered from British newspapers, and by his particularly intimate acquaintance with that portion of the war in which the ship was then engaged. He knew exactly the route she had taken in the north Maldives. Reports from the headmen of the islands visited had already been duly forwarded, and it was found that in some cases there was a suspicion that the *Raven* was an enemy ship carrying Germans in disguise.[349]

The Captain of the French Cruiser *Pothuau*, Commander C.R. Samson, DSO, RN, Flight Commander T.H. England, DSC, RN, and myself [Lt Jenkins, RNR, *Raven*] visited H.H. the Sultan [Muhammad Shamsuddeen Iskander III] the next morning, and arrangements were made with him for a search to be made amongst the Islands for the missing Officers and Seaplane. H.H. the Sultan promised to render all the assistance he could in the matter.[350]

Before the departure of the seaplane carrier the Sultan sent as gifts a young bull, two goats, two turtles, many coconuts, bananas and limes, and a plentiful supply of fresh fish and eggs-all very welcome. And each officer was presented with a mat of native workmanship made of fibre very finely woven and of designs strongly influenced by the proximity of India.[351]

Thus laden with tokens of high favour, *Raven* and *Pothuau* continued on to Colombo, where they arrived on 25 April.

At Colombo all the usual tasks of coaling, taking on supplies, tuning up the floatplanes, writing reports were undertaken. Between 4 and 5 May, a small party comprising Flt Cdr Clemson, four mechanics, and Sopwith Baby N1014, was transferred to the Australian cruiser HMAS *Brisbane*,[352] for an expedition to the equator (see below). On their return to Colombo on 13 May they learnt that Smith and Meade had returned from the dead.

On 6 May a boat rowed alongside, and two bearded and curiously dressed figures came up the accommodation ladder. They wore tarbooshes, white duck trousers and scarlet jackets with white facings. They looked like the leaders of the local band going round for subscriptions, but they were not. They were Guy the pilot and Willie the observer of the missing seaplane, and the adventures of the two officers must be taken on their own responsibility, since they were the only eye-witnesses of their exploits. Nor is it necessary that I should attempt to reduce to order the hilarious and incoherent versions which both of them gave simultaneously within a few minutes of their return to the ship. After the first excitement had died down each of them wrote out an account.[353]

Flt Lt Smith's was in a letter to his father which, by devious routes, reached Rudyard Kipling who used it as the basis for his story *A Flight Of Fact* in 1918.[354] Lt Meade's account appeared in another magazine, also published in 1918, under the heading *Our Seaplane Adventure*. The following brief account draws mainly on Meade's version.[355]

Shortly after taking off they had flown into a sudden wind storm and been blown off course. With darkness approaching, and fuel running low, Smith decided to put the Short down by a 'long black shadow of some island, surrounded, as they all are, by coral reef and lagoon.' They landed safely but immediately stranded on a coral reef. An attempt to send a wireless signal, running an aerial from the wing-tips and propeller, failed as 'the spark we obtained was miserably poor and nothing came in reply.' Around midnight the tide lifted them off and, restarting the engine, they taxied about trying to find and island to land on. Meade was up on the centre section firing off flares to provide some illumination. He was down to a single flare when Smith decided to risk all and taxy full out trying to cross the reefs into the main lagoon. Scraping over some coral they made it guided by the light of the last flare. Once in the lagoon they were lucky to locate a small island briefly illuminated by lightning. Tying the Short to a palm tree they 'spent a miserable night lying along the lower planes in an attempt to get some little shelter from the driving rain.'

In the morning the island was found to be small and uninhabited, with no water or food other than a few coconuts. Meade having 'sacrificed' his tunic to attempt to keep the magnetos and plugs dry, their other clothes were

'Two curiously dressed figures came up the accommodation ladder ...' Flt Lt G.D. Smith and Lt W.C.A. Meade on their return to *Raven* in the uniforms of honorary members of the Sultan of Malé's royal body guard. *(Kerry Family Collection)*

Above left: Smith and Meade surrounded by crew members of *Raven*.

A general view of the palm shed/hangar and Short 8018 on Fiale island, 8 May 1917. The torn fabric in the rear fuselage shows where the flotation balloon was removed when Smith and Meade swam ashore a few weeks earlier. Other than this the Short is in remarkably good condition after its adventure. The port side pillar mount for a Lewis gun is clearly seen, there is also a similar mount on the starboard side. *(EMK)*

hung in a palm frond shelter in the hope they may dry off. When the rain finally ended in the afternoon, they put in action a plan 'to make for the main island of the group, keeping well in sight of the intervening islands in case of accidents.' Always assuming the engine could be started and the damaged machine could be flown. Both floats were taking on water, an aileron was damaged and the wings so waterlogged, they had to puncture the fabric to drain them.

On the last charge of compressed air, used to turn over the engine, it started. Once the engine had started they had no time to collect their clothes. So, with two naked airmen onboard they flew over the atoll searching for an inhabited island. Once the engine had started they had had no time to collect their clothes. Passing over some local fishing boats, whose crews passed the word to the Sultan about a 'Strange and loud-voiced bird!' Their fuel ran out as they were approaching an occupied island, coming down close to the shore the wind and tide began carrying them further away. This time, with the dark approaching again, they decided to abandon the Short and swim for shore, which Smith estimated to be two miles away. Smith had a cork waistcoat available, but Meade had to remove a flotation balloon from the rear fuselage. Exhausted after a long struggle in the sea both made it to land.

On the island they found a few palm leaf huts, piles of coconuts, and a few scrawny chickens. One of these they killed and roasted over a fire that had been left burning. They were visited during the night by some natives who had landed from one of their boats.

> Through the palm grove there came a flicker of light and the low murmer of voices. A moment later two or three dark forms emerged into the moonlight and approached the huts.
>
> We got up.
>
> "*Salaam!*" Guy said, in his friendliest tones.
>
> There was one wild shriek of terror. The crazy lamp was dashed to the ground, where the coco-nut oil flared up for a moment and then died out, and back into the intricate mazes of the palm grove dashed three panic-stricken forms.

Come the dawn, the Short was barely visible in the distance, but still afloat. They started to build a raft, but were eventually able to persuade some native boats to permit them on board and, after much gesticulation, to take the seaplane in tow. It was landed on the beach of an island, later identified as Fiale (Feeali), on the north east side of Nilandhu (Faafu) Atoll, some 30 kilometres south of Ari Atoll.

Surrendering themselves to the headman, they were taken into the care of the village bachelors and accommodated in their hut, and provided with loin cloths and a few other rags of clothing. Over the next few days, although always being watched, a wary friendship was established. They exhibited their machine and gained a great prestige by giving their friends shocks with the wireless battery. A move which also ensured it would not be tampered with, the islanders erecting a rope barrier around it. By day they bathed; in the evenings they had sing-songs interspersing ragtime and music-hall tunes with the drowsy chants contributed by native talent. Finally, after several days, they were put in a large fishing dhow, carrying a crew of eight. The headman of the island was in charge, setting sail for Malé some 120 kilometres from Fiale, to hand them over to the Sultan.

At Malé they were met by the Sultan's secretary. From him they learned of *Raven*'s visit and that the Sultan had sent instructions that a thatched shed was to be built over the plane to protect it from the weather, until such time as it could be removed. Most importantly, they were relieved to hear that they would soon be sent to Colombo. 'Our appearance in the skimpy attire of the natives of Fiale probably grieved the Sultan, for a few hours after our arrival we were ordered into the presence of the royal tailor, who promptly provided us with white duck trousers, cut on a pattern that would scarcely past muster before American eyes. We were also notified that we had been made honorary members of the royal body guard, and that as a further mark of distinction we were to wear the guard's coat, a heavy, highly ornamental garment of violent red hue.'[356] They left Malé aboard a sea going dhow and, as we have seen, reached Colombo on 6 May.

Before *Raven* could sail to retrieve Short 8018, a major change in command became effective. On 8 May Wing Commander C.R. Samson, left the ship to return to the UK. He had been replaced in command of the East Indies and Egypt Seaplane Squadron at Port Said by Wing Cdr C.E. Risk on 25 April. Also leaving at this time were Flt Cdr T.H. England and Captain Robert R.E.C. Knight-Bruce. Both Samson and England visited the EIESS island at Port Said on 22 May on their way back to the UK.

On his return to England, Samson assumed command of Great Yarmouth, and later Felixstowe, naval air stations. The stations conducted anti-submarine and anti-Zeppelin patrols over the North Sea using single engined Shorts and Sopwith floatplanes and large twin-engined flying-boats. Samson experimented with a Sopwith Camel scout taking off from a platform mounted on a lighter towed at high speed behind a destroyer. Always eager to lead from the front, he made the first attempt to takeoff. The Camel tumbled over the bows to be run over by the lighter, Samson barely survived this experience. Using an improved version of the platform Lt S.D. Culley safely took off and destroyed Zeppelin L53 on 11 August 1918. A year later Samson left the Royal Navy to accept a permanent commission in the Royal Air Force with the rank of Group Captain. Over the next ten years he commanded RAF units in the Mediterranean, a fighter group at Kenley and, in 1926, became Chief Staff Officer, Middle East Command at Cairo. Never one to happily fly a desk, in 1927 he organised and led a flight of Fairey IIIF biplanes across Africa from Cairo to the Cape of Good Hope. Promoted Air Commodore in 1922, Samson was placed on the retired list on account of ill health in 1929. He had been awarded the DSO and bar (for services in Egypt 1916/17), the AFC and been appointed CMG in 1919. His foreign orders included appointment as a *Chevalier de la Legion d'Honneur*. Charles Rumney Samson died at his home on 5 February 1931, leaving his wife and a son, and a daughter from his first marriage.

England returned to the UK, arriving at the beginning of June, he was promoted Squadron Commander on 30 June 1917. He served aboard *Nairana* and, from February 1918, he was at Calshot. Captain Knight-Bruce disappears from the EIESS, although there is no record of his ever being attached to the EIESS. He is recorded as serving with 47 Squadron, RFC, in Salonika from August 1916 until August 1917, and attended an Aerial Gunnery course at Abbassia, Egypt, in June 1917. So, his sojourn with the EIESS is a bit of a mystery.

Raven sailed from Colombo at 09.00 on 6 May in company with *Diana*.[357] The two ships proceeded to Nilandhu Atoll, *Raven* entering the lagoon and anchoring off Fiale island at 12.00 two days later. *Diana* remained on patrol outside the atoll.

Guy Smith now picks up the story.

As we drew near the island, once again we saw the natives gathered on the shore apparently greatly excited. We put off in a boat, and there was a hostile demonstration on their part until we got near enough for them to recognize Meade and me. Then they began to shout with joy. I had on my best uniform and the men, women and children alike would keep rushing up to me and feel the cloth. Other men in the party were similarly clad, but the Maldivians took no liberties with them. They figured, however, that Meade and I were in a measure their property, and their delight was unbounded. We brought forth our presents, chiefly knives, mirrors, razors, candy and cigarettes, and the pleasure of children over new toys was no keener than theirs.

Short 8018 being prepared to leave the island and return to *Raven* out in the atoll. *(Kerry Family Collection)*

With great gusto and much beating of tom-toms, the headman at last led us to the cache of our seaplane. The machine had been guarded well. There was scarcely a possibility that anybody would steal it during our absence, but the natives had busied themselves with building a strong fence [and shed[358]] about it, made of cocoanut saplings and palm leaves. They had a small gate in one side, and this had been fastened with a rope attached to a gigantic padlock. Where that padlock came from I have no idea. Probably it had been a treasured island possession for many years. Anyway, there was no key for it, which did not perturb the natives in the least. They unfastened the gate by the simple expedient of cutting the rope.

We had to demolish one side of the [shed] to get the airplane out, but with the help at our disposal had little difficulty in taking it to the ship and having it hoisted aboard. Then we bade farewell to our friends and sailed off towards the west. As far as we could see them, and we looked back through a glass, the natives of Fiale stood on the shore of their little island and watched our ship.

The two ships immediately proceeded to Colombo, arriving during the afternoon of 13 May. The Short was beyond local repair but, once returned to Port Said, was quickly returned to service. Short 8018 had other adventures ahead of it.

HMAS *Brisbane* smoking away during the early 1920s. One of four Town-class cruisers to serve in the Royal Australian Navy and the first of two to be built in Australia. She spent most of her war 'East of Suez'. The Baby was probably located on a platform built aft of the funnels. *(State Library Victoria, gr001830)*

HMAS *Brisbane* and the EIESS, Colombo Detachment

As mentioned above Sopwith Baby N1014 was temporarily attached to *Brisbane* in early May. It was stowed on a temporary platform, just forward of the mainmast upon which a boom was rigged to lift the machine on to and off the water. The intent being to conduct a series of ocean reconnaissance flights, along the equator from 75°E to 89°E, in the search for the elusive *Wolf*. A total of six flights were made, two each day on 9, 10, and 11 May. *Brisbane* returning to Colombo on 13 May.

Clemson flew a similar course for each flight at a height of 2500 to 3000 feet. Leaving the ship, which was proceeding east at 12 to 15 knots, he headed NE for 15 to 20 minutes, then SE crossing *Brisbane*'s course out to a similar distance then headed back to the ship. Each flight lasted between 1 and 1½ hours. With a morning and afternoon flight, a track from 120 to 150 miles wide and 200 miles in length was covered each day.

Clemson reported, 'The machine, in spite of its light construction stood up to its work well. Its weakest points being the main and tail floats, which,

Sopwith Baby N1014 on the hook from *Brisbane*, Flight Commander Clemson in the cockpit. There are two bomb carriers for 65-lb bombs under the fuselage and a pair of fixed Lewis guns above the centre-section, each with a 47 round 'hopper' installed. The twin Lewis were not a common feature on EIESS Babys, in addition to N1014 only N1028, N1036 and N1038 are known to have had them. *(Australian War Memorial, 305421)*

although thoroughly overhauled and white leaded after each flight, became more and more leaky at the seams. I would submit that these floats, especially the tail float, might with advantage be more strongly constructed.'

On 16 May, whilst still at Colombo *Raven* 'Discharged Sopwith N1014 & motor car & motor boat to depot. Flight Commander Clemson, Flight Sub Lieutenant Stevens, 1 PO & 8 mechanics discharged to depot.'[359] An RNAS report also noted, 'Two Sopwith seaplanes will be left at Colombo with the necessary personnel, so that they will be available for working from the cruisers.'[360] The second Sopwith, Baby N1016, was probably also delivered from *Raven*. There is no indication that the detachment had a formal existence, so for convenience I have labelled it the Colombo Detachment.

Flt Lt Popham, who had arrived in Colombo from Port Said on 23 June, described the set up in his diary.

Colombo Graving Dock in the early 1900's. The EIESS Colombo Detachment was probably based close to the shed on the right.

> Went down in the car, a large Wolsley, with Clemson, Stevens, and the hands. There are 1 Petty Officer Groucott, one LM and about 6 mechanics, also 2 or 3 men from the Kite Balloon section which is at a rest camp after being in Mesopotamia. One drives along the harbour side through crowded native streets with occasional temple shrines and whiffs of incense, down to Walker's iron and marine engineers works next to the graving dock.
>
> The machines are stored in part of a shed up a railway line between two buildings. The line is too narrow to allow of the passage of the machines without tails, so that they are kept disconsolately in two pieces. There are three [N]1038 new just come out, [N]1014 and [N]1016 (or vice versa?) without an engine at all. There are a few stores and things in another shed. It is an awful business getting them out. It is done by means of a large steam crane and a truck running on the line. The truck and the plane are then pushed down to the end of the jetty, the tail fitted on and the engine being tested, then hoisted over by the crane into the water.
>
> They are building a magnificent new shed the other side of the graving dock on the water's edge, about 80 ft long. There is also a very steep little slipway.[361]

After *Raven* and *Brisbane* left the Detachment had been barely employed. The latter had returned at the end of May for a few days and possibly embarked a Baby and crew.[362] 'Her Captain had been very keen on flying encouraged by a number of successful flights. They had rigged up platforms for the machine, and got things well teed up'.[363] But since then other than a few test flights it appears to have been largely unemployed. With the departure of *Brisbane* in early June the reason for the detachment largely disappeared.

On 25 June, Clemson tried out the newly erected N1038. It was difficult to get

Sopwith Baby N1016 at Port Said, before being sent to Colombo. Bomb carriers have been installed but no sign of any Lewis gun. *(YEORN 2014-80-113. Courtesy of The National Museum of the Royal Navy)*

started and in the middle of the struggle he was summoned to the telephone. Clemson came back fuming, it had been the SNO authorizing his flight. Final, he was able to get the engine running and, after being towed out beyond a line of buoys, was able to get off. On his return less than half an hour later, he taxied straight up to the jetty. He was covered in oil, and the blipping switch did not work. Popham commented that the normally quietly spoken Clemson 'was for the first time audible: probably he was deafened by the engine.' The Baby needed some work and Popham lost his chance to essay his first flight at Colombo. After a false hope on 3 July, ruined by ongoing engine problems and poor weather, he finally had the opportunity on 4 July, it did not start well and finished even worse.

Taxiing out on N1016 he ran over a buoy getting it lodged between the floats. That was cleared, but to get it restarted a crew member from an accompanying boat had to climb onto one of the floats and turn the engine over. He then got off alright after a long run.

> Circled round a bit through clouds at 1000 [feet] and then went down south to look for a minefield. It was a pity it was so cloudy as I couldn't get much of a view of Colombo or the country. Turned to come back and noticed the revs dropping slowly at first then more rapidly until she petered out altogether and I had to land. Turned into the wind towards land and across the swell. Hit the top of a big wave and bounced. Port float gave and when I settled down she started sinking. Got out on the tip of the starboard wing but after about 15 minutes she turned slowly over on her back. I climbed round onto the bottom of the starboard float as she did so.

He was about four kilometres from land, but was rescued by a local fishing boat, then passed on to the harbour pilot boat. Trying to tow the wreck, the floats broke off and were loaded onto the pilot boat. Then a tug boat arrived and an attempt made to haul the Sopwith on board, but it broke free and sank. Clemson and Stephens now arrived in the Detachment's launch and took Popham aboard. The position of the wreck was buoyed, but several attempts over the next few days failed to recover anything of use.

During Popham's brief absence orders had arrived to pack up and return to Port Said. Packing up the two remaining Babys took a few days. Only N1038 had an engine, although a Clerget was on its way from Port Said for N1014, but may have arrived after the Detachment had departed. They were still in Colombo 11 July, when Popham's diary ends, but probably sailed on 15 July on HMT *Somali*.[364] Popham was back in the air at Port Said on 14 August.

A Final Visit to Aden

Returning to *Raven*, who finally turned her stern on Colombo shortly after 08.00 on 17 May 1917. In company with *Brisbane* she made her way up towards Bombay where, on 21 May, *Brisbane* parted company and *Raven* joined a troop convoy headed for Aden. Escorted by the cruiser HMS *Exmouth* the convoy included HMT *Kinfauns Castle*,[365] HMT *Nile*, HMT *Akbar*, shortly to be joined by SS *Khosrou* (P&O Company). On passage the convoy ran into some rough seas, slowing to five knots for several days. *Raven*'s log book noting that the canvas hangars protecting the floatplanes required constant attention, although no damage was noted. By 31 May the seas had calmed sufficiently for *Raven* to tow a target for *Exmouth*'s seamen gunners to use for rifle practice. The following morning *Kinfauns Castle* and *Khosrou* left the convoy steering a course towards the African mainland. A few hours later *Exmouth* and *Raven* entered Aden harbour, the two troopships are not mentioned again.

On arrival, Flt Lt Burling who had assumed command of *Raven*'s flight, was instructed to report to General Stewart, GOC Aden. Taking advantage of the seaplane carrier's arrival, the general requested a flight to update his intelligence on up country Aden. Accordingly, Flt Lt Burling and Sub Lt Kerry on Short 8019 left *Raven* at 16.45 to reconnoitre the Lahej Delta, dropping two 16-lb bombs each on Waht, Darb and Subar camp, returning after an hour's flight. They observed small trenches and three gun pits at Waht, but saw no troops, Subar camp appeared unchanged from the previous visits.

Despite feeling that he would 'never step into an airplane again' following his Maldives experience, Guy Smith was back in the air on 2 June, also piloting Short 8019, taking Major Paige of the GOCs Staff in Aden on a brief reconnaissance over Aden and its hinterland. The major apparently enjoying his flight, being entirely satisfied with the information obtained. They were the last flights the EIESS made at Aden, which would have to wait until the end of 1917 to get a permanent aeroplane unit, a Half Flight of 114 Squadron, RFC.[366]

After this flight *Raven* sailed for Suez and Port Said, where she moored at 08.00 on 10 June. After her return, *Raven* languished at Port Said, moored in the Abbas Hilmi Basin, until early October when she moved down to Suez for a short refit. Returning to Port Said after ten days she once more picked up her moorings, until required for one final trip in November.

CHAPTER 15

Maintaining a Steady Course

Squadron Commander (Major, RMLI) E.C. Risk. Detail from a group photograph taken at Calshot in July 1914. *(FAAM, Travers 78-47, Air Stations 3392. Courtesy of The National Museum of the Royal Navy)*

The EIESS in 1917

During *Raven*'s long absence, *Anne* had effectively been retired, *Empress* returned from Genoa, the First and Second Battles of Gaza had been fought, and lost. Most importantly, for the men and ships, Wing Commander Charles Erskine Risk had assumed command of the East Indies and Egypt Seaplane Squadron on 25 April 1917.

Wing Commander Risk was a Royal Marine, he had been commissioned into the Corps on 1 September 1901. He was one of three RM officers on the first course at the new Central Flying School at Upavon, August to December 1912. Risk gained his Royal Aero Club certificate, No.303, on 1 October 1912. After qualifying he was appointed in Command of Felixstowe Air Station with the RNAS rank of Flight Commander.[367] He was also an old compatriot of Samson, having been with him during the early heady days in Belgium of No.1 Naval Wing, where as a Squadron Commander he had been put in charge of the armoured cars. Risk was in command of the RNAS armoured cars at Gallipoli. He was wounded at Anzac on 30 April 1915, with a bullet through his left thigh and invalided back to the UK. On recovery he was appointed to command Dover Air Station 28 July 1915 to 28 February 1916. He was then in command of Isle of Grain Seaplane Station, until 15 March 1917. Promoted Wing Commander on 31 December 1916, he was appointed to command of the EIESS.

After the flamboyance of Samson, Risk was the calm steady hand needed to guide the EIESS through the next year. Flt Lt Popham recorded his first impressions. 'Col. Risk, Wing Commander, a very typical marine, nice looking with a close sandy moustache, and a shy humorous manner. I instinctively mistrust but rather admire him. I do all marines—they inevitably frighten me.' Which probably tells us as much about Popham as it does Risk.

A poor quality image of the island base as it appeared during the latter half of 1917 with two Bessonneau hangars. *(EMK)*

The situation at the base naturally was Risk's first concern. Following a settling-in period, he prepared reports to the C in C, East Indies Station, Rear Admirals Wemyss and Gaunt (from 23 June 1917), and the Admiralty, which provide a comprehensive appraisal of the state of the EIESS in 1917.[368]

Risk started by comparing the number of machines approved by the Air Department, Admiralty, with the actual machines at Port Said: 'Short 184—Approved 12; Actual 7: Sopwith Baby—Approved 6; Actual 3.'[369] He then says that the EIESS 'is very much under-manned, there being only 51 Air Service ratings on the Station, including Master-at-Arms, Writers, Boat's Crews, etc. Of these 9 are detached at Colombo, leaving 42 at Port Said (living on board *Empress* and *Raven II*).' Officers were similarly lacking, only 13, plus one Warrant Officer, being on station. His recommended complement was of 33 Officers and 170 ratings (including CPO and PO).

Risk's July report includes descriptions of the physical state of the Base at this time. After mentioning the local conditions prevalent, he continues.

> [A]t home, the seaplanes are kept in well built sheds, mostly water-heated, fitted with electric light, good floors, etc., and every convenience. Slipways are provided, fitted with electric or petrol engines, for hauling seaplanes in or out. Motor boats are provided for towing seaplanes should they breakdown, etc.

> At Port Said most of the machines are kept in two very old and dilapidated hangars. The temperature inside these hangars is so high that the planes of these machines become absolutely distorted and warped. It is so bad that when one stands by the wing tip and

looks along the leading edge of the plane, one can see the leading edge curving in and out, instead of being uniformly straight.

When these planes become distorted, it is a sign that some of the ribs are broken or twisted. An examination then has to be made and the ribs repaired which means additional work for the carpenters and fabric ratings.

The floors of these hangars are of sand. The consequence is that with ratings working on these machines a certain amount of sand is always being deposited on and in the machine. This is, of course, exceedingly bad for the engines and makes it much harder to keep the machine, planes and controls clean and in working order.

Machines have to be man-handled in and out on a very impromptu slipway. This is rather a tedious job and necessitates the employment of a good number of ratings and Arabs.

Floats do not last out here half the time that they do at home. Carpenters are constantly employed repairing these floats. The heat of the Sun on the sides of the floats when in flight probably is the main cause of weakness.

By 1917 there were a growing number of discarded Shorts and Sopwiths, lying in odd corners of the island. Here is Flt Lt Burling posing with three Short fuselages, their flying days over but probably a useful source of spare parts. *(EMK)*

> Owing to the existence of plague rats on the Island several
> buildings have had to be pulled down. These buildings will
> be re-erected later. All this work has had to be carried out
> by Air Service ratings with the assistance of a few Arabs.
>
> The argument might be made that it has always been possible in the
> past, for the East Indies & Egypt Seaplane Squadron to CARRY ON
> with the small number of ratings provided. This is quite true; they
> have been able to CARRY ON and do most of the things they are asked
> to do, but this has been done at the expense of machines. I can to an
> extent CARRY ON at present, but it would be impossible to undertake
> any operations in which two Seaplane Carriers would be absent at the
> same time. This is also provided I do not keep all machines clean, but
> I consider this to be an extremely expensive way of working. The additional expenditure incurred by providing an increased complement
> would be more than covered by the saving made in the prolonged
> life of the machines, which now-a-days are very expensive articles.
>
> It is of the utmost importance that machines should be kept in
> a thorough state of cleanliness and efficiency, especially on this
> of all stations seeing that Seaplanes are constantly called upon to
> carry out flights of from 30 to 60 miles inland. An engine failure
> under these conditions practically means certain disaster.

His appeal did not fall upon deaf ears, although action was slow in coming. At the end of August the Admiralty informed the C-in-C Egypt that several actions had been taken to improve the situation of the EIESS. These included the allocation of five Renault (240hp Renault-Mercedes) Shorts and eleven Sopwith Baby Seaplanes, of which two Shorts, four Sopwith Baby and two Fairey Hamble Baby machines were already in transit. As were three Bessonneau hangars. The Admiralty also approved an increase of total complement to 156. This to include 34 Engineers, 28 Carpenters and 77 Aircraftsmen, the balance to include a small increase in Officers.

However, all was not ideal, and towards the end of the year Risk forwarded a further report principally concerning ratings provided as reinforcements. In it he emphasised the following points.

> During [work] on the coast, it is sometimes necessary to keep
> two Seaplane Carriers constantly employed in spotting and
> reconnaissance work. A large staff is, on these occasions, required
> at the base repairing machines as they return, and, at the same
> time, it is essential to have an ample staff on board the Carriers.
>
> The majority of the Officers and men of the Squadron are
> accommodated on the various Carriers and have to be transported

to and from the Depot three and four times daily. This necessitates a Petty Officer and one or two "E" ratings being allocated to the upkeep and repair of motor boats, which are manned by Aircraftsmen.

About 60 per cent of the total complement of the Squadron is now composed of Aircraft ratings, and a good deal of the engine work has to be performed by them, necessitating much more supervision than if performed by actually trained "E" ratings. In this respect I would point out that in many cases the Aircraft ratings appointed to this Station are only enlisted a few weeks prior to their arrival here, and have had no actual experience of work on an Air Station at all. It would appear that the "E" ratings who have arrived have had very little experience with the Sunbeam engine, which is largely used in this Squadron.

Due to purely local conditions, engines require a great deal more attention here than at home. For instance great trouble is experienced due to sand finding its way into the engines, necessitating their constantly being taken down, cleaned and overhauled.

For most of 1917 the EIESS was only able to operate a single seaplane carrier on the coast, *Empress*, usually with four Short 184's in her hangar. As preparations for the final Gaza battle proceeded, reinforcements of men, machines, and a new ship, permitted the EIESS to commit three seaplane carriers to support the attack – *Empress, Raven* and *City of Oxford*.

HMS *City of Oxford*

As early as December 1916 Admiral Wemyss, the C-in-C, East Indies Station, had requested replacing both *Anne* and *Raven* with a single ship having twice the speed, as their limited speed made them vulnerable to submarine attack.[370] Wemyss was clearly describing, probably at Samson's urging, one of the seaplane carrier conversions similar to *Ben-my-Chree* and *Empress*. Which was not quite what he received, for whilst *Anne* and *Raven* would soon be decommissioned, their replacement was more of the same.

The steamer *City of Oxford* (4016 tons gross; length 412 feet; 12 knots when new) was built on the Clyde by Barclay, Curle & Co. Ltd in 1882 for George Smith & Sons, Glasgow. She was employed on the company's passenger/cargo UK to India routes until 1901 when George Smith & Sons were taken over by Ellerman Lines. Shortly after the outbreak of the war the Admiralty requisitioned 14 merchant vessels for conversion to dummy battleships. On 28 October 1914, *City of Oxford* was purchased and sent to Belfast for the necessary alterations to be made by Harland & Wolff. These had been completed by early December 1914 and, as the dummy battleship *St Vincent*, she joined the Special Service Squadron in Scapa Flow.

SS *City of Oxford* probably on trials in 1882. *(University of Glasgow Archives & Special Collections: DC101-0138)*

For seven months she swung round a buoy and then, when it became apparent that no-one was being fooled, le ast of all the enemy, *City of Oxford* was taken in hand for reconstruction to a kite balloon ship. Again the work was carried out by Harland & Wolff, who began work on 16 July 1915. An extensive hydrogen generating and storage plant was installed, and one of the forward holds converted to house a balloon. The ship's foremast was removed and re-stepped abaft the bridge. A stump balloon mooring mast was fitted forward. Another special feature was the addition of removable anti-torpedo bulges fixed in sections, and arranged so that damaged sections could be removed and replaced without the need for the ship to be docked.

City of Oxford left Belfast on completion of conversion on 9 March 1916 and arrived at Dover on the 15th to provide spotting facilities for the monitors that were operating off the Belgian coast. In August of that year she was released from the Dover Patrol and joined the Grand Fleet and later became attached to the Battlecruiser Force for trials in connection with the development of balloon installations in battleships, light cruisers and destroyers.

Unlike *Anne* and *Raven*, *City of Oxford* was converted into a seaplane depot ship at a shipyard in Hull to Admiralty requirements before sailing to Egypt. The original passenger/cargo ship has been altered so much that it is barely identifiable. Arrangements for floatplanes included a semi-permanent forward bulkhead to the well deck with supports for canvas shelters on both fore and aft well decks. Seen here at Port Said with two Shorts on the forward well deck, the second possibly being engineless. *(via Stuart Hadaway)*

With the increasing use of kite balloons flying from capital ships, at the end of 1916 the Grand Fleet Aeronautical Committee recommended that she should be paid off. In March 1917 it was decided that she should be converted yet again, this time for a seaplane-carrying role. This final transformation began at Hull on 1 April and lasted to the end of June. The kite balloon equipment was landed and the forward well deck was modified for use as a seaplane servicing area when the ship was at sea or, alternatively, for carrying six large seaplanes when the ship was required to operate as a transport. The hydrogen plant was removed from the hold and the space cleared to allow for the accommodation of six Sopwith Babies. Both the stump foremast and centre mast were taken down and a new foremast, fitted with two derricks, was stepped in the middle of the forward well deck to hoist the seaplanes to and from the water. The anti-torpedo bulges were also permanently removed.

Allocated to the East Indies and Egypt Seaplane Squadron on completion of the conversion, she sailed to Sheerness to embark a number of cased floatplanes for the squadron. On 13 July she departed Sheerness for Port Said arriving on 7 August. For the next two and a half months she served as a depot ship for the squadron in lieu of *Anne*. The latter having been decommissioned the day after *City of Oxford* arrived in Port Said.

The Military Situation in 1917–1918

The Turkish armies in Palestine, Syria and Mesopotamia were reorganized in May 1917 becoming the Yildirim Army Group (*Yıldırım Ordular Grubu*). Essentially it comprised Turkish units, with small numbers of German troops and auxiliary formations, including air services, all under German command and staff.[371] The first commander in chief was *General der Infanterie* Erich von Falkenhayn who was replaced by *General der Kavallerie* Otto Liman von Sanders on 25 February 1918. Both were given the rank of *Müşir* (Field Marshal) in the Ottoman army. The final commander, for a few days following 30 October 1918, was Mustafa Kemal Pasha.[372] Throughout 1917 and 1918 all Yilderim units were subject to on going supply difficulties resulting from the incomplete tunnels through the Taurus mountains.

The Allied Army was also reorganizing after the failures at Gaza. General Murray was replaced on 28 June 1917 by Lieutenant General Sir Edmund Allenby. Allenby spent the summer months improving and extending the water and rail supply lines from the Suez Canal to Rafa, built by his predecessor, also preparing a new strategy. Gaza was a strongly held fortress, on which any frontal attacks would be difficult and costly, as the two previous attempts had demonstrated. The Turkish line from Gaza to Beersheba was held by a series of strongpoints, the most important being at Sheria and Beersheba itself. In a truly Machiavellian move, Allenby spread false information that any attack at Beersheba was a feint and that landings would be made on the coast in the rear of the Gaza defences. When, just before his main attack, an attack on Beersheba was made it was initially ignored by the Turco-German high command. Beersheba was captured and a few days later would be used as a springboard to outflank main Gaza line. As part of the final preparations for the assault on Gaza an artillery bombardment, including naval ships, commenced on 27 October, four days before the attack at Beersheba. The attack on Gaza, as expected, proved to be costly and difficult. However, the attack from Beersheba succeeded in breaking the Turkish left and on 7 November they withdrew from Gaza. The subsequent British advance continued into December, and on 9 December captured Jerusalem.[373]

Empress is the main protagonist in this chapter, so here she is moored at Port Said in the Sherif Basin at the nearest point to the EIESS island base. Extensive awnings are spread in an attempt to keep the ship habitable. *(EMK)*

The supporting role of the EIESS following Risk's appointment is admirably explained in this extract from C.E. Hughes' book.

> The part played by the RNAS, though it was of the highest importance, was always an indirect one. We have seen how they kept watch over the inundated tracts which formed part of the defences of the Canal, and we have seen that their reconnaissance and bombing raids were directed against numerous positions and districts of country which at the time the flights were made were out of reach of the RFC. As the Army advanced, the Naval Air Service advanced too, or, rather, it did not so much advance as compress its area of enterprise, for from first to last, all available points behind the Turkish front right up to Adana and the approaches to the Taurus tunnel were included within the field of its investigations. The forward march of the British troops merely meant to the RNAS that certain places which they formerly visited, photographed, and disfigured with high explosives, were transferred to the very efficient care of the RFC, who carried on the good work.
>
> In this way, by the beginning of 1917, every part of the Palestine coast as far north as Jaffa had dropped out of the RNAS list. El Arish had come into British hands on 22 December 1916, Rafa on 9 January 1917, and Khan Yunis, fourteen miles south of Gaza, by the end of February. North of Jaffa was still our province, and Tul Keram, Haifa, Beirut, Chikaldere, and Adana, were all inspected and dealt with on one or more occasions. With the exception of the attacks on the submarine supply base at Beirut, these expeditions were always carried out with the object of assisting the plans of the Army which was faced with the problem of capturing Gaza and Jerusalem.[374]

A somewhat pensive Flt Cdr J.C. Brooke surveying three 500lb bombs in the bomb dump on the island. The length of the cylinder surrounding the vanes and the extended fuse casing identify these as heavy case, Mark 1/N/, 500/550 lb bombs, with a bursting charge of 180 lb 'trotyl' (TNT) and a total weight of 550 lb. *(EMK)*

A Visit to Beirut

The first use of one of the three Model D Shorts as a single-seater was on 13 May. On board *Empress* were three Shorts, 8004, 8020 and 8075, with pilots Flt Cdr J.C. Brooke, Flt Lt H.deV. Leigh, FSL E.M. King, and observers Lt V. Millard, 2Lt A.D. Ferguson. The seaplane carrier left Port Said around midday the day before, escorted by French destroyers *Coutelas* and *Pierrier*, and was off Beirut at 04.14.

Empress' navigator was uncertain of his position as there was very heavy haze overland. So, 8004 (FSL King and 2Lt Ferguson) was hoisted out at 04.28 to reconnoitre the coast. They returned at 05.00 and signalled that the position was correct. Short 8020, being flown solo by Flt Cdr Brooke and carrying a single 550-lb bomb with a 15 second delay fuze, was immediately hoisted out. Fifteen minutes later 8075 (Flt Lt Leigh and Lt Millard) also set out. Both two-seater Shorts were carrying two 65-lb bombs with 25 second delay fuzes. The three machines set course for Beirut, at varying heights, and were over the harbour between 06.00 and 06.15. The prime target was any submarine in the harbour, the secondary target were petrol stores on the eastern side of the harbour. There was no submarine and all five bombs aimed at the petrol store dropped into the harbour. All machines returned to the ship between 06.25 and 06.46. Once they were hoisted in, *Empress* and her escorts returned to Port Said.

Quite apart from the weight of the bomb, the size also required some creative thinking. One of the 550 lb bombs is held up against the fuselage steadied by two wooden frames and attached to a bomb release protruding from below the forward cockpit. Flight Lieutenant Worrall is in the aft pilot's cockpit of Short 8020, there is an extemporised mounting for a Lewis gun attached behind the forward observer's cockpit. *(EMK)*

As a bombing raid it was a total failure. Notably, this was the only occasion a 550-lb bomb was dropped by the EIESS and it dropped short of the target. On the engineering side, 'No trouble was experienced with any machine, 225 HP Short No 8075 climbed much better than No 8004 (a 225 HP Short converted into a 240 HP Short).'[375] The single-seat Short 8020 (240 hp) was able to rise to 4000 feet before bombing, 8004 bombed from between 4200 and 4500 feet, whilst 8075 with the same bomb load as 8004 was able to rise to over 6000 feet before bombing.

The EIESS, however, had not finished with Beirut and would return in August and again in September.

The Raid on Tul Keram, 23 June 1917

The raid on the railway station at Tul Keram (Tulkarm) was a rare, and successful, co-operation between the RNAS and RFC. The Official History has the following to say.

> It became known from agents' reports at the end of April 1917 that large quantities of stores had been accumulated by the enemy at Tul Keram. The Vice Admiral of the East Indies and Egypt station was asked to organize a seaplane attack to bomb the railway and to destroy as many of the accumulated supplies as possible. Owing to a shortage of seaplanes, the raid could not be made until the 23rd of June.[376]

Wing Cdr Risk, well aware that some of the available Shorts were only fitted with extemporised Lewis gun mountings, requested:

> As the latest Intelligence Reports showed that there were seven hostile machines at El Ramleh, I submitted that this operation might be made a combined operation with the Royal Flying Corps; Tul Keram being bombed by seaplanes from the *Empress* whilst the aerodrome, sheds, etc, at El Ramleh were being bombed by the Royal Flying Corps.[377]

His proposal was approved and Risk visited Lt Colonel A.E. Borton, Officer Commanding 5 Wing, RFC. It was arranged that the RFC would have bombing machines over Ramleh between 05.00 and 05.30 on the morning of 23 June. The key detail was that a pre-arranged signal would be sent to *Empress*, waiting 5–10 miles off the coast near Tul Keram, from the RFC W/T station at Deir El Belah when the RFC machines had left the airfield. Thus co-ordinating times over their respective targets.

Empress sailed from Port Said at 17.00 on 22 June, escorted by the French destroyer *Hache*. In the hangar were four Shorts, 8004, 8019, 8020, and 8075, all as two-seaters. Wing Cdr Risk was aboard, he would fly as an observer, the other observers were Lt Millard, 2Lt Ferguson, and Lt Pakenham-Walsh. The pilots were Flt Lt Worrall, Flt Lt Barr, FSL Bronson, and FSL King.

The RFC W/T was received shortly after 04.00, and at 04.15 the ships were off the Palestine coast, seven miles west of the Nahr Iskanderun. Tul Keram lies approximately 15 kilometres inland from the mouth of the wadi and a little south. All the Shorts were heavily loaded; 8004 (Worrall and Risk) had two 65-lb and four 16-lb bombs and two Lewis guns, as did 8020 (Bronson and Millard), 8019 (King and Pakenham-Walsh) had two 65-lb and eight 16-lb bombs but only a single Lewis gun, and 8075 (Barr and Ferguson) whilst only carrying two 65-lb and one 16-lb bombs also was loaded down with a Lewis gun, an aerial camera, and a Very pistol with 12 flares. Starting at 04.23 all four Shorts were hoisted out; 8075 failed to rise and was hoisted in. The remaining three were airborne by 05.00, although 8004 required a long run before taking off.

The three proceeded inland, flying in a single line at intervals of 800 yards from each other, 8004 leading, followed by 8019, then 8020. FSL King in 8019 was the first to drop his bombs, four 16-lb bombs hitting the station the remaining bombs falling around the station. The other two machines dropped almost simultaneously, with similar results. Fires were burning in the station as the machines flew back towards the coast. The Shorts were not fired upon during the attack, returning to be hoisted in by 06.08. *Empress*, and *Hache*, then proceeded immediately back to Port Said arriving at 17.15 the same day.

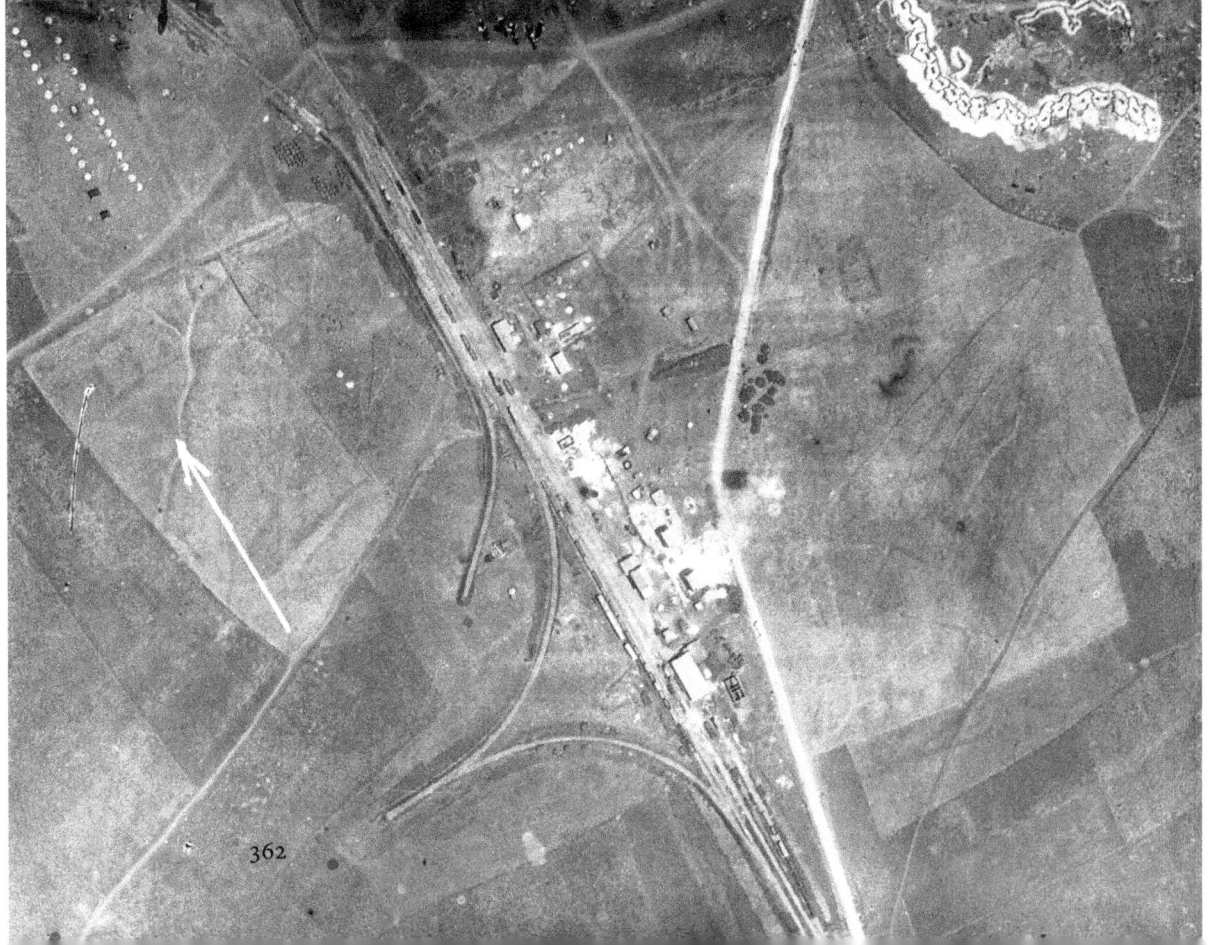

Bombing Tul Keram station, 23 June 1917. Bombs can be seen exploding in the sheds adjacent to the road. *(EMK)*

BE2e A2776 of 14 Squadron, RFC, somewhere in Palestine 1917. There is another BE2 and a Bristol Scout in the portable tent hangar behind. *(Paul Hare Collection)*

FSL King's logbook recorded the flight in some detail. 'From HMS *Empress*, hoisted out about 20 miles N of Jaffa to bomb Tul Keram rly. stn. 12 miles inland. No hostile fire. One good hit for certain, several other doubtful. Dropped 2/65 lbs & 8/16 lbs. Fired Lewis gun at & dispersed enemy mounted – about 300 men.'[378]

According to the Official History, 'In co-operation with this attack, seven aeroplanes [BE2e's from 14 Squadron, RFC, at Deir El Belah] bombed the German aerodrome at Ramleh to distract the attention of the enemy pilots from the Tul Keram attack. Seventy-three bombs of 20-lb or 16-lb weight were dropped, some of which were seen to fall on the aerodrome.' The operation was not significant enough to warrant recording in the squadron's post-war summary of events.[379] Raids on Ramleh appear to have been common enough occurrences not to warrant special mention.

The Raid on Adana, 15 July 1917

It would mid-July before *Empress* sailed again. But that does not mean her machines were inactive. They were making regular anti-submarine and convoy escort flights from within the harbour. Tasks that would become very familiar in 1918. Typical of these was a one hour flight on 12 July by Flt Lt M.C. Wood and Lt Kerry on Short 8019. Taking off from the outer harbour of Port Said at 09.18 they set a course to the north from the lighthouse. They had two 65-lb and eight 16-lb bombs, but no Lewis gun. Reaching the end of the swept channel at 09.42, they searched the immediate area for mines or submarines. Finding nothing, other than several drifters and motor launches going about their business, they returned to the swept channel and returned to land in the outer harbour at 10.20. An all too typical patrol and report.

Shortly after this flight, *Empress* was preparing to return to up the coast, with orders to 'Bomb the two important German cotton factories, situated to the N.W. of Adana, and to destroy crops, etc., in the district.' When leaving Port Said at 05.00 on 14 July she had four Shorts, 8004, 8018, 8019, and 8020, aboard – including all three Model D bombers. Also pilots Flt Lts Bronson, Burling, Leigh, Smith, Wood, Worrall, and FSL King, and observers Wing Cdr Risk, Lt Kerry, Capt Kempson, Lt Millard, Lt Pakenham-Walsh. She set out with the French destroyer *Voltigeur* as escort, but the latter suffered an engine breakdown and had to head into Famagusta, Cyprus, for repairs. Proceeding without an escort, *Empress* arrived off Karatash Burnu (Karataş Burnu), on the northern side of the Gulf of Alexandretta and due south of Adana, at 04.00 on 15 July. Between 05.00 and 05.15 all four Shorts were hoisted out, they were being flown as two-seaters, carrying in addition to their bomb load *(see Table)* a Lewis gun. All machines got away within minutes of being hoisted out, and headed inland to Adana about 50 kilometres from the coast.

Flt Lt Wood, Lt Pakenham-Walsh	8004	2 65-lb & 6 Petrol Bombs.
Flt Lt Smith, Capt Kempson	8018	2 65-lb & 8 16-lb Bombs.
Flt Lt Burling, Lt Kerry	8019	2 65-lb & 8 16-lb Bombs.
Flt Lt Leigh, Lt Millard	8020	2 65-lb, 2 16-lb & 6 Petrol Bombs.

Note: The explosive bombs were intended for the two cotton factories and the petrol bombs for any crops located.

Guy Duncan Smith provides a typical account of the raid. Whilst generally accurate, parts should be taken with a pinch of salt.

> We were called that morning in pitchy darkness. A raw, cold morning it was, too. The sea had been unusually rough for several days, and on the previous day all of us had been seasick. In spite of this, however, we had gone thoroughly over our machines, and tested our machine guns and bomb racks. With the first streak of dawn we started out from the mother ship, eight of us.

Not the best quality but it shows an interesting modification to Short 8018 following its return from the Maldives. There is a plain fabric patch repair by the tailplane but more interesting is what is going on in the forward cockpit. The rounded top to the fairing has been cut off and a flat panel installed across two stub walls. Into this panel has been fitted a ring mounting, possibly a Scarff ring, for a Lewis gun. Short 8018 was only briefly flown as a two seater in this configuration, the only record being on 15 July 1917 during the bombing of Adana. All 8018's subsequent operational flights were as a single-seater. *(EMK)*

I was first into the air, and circled around waiting for the others and keeping a lookout for submarines. When at last I saw they would be under way in another minute or two, I headed in the direction of the coast, which was still invisible in the feeble light of the morning. I picked up the shore line quickly, however, and crossed it at about 1,000 feet.

We were rather heavily loaded with bombs, machine guns and ammunition, but the engine was revving well and climbed splendidly. Heavy clouds hung unusually low, about 2,000 feet, so we climbed up through them and came out into bright, warm sunshine, under the clearest of skies.

Setting the course by compass, I kept the machine's nose well up, sighting the earth occasionally through rifts in the clouds. A river [Seyhan] runs through the city of Adana right down to the coast, and once in a while I could catch a glimmer of it. At last I glimpsed the city itself, so throttling my engine right back, I volplaned through the cloudbank, and then straightened out in the murky atmosphere to look for the objective.

Straight ahead and right in the line of our flight, were two long lines of Turkish tents. This encampment was not my objective, but while I had had no orders to bomb it I had had no orders not to. So I let loose one small bomb and saw it explode between the tent lines. The Turks were taken completely by surprise. They came rushing from the tents in wild disorder, but we did not stop to see what happened next.

The other machines had followed us down through the clouds, and we went speeding in over the city in line formation. If the Turkish soldiers had been surprised, even greater was the surprise of the people of Adana. We came in over them at low altitude with our engines throttled down, and we could see the population rushing about in a veritable panic.

Our instructions had been to bomb the barracks and destroy storehouses. We picked out the barracks [his flight report clearly states 'cotton factories', but barracks make a better story!], where soldiers were running madly about, occasionally stopping to fire their rifles at us. Officers on horseback dashed at a wild gallop up and down the streets. We dropped bombs upon the barracks and storehouses, doing a lot of damage, although all the bombs did not find their mark.

Bullets from rifles and machine guns time, kept whizzing by us all the time, for we were flying at low altitude. One machine, piloted by

Flight Commander [Burling], with Lieutenant [Kerry] as observer, seemed to be skimming just above the housetops.[380]

Concluding that our share of the day's work had been done, I was just rounding the machine for the return trip when my observer signalled to me that he had one big bomb left. So I made another turn and flew back over the target again. The observer let the bomb go, and we had the satisfaction of seeing it crash squarely through the roof of a factory, which was badly wrecked in consequence.

Relieved of her load of bombs, the old bus seemed to shake herself as though glad to be free, and simply jumped on her way to a higher and safer altitude. On the way back we passed villages, farms and herds of cattle, and a few groups of soldiers whom we machine-gunned a little. I recollected suddenly that we had neglected dropping our pamphlets. On these raids it is customary to take along pamphlets prepared in the language of the district, setting forth the allies' cause and urging all who read them to come to our support.

To remind the observer, I tore a small bit of paper off my map, held it up so he could see it, and then tossed it over the side. He took the hint with a smile, undid his bundle on the floor of the fuselage, and hurled the pamphlets overboard above the next town we saw. The pamphlets spread out in all directions and then floated slowly down like a shower of confetti.

Adana, 15 July 1917. Bombs exploding on and around the large cotton factory. *(EMK)*

Out over the sea again, we sighted the ship, a tiny dot on the horizon. Our plane was the third to return. It was still early morning, and we were home again in time for breakfast. Pilots and observers spent a busy day comparing notes, relating experiences, and estimating the damage done.

Wing Cdr Risk's official report was more matter of fact.

> The photographs taken do not in any way show the amount of damage that was undoubtedly caused by the bomb-dropping. The reason for this is due to the fact that only a few of the first photos taken were at all clear, owing to low-lying clouds, etc. I have therefore had a list made showing where each bomb dropped—this being checked by observers from the other machines. From reports given to me by Pilots and Observers, I consider that great damage has been done to both these factories, especially the larger one.[381]

Adana, 15 July 1917. Bombs exploding close to the small cotton factory. *(EMK)*

The list of bombs dropped is best summarised as follows. Burling and Kerry had the best results with both 65-lb bombs hitting the larger cotton factory close to the main entrance and all the 16-lb bombs striking the roofs of both cotton factories and an adjacent tobacco factory. Smith and Kempson's bombs, after hitting the army camp, dropped into the railway station yard and on or close to the smaller factory buildings. Their final 65-lb bomb dropped onto the roof of one of the factories. Leigh and Millard mostly just missed the factories, except for one 16-lb bomb which hit the roof of the large factory. Their petrol bombs were dropped on the factory, igniting a fire, another setting some threshed corn on fire. Finally, Wood and Pakenham-Walsh managed to hit the small factory with one of their 65-lb bombs, whilst the petrol bombs started a fire in the station yard and 'set light to crops and wood close to a granary.' Pakenham-Walsh also recorded in his report that, 'At about 1000 feet fires and smoke were observed in both factories when machine left.'

All machines returned safely, having been subjected to rifle fire throughout their flights, and had been hoisted in by 07.00. *Empress* now proceeded at full speed back to Port Said arriving at 05.15 on 16 July. As a precaution, Short 8020 (Worrall and Pakenham-Walsh) was sent off 10 miles out from the harbour to scout for submarines, circling around *Empress* at two miles distance until the ship was safe in harbour. Worrall brought the Short down within the harbour and taxied over to the base.

Back to Beirut

Empress made two further visits to Beirut prior to the next Gaza battles. What was the interest in the harbour? In Operation Orders the prime instruction was always to search for evidence of a submarine, secondary was the destruction of stores. As the only established port on the Palestine coast, the harbour at Beirut had several warehouses, so the concern about stores was reasonable. Since the beginning of the war passing French and British warships had reported a submarine in Beirut, probably glimpses of the half submerged wreck of the *Avnillah*. Intelligence agents in the area may have seen and reported submarines as well. By 1917 these reports must have raised many an eyebrow at headquarters, but always had to be investigated—just in case. It should be noted that a few submarines did visit Beirut, but usually for less than 24 hours to re-water and obtain some supplies, including small quantities of fuel oil from barrels. These visits appear to have been late in 1917 or early 1918, earlier visits are unknown. The only confirmed visit was by UB-66 (*Kptlt* Fritz Wernicke) on 10 January 1918.[382]

Empress' first return to Beirut was on 17 August. She sailed from Port Said at 12.45 on 16 August, having on board four Shorts, this time 8004, 8018, 8019, and 8022. Pilots were Flt Lts Bronson, Burling and Worrall, and FSL King, with Wing Cdr Risk and Lt Pakenham-Walsh as observers. Shorts 8018 and 8019 were described as 'bombing seaplanes' and flown as single-seaters. She sailed with a French destroyer as escort but, as in July, it broke down and had to return to Port Said shortly after sailing. *Empress* was in position, 22½ miles (per Log Book) off Beirut, by 04.05 on 17 August, and commenced hoisting out the Shorts half an hour later.

U-boat UB-66 (*Kptlt* Fritz Wernicke) on 10 January 1918. Moored alongside storage shed 'A' apparently taking on fuel from the barrels in the background.

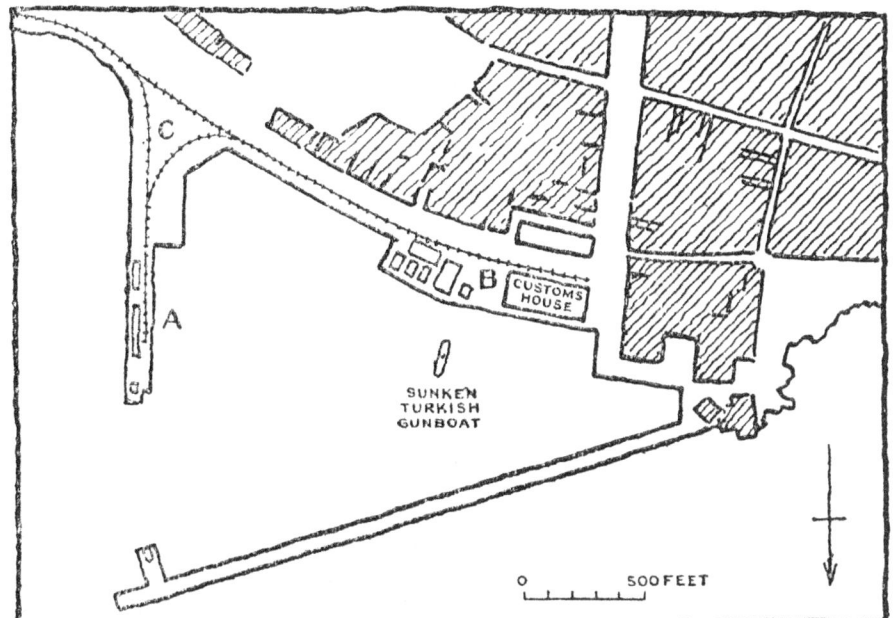

Sketch map of Beirut showing storage shed 'A', Customs House and storage sheds 'B', and potential storage at 'C'. All were targets for Empress' visit on 17 August 1917. *(Hughes, Above and Beyond Palestine)*

Flt Lt Worrall, Wing Cdr Risk	8022	2 65-lb and 6 16-lb bombs. 1 Lewis gun and 4 hoppers.
Flt Lt Bronson, Lt Pakenham-Walsh	8004	2 65-lb and 6 16-lb bombs. 1 Lewis gun and 4 hoppers.
Flt Lt Burling -solo-	8019	36 16-lb bombs.
FSL King -solo-	8018	36 16-lb bombs.

Wing Cdr Risk's orders were for the two-seaters to proceed in company to Beirut and scout out any submarines. If a submarine were spotted it was to be attacked immediately. If not then the two-seaters were to wait for the bombing seaplanes to attack before dropping their own bombs. The chief objective were storage sheds marked A in the accompanying sketch map, followed by sheds at locations B and C, finally 'any feluccas lying in the harbour. I would draw the attention of all Pilots and Observers to the fact that great care is to be taken that bombs are not dropped on any part of the residential quarters.'

There was a light wind and smooth sea which did not make the take-off easy for the heavily loaded machines. Worrall and Risk were first away at 04.32 and headed for Beirut without waiting for Bronson and Pakenham-Walsh, who had difficulty rising and had to dump four 16-lb bombs before they could get away at 04.45. The two bombers got away within minutes of Bronson and Pakenham-Walsh.

Wing Cdr Risk reported of his flight,

> Machine no 8022 was the first machine to arrive at Beirut. It was quite obvious that the arrival of this machine was absolutely a surprise, as one boat was seen to be pulling vigorously across the mouth of the harbour, going from the lookout tower towards objective "A". Afterwards it was noticed that a flag (probably warning flag) was flying from this lookout tower. As it was thought very likely that machine-guns were stationed here, 2 trays were fired from our Lewis gun, and several hits were observed.

They then circled around watching and recording the bombing results before making their own attack on objective B. Both 65-lb bombs hit the building, exploding and starting internal fires. Four of the smaller bombs damaged small sheds and two feluccas, the two remaining bombs are unrecorded. Burling dropped 16 of his bombs close to objective A, but claimed no hits, the remainder were dropped on objective B, claiming 16 of the 20 were through the roof. In his log book, King later recorded, 'From *Empress* 25 miles W of Beirut. Carrying 36/16-lb bombs. Twenty hits on one big store shed on jetty, setting the place on fire. Four others through roof of another shed, but did not explode. Bombs dropped from 400 to 1100 ft altitude.'

Bronson and Pakenham-Walsh were the last to return to *Empress*, having lingered to record the results of the bombing before making their own attack.

A general view of the harbour with a bomb exploding close to storage shed 'A'. The Customs House 'B' is at the top centre of the image and area 'C' at the lower left. The old gunboat *Avnillah* is resting on the bottom in the middle of the harbour. *(EMK)*

A 65-lb bomb exploding in the large Customs House storage shed. *(EMK)*

'Direct hits were obtained [on objective A] and fire started on the quay, in some railway trucks, and other combustible material.' On objective B, 'Numerous hits were obtained, causing internal explosions and fire. This objective was blazing fiercely and to all appearances will be totally destroyed by fire. The whole of the centre shed was unroofed and holes were numerous; fire could be plainly observed inside.' They then dropped their 16-lb bombs on Objective A and one each 65-lb bomb on objectives A and B.

Once Bronson and Pakenham-Walsh had been hoisted in, *Empress* proceeded at speed back to Port Said, anchoring there at 08.38 the following morning.

Wing Cdr Risk concluded his report of operations as follows—the capitals are his.

> NOT A SINGLE BOMB FELL IN ANY PART OF THE RESIDENTIAL QUARTER.
>
> It might be noted with interest that as the seaplanes returned to HMS *Empress* it was observed that in many cases the occupants were assembled on the roofs and verandahs of their houses watching the attack.

Whilst waiting at Port Said for the next call, *Empress* and her aircrew were able to engage in some mirror bombing practice, testing out some new Shorts and Sopwiths, including those delivered by *City of Oxford*, and some important spotting exercises with a monitor of the Royal Navy. The preparations for the battle at Gaza were proceeding and, as mentioned above, the bombardment would involve a naval squadron on the Turkish flank. It was therefore important that the ships and the floatplanes were able to practice working together.

Mirror Bombing was designed to permit pilots to practice bomb dropping without using dummy bombs.

> Method: The stand is placed on a large square of canvas and the pilot is ordered to fly at a certain height and to attack this as the target; but instead of releasing a bomb he informs the (two) observers at the mirror either by firing a Very's Light or other visual signal, or by wireless, of the moment when he would have dropped it. By means of their observations, the observers at the table can immediately inform the pilot by signal where this bomb would have dropped in regard to the target. The pilot then turns and repeats his attack as many times as is considered necessary.[383]

FSL King has brief mentions of the bombing practices and flying the new machines in his log book. Most of the mirror bombing practices were made on Sopwith Baby N1038, flying well after its return from Colombo. In October, he was able to try out Fairey Hamble Baby, N1210, noting that it was 'slightly slower than the Schneider!' This despite having a 130-hp Clerget engine. A few days earlier he had tried out one of the new Shorts, N1262, a Robey-built 184 fitted with a 240-hp Renault-Mercedes engine, which he declared 'A1'.

On 16 September, *Empress* sailed down the Suez Canal, past Ismailia, to anchor in Lake Timsah. She would spend a week here, exercising spotting with the monitor *M.31,* also ensuring that all W/T sets were tuned and working. Four Shorts were employed, 8021, 8022, N1090, N1263, with pilots Flt Lts Bronson, Burling, Wood, Worrall, and FSL King and observers Wing Cdr Risk, Lt Kerry, Capt Kempson, Capt Millard, 2Lt E.A. Newton, Lt Pakenham-Walsh. The Shorts had a Sterling W/T and Receiving set installed, but it took several flights before the sets could be properly tuned into the monitor. Toward the end of the stay several successful shoots were carried out.

Shortly after her return to Port Said, *Empress* was instructed to proceed once more to Beirut. This time, in addition to bombing, she was to provide spotting for the cruiser *Grafton*, an old friend from the Aegean. She sailed from Port Said at 12.49 on 26 September, having on board four Shorts, 8018, 8019, 8021, and N1090. Pilots were Flt Lts Bronson, Burling, Worrall, and FSL King, with Capt Millard and Lt Pakenham-Walsh as observers. Shorts 8018 and 8019 were once again to be flown as single-seaters. She sailed with French destroyer *Arbalète* as escort. The two ships arrived at a position 25 miles west of Beirut at 05.04 the following morning. *Grafton* was sighted approaching a few minutes later.

On leaving Stavros (Chapter 12) Captain Edgar Grace, and *Grafton*, had taken command of the 2nd Detached Squadron, based at Kephalo. They remained there until early June when was *Grafton* was ordered to Malta for a refit. On 11 June, she was torpedoed by the German U-boat *UB-43* (*Oblt* Horst Obermüller), 150 nautical miles (280 km) east of Malta. *Grafton*'s anti-torpedo bulges proved effective and damage was limited, allowing her to be safely brought into port at Malta under her own power with no casualties. Following a complete refit *Grafton* was ordered to Port Said, arriving there on 7 September.

Returning to Beirut on 27 September 1917, this photograph shows unrepaired damage to the Custom House shed and to some smaller sheds. *(EMK)*

On the morning of 27 September Short 8018, with FSL King piloting, was hoisted out. Carrying a 65-lb and 16 16-lb bombs the Short got away at 05.11, sighting *Grafton* in the distance, before setting a course toward Beirut.[384] Flt Lt Burling had been meant to accompany FSL King, but 8019 was unable to start. In his log book King recorded; 'Hoisted out at 25 m W of Beyrout [sic]. Dropped 1–65 lb & 16–16 lb bombs on sheds there from 500, 600 & 900 ft—all direct hits. No AA.'

His Report of Flight is a little more explicit:

> Shed A on eastern mole was found quite intact as far as aerial observation goes, only several small holes made by bombs last raid being visible in roof; this probably was due to the fact that the bombs exploded inside shed.
>
> Shed B. The front portion of this shed has been repaired and thoroughly retiled, whilst the rear half was evidently also under repair, only about a third part still requiring new tiles in the roof.
>
> Railway Offices, opposite B shed, which were damaged previously have also been thoroughly repaired and retiled whilst the broken skylights have been replaced.
>
> Bombs. All bombs dropped scored direct hits, four 16-lb hitting the rear shed of B, whilst twelve 16-lb and one 65-lb hit B.
>
> No AA gunfire was encountered, nor were there any signs of guns themselves.
>
> Numerous propaganda pamphlets were dropped in different parts of the town.

He also noted numbers of natives were observed to run out of the building and disappear up the side streets.

Grafton firing on Beirut harbour, 27 September 1917. Eight of the ship's 6-inch guns are visible in this aerial view, the four remaining were on a lower deck. The port side anti-torpedo bulges are clearly seen in the sunlight. *(EMK)*

Storage shed 'A' on fire from *Grafton*'s shooting, with a ricocheting shot just passing over the shed. *(EMK)*

He returned and was hoisted by 07.01. Following his verbal report, a signal was sent to *Grafton* detailing the results of the bombing and the state of Shed A, the target for the cruiser's guns. At 07.45 *Grafton* signalled she would be on position, 20 miles west of Beirut, to commence firing at 10.00.

Two machines were sent off at 09.30: N1090 (240hp Renault-Mercedes engine), Flt Lt Worrall with Lt G.H. Pakenham-Walsh as observer. In addition to wireless equipment, the Short was carrying a single 65-lb and four 16-lb bombs, and a Lewis gun. 8021, Flt Lt Bronson with Lt Millard as observer, also with wireless equipment, two Lewis guns but just four 16-lb bombs. They were to act as back up to the first machine and to take photographs of the bombardment. Worrall wrote in his report that after bombing the lighthouse he signalled *Grafton* to open fire at 09.56. '14 OK's sent. Some appeared to pass through building without exploding.' The cruiser's log book records she came to a stop at 10.05, opened fire three minutes later and ceased fire at 10.24, then got underway again. Both machines were ordered to close *Grafton* but only Bronson landed and 'conferred as to damage done.' Both machines were aboard *Empress* by 11.40.

In his report Wing Cdr Risk comments that the work done to repair the warehouses since the earlier attack 'points to the fact that these sheds are of importance and are required for use by the enemy.' Regarding *Grafton*'s bombardment, 'the fire was accurate, numerous hits being recorded and a fire started. The majority of the projectiles, however, seem to have pierced the building and burst in the sea beyond.' He again emphasised no damage was caused in the adjacent residential areas.

All machines recovered, *Empress* returned to Port Said, arriving at 07.55 on the 28 September.

Facing page: Short N1090, Flt Lt Worrall with 2Lt G H Pakenham-Walsh as observer, engine just stopped and gliding in toward *Empress* off Beirut 27 September 1917. Pakenham-Walsh is standing by hook on whilst Worrall has his head in the cockpit ensuring the engine is closed down. N1090 was fitted with a 240hp Renault-Mercedes engine and side radiators, these were found to block the pilot's landing view and later builds reverted to the box radiator. It also has a factory installed bomb rail, and a Scarff ring in the observer's cockpit. *(EMK / Puke Ariki. A66.610)*

A Final Visit to Adana and Chicaldere, 9–11 October 1917

Empress next sailed from Port Said at 05.15 on 8 October, escorted by French destroyers *Arbalète* and *Hache*, bound once more for Adana and the Chicaldere railway bridge. It would be an eventful and costly trip. Aboard were four Shorts; 8018, and 8019, both as single seaters, with 8021 and N1091 two-seaters. Pilots were Flt Cdr Clemson, Flt Lt Leigh, Flt Lt Popham, Flt Lt Stevens, Flt Lt Wood, and observers Capt Kempson, 2Lt Newton. A subsequent shortage of qualified observers led to Ldg Mech A. Prince volunteering for one flight.

Arriving off Karatash Burnu at 06.00, 9 October, *Empress* stopped and immediately commenced hoisting out the Shorts in the following order. Both two-seaters also carried a single Lewis gun.

To attack Adana:		
Flt Lt Wood -solo-	8018	3 65-lb and 16 16-lb bombs.
Flt Lt Leigh, 2Lt Newton	N1091	3 65-lb and 8 Petrol bombs.
To attack Chicaldere railway bridge:		
Flt Lt Stephens, Capt Kempson	8021	2 65-lb and 4 16-lb bombs.
Flt Lt Popham -solo-	8019	3 65-lb and 16 16-lb bombs.

There was a slight swell which made the take-off of the heavily loaded machines difficult. Both two-seaters had to jettison a 65-lb bomb in order to get off. Flt Lt Wood took off at 06.40 followed two minutes later by Leigh and Newton on the second Short. Both machines setting course for Adana. The two machines for Chicaldere rose from the water at 06.50, and headed towards the bridge.

All was not well with 8018. It was observed from N1091 to be having difficulty climbing above 700–800 feet, Flt Lt Wood turned back towards the ship dropping some, but not all, of his bombs. He landed about twenty yards from *Empress*, then there was a sudden explosion as the remaining bombs went off. When the smoke cleared 'nothing could be seen but small wreckage and the body of Flight Lieutenant Wood. The body was brought on board with all possible despatch but life was extinct.'[385]

After carrying out a submarine patrol around the ship whilst the remains were gathered in, N1091 landed and was hoisted in at 08.30.

Unaware of these events, the remaining two machines continued to the Chicaldere bridge, arriving shortly after 07.30. 8021 bombed first. Stephens' first bomb was dropped on a train '3 miles W of bridge, fell alongside train about 10 yards N. Train stopped.' His next bomb hit and demolished a guard house at the west end of the bridge, his 65-lb bomb missed the bridge but demolished the guard house at the east end of the bridge and set fire to some tents. His remaining 16-lb bombs appear to have caused not further damage. Stephens and Kempson then circled to watch, and protect, Popham in the single seater.

Popham made two runs over the target. On his first run he let go four 16-lb bombs at the western guard house, and eight at the bridge, scoring no hits on this difficult target. Coming around for another run, one 65-lb bomb was aimed at the guard house, and two at the bridge which passed through the steel work to explode in the river, his final four 16-lb bombs may have scored a hit or two.

Chicaldere railway bridge 9 October 1917, looking to the south-west. The smoke is from burning tents at the eastern end of the bridge caused by bombs dropped by Flt Lt Stephens and Capt Kempson on Short 8021. Captain Kempson then took this photograph of two 65-lb bombs dropped by Flt Lt Popham, which passed through the bridge to explode in the river. The sandbanks in the river are very different from those in the photographs taken the previous December. *(EMK)*

Throughout the attack both machines were subjected to heavy and accurate rifle fire, also some shrapnel from field guns in the area. There was one AA gun making accurate shooting against 8019 whilst circling at 4000 feet prior to attacking. Both machines returned to the ship with damage. The radiator of 8019 was shot through, the engine overheating due to loss of water. 8021 suffered damage to the starboard main float, and the lower main petrol tank was pierced in several places, the elevator was damaged and the tailplane struts shot through. Both may have been lucky to make it back. Short 8019 could not be repaired on board, and donated its tailplane and elevator to 8021, which was flying again the following day.

The remains of Flight Lieutenant Melville Cornelius Wood were buried at sea at 10.00, with Commander Drury reading the burial service in the presence of all available officers and men. *Empress* then proceeded to Famagusta to take on coal.

Flight Commander Clemson decided that the Adana part of the operation should still be attempted. Then, if possible, to visit Haifa on the way back to Port Said. So, following coaling, *Empress* returned to Karatash Burnu on the morning of 11 October. Only two Shorts were now available, 8021 and N1091. Flt Cdr Clemson with 2Lt Newton were to take the former, and Flt Lt Leigh with Capt Kempson the second.

Loaded with two 65-lb and four 16-lb bombs plus a Lewis gun, 8021 was hoisted out at 06.29. Nine minutes later, after an extended run, it was able to get off and immediately proceeded inland towards Adana. The principle target was to be the locomotive shed. The second Short, with three 65-lb bombs and four 16-lb, initially refused to start. It finally got off almost half an hour after Clemson and Newton had headed inland. Climbing to 1000 feet, it was forced to return with engine trouble and was hoisted back aboard.

Commander Drury takes up the story.

> After the first machine had been absent from the ship for two hours, orders were given for *Empress* to proceed inshore to within five miles, and a course parallel to the land was set. A destroyer (*Hache*) was ordered in as close as safe navigation allowed, and up to 11.15 am a thorough search was made of the shore, but no traces could be seen of the machine or its occupants.

Short N1091 had been repaired and was hoisted out at 09.18 with Flt Lt Leigh and Ldg Mech Prince aboard, to search for the missing machine. It carried no bombs but had a Lewis gun. Rising to 1000 feet, the engine again failed, and they were forced to return with engine trouble and was hoisted back aboard. Back on board it engine was found to be damaged beyond repair. During their brief flight they had been able to overlook the salt lake just inland from the coast, but could see no sign of the missing Short.

As no further operations were possible, after five hours had passed, twice the endurance of the Short, Commander Drury reluctantly 'considered that as a most thorough search had been made of the coast I was only unnecessarily endangering the ship by remaining, and orders were given to set course for Port Said.' *Empress* arrived at Port Said shortly after noon on 12 October.

A Turkish Communique, dated 12 October, offered one version of the story.[386] 'On 11.10.17 aeroplane flying over Alexandretta [sic] was brought down. The pilot is badly wounded. He was made prisoner, and the aeroplane, in a serviceable condition, has been captured.' From intelligence and diplomatic sources, a slightly different version emerged over the next few months. It appears that Clemson and Newton were able to bomb the locomotive sheds at Adana, but during the attack rifle or machine gun fire punctured the main fuel tank which rapidly drained. Lucky to escape a fire in the air Clemson was forced to bring the Short down on the land just outside Adana. Such a landing, at the very least, would have torn the floats off the machine, but both men were uninjured in the landing. Clemson was setting the Short on fire when some local troops arrived. According to Captain Wedgwood Benn, Clemson was 'sufficiently pugnacious in his bearing to force the Turks to shoot him through the body.'[387] Both men were taken prisoner and were released at the end of the war.

CHAPTER 16

From Gaza to *Goeben*

Naval Participation in the Third Battle of Gaza

The previous chapter included a brief outline of the military situation before and during the Third Battle of Gaza. This chapter looks at the naval contribution to the battle, particularly the role of the ships and floatplanes of the East Indies and Egypt Seaplane Squadron.

During the summer of 1917 not only was the Army command changed but also that of the Royal Navy. As previously noted, Vice Admiral Wemyss was succeeded as Commander-in-Chief, East Indies and Egypt Station by Rear Admiral Ernest F.A. Gaunt on 23 June 1917, who remained in the post until 1 August 1919.[388] Shortly afterwards, on 6 July, Rear Admiral Thomas Jackson was appointed to a newly created post as Flag Officer, Egypt & The Red Sea, taking up his duties on 20 July. As Admiral Gaunt's command no longer included Egypt, Admiral Jackson was in command of the Royal Navy ships during the bombardment of the Gaza flank.

General Allenby's plan was to storm Beersheba, on the eastern flank of the Turkish position, and to follow up with attacks on the Turkish centre, and Gaza. He specially requested that the naval forces should make feint landings to the north of Gaza at the same time as the Beersheba attack. Admiral Jackson, with whom General Allenby consulted, had available a motley collection of bombarding ships. These ranged from the monitor *Raglan*, with two 14-inch guns, and a newly fitted single 6 inch gun salvaged from the small monitor *M.30* sunk at Long Island in the Gulf of Smyrna in May 1916, several small monitors with either a single 9.2-inch or two 6-inch guns, two Insect-class gunboats, *Aphis* and *Ladybird*, also with two 6-inch guns, the cruiser *Grafton* (twelve 6-inch) and *garde-côtes cuirassé* (coastal defence cruiser) *Requin* (two 274 mm and numerous smaller guns). Support ships included two British destroyers, *Comet* and *Staunch*, five French destroyers, *Arbalète, Coutelas, Fauconneau, Hache,* and *Voltigeur*, several transports, trawlers and drifters, and a few X-lighters.[389]

As protection for the bombarding ships a protected anchorage was laid off the Palestine coast near Deir el Belah. The front line was just a few kilometres further along the coast. 'The anchorage was protected by a net which ran parallel to the shore and was distant two miles from it. The gaps between the ends of the net and the shore were patrolled by trawlers and drifters.'[390] Submarine detection was in its infancy, mostly reliant upon the Mark 1 Eyeball, and the gaps would prove to be distinctly permeable.

The Army's bombardment commenced on 27 October. The land based heavy artillery consisted of 68 medium and heavy guns and howitzers; 60-pounder (5-inch) and 6-inch field guns; and 5 and 6-inch howitzers. They were firing on pre-registered targets, and with assistance of RFC and balloon observation. The naval bombardment commenced 30 October, one day before the attack on Beersheba. The naval bombardment relied on visual observations from the ships, and spotting provided by floatplanes from the three seaplane carriers of the East Indies and Egypt Seaplane Squadron.

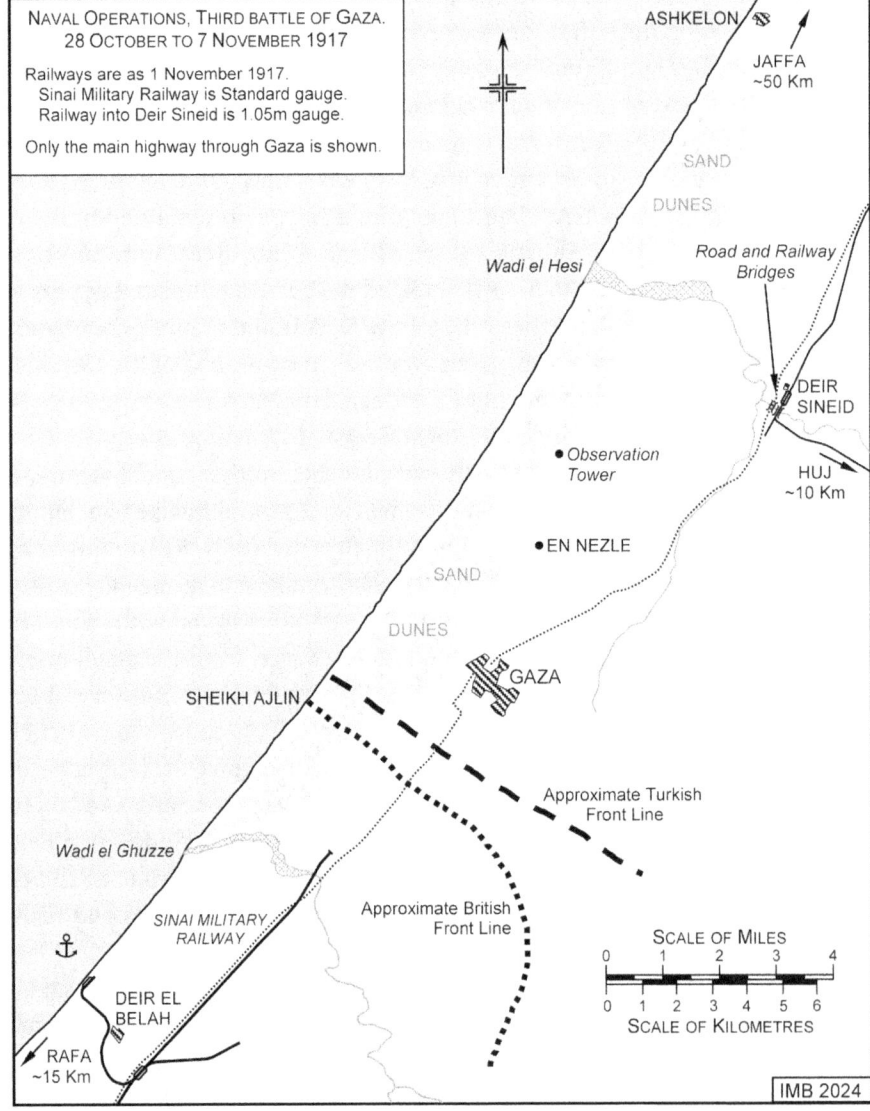

Map of operations detailed in this chapter. The railway line is shown as it was before the Third Battle of Gaza. After the battle the two lines were joined and the Turkish track widened to standard gauge. *(Based on Naval Operations, Vol 5)*

The 14-inch gun monitor *Raglan* at Port Said. For service off Palestine the monitor's AA armament had been improved, the two 12-pdr guns had been moved from a lower deck onto the forecastle deck abreast the main turret on HA mountings, adding to the single 3-pdr and 2-pdr fitted aft, and there also appears to be a machine gun mounted in the eyes of the ship. Following service at Gaza *Raglan* returned to Imbros in late December 1917. *(via Stuart Hadaway)*

Spotting For The Monitors

Wing Cdr Risk remained at Port Said when *Empress* headed up the coast to Adana at the end of September. His task was to prepare the EIESS for its coming role off Gaza. The *City of Oxford* had to be made ready for its operational debut, and *Raven* awakened from her slumber. For the former this required little more than testing hoisting arrangements and rigging shelters for the floatplanes on the foredeck. In *Raven*'s case there was no Prince Charming, just plain hard work. Possibly with Gaza in mind, she had been sent down to Suez for a quick dry docking and some remedial maintenance.

At 10.00 on 28 October Short 8019 was taken aboard the monitor *Raglan*. Shortly afterwards, Flt Lt Burling and Capt Kempson, with six mechanics, also joined the monitor. *Raglan* left Port Said at 18.00 and arrived off Gaza at 05.00 on 30 October, the monitors were very slow ships. She then proceeded to a bombarding position off Wadi el Hesi, where she was joined by *City of Oxford* at 08.30. *City of Oxford* had sailed at 18.00 on 29 October, with Shorts N1090, N1091, N1262, Flt Lts Leigh and Smith, and observers Capt Millard, and Lt Ferguson.

North of Gaza there was a ridge of low coastal hills and dunes which restricted the view from the sea, leaving little visible but a narrow strip of the maritime plain. But about 15 kilometres north of the town, the Wadi el Hesi had cut a small cleft through the hills and, from a ship off the wadi mouth, the railway station of Deir Sineid and the road and railway bridges over the wadi could be observed about eight kilometres (five miles) inland.

At 10.20 on 30 October *Raglan*'s Short was hoisted out, with Flt Lt Burling and Capt Kempson commencing the first RNAS flight in support of the Third Battle of Gaza. They were tasked with spotting *Raglan*'s 6-inch gun on to the railway station. Fire was opened at 10.50 and was slowly corrected on to the target, then firing 20 rounds rapid. The station buildings were set on fire, 'A large ammunition dump was then observed near the station and the 6 inch gun was switched on to it. At about the 8th round a direct hit exploded the dump which continued exploding for 35 minutes, demolishing the Railway Station and tearing up many yards of line.'³⁹¹ The 6-inch was then ranged on to the railway bridge, hitting the bridge after three shots the order for 20 rounds rapid was again given. The bridge was hit several times, damaging the track and bridge.

Returning to the ships shortly after 12.00 it was quickly evident that 8019's floats were leaking, and the Short was taxied over to *City of Oxford* and hoisted inboard. The floats were later replaced, but the afternoon spotting flight for *Raglan* was carried out on N1262 by Burling and Kempson. Back over Deir Sineid the 14-inch guns were directed against the railway bridge over the wadi. By 16.30 the railway bridge had been demolished and the guns were being registered on the road bridge. Before this could be completed the Short was interrupted in its work.

Around 16.40 Short N1262 was attacked by a 'Halberstadt Scout,' more likely an Albatros D.III (OAW) from FA301.

> [The] seaplane, heading towards the ship, was followed down by the Scout to within 800 feet of the water firing at the Seaplane from a gun firing straight ahead. The Observer of the Seaplane, Captain W.R. Kempson, opened fire on pursuing Scout at close range and let him have 2 trays from Lewis gun. Scout still followed; Seaplane was then turned sharply to the right, enabling Anti-Aircraft guns from HMS *Raglan* to drive off the Scout which had until then been too near the Seaplane for AA-fire from ships.

The 'Halberstadt Scout' that attacked *City of Oxford*'s Short N1262 on 30 October 1917, and again on 6 November (N1091), was more likely to have been an Albatros D.III(OAW) from FA301, similar to this from FA300. Albatros D.III, D.636/17, had been forced down on 8 October 1917 and captured intact. It is seen here with 1 Squadron, Australian Flying Corps, whose members had recovered the machine and moved it to their airfield where repairs were carried out returning it to flying condition. For a photograph of the Albatros with FA300 see Chapter 7.

> Seaplane was hit in about 35 places, in fuselage and top centre section; one elevator control was shot away and Captain Kempson obtained slight splinter wounds in the thigh.
>
> The floats of the Seaplane were so badly holed by machine gun fire that, after landing, it became waterlogged and turned over alongside HMS *Raglan* before it could be hoisted inboard. The wrecked machine was eventually got on board HMS *Raglan* and subsequent concussion, caused by gun-fire of 14 inch turret, blew the machine to pieces.

Flt Lt Burling and Capt Kempson, whose wound did not prevent him flying again the next day, had not finished working from *Raglan*.

At 10.25 on 31 October, Flt Lt Smith and Capt Kempson were air borne on N1090 from *City of Oxford*. They spotted *Raglan*'s 14-inch guns onto the road bridge near Deir Sineid, returning to the ship by 12.15. Later in the afternoon,

Flt Lt Leigh and Lt Ferguson repeated the exercise on N1090. Leaving at 14.00, they too spotted *Raglan*'s big guns onto the road bridge for an hour from 14.20 before being ordered to return to the ship. Ferguson reported that the 'Bridge across Wady Hesy is a heavy stone bridge of four span three pier construction. Centre pier and East parapet were damaged considerably but bridge appears to still be in a serviceable condition.'

Shortly after their return, *Raglan* with Burling and Kempson on board, proceeded to the anchorage off Gaza. Where Flt Lt Burling and Capt Kempson went ashore to consult with the RFC, specifically regarding the country around Deir Sineid. They later rejoined *Raglan* which was restocking ammunition at the anchorage. The monitor left the anchorage and proceeded to Wadi el Hesi in the evening of 1 November, returning to her firing position off the wadi at 09.00 the following morning.

City of Oxford had sailed for Port Said on 31 October, but was ordered to be at the Gaza anchorage 07.00 on 1 November, where she was instructed to have a seaplane available for spotting that afternoon. At 15.34 Short N1090 was airborne with Flt Lt Smith and Capt Millard, they had been instructed to locate two field guns and to spot the gunboats *Aphis* and *Ladybird* on to the target. Whilst the guns were located, close to an observation tower half way between Gaza and Wadi el Hesi, spotting was impossible as too many other guns were targeting the same area. They returned to the ship at 17.35. *City of Oxford* now proceeded towards Port Said, meeting *Empress* at 05.40, 2 November, to transfer Flt Lts Leigh and Smith to that ship, then on to Port Said arriving at 18.00. Less than 24 hours later she was away again bound for the Wadi el Hesi.

Raven had returned to Port Said from Suez by 15 October, taking on stores and preparing to receive floatplanes. But it would be the end of the month before Shorts 8022, and N1263, with Flt Lt Bronson, Flt Lt Worrall, and observer Lt Pakenham-Walsh, arrived on board. She then sailed at 17.30 on 31 October, escorted by the armed yacht *Managem* and trawler *Veresis*,[392] to be off Wadi el Hesi the following morning. In *Raglan*'s absence, *Requin* had arrived as the bombardment ship.

On 1 November, *Requin* requested a spotting flight to commence at 09.30. Short N1263 was hoisted out, Flt Lt Bronson and Lt Pakenham-Walsh, but before it was able take-off the two 65-lb bombs it carried had to be released into the sea. The Short was over Deir Sineid at 09.50 to commence spotting for *Requin*, continuing until 11.20. *Requin* was mainly firing with her secondary 100-mm guns. Further damage was caused to the railway and road bridge.

French *garde-côtes cuirassé* Requin commissioned in 1888 and completely rebuilt and rearmed in the late 1890's she was well past her prime but served usefully in the Eastern Mediterranean throughout WW1. This study, taken near Ismailia, shows the two 274mm guns that were put to good use off Wadi el Hesi in November 1917. *(Gallica)*

It was observed that an embankment had been built adjacent to the destroyed railway bridge and track had been laid across it. A second flight commenced at 14.15 using Short 8022, with Flt Lt Worrall and Lt Pakenham-Walsh. Spotting commenced for *Requin*'s 274-mm and 100-mm guns at 14.30, but the cruiser appeared not to be receiving all the signals. So, Worrall headed back to the ships to enquire if the signals were being received, *Requin* replying YES.

Requin may have been coping with a more serious problem at this time. She had anchored just within range of the Turkish guns and was twice hit; one shell exploded on the mess deck killing nine and wounding 29.

Spotting was resumed but at 15.18 the cruiser signalled the Short by searchlight that her W/T was out of order, and the Short returned to *Raven*. At 16.15, she signalled *Raven*, 'Many thanks to Pilots and Observer for Spotting.' During both flights, hostile aircraft were observed in the vicinity but did not interfere with the spotting.

An additional problem was that signals from *Grafton*, who was proceeding from Gaza to Wadi el Hesi, were jamming the Short's W/T. The cruiser was involved with the simulated landing requested by General Allenby.

> At Deir el Belah a party of the Egyptian Labour Corps were marched down to the beach within full view of the Turks on the heights above Gaza, and there embarked in a fleet of small craft specially brought up from Port Said. As the light waned the party in the boats moved off northward as if to be landed north of Gaza; but as soon as it was dark they returned to Deir el Belah and quietly went ashore again, though, to keep up the illusion, a procession of small vessels showing lights occasionally steamed northward past Gaza. The *Grafton* and two little river gunboats cruised off the [Wadi el] Hesi to prevent any Turkish reserves from crossing it.[393]

On the morning of 2 November, *Grafton* signalled *Raven* to prepare two machines to spot for her and *Raglan*. Shortly afterwards, Flt Lt Burling and Capt Kempson transferred from the monitor to fly one of the Shorts. The ensuing chain of events is best told by *Raven*'s report.

> At 1045   Signal received from *Grafton* and *Raglan* to hoist out seaplanes.
>
> At 1050   No. [N]1263 (Pilot Flt Lt Bronson, Observer Lt Pakenham-Walsh) was hoisted out, but returned on account of wireless generator breaking.
>
> At 1100   No. 8022 (Pilot Flt Lt Burling, Observer Capt Kempson) was hoisted out, but visibility was too bad to continue spotting, so machine returned.

| At 1210 | No. 8022 (Pilot Flt Lt Bronson, Observer Lt Pakenham-Walsh) was hoisted out to spot for *Grafton* against enemy entrenchments in Sector J8 Askalan [Ashkelon]. Several direct hits on entrenchments were registered. |

| At 1443 | No. [N]1263 (Pilot Flt Lt Burling, Observer Capt Kempson) was hoisted out to spot for *Raglan* against Railway Embankment at Deir Sineid. |

| At 1535 | No. 8022 (Pilot Flt Lt Bronson, Observer Lt Pakenham-Walsh) was hoisted out to spot for *Grafton* against Railway Embankment and Bridges at Deir Sineid. Spotting was carried out until exhaust pipe broke adrift from manifold, causing exhaust flames to blow back into the faces of the Pilot and Observer. As the machine was in danger of catching fire, Pilot was forced to shut off petrol and landed at 1612. This machine is out of commission as repairs could not be done on board. |

| At 1630 | Signal sent to *Grafton* that one machine was available for service. |

| At 1650 | Signal received from *Grafton* to change over serviceable machine with damaged one [N1262, see above] on board *Raglan* and then proceed to Port Said. |

| At 1710 | No. [N]1263 was hoisted out, towed over to *Raglan* and hoisted inboard. As the machine aboard *Raglan* was without floats, fuselage and wings badly damaged, and as it was now dark, it was impossible to transfer this machine. |

Raven now proceeded to Port Said, arriving off this port at 8 a.m. on November 3rd.

Although not supported by flights from *Raven* on 2 November, and despite her damage and losses, *Requin* was active throughout the day remaining continuously under fire. Her determination did not go unnoticed. 'The old French battleship expended all her ammunition, and sailed at 11.0 p.m.; as she steamed off she was loudly cheered by the rest of the little squadron.'[394] *Requin* returned from Port Said to resume bombardment duties on 4 November.

The Wadi el Hesi was *Raven*'s last voyage for the EIESS, but she would sail on for many years yet. She was decommissioned 31 December 1917, spending the remaining months of the war as *Ravenrock*, a fleet collier, stores carrier and troopship until January 1921. Sold and returned to mercantile service 1922. She was sold to a Japanese company in 1923 and survived almost to the end of WW2. As *Kenei Maru* she was sunk at Saigon on 12 January 1945 by US carrier aircraft.[395]

Damage to *Requin* from counter battery fire on 1 November 1917. *(forum.pages14-18.com)*

The Baby's Last Hurrah?

The Sopwith Baby, in its various versions, served the RNAS and RAF from early 1915 to the end of the war. Obsolescent almost before it entered service it nevertheless saw active service throughout, albeit finally on local submarine patrols. Their last operational use, other than patrols, may have been during the Gaza battle.

Empress was at Port Said on 31 October and 1 November when she took on board four Blackburn-built Sopwith Baby's (110-hp Clerget), N1028, N1036, N1038, and N1129, and two Parnall-built Fairey Hamble Baby's (130-hp Clerget), N1209, and N1210.[396] She was joined by pilots Flt Lt King, Flt Lt Popham, and Flt Lt Stephens. *Empress* sailed at 22.30 on 1 November, with two French destroyers, *Coutelas* and *Hache*, as escorts and proceeded up the coast towards Gaza. As we have seen, the following morning she took on board Flt Lts Leigh and Smith from *City of Oxford*. Steaming at full speed, at 09.00 on 2 November *Empress* was 20 miles of the coast adjacent to El Haram, some ten miles north of Jaffa.

The first target was a railway bridge at Jiljulie over the Wadi Kana, some ten miles inland from El Haram. The initial plan was to attack with all six machines in two waves. On the day only four Baby's were put on the water between 09.03 and 09.25. Of these one, N1038 (Flt Lt Smith), immediately sank by the tail and turned over. As the arming vanes on the two 65-lb bombs had been wound open by the wind it was too dangerous to attempt salvage, so the Baby was sunk by gunfire. Three other machines, N1209 (Flt Lt King),

Sopwith Baby with two over wing Lewis guns at the island. The pilot, Flt Lt A.W. Clemson, is surrounded by members of the Egyptian handling party. The Blackburn trademark on the float, but that does not help identify the specific Baby as all the 'N' serialled machines were built by Blackburn. There is a clear wind screen with a metal surround suggestion either N1028, N1036, or N1038, all of which were aboard *Empress* off Gaza on 2 November 1917. Unlike Clemson who had been taken prisoner on 11 October.

N1210 (Flt Lt Popham), and N1129 (Flt Lt Leigh), were given a pre-arranged signal and headed inland around 10.00 within sight of each other. Each machine was carrying two 65-lb bombs.

From Flt Lt King's logbook.

> From *Empress* 20 miles West of El Haram, about 10 m north of Jaffa in company with 3 other Schneiders (Leigh, Popham & Smith) one of which capsized in water (Smith) alongside ship & was later sunk by gunfire from *Empress*. Another (Popham) turned back after crossing the coast with dud engine. Proceeded inland about 10 miles to Jiljulie Rly bridge & dropped two 65 lb bombs from 600 ft. One failed to explode owing to fusing wire carrying away & other just missed bridge, falling in wady. Cross wind made bombing very difficult. Desultory rifle fire encountered. Sea V. difficult to rise off.

Flt Lt Leigh and Flt Lt Popham, the latter making no mention of engine trouble, reported dropping their bombs close to the bridge. But, 'although the bridge itself was not actually hit, the line on both N and S sides of the bridge was hit in several places.'[397]

Landing and recovery arrangements were detailed in the original plan.

> The attention of Pilots is again drawn to the fact that they should land as close to *Empress* as possible. Also to the fact that when machines are returning *Empress* will lie BEAM ON to the wind and NOT head to wind (so as to cause a lee for machines) and will fly a flag from the yardarm on the side on which machines should land, i.e.,

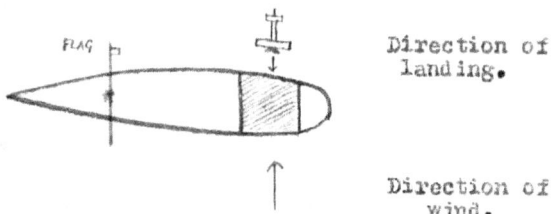

This is a very early example of the technique of using the ship to provide an area of relatively smooth sea for the floatplanes to land on. Similar methods were widely adopted after the war and through WW2 by all navies operating floatplanes.

After recovering the three returning machines *Empress*, at 11.00, proceeded north at full speed towards Haifa, arriving at 13.45. The five remaining machines were hoisted out, in the following order, between 13.50 and 14.18. N1209 (Flt Lt King), N1129 (Flt Lt Leigh), N1210 (Flt Lt Popham), N1036 (Flt Lt Stephens), and N1028 (Flt Lt Smith). Each machine was again carrying two 65-lb bombs, except N1036 which had two 16-lb bombs. They were briefed to bomb an olive oil factory near Haifa.

Flt Lt Stephens had orders to drop his two small bombs first 'in order to give the Christian employees time to evacuate the factory and take cover.' At 14.30 he dropped his bombs, 'which fell in the vicinity of the factory about 50 yards short. Workers seen to evacuate factory.' The four other machines dropped their bombs 10–15 minutes after Stephens, claiming several hits.

Flt Lt King again.

> From *Empress* 8 miles West of Haifa, proceeded to bomb oil factory at head of bay, leading 4 other machines (Smith, Stephens, Popham & Leigh). Scored two direct hits with 65 lb bombs from 400 ft & saw one machine (Popham) glide onto water with engine trouble. Other machines were circling round it, so I immediately returned to inform the ship, after seeing pilot was all right. Stephens came down alongside disabled machine & Popham swam to him, then Stephens, unable to get off again, taxied out to sea. French destroyer was sent in & eventually picked up both pilots when their machine was in a sinking condition near Mt Carmel. The first machine had to be abandoned & was probably captured intact.

On learning from Flt Lt King of the machine down:

> A signal was made to the French destroyer *Coutelas*, who immediately proceeded towards Haifa. She returned about 6.30 p.m. making the signal "All saved, machines lost". It appears that Flight Lieut. Popham's engine suddenly "cut out" and that he made a safe landing in the bay. Flight Lieut. Stephens saw that this had occurred and after dropping his warning bombs near the factory, he returned and landed alongside Flight Lieut. Popham. The latter, after having vainly tried to set fire to

Sopwith (Blackburn) Baby N1028 on the island date unknown. N1028 was one of two surviving Baby's following the 2 November 1917 operations from *Empress*.

A Parnall-built Hamble Baby similar to N1209, the other survivor of 2 November 1917. Believed to be N1194, if so it has a 110hp Clerget but is otherwise identical to N1209 and N1210. It has the Fairey patent camber-changing wings, ie; ailerons combined with trailing edge flaps, and Fairey designed tail float, but retained the rounded fin of the original design as opposed to the square cut Fairey rudder. The flying boat in the background is a Curtiss H.12.

his machine, swam to and boarded the other machine. Flight Lieut. Stephens then tried to proceed, but owing to the weight on the floats, they were more deeply immersed, and the spray thrown up was sufficient to break the propeller. One of the machines was eventually destroyed by gunfire from *Coutelas*.

I regret to report that Flight Lieut. Stephens was slightly injured in the leg by a splinter of wood from a broken strut, and that Flight Lieut. Leigh is also slightly injured in hands, head and body, as a result of a landing in a confused sea.

Four direct hits by 65-lb bombs were obtained on this factory at Haifa, and it is considered that great damage was caused.[398]

So, by the end of the day just two Baby's remained available, N1028 and N1209. N1038 had been sunk by gunfire in the morning; N1036's propeller had smashed, also breaking longerons and engine bearers. Before abandoning it, Stephens had 'Removed examination doors of main floats so as to destroy machine, which sank by the nose.' N1129 had capsized on landing when the tail float was damaged, although recovered it was never to fly again; N1210 was sunk by gunfire from *Coutelas*.

Not surprisingly, a third raid on the station and rolling stock at Tulkeram was cancelled. *Empress* returning to Port Said by 08.00 on 3 November. Here they found *City of Oxford* waiting for pilots, Flt Lts King, Popham, and Smith, before returning to the Wadi el Hesi. Sailing at 15.30 *City of Oxford* arrived at Wadi el Hesi at 09.00 on 4 November, remaining until the 8th.

One More Unto The Wadi

The military situation on 4 November was as follows. Beersheba had been captured on 31 October. Water remained a problem despite the wells having been taken undamaged, and the advance from Beersheba towards Tel el Sheria accordingly slow to develop. The assault on Gaza commenced on the night of 1/2 November, and would continue until 7 November when the Turkish defence crumbled following the capture of Tel el Sheria. By 4 November the main defences of Gaza had been captured and the British troops were attempting to advance along the coast towards Wadi el Hesi. The advance was held by strong Turkish batteries around En Nezle (or El Nuzle), nestled in the ridge of coastal hills just north of Gaza. Over the next few days, the role of *City of Oxford*'s Shorts (8019, N1090 and N1091; N1263 was already on *Raglan*) was to continue to provide spotting for *Raglan, Requin*, and several smaller monitors and gunboats, as they attempted to destroy these batteries and keep the Deir Sineid road and railway bridges unusable.

On 3 November, following the departure of *Raven* and before *City of Oxford* returned, N1263 on *Raglan* was put to use. Flt Lt Burling and Capt Kempson were ordered to spot for *Grafton* on trenches south east of En Nezle. Arriving at 06.25 they were ordered by searchlight to return to *Raglan*, 'owing to sun coming up over land and obscuring point of aim.' A second flight was made to spot *Raglan*'s big guns onto the road bridge and railway embankment at Deir Sineid. Several hits were made before a sandstorm reduced visibility, and the Short was recalled.

On the morning of 4 November they were spotting for *Raglan*'s main armament, and the single 6 inch gun. Flt Lt Burling and Capt Kempson, Short N1263, had left *Raglan* at 10.00 and started spotting for the 6 inch gun shortly afterwards. They had made several hits on the road bridge at Deir Sineid but, at 10.40, were attacked by two landplanes, a scout and a two-seater. Kempson was able to drive off the scout whilst retiring towards the ships, firing off all five hoppers of Lewis gun ammunition they were carrying. The two-seater was eventually driven away AA fire from the ships and by the 'machine gun of second machine on the water.' *City of Oxford* having arrived at 09.00, this was N1091,[399] Flt Lt King with Lt Ferguson as observer. They had been hoisted out at 10.30, then as King's log book explains…

> Was just preparing to start engine when two Huns appeared overhead chasing down Short [N]1263 with Burling & Kempson. Huns vigorously archied by all ships & one sheered off. Other hung around for a good while & machine gunned us, whilst Ferguson let back with one hopper full. Hun retreated & we got off & carried on spotting for HMS *Raglan* onto rly. embankment & road bridge across Wady El Hesy.

Burling and Kempson returned to *City of Oxford* at 11.00 'as the weather was unsettled and handling seaplanes from HMS *Raglan* was a risky operation at the best of times.'

King's Report of Flight continues:

> 11.00: Difficulty was experienced in picking up target owing to heavy ground mist; this necessitated flying inland to a point immediately above Wady El Hesy and between coast and target. At first, firing was erratic but soon became steady and three direct hits on embankment which damaged it considerably.
>
> 12.00: Observed *Raglan* signalling by searchlight, and as signal was not understood, landed alongside. HMS *Raglan* then signalled "Please wait ten minutes till we get better position."
>
> 12.10: *Raglan* signalled "We are all ready". Airborne.
>
> 12.30: Proceeded with spotting on Road Bridge. Fire on Road Bridge was very good, the spotting corrections being mostly 'OK' and 'Just ON'.
>
> 13.07: *Raglan* signalled Go Home.

King and Ferguson landed alongside *City of Oxford* at 13.25 and were hoisted in ten minutes later. Both *Raglan* and *City of Oxford* were congratulated that evening by Admiral Jackson on destroying the new railway embankment.

Morning fog delayed spotting on 5 November until early afternoon when visibility improved enough to attempt spotting. At 13.30 Short N1091 with Flt Lt Popham and Capt Millard set off to work with monitor *M.29*. Signalling and visibility hampered the shoot, but Capt Millard was able to direct the monitor's two 6-inch guns on to some trenches west of En Nezle. They then worked briefly with *Grafton* before clouds obscured the target, returning to *City of Oxford* at 15.55. At 15.30 Flt Lt King with Capt W.R. Kempson set out on Short 8019. 'Intended spotting for Raglan's 14 inch, but thick clouds were at 800 ft over land, so had to chuck it. Hun machine came over ship & bombed it, but did not hit.'

The next two days had slightly better weather and a number of successful shoots were made, although not without some interference. On 6 November, with the coast still covered in low cloud, *Raglan* was relieved by *Requin* and Flt Lt Burling, Capt Kempson, their six mechanics, and Short N1263 were welcomed back 'permanently' to *City of Oxford*. *Raglan* proceeded to the anchorage off Gaza to re-ammunition and returned with 24 hours.

However, returning to 6 November. 'Information was received from Army that Turks were retreating Northwards along road leading through Deir Sineid, and that large convoys were situated at Deir Sineid Junction up Wadi el Hesy. At 09.30, *Requin* took up firing position off Wady el Hesy, and was spotted for by Seaplanes for the best part of the day.'

Short N1091 was the machine in which Flt Lt Smith and his observer 2Lt Pakenham-Walsh had the combat with two German machines, a single-seater and a two-seater, on 6 November 1917. The side radiator on this 240hp Renault machine is clearly seen although the Scarff ring is mostly hidden in shadow.

When the clouds lifted, Flt Lt Smith and Lt Pakenham-Walsh were first off on N1091 at 10.00. They were tasked to work with *Requin* against the Deir Sineid railway embankment. Smith wrote about the flight in his memoir.

> I was keeping an eye on the battle, and a wonderful sight it was. On one sector a barrage fire was in progress at this particular moment, preceding an infantry charge. With joy I saw that the Turks were preparing to retreat. The observer and I were so busy with our eyes on the panorama that we paid no attention whatever to the Hun airmen. Suddenly came right in our ears the rat-tat-tat-tat-tat of a machine gun, with bullets whizzing by our heads.
>
> Behind us, not more than thirty feet away and driving straight at our tail, its gun spitting a regular lead hail upon us, came a Hun scout single seater. It was a ticklish situation, for scouting and bombing planes, such as ours, are not built for air-battling, being too heavy for the acrobatics by which it is possible easily to evade an attacking enemy. However, we had one advantage over our adversary. He was alone, and must do his own piloting and firing. To get our range he had to aim not the gun but the airplane at us, and so must be coming straight on. In our machine the observer handled the gun, which was so mounted that it could be swung to fire at any angle. Our chance for escape lay in quick manipulation of the airplane that would take us out of his range. [He presents a remarkably sanguine view their combat chances, but he did have the advantage of hindsight.]
>
> I gave the controls a quick twist that turned her with her planes on the vertical and dropped in a steep spiral. The Hun shot over

the top of our upward wings, missing them by a hair. Then I straightened out and started in pursuit. Meantime my observer had dropped his glasses, swung the machine gun into position and turned loose such a burst of fire that the lone German quit cold.

They had been flying at 3000 feet and the combat drove them down to 1500 feet. But, having seen off one opponent, they had no time for self congratulation as a second machine came into the attack, a two seater Rumpler or AEG.

We had a running fight with this fellow. In an air battle every third bullet fired is a 'tracer'—that is, it leaves a trail of smoke so that one can tell just how close one has come to the target. We could follow in this fight not only our own 'tracers' but the Hun's as well. The observer of the enemy plane was standing up manipulating his machine gun, and we could also see the head of the pilot. But accurate aim from an airplane is impossible, and all we could do was to fire in their general direction. We possibly did some damage on this occasion, but if we did we never were able to find out.

Climbing back through 2500 feet they recommenced spotting for *Requin*. The two German machines remained in the vicinity and made two more attacks, attempting to interrupt the spotting. However, by keeping closer to the ships, at the cost of a better view of the fall of shot, the ships and their anti-aircraft fire, added to Pakenham-Walsh's Lewis gun, kept off the two German machines. *Requin* made several hits on the embankment and nearby road bridge. The Short returned to *City of Oxford* at 12.02 after an eventful two hours.

The second flight commenced at 13.30, Flt Lt Popham with Lt Ferguson on N1091, which was clearly undamaged from its earlier encounter. They were

Although the Rumpler C.I remained in service with *FA300*, it is possible that the German two-seater encountered over the Wadi el Hesi in early November was an AEG C.IV from *FA301* similar to this one photographed at Huj earlier in 1917. Initially disliked as it was not as reliable as the Rumpler C.I, the AEG quickly became one of the more widely used two-seaters by both the German *Flieger-Abteilung* and Turkish *Tayyare Bölük*. (LoC)

able to direct *Requin*'s guns onto the embankment until 14.25 when it was chased off by 'a hostile aeroplane. Enemy machine was fired on by AA Guns from the ships and by machine gun from seaplane, being finally driven off by 14.40.' Spotting was then resumed, this time on the road bridge, until the Short returned to the ship at 15.00. A few minutes later, refuelled and with a fresh pilot, Flt Lt King, it was off again. 'Spotting for French cruiser *Requin* onto bridges. After few shots our W/T broke down so carried on signalling with lamp. Saw several Hun machines about but we kept well away & were not attacked.' A busy day for N1091, but it appears to have emerged unscathed.

By the morning of 7 November, 'The Turks were apparently now trying to check our advance Northwards along the coast by placing field guns North of Deir Sineid, and by occupying trenches, previously made, South of Ashkelon, and *Requin* was subsequently spotted for on to these targets.' It should not be thought that the Turkish Army was a defeated army. Forced out of Gaza, mainly by overwhelming artillery fire, it made a skilful withdrawal covered by several stubborn rear guard actions.

Each flight made from any ship in the EIESS was numbered in a simple chronological sequence. *City of Oxford*, Flight 1, was on 30 October, the flights made on 7 November numbered 16 to 21, although records are confused and incomplete for this day. Flight 16, N1091, with Flt Burling and Capt Kempson, was airborne at 06.55 to spot *Raglan* on to Deir Sineid, they found poor visibility owing to low clouds and returned to the ship at 07.38.

Short N1263 on *Raglan* 2–6 November 1917. From a batch of twelve Short 184s built by Robey's and fitted with 240hp Renault or 240hp Sunbeam Gurkha engines, probably all were completed with side mounted radiators. Both N1262 and N1263 were delivered to the EIESS with a 240hp Sunbeam Gurkha. *Raglan* had been fitted with a 6-inch gun salvaged from a sunken small monitor, its camouflaged turret can be seen aft of the tripod mast, reducing the deck space available for hosting a floatplane.

Flight 17, Flt Lt King and Capt Millard on 8019, set out half an hour after N1091 and flew up the coast to Ashkelon to work with *Aphis* and *Ladybird*. They found conditions every bit as difficult and returned to the ship by 08.30.

Flights 18 and 19 cannot be positively traced, but appear to have involved N1091 and N1090. Flight 18 spotted both *Raglan* and *Requin* onto the usual targets in the Deir Sineid area, Flight 19 there is no information on.

Earlier in the morning Flt Lt Burling and Capt Kempson had been called over to *Raglan* to be briefed on a new target at the railway station at Julis, north east of Ashkelon. A large convoy was reported and it was the last point on the railway within reach of guns from the sea. The monitor *M.15* (one 9.2-inch gun) was allocated for the shoot. Returning to *City of Oxford* the two set out on N1263 at 11.54 (Flight 20). Spotting was difficult and had to be carried out below clouds lying at 1200 feet, the station being around 100 feet above sea level. Hits were made on the village and station. After about an hour spent spotting they returned to *City of Oxford* at 13.45. Meanwhile, at 14.08, N1091 was sent off with Flt Lt Popham and Lt Ferguson, to continue spotting for *M.15*, later joined by *Raglan* and *Requin* (Flight 21). Shortly after take-off they were attacked 'by hostile aeroplane and withdrew to within range of ship's guns. Hostile machine retired.' By 14.20 they were able to commence spotting for *M.15*, at 14.45 they were in communication with *Raglan* and *Requin* and began spotting for them. Another hostile machine was sighted at 16.15 and spotting had to be ended ten minutes later was they were getting low on petrol. The Short returned safely and was hoisted in at 16.40.

At 19.00 Flt Lt Burling and Capt Kempson again transferred to *Raglan* 'to arrange spotting for bombardment of *Raglan*', presumably on the following day. However, plans were 'abandoned owing to the weather, and HMS *Raglan* was ordered to Port Said.' Burling and Kempson were returned to *City of Oxford* which proceeded up the coast to be off Ashkelon by day break.

On 8 November the breakthrough at Gaza was well underway.

> The previous evening at 17.00, our guns and some cavalry had advanced along the coast as far North as Wady el Hesy, although the Wady was occupied by Turks about 5 miles inland. Our troops continued advancing Northwards along the coast in strength and were being threatened by shrapnel from the Turks who were just over the sand hills a few miles inland. All ships then took up a firing position along the coast from Wady el Hesy to a few miles North of Ashkelon with a view to shifting Turks further inland.

Four flights sent out from *City of Oxford* between 11.45 and 13.25, the last flight being hoisted out as the first was returning. They were to direct fire of the ships onto positions L9, L15, and L20. These were all located west of the railway, north of Deir Sineid.

Starboard side view of Short N1090 being hoisted aboard *City of Oxford* off the Wadi el Hesi during October/November 1917. The Short was on *City of Oxford* throughout the Gaza campaign then transferred to *Empress* in January 1918 for operations against *Goeben*. (Australian National Maritime Museum, Sub Lieutenant H.B. Buck album)

First of these was Flt Lt King with Capt Millard on N1091, who were airborne at 11.45, returning at 13.25.

> From *City of Oxford* off Ashkelon. Spotting for *M.15* onto villages inland & north of Wadi El Hesy. Saw large columns of our troops, batteries, ambulances, camels, etc, etc, advancing north up the beach & past the Wady – a wonderful sight – & others skirmishing inland. Whilst we were a couple of miles inland & only 1600 ft up we were archied rather too accurately so I promptly returned to the beat up & down the coast. *Requin* was doing good shooting onto villages. Saw one huge column of smoke, about 1000 ft high, from a dump evidently exploded by retreating Turks.[400]

The explosion was witnessed by war correspondent W.T. Massey at Gaza.

> I was standing on Raspberry Hill, the battle headquarters of XXIst Corps, when I heard a terrific report. Staff Officers who were used to the visitations of aerial marauders came out of their shelters and searched the pearly vault of the heavens for Fritz. No machine could be found. Some one looking across the country towards Atawineh saw a huge mushroom-shaped cloud, and then we knew that one enormous dump at least contained no more projectiles to hold up an advance. This ammunition store must have been eight miles away as the crow flies, but the noise of the explosion was so violent that it was a considerable time before some Officers could be brought to believe an enemy plane had not laid an egg near us. The blowing up of that dump was a signal that the Turk was off.[401]

Next up were Flt Lt Smith with Lt Pakenham-Walsh on N1263 to spot for *Ladybird* and *M.31*. They made two flights the first leaving at 12.00 and returning at 12.57, the second between 13.25 and 14.22. On the first flight no targets were observed and when the W/T failed they returned to the ship. It was discovered that the aerial wire had been bumped off the spindle and shorted out the W/T set. Whilst this was being repaired, Flt Lt Popham and Lt Ferguson were hoisted out on 8019 to spot for *Requin* on trenches and camps near Burbera, south east of Ashkelon. Airborne at 12.20 they were in contact with the cruiser fifteen minutes later. Hits were made on the road and trenches in the area and later the target was switched to a camp, making at least four direct hits. The Short was engaged by anti-aircraft guns, shrapnel from one near miss piercing the port float. The Short landed safely and was quickly hoisted onboard at 13.42.

The final flight of the day, and the EIESS's participation in the Third Battle of Gaza, was Flt Lt Smith and Capt Millard's second flight, with the radio repaired. Airborne at 13.25 they were unable to communicate by wireless or signal lamp so Flt Lt Smith decided to land alongside *Ladybird*. He got down safely and was told that the two gunboats has received orders not to fire south of a line running E-W through Burbera, which was where they had been briefed to direct the shoot. The sea had become rather rough and in trying to get off [a float] strut broke and propeller cut through float. Quickly taken under tow by *ML31* which 'came alongside and took us in tow in a most seamanlike manner, otherwise seaplane would have quickly foundered.' The ML delivered the Short safely back to the ship. *City of Oxford* returned that evening to Port Said.

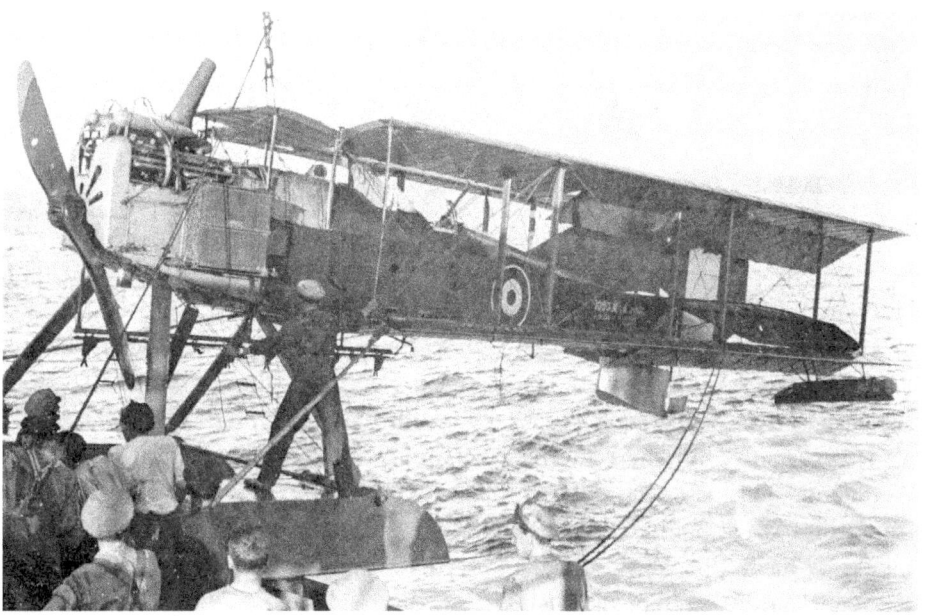

Short N1090 wings folded and being brought aboard *City of Oxford* off the Wadi el Hesi during October/November 1917. The 240-hp Renault engine and side radiator are clearly shown, as are the Scarff ring for the observer. The short angled tube above the aft end of the bomb rail is a guide for the W/T aerial with a bob-weight to keep it extended during flight. There is a rather unusual presentation of the serial, compare with the image of the starboard side of N1090. Perhaps the sign writer at the Short factory was having a bad day. *(Australian National Maritime Museum, Sub Lieutenant H.B. Buck album)*

Back at Port Said, Smith recalled:

> Several of us were summoned to the commanding officer's quarters. And again we looked at each other apprehensively, anticipating a strafing. The CO greeted us with a stern and disapproving mien. Then he began to smile widely, shook each by the hand and said, holding up a document: 'I can't read French, but I can just make out enough of this paper to know that you have been awarded the French *Croix de Guerre, pour la belle audace* displayed by you in the operations at Gaza.'

It would be several months before they were presented with the medal.

A cheerful group of pilots and observers returning from Gaza on *City of Oxford*.

Back row L-R: Flt Lt King, Lt Ferguson, Flt Lt Burling, Lt Pakenham-Walsh, Flt Lt Popham.

Front row L-R: Capt Kempson, Flt Lt Smith, Capt Millard. *(EMK)*

I mentioned earlier that the anchorage at Deir el Belah would prove to be distinctly permeable. During 11 November *UC-38* (*Oberleutnant* Hans Hermann Wendlandt) was able to pass through the line of trawlers and drifters to enter the anchorage. Once there he torpedoed *M.15* and the destroyer *Staunch*. Twenty six men were lost from *M.15* and eight from *Staunch*. The U-boat escaped, to be sunk 14 December off Corfu by French destroyers, 9 men were lost but the remainder of the crew survived, including *Oblt* Wendlandt. The losses caused Admiral Jackson to withdraw the bombarding squadron to Port Said, ending naval support for the Gaza battle. However, the Army having broken the Gaza line was now advancing too swiftly for naval bombardment to be of much assistance.

An Australian airman passing through Deir Sineid on 5 December 1917 recorded this impression of the work of the naval guns.

> Had a pretty rough trip but woke up at railhead [at Deir Sineid] this morning after travelling back to Rafa and out on this line. I visited the ammunition trucks belonging to Jacko which were blown up by a direct hit from a monitor. There were thousands of shells laying all over the place. The holes made by the 12″ shells along the line and in the Wadi go to show what splendid shooting the Navy have been doing.[402]

Back at Port Said there was no immediate employment for the two remaining seaplane carriers, *Empress* and *City of Oxford*. However, in January events at Gallipoli required the urgent attendance of *Empress* and, a few days later, *City of Oxford* commenced preparing for a visit to the Red Sea.

Two Lone Ships

The *Kaiserliche Marine's Mittelmeerdivision* was established in response to the First Balkan War 1912/13. In 1914 it comprised the battlecruiser SMS *Goeben* and light cruiser SMS *Breslau*, both modern ships, under the command of *Konteradmiral* Wilhelm Souchon.[403] The outbreak of the war found both ships visiting Pola in the Adriatic. Not wishing to be trapped Souchon immediately ordered the ships to make full speed down the Adriatic and into

SMS *Goeben* from a pre-war post card. A modern battlecruiser completed in 1912, ten 280mm, ten 150mm guns, and twelve 88mm AA guns, 27 knots. In Turkish service it became *Yavuz Sultan Selim* (usually *Yavuz*). (M.L. Carstens post card)

the Mediterranean. The ensuing events and chase of the *Goeben* and *Breslau* play no part in this story. Suffice to say that they entered into and through the Dardanelles on 10 August, just ahead of pursuing Royal Navy ships. On 16 August both ships were transferred to the Turkish Navy as *Yavuz Sultan Selim* (usually *Yavuz*) and *Midilli* respectively. They retained their German crews and Souchon appointed commander-in-chief of the Ottoman Navy. Their wartime exploits were mainly in the Black Sea against the Imperial Russian Navy but in early 1918 they emerged, suddenly, out of the Dardanelles.[404]

In September 1917, *Vizeadmiral* Souchon (he had been promoted in 1916) was recalled to Germany. His replacement *Vizeadmiral* Hubert von Rebeur-Paschwitz decided to mount a foray into the Aegean, hoping to draw Allied ships away from Palestine, where the Ottomans were under pressure. To quote *Kapitänleutnant* Herman Baltzer, who in action commanded *Goeben*'s No. 3 turret:

> The question of risk had an important bearing on the objective, since because the ships were Turkish vessels, it was necessary to obtain the sanction of the Turkish authorities to any operations in which they might take part, which would not have been the case had they been considered to be German ships.
>
> One objective which had been considered by the German Staff was to block the Suez Canal by sinking the *Goeben* in it, but as this would have entailed the loss of the ship, it was thought unlikely the Turks would agree.
>
> The objective on 20th January [therefore] was twofold; (a) To destroy the British monitors and other craft watching the entrance of the Dardanelles and to do as much damage as possible to the Naval base in Port Mudros; (b) to raise the morale of the Navy, especially the German crews.[405]

The British ships in the area were mostly intended for shore bombardment, convoy escort and general mine laying and minesweeping. However, two pre-Dreadnought battleships, *Agamemnon* and *Lord Nelson*,[406] were based at Mudros to provide big gun support should *Goeben* and *Breslau* emerge from the Dardanelles. The defence was also heavily reliant on minefields. Concerning which the Germans believed they had had an intelligence coup.

> In the early days of January, 1918, a chart was obtained from some small craft in the neighbourhood of Rabbit Island, on which were drawn certain lines which the German Intelligence Staff believed to represent the position of various lines of mines. This information, coupled with the reconnaissance reports received from their aircraft, made them confident that they were acquainted with the position of all minefields.[407]

SMS *Breslau* from a pre-war post card. Light cruiser completed in 1912, lead ship of a class of four, twelve 105mm guns rearmed in 1917 with eight 150mm guns, 27 knots. In Turkish service it became *Midilli*. (M.L. Carstens post card)

The captured chart that showed that there were more minefields than they had realised. It appeared, however, to show that there was a gap between them, but they did not realise that it was only a rough indication rather than an exact plan. Aerial reconnaissance had told the Germans that *Lord Nelson* was not at Mudros. She was taking Rear Admiral Arthur Hayes-Sadler, who had just assumed command of the Aegean Squadron, to meetings in Salonika. To the German staff, based on the information available, the time appeared right for a sortie.

The ships left Constantinople at noon on 19 January 1918, commanded by *Vizeadmiral* von Rebeur-Paschwitz. Accompanying *Goeben* and *Breslau* were four Ottoman destroyers; *Basra, Samsun, Muavanet-i Milliye,* and *Nümune Hamiyet*.[408] After conducting exercises in the Sea of Marmora, they proceeded through the Dardanelles, passing through the Nagara net about 04.00,[409] 20 January, and Seddel Bahr at 05.40. Leaving the destroyers behind to await further orders, as they were too slow to accompany them, *Goeben* and *Breslau* proceeded on a westerly course. Just north of Mavro island they turned to the south west. Shortly afterwards, at 06.10, *Goeben* struck a mine abreast the bridge the port side. The damage was minor and did not change the plan. Around 06.30 the two ships turned to the north towards Imbros, *Breslau* leading.

A patrolling British destroyer, *Lizard*, sighted the two ships at 07.20. She challenged by light signal, on receiving no reply she immediately flashed the signal GOBLO on full w/t power which, cutting through enemy wireless jamming, warned that *Goeben* and *Breslau* were out.

At 07.40, *Breslau* fired on *Lizard*, driving her away to the north. The cruiser then commenced firing on the monitor *Raglan*, which replied with

SORTIE BY *Yawuz Sultan Selim* (EX *Goeben*) & *Midilli* (EX *Breslau*) ON 20TH JANUARY 1918.

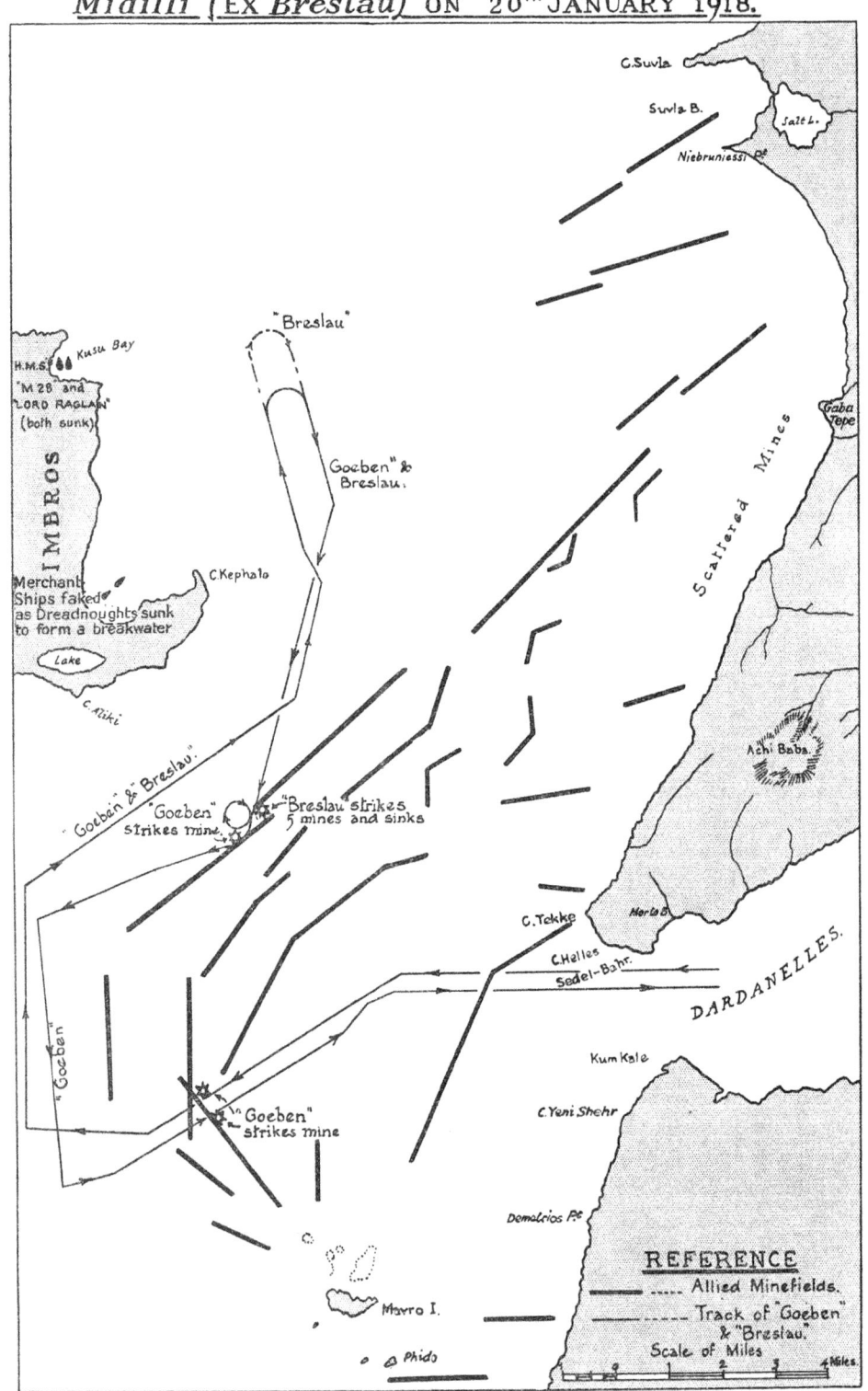

Map of the sortie of *Goeben* and *Breslau* 20 January 1918. Based on the maps in the Official History but omitting movements of British ships. (*Smoke on the Horizon*, fp.256.)

her single 6-inch and twin 14-inch guns. She did not have the time to find the range before *Breslau*'s guns put *Raglan*'s spotting top out of action. The small monitor *M.28*, one 9.2-inch gun, barely had time to open fire before both ships were struck by fire from both *Goeben* and *Breslau*. Within minutes *Raglan* had sunk and *M.28* was on fire and shortly blew up. British destroyer *Tigress* now came up, but the destroyer was quickly bracketed by *Breslau*'s guns and was forced off to the north to join *Lizard*.

M.28 at Mudros ca.1917 with Ark Royal in background. Both M.28 and Raglan were sunk by Goeben and Breslau in Kusu Bay, Imbros, on 20 January 1918. (from an old post card)

Just one of many RNAS flights on this day was made by Flt Lt J.W.B. Grigson flying a Sopwith 1½ Strutter of C Squadron from Imbros.

> On the 20th January 1918 the *Goeben* and *Breslau* appeared off Imbros before dawn and sank the monitors *Raglan* and *M28* in Kusu Bay. There was complete W/T jamming by the German and Turkish stations. The Imbros aerodrome was, I believe, the only station to receive the patrolling destroyers' report. My observer and I were permanently detailed for W/T reconnaissance in a Sopwith in the squadron scheme which had been prepared for action with the *Goeben*. We were over the enemy before they finished firing at the monitors. The W/T, was useless owing to the jamming. Having dealt with the monitors the enemy turned to pass between Kephalo Head and the minefields off the Peninsula. Off Kephalo the DH bombers appeared and the enemy began to zig-zag to avoid bombs. The *Breslau* went too far and struck several mines. *Goeben* tried to take her in tow but she sank in a few minutes. We immediately returned to Gliki to report. When we came out again the *Goeben* was heading into the entrance to the Dardanelles and was met by the Turk destroyers and torpedo boats. The enemy light craft attacked the *Tigress* and *Lizard*, our patrol destroyers, who were picking up *Breslau* survivors. The rescue was abandoned and the Turks were chased off. We had to return to Gliki to refuel. Later the *Goeben* ran aground at Nagara Point in the Narrows.[410]

Having sunk the monitors the two Turkish ships had reversed course planning to repeat their success against the ships at Mudros, including *Agamemnon*. *Kapitänleutnant* Baltzer takes up the story.

> Soon after turning, *Breslau* being astern, they were attacked by aircraft, and as the shells from *Goeben*'s anti-aircraft guns were falling about *Breslau*, the latter was ordered to take station ahead. Whilst overhauling *Goeben*, *Breslau* struck a mine and *Goeben* turned to take

her in tow. Hawsers were passed, but whilst passing them many mines were seen and were only avoided by *Goeben*, thanks to an officer on Breslau's forecastle hailing and pointing out their direction. The mine appears to have struck *Breslau* under her stern as both screws and rudder and the stern of the ship were blown off, and the officers on board *Goeben* could see right into the interior of the ship. It was soon apparent that *Breslau* could not float, and so hawsers were cast off and *Goeben* proceeded alone, still with the intention of attacking Mudros. *Breslau* struck three more mines and sank, whilst the torpedo boat destroyers were going to her assistance.

After leaving *Breslau*, *Goeben* still intended to attack Mudros but also struck a mine. Again on the port side a little forward of her midships No.3 turret. *Vizeadmiral* von Rebeur-Paschwitz abandoned the attack and set course for safety in the Dardanelles. Now under attack by aeroplanes from Imbros, *Goeben* attempted to return along the same track she had followed when leaving the Dardanelles, but was a little off course. At 09.48 she struck a third mine, from the same mine field as her first, this time on the starboard side abreast turret No.3. Still able to steam she entered the Dardanelles. As she approached Nagara Point, two planes were observing her movements and she was seen to turn suddenly and unaccountably towards the land, and run fast aground.

> On approaching Nagara Point, the Captain noticed, but too late, that there was a mistake on the chart with regard to the net buoys. The Captain believed the easternmost buoy to be the end buoy of the net and was unaware that an additional buoy, not shown on the chart, had been laid to mark the shoal off the point, with the result that when endeavouring to pass eastward of the marker buoy Goeben took the ground…[411]

The two British destroyers had meanwhile, around 09.30, engaged the Turkish destroyers which had gone to the assistance of *Breslau*. Hits were made on *Basra*, but they did not explode, and the Turkish destroyers were recalled by a W/T message from *Goeben*. Following a short stern chase *Tigress* and *Lizard* came under fire from the shore batteries and, being close to shallow minefields, turned back to rescue survivors from *Breslau*, picking up 14 officers and 148 men from a crew of 354.

Losses on the British side had been no less grievous. *Raglan* was sunk with the loss of 127 lives. There were 93 survivors, including her commanding officer Cdr Henry F. Chevalier, Viscount Broome, who had been wounded when the spotting top was hit. *M.28* suffered 11 of her crew killed, including her commander Lt Cdr Donald Priaulx MacGregor, who was killed when the 9.2-inch gun was hit. There were 55 survivors from *M.28*.

Night Bombing *Goeben*

Goeben ran aground near Nagara Point around 11.30, 20 January 1918. Over the next few days she was subjected to near constant air attack from aeroplanes and floatplanes of the RNAS at Imbros.[412]

Royal Naval Air Service aircraft, plus some machines from the Royal Hellenic Naval Air Service, Flight H.2, stationed at Mudros, flew 270 sorties against her, dropping 15 tons of bombs. Strong winds, low clouds and effective anti-aircraft fire meant that only two hits were scored 'one on the net shelf and one on the funnel casing, both causing but little damage.'[413] Even if more hits had been obtained, the 65 and 112 pound bombs used could have done little damage. Two torpedo carrying Short 320's, and some 230-lb bombs, were brought from Otranto aboard HMS *Manxman*. The Official History noted that she arrived at Mudros on 25 January. Adding, 'but there was an unaccountable delay in sending up her seaplanes, and, on the 26th, the day of the *Goeben*'s departure, the wind freshened and the sea was judged too choppy for the torpedo-loaded seaplanes to get away.'[414] Also involved was *Empress* from Port Said.

Summoned from Egypt, *Empress* took onboard four 240hp Renault-Mercedes Shorts, N1090, N1581, N1582, and N1590, with pilots Flt Lts Bronson, Leigh and Worrall, and observers Capt Kempson, Lts Ferguson, and Pakenham-Walsh. Leaving Port Said by 14.00 on 21 January, she proceeded to Milo for coal, then on to Mudros arriving at 17.00 on 24 January. There, orders were received to 'attack *Goeben* with two seaplanes by night.'[415] The first attacks were made that same evening, starting at 22.30. None of the pilots, or observers, had made night flights before.

First to test the night air were Flt Lt Worrall with Lt Pakenham-Walsh on Short N1581. They had two 65-lb bombs, a wireless transmitting and receiving set, and some Very's pistol and flares, but no Lewis gun. Taking off at 22.30

Empress at Mudros early 1918, with the airship shed in the background.

they first set course for Tenedos, to get their bearings, then north towards the mouth of the Dardanelles. Continuing north to Nagara they bombed *Goeben* at 00.15, returning to *Empress*, by way of Imbros, and landed at 01.15. They were followed by Flt Lt Bronson and Lt Ferguson on N1582. They carried the same load as the first machine with the addition of an Aldis signalling lamp. Airborne at 23.05 and, using Imbros as a navigation guide, they bombed *Goeben* at 00.25, landing back at *Empress* at 01.30.

Both crews had similar experiences, the following quotes are from Lt Pakenham-Walsh's report. 'Weather—Slightly misty. Wind slight from the north. Bombs were dropped in one run from stern to bow from 3000 feet, one bomb striking the bow and the other falling over.' Lt Ferguson reported both of their bombs missing on the port side by 100 and 150 feet. *Goeben* was blacked out, but 'Rather heavy anti-aircraft fire was experienced, but shooting was rather low at first, improving as we left. Three searchlights were sweeping the Narrows. No craft were seen alongside or near *Goeben*.' Neither crew mentioned if special arrangements had been made for, or of any difficulties in, landing in the dark.

The following two nights the weather was unsuitable, but 'Early on the morning of the 27th a Camel reconnoitred the Straits in adverse air conditions and reported the *Goeben* still in position. A raid was made in the afternoon by five DH.4's, but it was found impossible to make headway against the wind.'[416] Nonetheless, *Empress* was instructed to make another night attack over the

Goeben was aground on Nagara Point from 20 to 25 January 1918. Although subject to continual bombing attacks day and night only minor damage was caused by the small bombs available. (Finke post card)

Short N1582 set out to bomb *Goeben* on the night of 27/28 January 1918 based on an inaccurate report that the battlecruiser was still aground. It came down near Nagara and was captured intact, the crew Flt Lt Bronson and Lt Pakenham-Walsh were uninjured and made prisoner. The Short is seen here at the Naval Aircraft School *(Bahriye Tayyare Mektebi)* at San Stefano later in 1918. *(via Ole Nikolajsen)*

27/28th. Four floatplanes were also to be sent off from Mudros, the intent being to time the night attacks to act as a diversion for submarine *E.14* (Lt Cdr G.S. White) which had been sent out that afternoon to torpedo *Goeben*.

Accordingly, Flt Lt Bronson and Lt Pakenham-Walsh took N1582 out with a single 230-lb bomb, W/T, Very's flares, an Aldis lamp and this time a Lewis gun. Short N1582 was airborne at 22.25, but did not return. A search was flown the following morning, but no signs of the missing Short were found. A Turkish official message was intercepted on 31 January stating that, 'An English seaplane was forced by machine-gun fire to land near Nagara, and the crew were captured uninjured.'[417] The Short cannot have been too badly damaged as it was transferred to the Naval Aircraft School (*Bahriye Tayyare Mektebi*) at San Stefano, and may have been test flown later in the year. Both Bronson and Pakenham-Walsh were released at the end of the war. Bronson arrived at Alexandria on 24 November, and was in the UK before Christmas.

Of the four Shorts sent from Mudros after midnight, the weather conditions forced three to return without bombing. The remaining Short 'came down to 1600 feet and was subjected to severe anti-aircraft fire. He could not find the *Goeben* although he saw small boats close at hand.'

The fate of *E.14* is best described by Lt Cdr White's citation for the Victoria Cross.

> *E.14* left Mudros on the 27th of January under instructions to force the Narrows and attack the *Goeben*, which was reported aground off Nagara Point after being damaged during her sortie from the Dardanelles. The latter vessel was not found and *E.14* turned back. At about 8.45 a.m. on the 28th of January a torpedo was fired from *E.14* at an enemy ship [*İntibah*, tug/minelayer]; 11 seconds after the torpedo left the tube a heavy explosion took place, caused all lights to go out, and sprang the fore hatch. Leaking badly the boat was blown to 15 feet, and at once a heavy fire came from the forts, but the hull was not hit. *E.14* then dived and proceeded on her way out.

Soon afterwards the boat became out of control, and as the air supply was nearly exhausted, Lieutenant-Commander White decided to run the risk of proceeding on the surface. Heavy fire was immediately opened from both sides, and, after running the gauntlet for half-an-hour, being steered from below, *E.14* was so badly damaged that Lieutenant-Commander White turned towards the shore in order to give the crew a chance of being saved. He remained on deck the whole time himself until he was killed by a shell.[418]

In all nine crew members, from a crew of 30, were rescued and taken prisoner.

Come the morning of 28 January and early flights discovered that *Goeben* had gone. The following day a reconnaissance of Constantinople was made by Flt Cdr Lionel A. Hervey with Observer Sub Lt S. Chryssidy of the Greek Navy as observer, flying DH4 N6420, of G Squadron, 2 Wing at Mudros. They found *Goeben* 'lying near the inner of the two bridges spanning the Golden Horn by the arsenal. The largest of the arsenal dry docks was seen to be flooded and open.'[419]

What the Camel pilot actually saw on the morning of 27 January is unknown, as *Goeben* had got off the sand bar the previous day. The pre-dreadnought *Torgut Reis*, and other ships, had used their propellers to blast away some of the sand holding *Goeben* which was able to pull off early on 26 January. Still able to steam she had been escorted to Constantinople by *Torgut Reis*.

Kapitänleutnant Baltzer has a final comment on the bombing. 'At first, so I have been informed by a Turkish officer, they were much alarmed by the British bombing operations, and all hands took cover on the approach of the aircraft, but the poor results of the first attacks encouraged them, so that subsequently, on the approach of aircraft, boats were manned to collect fish.'

Goeben, properly *Yavuz Sultan Selim*, was slowly repaired and refitted. It was not until 1931 she was commissioned as the flagship of the Turkish fleet. She was decommissioned from active service on 20 December 1950 and stricken from the Navy register on 14 November 1954. The Turkish government offered to sell the ship to the West Germany in 1963 as a museum ship, sadly the offer was declined. Finally in 1971 the ship was sold for scrapping, she was towed to the breakers on 7 June 1973, and the work was completed in February 1976. Several parts of the ship have been preserved, including three of her screws (one of which is displayed at the Istanbul Naval Museum/*İstanbul Deniz Müzesi*) and her foremast which is located at the Turkish naval academy/*Deniz Harp Okulu*.

But to return to *Empress* in January 1918. She remained at Mudros until 8 February. Making just two anti-submarine patrols on 30 January. Finally released she sailed for Port Said, by way of Milo for coal, arriving 12 February.

CHAPTER 17

The Final Cruise

When *Empress* returned to Port Said she was to remain there until the end of April 1918. *Raven* had been paid off on 31 December 1917. Only *City of Oxford* was actively employed. To her fell the final cruise of the East Indies and Egypt Seaplane Squadron, a final visit to the Red Sea.

The Red Sea in 1918

The situation in the Red Sea had changed significantly since the EIESS's last visit in January 1917. Captain Boyle, writing anonymously in *The Naval Review*,[420] provided the following summary.

> Hejaz: No Turks within 70 miles of the coast.
>
> Asir: Our Allies, but inactive. No Turks near the coast.[421]
>
> Yemen: Turks in occupation of coast towns. The Imam (ruler of Yemen) "sitting on the fence."
>
> Political Administration: Hejaz affairs dealt with direct from Cairo; Asir and Yemen from Aden.

At this time the SNO, Red Sea Patrol, was Captain H.A. Buchanan-Wollaston, who was appointed to *Fox* on 24 October 1917 to relieve Boyle. The squadron itself remained a revolving mixture of older ships, most we have encountered earlier, some new.

The main impetus of the Arab Revolt had long since moved north and towards Damascus. Isolated Turkish forces remained in Medina, throughout Asir, Yemen, and around Aden. The local Arab sheikhs were either neutral or lukewarm in their support of the revolt. The most active support for the revolt came from the Idrisi in Asir, between Mecca and Yemen.

Asir was an ill defined area lying between the Hejaz and Yemen. At this time the British defined it as between Lith, some 120 kilometres south of Mecca, and Loheia. A stretch of coast some 550 kilometres in length. Harbours suitable for local trading dhows could be found at several locations, principally Lith, Jizán, and Loheia, but for modern ships these were just an open roadstead.

The War in Asir

The only available portrait of Sayyid Muhammad ibn Ali al-Idrisi, date unknown.

Sayyid Muhammad ibn Ali al-Idrisi, was born at Sabya in Asir in 1876. He studied in Mecca, at al-Azhar in Cairo, and with the Senussi at Kufra, before spending a period with his Idrisi relatives in Egypt and the Sudan. In 1905/06 he returned to Asir and in the following year led a successful revolt against the local Turkish administration. From 1908 the Idrisid Emirate of Asir was a factor in the politics of Arabia. However, owing to religious differences, there was no love lost between al-Idrisi and the Sharif of Mecca. He had more success with the Italians, siding with them in the their 1911/12 war against the Turks. In May 1915 he signed a Treaty with the British Resident at Aden, publishing a proclamation denouncing the Ottoman state and urging Arab independence, receiving in return certain trading privileges.

By June 1915 al-Idrisi had a following of up to 12000 irregulars but failed to capture the port of Loheia (Al Luḥayyah). The Red Sea Patrol had attempted to support al-Idrisi, but only succeeded in setting part of the town on fire. Two years later he announced new plans to capture Loheia, and requested that a British ship should be sent to Hodeida (Al-Hudaydah), further down the coast, to prevent the Turks from sending reinforcements to the former place. This demonstration had to be called off when the ship had to be diverted to a more urgent task. Not surprisingly, the attack on Loheia failed to materialise. One would have thought by now that the Idrisi had little faith in British assistance. But a further request for assistance was made when *Fox* visited Jizán on 7 January 1918.

Flagship of the Red Sea Patrol, *Fox* at Loheia with awnings spread, although the forward 6-inch gun has been left open, just in case. *(EMK)*

Al-Idrisi, then residing at Sabya a few kilometres inland, north of Jizán, requested naval assistance for another attempt to capture Loheia. An understanding was reached with Captain Wollaston, by which the armed steam launch *Kamaran* (CPO W.H. Duke) would escort Idrisi dhows carrying ammunition and provisions to the front.[422] When the assault took place, *Fox* was to shell Loheia fort and the principal Turkish positions to the north and to send a landing party ashore when requested by al-Idrisi.

Captain Boyle describes the events which followed.

Sayyid Mustafa el Idrisi in 1922 from a photograph by T.D. McLeish. *(Detail from Peoples of All Nations, Vol 1, p.187, 1922)*

> Sayyid Mustafa el Idrisi comes to the fore now. He is a distant relative of Sayyid [Muhammad]—the Idrisi, and was acting as a Political Agent for us, under the wing of the political representatives at Aden. A splendid type of Arab gentleman, this Mustafa; a great friend of the RSP [Red Sea Patrol], and a most zealous worker for the Allied cause. He quickly won the Idrisi's confidence in us and in himself, and it was largely his forceful character that eventually got the Idrisian troops to move. Mustafa had a delightful sense of humour; on one occasion; after a very strenuous sojourn ashore, he returned aboard the ship accommodating him, was ushered by the interpreter into the Captain's presence, and, after the usual exchange of greetings, caught sight of his face in the glass and burst into a most hearty and infectious laugh, with a remark which set the interpreter laughing even more boisterously. Now Mustafa's complexion is of a dark walnut hue. At last the interpreter was able to translate the joke: "Sayyid Mustafa says he is sunburnt."

> The Idrisi's tribesmen started early in January 1918 to move down the coast, picking up others en route and with them, mercenaries from the hills, of the Hashid and Bakil tribes. The advance was slow, and after crossing their border and coming a few miles within the Yemen country, it grew yet slower, ceasing for the time being just within sight of Loheia, their first objective. Loheia was the northernmost town in Turkish occupation. The RSP carried spare ammunition and kept at least one ship on the right flank.

> It was a time of political conferences, vows of friendship, much palaver, many promises, but not a movement. One of the troubles was lack of water between the halting place and the objective – a good many arid miles. The RSP however, undertook to supply the water and the Arabs got going again. The ships lent encouragement by some demonstration bombardments of unoccupied parts of Loheia, including an old Castle. The *Fox* with her 6-inch guns was able to reach a spot called el Atn [El Attan or 'Atn] some 5 miles inland; this was important as the only decent wells were there—the fresh water supply

of Loheia. At a later period the *Espiegle* found that by anchoring in a spot which was well inland according to the chart, she could also bombard el Atn.

Obviously el Atn must be taken, and Loheia's principal water supply cut off. Till this was done the Idrisian troops were chiefly dependent on the RSP for water; and herein the RSP held a trump card after a certain point down the coast had been reached; for if the ships moved on our Arab allies had to follow, and thus was punctuality forced upon them for a time and plans were adhered to.

The Navy played no part ashore in the operation beyond visiting, suggesting plans, furnishing spotting and signal parties.

The day came when the Idrisians took el Atn, though they did not hold it long. There were several wounded; and this seems indeed remarkable, for it afterwards transpired that there was no resistance. It is only necessary to see these Arabs attacking and to know their marksmanship to understand how there came to be wounded. The wounded were treated by the ships medical officers.[423]

Idrisian troops had attacked Turkish positions near Loheia on 13 January, whilst *Fox* shelled el Atn. That day *Fox* and *Clio* (Cdr A.W. Lowis) also bombarded Loheia.[424] The next day, *Fox* again bombarded el Atn but the Idrisi failed to advance. On the 15th it was decided to halt all attacks against Loheia until el Atn and its wells had been occupied by the Idrisi. So, on the 16th el Atn was again shelled, by *Fox* and *Perth* (replacing *Clio*), as a preliminary to another failed Idrisi attack. It was now agreed that all bombardments would cease until such time as the Idrisi were sufficiently strong not only to take but to hold el Atn and Loheia. It was not until 16 February that al-Idrisi made a full-scale attack against Loheia. Capturing el Atn and Loheia the next day.

The Cadmus-class sloops were an anachronism even when first commissioned. This is either *Clio* or *Cadmus* in New Zealand waters circa 1904/05. By the early 1900s it must have been a rare experience to stop engines and hoist sails, and here we have all sails set including studding sails on the foremast. By 1914 the sails were left in the past but the ships still had much work ahead of them.

There is no doubt that the Arabs would have achieved nothing without the RSP. Besides the aid already indicated, its tremendous moral and potential support must not be forgotten. The policy was that the movement should be the Idrisi's and his the victory, if he could gain it. He did gain it, but not till the middle of February when Loheia fell, a few prisoners being taken but the majority escaping inland. Most of the Turkish garrison took to the hills or went to Zohrah, some 20 miles inland, where probably the civilians accompanied them.

The *City of Oxford* now appears on the scene, with a flight of seaplanes under Flight Commander Leigh. It was hoped that with her there would be sufficient incentive to the Idrisi and his followers to continue their activities against the Turks. But the effect of aerial operations was rather disappointing, as the Arabs seemed to prefer that their job should be shifted to the shoulders of the Air Force. The direct result of the air operations was the clearance of Turks from the hills and all the vicinity of Loheia. They also made a successful bombing attack on a fort at Zohrah, said to be the new Turkish Military H.Q.

The indirect result [of the bombing attacks] was the transfer of prestige from Turks to the Idrisi, to whom a number of wavering tribes sent important hostages in token of allegiance; according to some of these, warfare from the Heavens as well as on earth was not to be opposed. The Idrisi's increased confidence in and friendship with the British was very marked after the good work of the seaplanes.

City of Oxford at Port Said in 1918 with shelters rigged over her fore deck.

City of Oxford's Red Sea Cruise

City of Oxford left Port Said on 27 January 1918, sailing through the canal to Ismailia. On 9 and 10 February Shorts 8020, 8022, N1263, and N1639, were flown down from Port Said to the ship at Ismailia. The ship moved down to Suez on 11 February, where she was joined by pilots Flt Cdr H.deV. Leigh (Flight commander), Flt Lt G.D. Smith, Flt Lt E.M. King and FSL G.A.A. Pennington with observers Obs Lt R.C. Kennedy, RN, Capt V. Millard, and Lt A.D. Ferguson.

Guy Smith nearly missed the trip, making some allowance for his story telling.

> I had a brief furlough after arriving at Port Said and, with some of the other officers, went on a little pleasure trip to Alexandria. We had just begun to enjoy our vacation when orders came for us to return to the base without delay. To my dismay, I found on reaching Port Said that my ship had started out on another southern trip. And here was I, left behind. Hurriedly hunting up my commanding officer, I said:
>
> "We would never have received a hurry-up order if some special stunt wasn't on. I don't want to be left out of it."
>
> "Yes," he answered, "something special is on, and you'll be needed. If you get ready without delay, you can catch the train to Suez and go aboard your ship there."
>
> Get ready! I was ready that very instant. So were the other fellows who had been with me at Alexandria. We certainly were not going to miss anything if we could help it. From Port Said to Suez is 100 miles of the most horrible train ride one can imagine. The railroad itself is bumpy enough to make one seasick, the cars are none too comfortable, and over the whole route one must endure suffocating heat and dust and sandstorms. However, our little crowd was used to hardships by this time, and we made that journey pleasantly enough. Not a word of complaint was heard. We occupied most of the time conjecturing what our next job would be and picking out our probable destination. The only information we had was that we were headed for Arabia. But one can make a lot of guesses concerning a probable Arabian destination and still not come within hundreds of miles of the right one.

Flt Cdr Humphrey deVerd Leigh, pilot and commanding the RNAS detachment aboard *City of Oxford* February/March 1918. Seen here at the island with a natural aviator in hand. (EMK)

They departed Suez on 13 February, on what would be the last cruise of the EIESS, proceeding down the Red Sea to Port Sudan then on to Loheia. Arriving on 22 February, anchoring off Kotama Island, an island in the outer roadstead of Loheia, in company with *Fox* at 20.20. *City of Oxford* now came under the orders of Capt Buchanan-Wollaston, SNO Red Sea Patrol. The following morning the two ships changed anchorage to Loheia. According to the Red Sea Pilot, 'The best anchorage for large vessels is about 4 or 5 miles from Loheia, in from 7 to 9 fathoms, NNE of Urmek island.'[425] There is no reason to doubt this was the anchorage used.

Phoenix Dynamo built 240hp Renault Short N1631 from the same production batch as N1639 which was aboard *City of Oxford* February/March 1918. The radiator is now back above the engine as the side radiators were unpopular with pilots who found they interfered with their view when landing. N1631 was delivered to Isle of Grain for trials in June 1917, but damaged beyond repair at Calshot in October. N1639, however, had an active career with the EIESS from November 1917 and the RAF until March 1919. *(via Colin Owers)*

'Acting under Orders received from the Senior Naval Officer, Red Sea Patrol, from 22nd February to 18th March inclusive, weather permitting, daily flights were made as required, being mostly bombing flights carried out against the Turkish positions on Jebel al Milh, and the hostile Arab positions around the Jebel as far East as Zohrah.'[426] It was hoped *City of Oxford*'s aircraft would encourage al-Idrisi and his followers to continue operations against the Turks. In the main operations were over the Loheia–Zohrah district in support of efforts to eject the Turkish forces holding the heights of the Jebel al Milh nine miles inland from Loheia. The Jebel is only 150 feet high but, rising out of the flat plain, it dominates a wide area.

Before examining the actual flights, a few notes from the final report prepared by Flt Cdr Leigh throw some light on the operational conditions experienced. 'Except on one or two special occasions all the flying at Loheia was carried out after the heat of the day. Early morning flying was found impracticable owing to heavy ground mists, and in the heat of the day the atmosphere was found to be too rough.' He also comments 'No difficulty was experienced at Loheia in either taking off or landing, as the anchorage was well protected from the open sea by reefs and islands.' Other reports contradict this, mentioning a few occasions where flying had to be cancelled due to rough seas.

The weather was not their only problem, the available maps were next to useless. 'Photographs were taken on all bombing flights of the country over which we were operating with a view to mapping same. The existing Map—South West Arabia, SANAA, Sheet 1/253,440—was found to be totally inaccurate, and mentioned very few places.'[427]

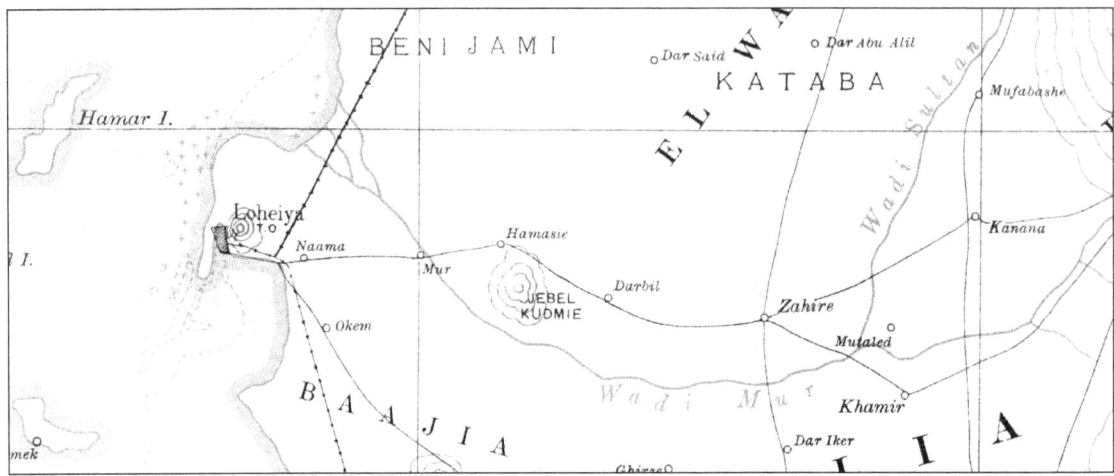

The area of operations over Loheia as shown in a detail from a map prepared for the Geographical Section, General Staff in 1915, SW Arabia, Sheet 2—Sana'a. In his report Flt Cdr Leigh commented 'The existing Map—South West Arabia, SANAA, Sheet 1/253,400—was found to be totally inaccurate.' The telegraph line shown on the map was working up to 1914 but had since fallen out of use.

Flying Over The Jebel Al Milh

On the evening of 22 February, Flt Lt Smith and Capt Millard made the first foray over Jebel al Milh. Short 8022, carried no bombs, but Millard had a copy of the inaccurate official map and a sketch map provided by al-Idrisi. Hoisted out at 17.00 they returned within the hour, having flown over and photographed the Jebel and Zohrah. Millard's observations provided the first accurate information available. In addition to noting a number of small camps and gun positions, he reported on both the Jebel and the town of Zohrah.

> Jebel al Milh is a naturally strong position and appeared well fortified on all flanks. The sides of the jebel are steep and rocky and afford good natural cover to the enemy.

> The large town Zohrah appeared to be about where Khamir is on the SW Arabia map, and Zahire appeared only a small village. About 30 camels and 20–30 men were observed here. Zohrah [Khamir?] was a large town with three lots of white stone buildings. Practically no movement here, only a few men seen. Zohrah [Khamir?] was situated about a mile South-east of the junction of two Wadys running West and East past the town, then turning Northwards.

Flt Lt Smith recalled:

> I was detailed for the first flight over this country, with orders to drop no bombs, but to make thorough reconnaissance, and then return. We flew over the fortified mountain of Jebel al Milh, and circled around it a few times, my observer taking numerous photographs and copious notes. The Turks exploded a few shells around us, but none came close enough to cause us any uneasiness. We went about twenty miles inland, passing over a village every mile or two.[428]

Regarding the 'shells', Millard noted:

> The plane was under considerable fire from rifles on Jebel al Milh. The plane evidently surprised the enemy who came out of their tents and village huts, opening fire with their rifles using black powder. On the return journey they fired about 8 rounds from the mountain guns at the machine.

Over the following four weeks operational flights were made whenever the weather permitted. The bombing or reconnaissance flights over the Jebel al Milh and surrounding area were, apart for a few minutes at the start, entirely over land and mostly enemy held territory. Pauses in the flights whilst a welcome break to the pilots and observers, permitted the vital work of the airframe and engine mechanics to continue without a break to ensure the Shorts were available when required. Details of the flights are very similar and would soon become repetitive. So, the following text picks a few flights to follow in more detail.

On 23 and 24 February all four Shorts made afternoon bombing flights over the Jebel al Milh and the surrounding area. Flt Lt King recorded his first flight in his log book, in a typically lengthy entry. On 23 February, piloting Short 8020 with AM Perry as observer, taking off at 17.55, they 'Flew in from *City of Oxford* to Jebel and dropped 2–65 lb and 6–16 lb bombs on trenches on Jebel and villages at foot. One 65 lb failed to go off, but 1–65 & 5–16

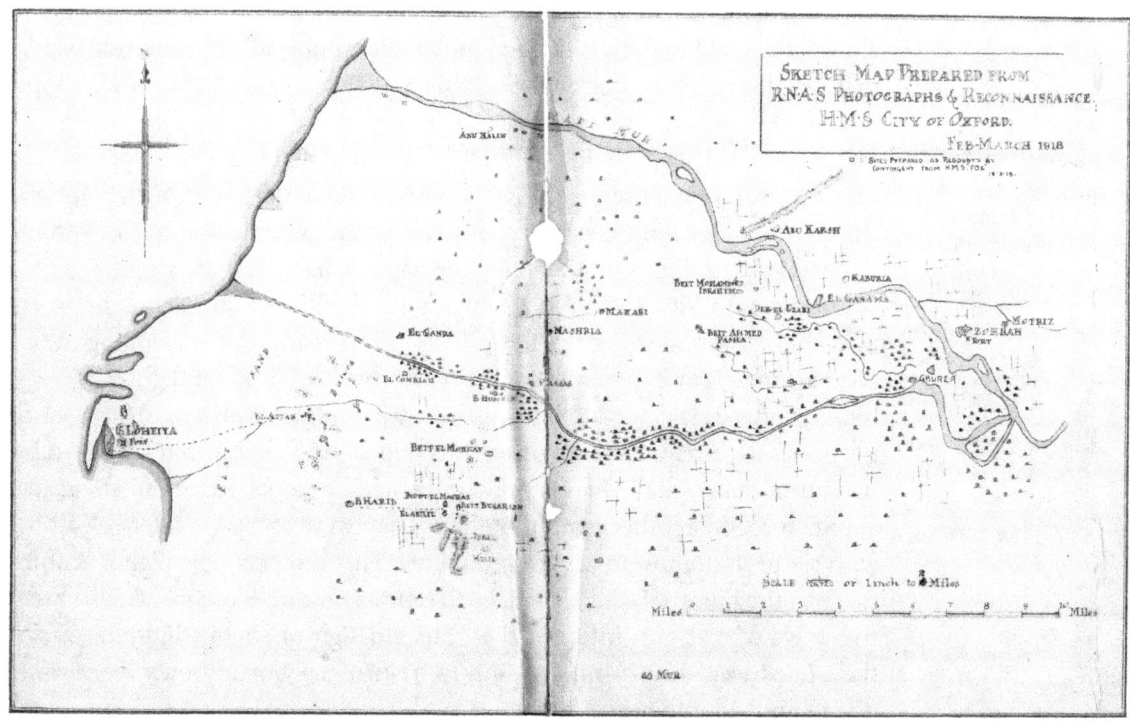

The same area shown in a map prepared aboard *City of Oxford* using aerial photographs and a limited land survey. *(TNA, AIR 1/1708/204/123/75)*

The town of Loheia (Al Luhayyah) from 4000 feet on 22 February 1918. The old town walls and town fort are seen, the Al-Zayla'i Mosque lies just outside the walls. Both town and mosque are overlooked by a Turkish fort on an isolated jebel just behind the town itself. The fort was bombarded by Italian cruisers *Calabria* and *Puglia* during the Italo-Turkish War in January 1912, and was abandoned by 1918. *(EMK)*

were direct hits on villages. No sign of enemy & no fire encountered. Landed alongside ship after dark with good moon. No trouble at all.' It was noticeable that fire discipline was stronger on this and subsequent flights, and troops remained under cover. Coverings were also provided for the gun pits.

Capt Millard, again flying with Flt Lt Smith, noted that it was too dark to see any results from the bombing and 'It was too dark to see whether Idrisi's troops were attacking. Visibility was very poor, as machine left at sunset.' Duly noted, the flights on the following day departed mid-afternoon.

On the 24th Flt Lt King made two more flights over the Jebel, between 14.35 and 16.39 returning briefly to the ship to rearm. He was flying Short 8020 again, with CPO H.M. Goodwin as observer. The first flight was 'very bumpy indeed. Dropped 8 bombs [two 65-lb and six 16-lb], mostly on Jebel, 1–65 & 4–16 lb bombs being direct hits. No sign of enemy, but Idrisi Arabs had burnt many small villages previously held by Turks. Returned alongside ship & took off another load of bombs.' They were off again, after reloading with six 16-lb bombs, in under 15 minutes. This time he 'flew higher & thus escaped the bumps. Landed all 6–16 lb bombs on ridges of Jebel & also fired two trays of m/c gun into trenches. Still no sign of enemy, but afterwards their guns were seen from the ship to be firing. Evidently they were lying doggo until after the m/cs had left.'

Three other machines made just a single flight each. After bombing mainly the Jebel they returned with similar reports; this from FSL Pennington and Lt Ferguson on 8022: 'No hostile troops was [sic] observed in village or on Jebel. All tents and shelters previously seen had been removed, mud huts alone remaining. Atmosphere was very disturbed at 2,000 feet and made flying difficult; visibility good, slight haze.'

In order to update the existing inadequate maps, Captain Millard went ashore on 25 February to reconnoitre, from ground level, the local area. In the company of Sayyid Mustafa he visited El Attan. On a second visit ashore, on 18 March, he was also taken to El Comriah and to a position a mile west of Nashria. On both trips, taking bearings with a prismatic compass he was able to locate at least some villages and obtain their local names. Whilst he was ashore he was able to interview a deserter, Omar Khalil, an artilleryman who had been serving in Yemen for six years. Millard thought the deserter was reliable, as much of his information tallied with that from other sources.

Khalil stated that there were 400 Turks and 150–200 Arabs on and around the Jebel. They had four 75mm guns and two Nordenfelt guns, with ample ammunition. Two of the 75 mm guns were located on the Jebel, the other two and the Nordenfelts were situated in the villages to the north of the Jebel. Their headquarters was at Jarb, a Turkish post some 42 kilometres to the north, with which they were in communication by heliograph. Since the first attacks on 23 February the guns had been covered over with matting and blankets, the Turks hid in caves during the attacks. Immediately after the

A nice vertical of Loheia taken on the same flight. Left to right through the centre of the image, the old town fort, the mosque, and the Turkish Fort. *(EMK)*

first floatplanes had flown over, all the Arabs and their Sheikhs had deserted. However, the deserter asserted that the Turks had no intention of evacuating the Jebel. They were supplied with food from Jarb, and there was water on the Jebel itself.

There were no flights on 25 February due to rough seas in the anchorage, but three the following day. Two flights were mainly reconnaissances to help clarify names, places and locations for the sketch map then being prepared. But Flt Lt King, Short N1263 with Lt Ferguson as observer, made

> Another flight over Jebel al Milh to bomb Jebel if Turks holding it open fire in Idrisi Arabs, who are to attack building near foot of Jebel. Stayed over hill for 35 min dropping 2–65 lb bombs, one of which failed to explode and firing 300 rounds of m/c onto trenches. Turks didn't fire and Arabs didn't turn up. They are a lot of bally wasters, so we returned to ship.

Although the Idrisi forces were making attacks on the Jebel, it was not their type of warfare and these were not pressed with any enthusiasm. Moreover, when the Shorts arrived over the Jebel, they ceased attacking, being content to let the flying machines do all the work. However, Sayyid Mustafa was now informed he could expect no further aerial help until his forces were seen to be making a determined attack on the Jebel. This was ordered for the 28 February. In support, between 10.30 and 18.30, the ship maintained a regular patrol over the Jebel in support of al-Idrisi's forces. In total six flights were

The Turkish forces were established a few kilometres inland on the Jebel al Milh. This photograph taken on 22 February 1918, looking north-east, shows Turkish tents and some trenches and possible gun emplacements on the occupied jebels. The main encampment can be seen on the jebel centre of picture, whilst some more tents are on the ridge of the right hand jebel. *(EMK)*

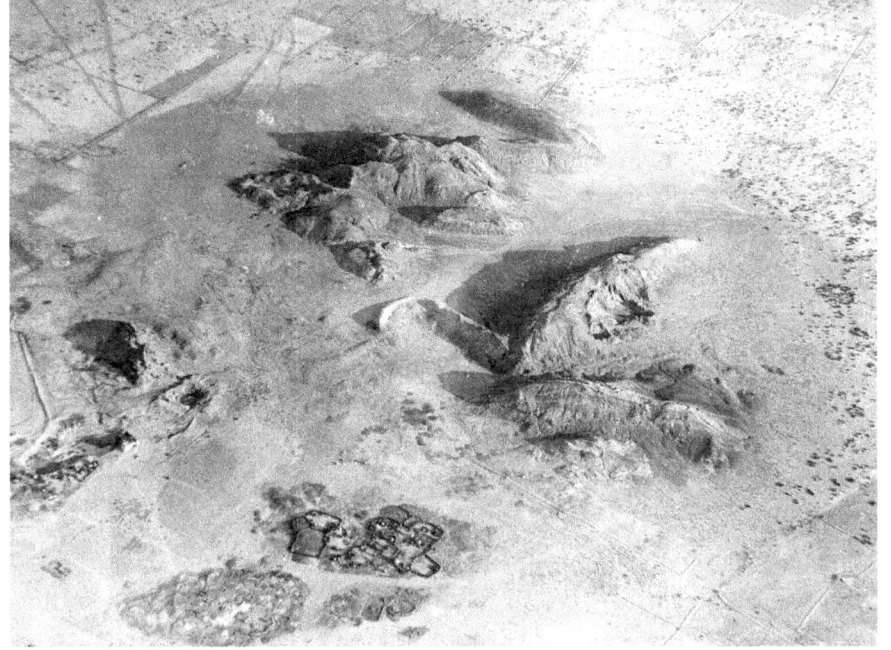

Taken on a different date this image provides an overall view of the Jebel al Milh looking to the south-east, with several Arab villages to the north of the Jebel. The Turkish forces occupied the two jebels just right of the centre of this image, experience of bombing has caused them to camouflage their positions. *(EMK)*

made on 28 February, each spending around 45 minutes over the Jebel. As one Short was landing the next was preparing to set out.

The first flight (FSL Pennington with Capt Millard on 8022) reported the Arab troops on the south-eastern end of the eastern spur. Millard, as always, prepared a detailed report.

> They were on the side of the hill under the lee of the top cliff. It is estimated there were not more than 400 to 500 Arabs. The Turks have made two stone barricades on the top of the Eastern Jebel, facing South-east, but no Turks were visible behind them. Just behind the stone barricades was a dark object suggesting a gun pit. In their present positions neither side are able to fire on each other with any effect. The other two spurs of the Jebel al Milh appeared quite deserted and no movement was observed in any of the three villages to the North of the Jebel. No gun or rifle fire was observed.

First of the bombers was N1639 crewed by Flt Lt Smith with Lt Ferguson. Each Short carried one or two 65-lb bombs and a Lewis gun. As they were landing Flt Lt King with Lt Kennedy was taking off on N1263, to arrive over the Jebel at 14.15 and remaining on station until 14.51. As he was landing at 15.00, Flt Cdr Leigh with Capt Millard were being hoisted out on N1639. The last two flights were made by FSL Pennington and Lt Ferguson on N1263 and Flt Lt Smith with Lt R.C. Kennedy on N1639. Whilst bombs were dropped, in preference to landing with them, little was achieved. Three of the machines were hit by rifle fire but no serious damage was caused.

The bombing raids continued on 1 and 2 March, three flights being made each day. Although one of these had to return with engine trouble before it could bomb the Jebel. One of the flights on 2 March flew past the Jebel and on to Zohrah. At the special request of Sayyid Mustafa this was to bomb the

Turkish Headquarters in a fort just outside the town, mainly with a view to impressing the hostile Arabs. Flt Cdr Leigh and Lt Kennedy undertook this raid, flying N1639, with two 65-lb bombs and a Lewis gun. Airborne at 16.50 they reached Zohrah just over half an hour later.

> Crossed the town steering SE and dropped 2–65 lb bombs on fort: 1st burst about 20 yards outside NW tower of fort: 2nd burst over fort about 100 yards South of it. Smoke from explosions covered whole of fort so no movement could be seen. No hostile fire was directed at machine. People were seen in town of Zohrah, but no sign of troops, camps or entrenchments.

Ultimately, the Arab forces appear to have melted away, the small gains abandoned, although the Turkish forces seemed unable, or unwilling, to benefit from this. From now until mid-March the Jebel was visited regularly. Occasionally, the nearby villages of Beit Ahmed Pasha and Abu Karsh, both 20 miles inland nearer to Zohrah, were visited instead.

Beit Ahmed Pasha received several flights on 4 and 5 March. The first, on 4 March, by Flt Lt King with Lt Ferguson as observer, on 8020. Taking off at 16.37 they returned at 17.53. King's log book records the flight, 'To a village— Beit Ahmed Pasha, dropped 2–65 lb on one place. One hit, other failed to explode. Not sure if right village as directions given for finding hopelessly wrong.' Flt Lt Smith, with AM Perry, on 8022 was also over the village in the evening dropping a single 65-lb bomb. His second bomb failed to release until past the village.

The following day King and Ferguson were back over the village on Short 8022, finding the 'village burnt to the ground and was still smouldering. White house remains. Proceeded to Abu Karsh. Dropped 2–65 lb bombs. One burst on Western outskirts of village, other fell short in Wady. FSL Pennington and Capt Millard were over the village at the same time. Millard reporting dropping two 65-lb bombs on the white house 'one exploding about 100 yards to east of it, the other failing to detonate. Smoke from a small fire was issuing from the scrub surrounding the house.'

Strong winds, heavy swells and poor visibility curtailed operations between 6 and 11 March. The respite permitting essential work on the Shorts, the engines of which were beginning to wear out. The mechanics were able to restore reliability to most of them, only that of N1639 proving eventually to be irremediable.

A single flight was made on 12 March, a reconnaissance of the Jebel al Milh. Visibility was poor and little useful information was obtained.

Two days later, three patrols were flown over the Jebel. A morning flight encountered low cloud and limited visibility and returned after half an hour. Flt Lt King and Capt Millard set out at 16.50 on 8020. Dropping their bombs on the Jebel, one fell short and the other failed to explode. On their return King had to land some way from the ship, with no petrol pressure, and be towed in. A later flight, by Flt Cdr Leigh and Lt Kennedy on N1639, also had to return with a missing engine.

Test flights for both machines the following afternoon achieved mixed result. Flt Lt King had to land 8020, with leaks from the pressure tank and the engine failing to run using the gravity tank. Further work solved the problem and, after a brief test flight on 16 March, Flt Lt King took it over the Jebel with Lt T.D. McLeish,[429] in the observer's seat. Shortly after take off they landed to 'replace bombs which had shifted.' His log book described the flight as a 'Test & photographic flight to the Jebel. Dropped 1–16 lb bomb on crest of Jebel, pin broke on other, but it exploded on contact.' On return King reported the Short as 'satisfactory.' Meanwhile, N1639 had ongoing engine problems, which continued work initially failed to resolve, but was declared operational by 20 March.

On 17 March, 8020 continued to be recalcitrant. Flt Lt King with Capt Millard had set out but 'landed after rising 20 ft as one aileron wire got stuck. Fixed it & got up then found fan had come off star[board] 65lb bomb. So dropped it in sea. Other bomb hit side of Jebel. Vis very bad.'[430] On their return to the ship Lt Walker, the Armament Officer, investigated and reported.

So many of the images used in this book come from the albums of 'EMK'. Here is the man, Flight Lieutenant Eliot Millar King piloting Short 8020 over the desert between Loheia and Jebel al Milh. *(EMK)*

> It is submitted that the fuse wire became detached owing either to excessive jolting in the heavy sea running at the time, or alternatively to a wave striking the wire. The fusing arrangements were all correct when the machine left the ship. I understand there is a down draught through the cells containing the bombs on this machine and this would account for the fan working off the spindle, instead of further on [ie; when dropped], as might be expected. Acting on Instructions received from Flight Commander Leigh the 65lb bomb gears have been removed, and the machine fitted to carry eight 16lb bombs.[431]

King later added a note in his log book concerning the one bomb they dropped on the Jebel. 'From information received after this attack have every reason to believe this 65 lb bomb laid out about 20 Turks & 3 mules, as one bomb fell in the middle of a party this day. Only two machines bombed Jebel same day, so not quite sure which machine it was.' The second flight was made by Flt Lt Smith with Lt Kennedy as observer on Short 8022. They set out about 15 minutes before Flt Lt King and reported dropping a 65-lb bomb on the western spur of the Jebel. King may have been unsure but, as was his wont, Smith claimed that it was his bomb that caused the damage, here is his version of the story.

> About 6.30 o'clock I was sent over with a few bombs to drop on their gun positions, as they had been keeping up a steady and annoying fire. All the information we had as to the positions had come from the Arabs, and so we were going largely on guesswork. However, I dropped a couple of bombs, one big one exploding on what we had figured as the right location. Two days later Sayyid Mustafa el Idrisi, the King's cousin, came aboard and told us a story related by a deserter. My bomb had dropped among a party of Turks, killing two officers, eighteen men and three mules. The man who brought the story had been close by at the time and barely escaped death himself. And that was why he had deserted.

City of Oxford's time in the Red Sea was coming to an end, and two flights on 18 March marked the end of bombing missions over the Jebel and Zohrah. Flt Lt Smith and AM Perry were first away at 16.30 flying 8022. Their orders were to bomb the Jebel then 'reconnoitre Town 5 miles SE of Jebel.' The town cannot be located on any contemporary maps, and is not mentioned in the Report of Flight. However, the single 65-lb was dropped on the western spur of the Jebel. They returned to the ship an hour after leaving.

One of the Short Model D bombers taxying at Port Said, the original caption suggests a date in 1918/19. If the date is correct, this is 8020 which was the only active survivor at this time. *(W. Toohey 825, via Colin Owers)*

Meanwhile, Flt Cdr Leigh and Lt Kennedy had departed at 16.53 on N1263 to bomb Zohrah. Taking their time, they reached Zohrah at 18.00, it is possible they struggled to climb as they did not cross the coast until 17.35 having attained 4000 feet. A 65-lb and two 16-lb bombs were dropped on the town. The 65-lb and one 16-lb fell on town, and one 16-lb bomb fell 50 yards East of town. They crossed the coast, seven miles above Loheia just ten minutes later and returned to the ship at 18.23.

On the morning of 19 March, *City of Oxford* and *Fox* proceeded north along the coast to Habil (Habl). On board were Sayyid Mustafa and Nasir ud din Ahmed, an Indian Political Officer and interpreter attached to al-Idrisi. The anchorage at Habil was an open roadstead but, fortunately, the weather was mostly kind for the next few days. Habil, located approximately half-way between Loheia and Midi. It was conveniently located for a land expedition to survey Turkish positions on Jebel Abs, some 40 kilometres due east of Habil. So, at dawn the following morning Sayyid Mustafa and Nasir ud din Ahmed set out with observers Capt Millard and Lt Kennedy on a round trip. The two observers were able to obtain local names and take bearings of various positions. Again these were transformed into a sketch map similar to that prepared of Jebel al Milh, any similarity to existing maps being serendipitous.

> On their return to Habil at about midnight they were personally presented to Sayyid [al-]Idrisi, who expressed great admiration for the work performed and stated that "the magnificent work of the seaplanes could not be exaggerated. That practically the whole of the Arabs in Yemen had already come over to him, and others were still daily surrendering". He also expressed himself very pleased with the demonstration flights made that morning.

The flights had been requested by Sayyid Mustafa, before he set out, as demonstrations over Habil and Midi, to impress the Arabs and in honour of Idrisi. Flt Lt King described Midi as 'the Headquarters of the King of the Idrisi, a smaller town than Loheiya.' Whilst the land party was riding inland, two Shorts set out during the morning of 20 March. First away at 09.46 were Flt Cdr Leigh and Lt Ferguson on N1639. For once the Short's engine behaved and they were able reach Midi about 20 kilometres further north along the coast at 10.01. After circling the town twice they returned to circle Habil before returning to the ship at 10.40. However, when called upon that afternoon, N1639 once again failed to perform. Flt Lt King recording in his log book, 'Was going to reconnoitre inland of Habil, but engine conked, so returned to ship.' This was the Short's final flight of the trip.

Flt Lt Smith with LM Robertson on 8022 left the ship at 10.05 heading straight to Habil before proceeding to Midi, circling the town whilst descending from 3000 to 800 feet. Smith then turned back towards Wadi Hairan, halfway to Habil, before returning to Midi whilst climbing to 4000 feet. Once again he spiralled down to 800 feet, then crossed the coast before turning back towards the ship, landing at 11.11.

Smith's memoir included an account of this flight in his typical style.

> We had been on the campaign for many days when word came that the King of the Idrisi wished to see an airplane. The King is all-powerful. He has among the natives the reputation of being very religious. It would be an offense for such a holy man to look upon the face of an infidel, so that he had never permitted a Christian to be brought into his presence until the Christian had been properly veiled. However, in the crisis of war he was willing to put aside some of his religious scruples and expressed a desire to welcome a flying man, even though the flying man were a Christian, so long as he remained in the air.
>
> Far and wide he sent an order summoning the sheiks, or tribal chiefs, to come to his city of Midi, as he had something to show them. The King is as wise as Solomon. He realized Turkish influence was such that many of the tribes were wavering in allegiance, and if he could show the sheiks the strange birds that had come to help him it might strengthen his hold on them, and perhaps put a little more "jazz" into their efforts in his behalf.

Returning from a bombing raid over the desert. *(EMK)*

Midi, 20 March 1918, probably one of the photographs taken by Cdr Leigh and Lt Ferguson on N1639. The main street, with adjacent market, and a caravanserai attached to the mosque form the centre of the town. The 'King's' house is possibly marked by the tower to the upper right of the image. *(EMK)*

My observer and I were picked to pay the visit to the King. Before starting we prepared a message to drop upon the town. It was addressed to the King of the Idrisi, and read: "With the compliments of the officers and men of the East Indies and Egypt Seaplane Squadron."

We passed over Midi at a height of about 4,000 feet. Turning to my observer, I asked him to hand me the message, which was carefully wrapped up and had a coloured handkerchief tied to it as a streamer. He passed it to me, but he lost a little time fumbling in his pocket for it, and as we were well over the middle of the town I feared we had lost out opportunity of dropping it with proper dramatic effect. However, I took the chance, slipping it through a hole in the fuselage. By the magic of good luck, it landed squarely on top of the King's house.

He was pleased beyond words. He probably pointed to the accuracy of my marksmanship to show that we could accomplish it with our bombs. I never undeceived him, but as a matter of fact, I didn't know which house was his, and if I had known and had taken deliberate aim would probably have missed it by a wide, wide margin.

During the afternoon of 21 March, Flt Lt Smith set out with Capt Millard on 8022 to reconnoitre and photograph the locations Millard had viewed from ground level the previous day. It was too late in the day for photographs when they reached the location of Jebel Abs, but two 16-lb bombs were dropped on Ranf, a location that may correspond to Wariah on the official map. Millard commented that Ranf was 'the Headquarters of the Arab tribes with the Turks, is about one mile South of Jebel Abs with a fort half a mile to the North East.' By now the sun was setting and further detailed observations impossible. They turned back to the coast at 18.25, landing alongside *City of Oxford* at 18.41.

Before leaving Habil on 22 March a final flight around Midi and Habil was made by Flt Lt King and Lt Kennedy on 8020. It was uneventful, lasting just under an hour. Flt Lt King's only comment being, 'Ailerons a bit funny coming out of bumps,' a little concerning given 8020's recent history. *City of Oxford* sailed from Habil at 14.15, again in company with *Fox*. Shortly after 17.00 the ships came to a stop and dropped anchor off Baäs Island, a tiny speck some miles north of Loheia. Here, Shorts 8022 and N1263 were hoisted out. When the latter failed to get off due to engine problems, just 8022 was sent

An aerial view of Port Said, looking south, in January 1918. The EIESS island is the most distant of the four in the harbour. The area to the centre left of the image has changed considerably from the Port Said map earlier in the book, the likely location of the EIESS/RAF Out Station is just out of frame. The de Lesseps statue (blown off its pedestal in 1956) is lower centre connected to the mainland by a broad multi arched bridge. *(EMK)*

off with Flt Lt Smith and CPO Goodwin, to reconnoitre, one last time, Jebel al Milh. Both ships then got under way proceeding south towards Loheia. Observing no activity on the Jebel, Smith and Goodwin returned to the ship, then approaching Loheia. Landing alongside he was instructed to 'Carry on until we pass islands.' So, taking off again, they scouted ahead of the ship until she anchored off Loheia, where he was able to land and be recovered.

At 11.15 on 23 March *City of Oxford* raised anchor and proceeded to Kamaran Island, anchoring at 16.00, probably off the old quarantine station in the Kamaran Passage between the island and mainland. At 16.42, Flt Cdr Leigh and Lt Ferguson were airborne on 8022 to carry out a demonstration flight over the mainland village of Salif and surrounding area. Landing at 17.15, they were quickly hoisted in and the ship was under way for Suez within 15 minutes. *City of Oxford* reached Suez on 29 March 1918 concluding a successful extended cruise.

Over the previous month *City of Oxford*'s four Short 184 floatplanes made 51 operational flights, and another eight test flights. A total of 51¼ hours were flown, and 63 16-lb and 52 65-lb bombs were dropped. Not surprisingly at the end of this period the machines were showing signs of hard use. Both N1639 and N1263 were out of service. Only the two older Shorts, 8020 and 8022, were even remotely operational, but were noted as 'not fit for inland flying until overhauled.'

It had been a busy and, at times, frustrating month. However, 'The bombing attacks caused considerable moral effect and most of the hostile Arabs at once deserted the Turks, and up to the time of leaving practically all the hostile Arab tribes around the Jebel Al Milh and Zohrah districts had deserted and come over to the Idrisi. As a result of this the Jebel Al Milh is now cut off and surrounded by the Arabs.'

Captain Boyle also mentioned that, 'It must be added that air observations and photographs enabled a number of corrections to charts and maps of the neighbourhood to be made, not with precise accuracy, to be sure, but enough to facilitate navigation considerably.'[432]

At the conclusion of his report on operations,[433] Flt Cdr H.deV. Leigh gave due praise for his hard working other ranks.

> All ranks worked under trying conditions extremely satisfactorily.
> I would specially bring to your notice the following ratings who often worked long hours and always with success:-
>
> Petty Officer S. Lay   (E)
> Leading Mechanic J. Stokes   (E)
> Leading Mechanic W. Robertson (E) [434]

The End of the Asir Campaign

In his account of the Red Sea Patrol Captain Boyle wrote that, with the departure of the *City of Oxford*, 'The Idrisian operations may now be considered at an end, for nothing more of any importance can be placed to their credit.' Whilst he was correct that nothing creditable could be added, the operations were not quite at an end.

Anglo-Idrisi cooperation which culminated in the fall of Loheia was amongst the most successful British naval operations of the war against Turkish Yemen. It not only gave great satisfaction to al-Idrisi, but also gave rise to considerable alarm among the Turks. Captain Wollaston recommended on 5 April the detailing of a military adviser to work with the Idrisi. The proposal was accepted and six weeks later an officer from the Sudan Government, Capt S.A. Tippets, took up his duties as liaison officer with the Idrisi.

However, the Turks were still in occupation of Jebel al Milh and, in early June, began concentrating troops around Beit al Pasha with a view to reoccupying Loheia. On 8 June el-Idrisi lost Beit Hussein and four days later he evacuated Loheia. On the 15th *Clio* shelled the newly occupied Turkish positions at Loheia. Captain Wollaston blamed the loss of Loheia on the faulty dispositions of Sayyid Mustafa and his refusal to listen to advice proffered by the Tippets.

During September and October 1918 renewed Idrisi attempts were made to reoccupy Loheia and British warships stood by for several weeks. At Sayyid Mustafa's request *Odin* fired ten rounds of lyddite at Loheia on 10 September and two days later Idrisi troops occupied the town but soon withdrew as the Turks continued to hold the wells at el Atn. On the 18th *Suva* fired seven rounds of 4.7-inch shrapnel at Loheia when it was discovered that the Turks were endeavouring to burn the town. *Espiegle* also assisting Idrisi operations in the vicinity.

Further British naval action was taken in October when the Turks again attempted to burn Loheia in retaliation for the casualties caused to them by the launch *Kamaran* whilst sinking two dhows. A British vessel replied by opening fire with shrapnel upon part of the town from where soon afterwards smoke was seen rising. The fort on the ridge overlooking the town was also shelled.

On 22 October 1918 the Idrisi made known his plan to attack Loheia yet again. This got underway on the 24th and the following day, in support, British ships opened fire in the direction of the Turkish gun positions. Idrisi troops made little headway and on the 30th the attack was called off. That constituted the final Anglo-Idrisi joint operation of the war for on 2–3 November messages were sent summoning the Turkish garrisons to surrender in accordance with the terms of the armistice.

CHAPTER 18

Royal Air Force Days and Ways

Final Days of the East Indies and Egypt Seaplane Squadron

With the success of the Gaza Battle, and the mobility of the RFC units, the need for the East Indies and Egypt Seaplane Squadron's seaplane carriers almost disappeared overnight. *Anne* had already been decommissioned and *Raven* would soon follow, *Empress* was not actively employed by the EIESS again. It fell to *City of Oxford* to make the squadron's last cruise, as previously related. Wing Cdr Risk had to find a new role for the floatplanes.

Previously, *l'escadrille de Port-saïd* had conducted patrols from Port Said. The first recorded submarine patrols by the EIESS were from 12 to 20 July 1916. As this report by Wing Cdr Samson shows:

> The presence of enemy submarines in the Eastern Mediterranean… necessitated the institution of a continuous coast patrol. This was carried out by means of a succession of flights, both by Sopwith and Short machines.[435]
>
> The total number of flights made in the period under review was 27, and the total time in the air 17 hrs 8 mins. On 15 occasions observers were taken up provided with the specially prepared anti-submarine maps of the harbour and coast. The opportunity was taken to test the qualities of the submarine detector, a form of binocular which by means of special prisms, cuts out all the polarised rays of light and thus removes the reflected glare of the water in order that the view may penetrate its depth.
>
> Since the patrol was instituted no submarine has been detected in the neighbourhood of Port Said and no casualties have been suffered by shipping. The total gross tonnage entered and cleared July 15th to 21st is stated by the Canal Company to be 323,730 tons.[436]

By late December 1917 the EIESS commenced regular anti-submarine and convoy patrols. Flt Lt King recorded what was probably one of the first of these in his log book. On 22 December, with Lt Walsh as observer on Short

N1649, he flew a patrol over a 17 ship convoy for over two hours. It was a precursor of many hundreds of similar patrols flown by the EIESS and its RAF successors.

In a report Wing Cdr Risk detailed the 22 December flights.[437] '[S]eaplanes were ordered to assist as escort to a convoy leaving the port. Three machines [N1649, N1582, and N1590] were detailed for this duty, and the convoy was escorted to a point 40 miles out.' Submarine patrols were also carried out on 28 December and 2, 3, 4, and 5 January, with no sightings reported. By the end of January 1918, Wing Cdr Risk was reporting on flights from both Port Said and Alexandria. Noting, 'On 15th January, two seaplanes were sent to Alexandria for the purpose of carrying out patrol work.' The Alexandria Detachment was established in the Outer Harbour at a location known as Kamaria Port. Separate from the main activity of the port, with easy access to sheltered water for floatplane operations, it was commanded by Flt Cdr Burling.[438] Squadron Commander P.L. Holmes commanded the Port Said Detachment, and had overall charge of the development of a system of

'Convoy leaving for England' states the original caption with a date 28 March 1918 on the photograph. This is therefore from one of the final patrols by the East Indies and Egypt Seaplane Squadron. Probably from Alexandria on Short N1581, piloted by Flt Cdr Burling with an unknown observer. *(EMK)*

Submarine and Convoy Patrols, putting his previous experience at Felixstowe to good use. There is even mention of a Port Said 'Spider Web' patrol in Capt King's log book.[439]

By the end of February 1918 Wing Cdr Risk was forced to report that the two detachments were in a parlous state.[440] Port Said had four Shorts and four Sopwith Baby's, but Alexandria just a single Short. Both detachments had only a single pilot available. As he explained:

> During the early part of this period HMS *Empress* was at Imbros for operations against the *Goeben* and whilst there carried out three patrols amounting to 3hrs 45mins. Upon her return, Pilots and Observers were transferred the same day to HMS *City of Oxford* for duty in the Red Sea, thus leaving Port Said and Alexandria very short of Pilots.
>
> The new Short Seaplanes with the 260 H.P. Sunbeam engines, which have been received here, are not reliable for long patrols on account of the valves. The new type valves have been demanded but have not yet been received. It would be possible to carry out very valuable patrols and convoy work from these Bases if sufficient Pilots and suitable Seaplanes were available.

The situation slowly improved through March and after the return of *City of Oxford*.

The *Croix de Guerre* Investiture

Before turning to the Royal Air Force there remains one final piece of East Indies and Egypt Seaplane Squadron, RNAS, business—the *Croix de Guerre* investiture.

The original letter announcing the awards was dated 29 November 1917.[441] The exact date of the investiture is elusive, but Flt Lt Popham appears in the photos so it was before he was invalided back to the UK in mid-April 1918. Flt Lt Smith gives us a clue when after being informed of the award he writes 'We were formally presented with our crosses a few months later.' So, a date in early April 1918, following *City of Oxford*'s return, appears plausible. Also, Lt A.D. Ferguson is wearing a GS officers cap with an RFC (?) cap badge as opposed to his usual Glengarry, perhaps due to his transfer from the RFC to RAF. Only seven of the ten flyers listed were present; Flt Lt Bronson and Lt Pakenham-Walsh, were guests of the Turkish Government from January 1918, and Flt Cdr Burling as the only pilot available at Alexandria may have been unable to attend due to the exigencies of service.

The investiture parade for the *Croix de Guerre* was held on the Base island. The medals being presented by *contre-amiral* Georges Varney,[442] as recalled by Flt Lt Smith:

The *Croix de Guerre* investiture at the island. EIESS officers and other ranks are formed up in their best uniforms around the island's parade ground with a well worn Bessonneau hanger as backdrop.

Contre-Amiral Georges F.C. Varney is presenting the first medal to Flt Lt H.V. Worrall. Wing Cdr C.E. Risk stands behind the *amiral* probably with details of the award to be read out for each recipient. The remaining recipients are, in order, Flt Lt G.D. Smith, Flt Lt A.E. Popham, Flt Lt E.M. King, Capt V. Millard, Capt W.R. Kempson, Lt A.D. Ferguson. The *amiral*'s flag officer (looking away) is carrying a tray with medals. There are two French buglers behind the flag officer. *(EMK)*

Two French buglers sounded the call and everybody lined up on the parade ground. Then the French Admiral advanced and, drawing his sword, said: "*Au nom du Président de la République Française je vous conféré le Croix de Guerre.*" He then pinned the cross on the left breast. The buglers blew another call, the Admiral saluted, sheathed his sword and marched away, while we stood at rigid attention.

From RNAS to RAF

The decision to create the Royal Air Force was made in late 1917, with the passing of the Air Force Bill in Parliament, and its receipt of Royal Assent on 29 November. The actual creation took months more of Staff work in the Admiralty and War Ministry. The Admiralty provided details of the new organization to all commands and Senior Officers in a series of Admiralty Fleet Orders (AFO) published in March 1918. Though, doubtless, rumours of the impending changes had been circulating for some time.

The key AFO was No.1391 published on 19 March 1918. The first two paragraphs of which read:

> In view of the impending transfer of the RNAS to the Air Force (after which the Air units serving with the Navy will be known as "Air Force Contingents"), the following outline of the organisation which will be adopted is communicated for general information and guidance. The date on which this organisation will come into force will be communicated later.

The general principle underlying the organisation is that Air Force Contingents will come under the respective Senior Naval Officers, and through them under the Admiralty, for operations, but that they will come under the Air Council for discipline and for maintenance of personnel, material, and buildings. The most important exception to this principle will be in the case of the Air Force Contingents with the Grand Fleet, which will come under Naval discipline.

The arrangement for the Grand Fleet and other home based Contingents were then laid down in detail. Those for Contingents on Foreign Stations, including Egypt, were less precise, noting that,

> [L]ocal requirements will involve some slight modifications, which can be settled as requisite for each command. The organisation settled for the Mediterranean, which may be regarded as typical, is as follows:–
>
> (a) There will be Air Force Group Officers attached to each Mediterranean sub-command (viz., Malta, Gibraltar, Adriatic, Aegean, Egypt), and a Senior Air Force Officer at Malta in command of the whole Air Force District.
>
> (b) The Air Force Group Officers will come under the local SNO's for operations (which may be ordered by them or by the Commander-in-Chief, Mediterranean), and under the Officer Commanding Air Force District, Malta, for discipline and maintenance.

Officers of the EIESS after the investiture gathered in front of one of the hangars.

Centre row, L–R : Capt V. Millard; Capt W.R. Kempson; Lt A.D. Ferguson; Sqd Cdr P.L. Holmes; Wing Cdr C.E. Risk; Flt Lt H.V. Worrall; Flt Lt G.D. Smith; Flt Lt E.M. King; Flt Lt A.E. Popham.

Standing in the back row, second from left, Lt Cecil Eldred Hughes, Somerset Light Infantry, and Captain Benn's replacement as EIESS Intelligence Officer. Also in the back row, second from right, Flt Lt H.deV. Leigh. The remainder of the officers are unidentified. *(EMK)*

(c) The OC, Air Force District, will allocate machines, personnel, &c,. at his disposal in accordance with the directions of the Commander-in-Chief, Mediterranean, and will act as advisory to him in all matters connected with Aircraft. He will be in direct communication with the Air Ministry as regards discipline and the maintenance of the total establishment of personnel and material approved for the whole Mediterranean command.

The RAF in the beginning was essentially an army unit, the RFC, with the RNAS tacked on. The RFC did not understand naval aviation and did not want to. The RAF evinced no change in that attitude. Therefore, with the exception of the RNAS fighter and bomber units serving on the Western Front, the naval aviators were largely left to get on with their job. They were aided in this by the fact that most senior Air Force Officers involved in naval aviation were ex-RNAS themselves. Essentially the old RNAS with new titles, uniforms, and ranks. At the front line most RNAS personnel took the attitude—they can call us what they want, but we are still naval airmen doing a naval job. So, let's get on with it.

From EIESS to Naval Wing, RAF, Egypt

The East Indies and Egypt Seaplane Squadron ceased to exist on 1 April 1918, becoming the Naval Wing, RAF, Egypt. But what of its pilots, observers, non-commissioned personnel and ships?

The fate of the remaining ships is easiest to answer. *City of Oxford* was at Ismailia on 1 April 1918. She was retained as a seaplane depot ship for the Egypt Group, Mediterranean Command, RAF, until the middle of September.

Empress in 1919 at Constantinople. *(FAAM, Carriers E064. Courtesy of The National Museum of the Royal Navy)*

Ordered home, she was paid off and decommissioned at Cardiff on 20 November 1918, and sold to Ellerman & Bucknall for general cargo duties. *City of Oxford* was broken up in Italy in 1924.

Empress had been inactive since returning from Mudros in February. On 25 April she sailed from Port Said to Gibraltar to carry out submarine patrols and escort of convoys in the Straits. On 5 August she proceeded to the UK for a brief refit Devonport Dockyard. On completion of this, on 29 September, she returned once again to Mudros, arriving shortly before the Turkish surrender on 30 October. She was part of the Allied Fleet which sailed through the Dardanelles to Constantinople on 12 November 1918. She remained in the Aegean area until October 1919 then being ordered to the UK to decommission. In November 1919 *Empress* was returned to the South Eastern and Chatham Railway, from 1923 the Southern Railway. In 1923 she was sold to a French company, *Société Anonyme de Gérance et d'Armement* (SAGA),[443] and operated on the Calais-Dover route on behalf of the *Compagnie des Chemins de Fer du Nord*. *Empress* was scrapped in 1933.

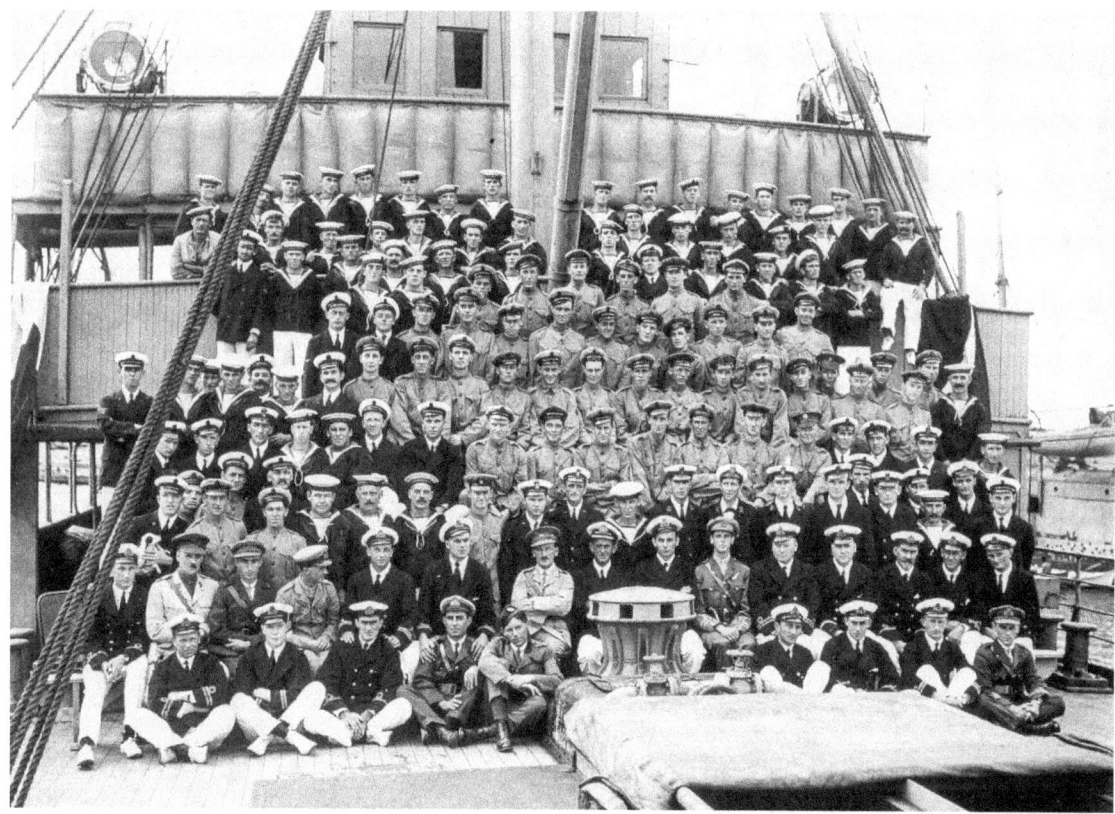

Empress departed from Port Said on 25 April 1918. This group photograph was taken before then, probably in December 1917 as among those identified is Lt Pakenham-Walsh who was made PoW in January 1918. As always the majority of officers and men cannot be identified, however some of the ex-EIESS pilots and observers are to be found. Front row (cross-legged): 4th from left Flt Lt H.V. Worrall, Lt A.D. Ferguson (with his usual Glengarry). Seated: 2rd from left Lt C.E. Hughes, Lt L.H. Pakenham-Walsh, Flt Cdr E.J.P. Burling, Flt Lt G.D. Smith, n/k, Wing Cdr C.E. Risk, Cdr Edward D. Drury (Captain of *Empress*), n/k, Capt V. Millard. *(EMK)*

Headquarters, RFC, Middle East, requested on 27 March information regarding 'the numbers, employment and trades of RNAS personnel... which make up the establishment. Details should indicate numbers, grading and employment of Officers, i.e. Numbers of Commanders, Pilots, Observers or Equipment Officers, etc: Numbers by trades of Other Ranks.'

Wing Cdr Risk's reply on 28 March listed 25 Officers and 199 Ratings in the Egypt Command. I have added known identities in [square brackets].

1 Wing Commander	[C.E. Risk]
1 Squadron Commander	[P.L. Holmes]
3 Flight Commanders	[E.J.P. Burling, H.deV. Leigh, n/k]
3 Flight Lieutenants	[E.M. King, G.D. Smith, H.V. Worrall][444]
5 Flight Sub-Lieutenants	[G.W. Morey, G.A. Pennington, G.J. Pilgrim, G. Waugh, n/k][445]
2 [RNAS] Observer Lieutenants	[A.C. Kennedy, S.C. Howes]
3 Military Officers attached for duty as Observers	[Capt W.R. Kempson, Capt V. Millard, Lt A.D. Ferguson]
2 Armament Officers	[Lt C.F.T. Walker, n/k]
1 Photographic Officer	[Lt T.D. McLeish]
1 Executive Officer, RN	
3 Warrant Officers	[WO.1 (Lieutenant, RAF) W.C. England – Station EO, rest n/k]

RATINGS:

Engineers	40
Riggers and Carpenters	37
Armament	3
Wireless	5
Photographic	5
Draughtsmen	2
Writers	2
Aircraftsmen and General Ratings carrying out miscellaneous unskilled duties	105. Total: 199.

A few days later, all personnel were transferred to the RAF with only a change in rank to show for it. Wing Commander Risk, for example, became a Lieutenant Colonel, RAF. The three Flight Lieutenants initially became Lieutenants, RAF. They were, however, within a month appointed (Temporary) Captains, RAF. The three military observers remained with the squadron, Kempson and Ferguson for the duration, Millard at least until July after which he appears to have commenced a series of armament courses whilst remaining in the Middle East. Flt Cdr Leigh and Obs Lt Kennedy joined *Empress* when she left for Gibraltar.

Now land based, the EIESS was divided into two detachments, one each at Port Said and Alexandria. Port Said remaining the main engineering base. Both Detachments were part of the Naval Wing, Royal Air Force, Egypt, and were referred to as the Port Said Squadron and the Alexandria Squadron. There were minor variations but all reports and communications used these formats.

At some stage numbers began to be applied to the units. On 6 June 1918, the Naval Wing, RAF, Egypt, was re-designated 64 (Naval) Wing, RAF, with Lt Col C.E. Risk in command. At the same time 431 and 432 (Seaplane) Flights were authorised; 431 Flt was the Port Said Squadron, but there is no confirmation that the Alexandria Squadron became 432 Flt. Major P.L. Holmes commanded 431 Flt and Major E.J. Hodsoll commanded at Alexandria with Captain Burling as second-in-command. On 6 October, the two Flights were combined to form 269 Squadron, RAF, with Major P.L. Holmes in command at Port Said. At the same time 270 Squadron was formed at Alexandria, with Short 184s and a Felixstowe F.3 flying boat, with Major E.J. Hodsoll commanding. However, from 29 September Captain Burling is signing the Weekly Reports as Commanding Seaplane Squadron, Alexandria. The flight/squadron numbers were never used in surviving documents.

Report on Present Position of RAF in the Mediterranean, August 1918

This report (AIR 1/649/17/122/409) covers the entire RAF presence throughout the Mediterranean theatre, and provides detailed reports of all the Groups—Aegean, Adriatic, Egypt, Malta, Gibraltar. The following extracts are mostly relevant to the Naval Wing, RAF, Egypt.

> The R.A.F. Mediterranean Command comprises all former R.N.A.S. units in the Mediterranean. All R.A.F. units in the command are working with or carrying out operations for the Navy; all units are therefore under the orders of the Admiral or S.N.O. of the Naval Command in which situated for operations, and ultimately under the British Commander in Chief, Mediterranean.
>
> Other R.A.F. units in the Mediterranean which are working with the

Army—as for instance those at Salonika or in Palestine—belong not to the Mediterranean but to the Middle East Command.

The operations carried out are almost entirely in connection with the anti-submarine campaign; and may broadly be divided into two classes: (a) anti-submarine patrols and direct attack of submarines at sea, (b) bombing of enemy ports and depots, particularly submarine bases.

Included under (a) protective escort work with convoys, of which a considerable amount is carried out with convoys arriving at or leaving those ports where seaplane stations are situated—i.e. Gibraltar, Malta, Alexandria, Port Said and Taranto.

EGYPT GROUP.

(a) Organisation

37. This Group is organised as one Wing, No. 64 Wing, R.A.F., and consists of seaplane stations at Alexandria and Port Said, and a Kite Balloon base at Alexandria. A new K.B. base is also about to be established at Port Said.[446]

The island base at Port Said, looking south, as it was in 1918. There are now three Bessonneau hangers, with a Short outside the centre hanger. The parade ground was the small area of clear space next to the northern hanger, and is surrounded by multiple makeshift offices, stores, and workshops. Most accommodations had moved off the island at this time. The Egyptian Coastguard continues to use part of the island as a shipyard for the coaling barges and tugs. *(EMK)*

38. Although forming part of the Mediterranean R.A.F. command, and coming under the Rear Admiral, Egypt for operations, the group has recently been placed for detail administrative purposes under the control of G.O.C. Middle East, whose Headquarters are at Cairo.

That is to say, all details of administration, discipline and supply are dealt with by Middle East Headquarters, while all large questions of policy, proposals for new programmes, increases of material or personnel, etc, are referred from Middle East to G.O.C., Malta. This procedure works well in practice, the group fitting in easily with the existing R.A.F. Middle East organisation, and much time being saved in dealing with small questions of administration which would otherwise have to be referred to Malta. The headquarters of the Group or Wing are situated at Alexandria.

(b) Equipment.

39. The policy originally laid down for this group consisted of a squadron of F.3 Large America seaplanes at Alexandria, and two flights (six machines each) of Short seaplanes at Port Said, together with a reserve at the latter place of Short seaplanes for any aircraft carriers attached to the Egyptian Command.

The Alexandria Seaplane Station was established at the western end of the Outer Harbour at a location known as Kamaria Port, seen in this view looking northeast in 1918. There are five Bessonneau hangars, a slipway, but only a single Short is visible. Mex Road is off picture to the right of the warehouses. (Richard John Grandin - The Western Front Association)

A general view of the Alexandria base showing two of the hangars with ubiquitous aircraft transport crate 'buildings' and several Crossley tenders alongside. The Short is under the only derrick available to lift machines onto the water, the slipway has yet to be constructed. The Short rests on the ground but a ground transport dolly is to its left. *(EMK)*

Owing, however, to the slow rate of output of F.3 seaplanes from Malta,[447] it is not anticipated that any of this type will be available in sufficient numbers to equip a squadron at Alexandria until well into next year, for this reason, therefore, the policy being worked to at present is one squadron (two flights) of Short seaplanes at both Port Said and Alexandria. Apart from a few obsolescent seaplanes of the "Baby" type all the machines in both squadrons are of the Short "184" type, the average strength at present being 6 machines at Alexandria and 12 machines at Port Said. The reserve for aircraft carriers at the latter place is not being held as there are no longer any aircraft carriers working in these waters.

40. Port Said—originally formed only as a shore base for a number of seaplane carrying ships which were at one time working from there—has not been found suitable as a base from which to carry out active patrol operations due to the narrowness and congested state of the harbour, and lack of accommodation and room for expansion in the site originally selected. This site was a long way from the mouth of the harbour, and in order to rise, seaplanes had either to risk fouling shipping, etc, in the harbour or else taxy right to the open sea thereby entailing unnecessary wear and tear of floats and engines.

41. For these reasons it has been put forward by G.O.C. Mediterranean and concurred in by Commander in Chief and Admiralty to replace the existing seaplane squadron at Port Said by a squadron of D.H.9 aeroplanes specially fitted (as in the case

of the Malta aeroplanes)[448] for over sea work. An aerodrome is available, and is now being prepared, and it only remains for Air ministry to make provision for the necessary machines.

Meantime, to get over the difficulty of operating seaplanes mentioned above, a temporary advance site has been selected at the mouth of the harbour, where a "war flight" of six machines is maintained; the old site being retained as an erecting and repair base only.

(c) <u>Personnel</u>.

42. An establishment for the seaplane squadron at Port Said allowing for 31 Officers and 205 men was drawn up by Air Ministry, and the same establishment (pending the provision of the F.3 squadron mentioned in paragraph 39) is being worked to at Alexandria.

There is at present a slight shortage of seaplane pilots and observers at both Alexandria and Port Said and of Balloon officers at Alexandria; but beyond this (which it is anticipated will be made good very shortly) all units are practically up to full strength.

(d) <u>Operations</u>.

43. The operations of this Wing consist entirely of anti-submarine and mine-field patrols, and the escort of convoys arriving at or leaving Egypt. The latter work is more important because submarines do not as a rule operate close to the Egyptian coast (i.e. within reach of our seaplanes) except when convoys are available for attack.

A Short being hoisted in at Alexandria after a flight, the serial is unknown but it is named FLAGSHIP.

In other words, submarines are not met with "on passage", and therefore the escort of convoys not only forms an important protection for them, but also gives the best chance of meeting and attacking submarine.

Kite Balloons are also employed in local defence, both in mine-sweeping and in the local protection of in-coming and out-going conveys.

Kite Balloons, whilst not strictly part of this story, were an important part of the convoy escort system. A number of the Flower-class sloops,[449] based at Alexandria, were fitted with winches and a small balloon deck, it is possible that some trawlers were similarly fitted. The Report has the following to say about them.

50. <u>No. 2 Balloon Base, Alexandria</u>.

    This Station is situated at Mex Road,[450] just inside the breakwater, and consists of a canvas camp in which the officers and men are housed in 60' × 20' Pollard standard tent huts. The Commanding Officer has, however, quarters in the adjoining villa, belonging to the Medical Quarantine Staff, adjacent to the Balloon Station.

51. The balloon facilities consist of six sheds and gas-making facilities consist of three, type "A", Silicol Plants and two Compressors (Reavell type) together with 900 gas bottles.

No.2 Balloon Base was established at Alexandria during 1918, located along Mex Road close to the Seaplane Station although its exact location has not been determined. It provided kite balloons for convoy escort ships, mostly specially equipped *Flower*-class sloops. Five of the six balloon sheds are shown, the most distant one with a nurse balloon outside. But what is happening here? There are five men in or around the balloon basket to weigh it down whilst a reduced ground party appear to be manoeuvring the balloon, possibly out of the base to a waiting ship at the nearby breakwater. *(EMK)*

52. The Naval Section of this Base consist of 25 officers and about 105 ratings, together with about 30 odd ratings, detailed to this Base by G.O.C., Middle East. There is also included in the camp a section of ex R.F.C., officers and men of whose numbers no exact particulars are to hand.

Submarine and Convoy Patrols

Between 22 December 1917 and 31 March 1918, the EIESS had flown 199.5 hours on patrols from Port Said and Alexandria. Over this period only a handful of Short 184s were available and even fewer pilots and observers. From 1 April to 16 November 1918 the Naval Wing, RAF, Egypt, completed 1632 flying hours on patrol and convoy escort. Almost 3000 ships, mostly in convoys, were escorted into and from the two ports. Patrols were usually between two to three and a half hours in duration, although shorter patrols were not infrequent. The longest patrol recorded was 4 hours 25 minutes by Short N2905 on 17 July, and there were at least eight more patrols of over four hours duration. Pilots and observers are not named in the surviving weekly summaries. The majority of the patrols were convoy escort, relays of Shorts covering inbound and outbound convoys up to 50 miles out from the coast. Submarine patrol areas were established, including Patrol Area (port), Patrol Area (starboard), and a Spider Web from Port Said, there were probably similar patrol areas from Alexandria. As well as the floatplanes there were numbers of minesweepers and trawlers operating in the same areas. In addition each convoy had its own escorts, usually French or British destroyers, although Italian and Japanese ships also took part.[451]

The numbers of personnel available after the creation of the RAF are difficult to determine. None of the weekly reports from Port Said and Alexandria contain pilot or observer information. That individual Reports of Flights, which would contain this information, continued to be prepared is confirmed from a handful of surviving examples. From group photographs, taken at the end of the war, the two units each appear to have comprised up to ten pilots and observers each. But in each case only a few familiar faces can be identified.

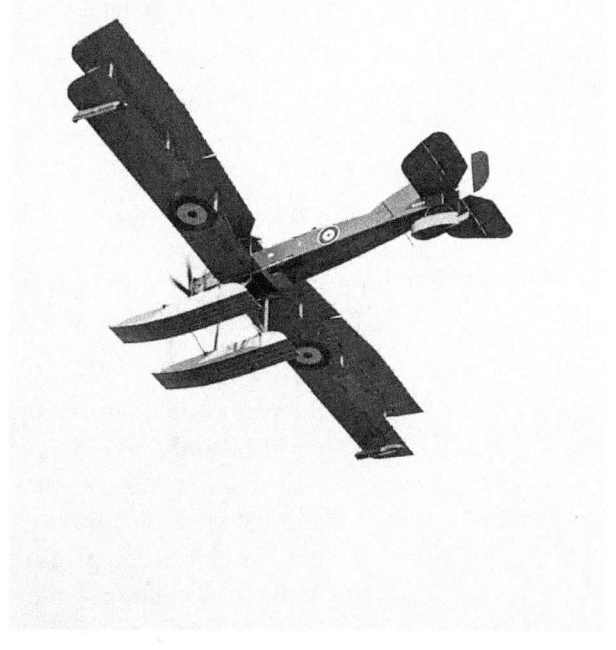

Short 184 Model D, 8020—a survivor. One of three Model D single-seat bombers supplied to the EIESS in early 1917, 8020 flew its first operation for the EIESS on 22 March 1917 and its final one exactly one year later. Unfit for further operational flying it was retained as a 'School Machine' at Port Said until crashing on 15 July 1918. *(EMK)*

The actual number of machines available at any given time is also problematic. The Weekly Reports provide an indication, for example:

> Week Ending, 27 April—Port Said, 12 Shorts (including some familiar from EIESS operations), and 1 Fairey Hamble Baby. Alexandria, 2 Shorts.

> Week Ending, 11 May—Port Said, 1 Short and 1 Sopwith Baby. Alexandria, 1 Short.

> Week Ending, 6 July—Port Said, 3 Shorts and 1 Sopwith Baby. Alexandria, 5 Shorts and one Sopwith Baby.

> Week Ending, 21 September – Port Said, 4 Shorts and 1 Sopwith Baby. Alexandria, 4 Shorts.

> Week Ending, 9 November – Port Said, 4 Shorts and 4 BE2e. Alexandria, 5 Shorts.

Clearly, 27 April was an anomaly, and the actual number of available machines remained low throughout the remainder of the war. The addition of the BE2e at Port Said was a harbinger of change from floatplanes to landplanes that was in hand at the time. The existing seaplane squadron at Port Said was to be replaced by a squadron of DH9 aeroplanes specially fitted for over water work. Although none had been supplied by the end of the war.

In his month's end review for April 1918, Lt Col Risk noted:

> During the latter part of the month, HMS *Empress* was sent away from the station with a full complement of the four best machines, leaving Port Said and Alexandria very short handed, especially in skilled "E" ratings, and two semi-efficient machines, 240 horse-power Sunbeam Short at Port Said and one 240 horse-power Renault at Alexandria, both machines being unsuitable for long submarine patrols. The result of this was that during the last week of the month it was practically impossible to carry out patrols or escorts.[452]

Two reports from the end of June provide a snap-shot of the situation.[453] On 22 June, Lt Col Risk summarised the situation thus:

At Port Said:	13 Short Seaplanes (of which 4 are supposed to be in reserve for City of Oxford for Seaplane Operations).
	3 Baby Seaplanes.
At Alexandria:	10 Short Seaplanes.
	2 Baby Seaplanes.

Of these, there are only 3 Short Seaplanes and 3 Baby Seaplanes ready.

The report of machines available at Port Said engineering base on 29 June details a sad story.

Description	Number	Engine	Type	Condition of Machines, etc.
Short Seaplane	8649	150 Sunbeam	School Machine	Machine being overhauled.
" "	8020	240 Sunbeam	" "	Under repair.
" "	8019	" "	2-Seater Reconnaissance	Engine down. Machines to be Deleted
" "	8022	" "	" "	" "
" "	N1679	260 Sunbeam	" "	Engine down, faulty valves.
" "	N1827	" "	" "	Engine down, faulty valves.
" "	N1597	" "	" "	Engine at Abassia Repair Depot.
" "	N2792	" "	" "	Engine at Abassia Repair Depot.
" "	N2823	" "	" "	Engine at Abassia Repair Depot.
" "	N2811	" "	" "	Engine at Abassia Repair Depot (from m/c 2791). New landing chassis and planes to be fitted.
" "	N2791	" "	" "	Engine (from m/c 2811) ready for fitting in machine.
" "	N2812	" "	" "	Fabric on fuselage being repaired. Engine at Abassia Repair Depot.
" "	N1839	" "	" "	Engine being dismantled from machine for examination.
" "	N2847	" "	" "	New machine being erected.
" "	N2849	" "	" "	New machine, in case.
" "	N1091	240 Renault	" "	Engine down, waiting for spare parts. Machine being overhauled and new under-carriage fitted.
Sopwith Baby	N1128	110 Clerget	Small Bombing Machine	Fuselage to be trued up.
Hamble Baby	N1209	130 "	" " "	Recommended for deletion. Engine suitable for a spare.
Sopwith Baby	N2073	130 "	" " "	Ready.

Lt Col Risk's 22 June report continues.

> As far as possible, I try and arrange for seaplanes to escort convoys, both inward and outward bound, to a distance of at least 40 miles. This is the utmost we can do with the present type machines, owing to the unreliable engines (260 HP Sunbeams) even when fitted with the new type valves.
>
> Three of the Sunbeam-Shorts at Alexandria are fitted with two magnetos only, these being of a condemned type. (N.B. These 260 HP Sunbeams should <u>all</u> be fitted with 4 magnetos.) Also the above three engines are not even fitted with the extra brackets for the other two magnetos, so that it would be impossible to fix them even if they arrived on the station.
>
> The other Short Seaplanes are mostly fitted with Renault-Mercedes engines, 240 HP. This is a very reliable engine, but they have ceased to exist in England and spare parts are very hard to obtain. Also, the machines themselves are very old and "soggy" and will not lift any useful weight.

He concludes by making what amounts to a plea for Felixstowe flying boats to be urgently sent out to Egypt. However, none were available at the time.

U-boat Operations off Egypt

The sinking of the old French cruiser *Amiral Charner* on 8 February 1916 by *U-21* (*Kapitänleutnant* Otto Hersing) sent shock waves through the naval community. Up to that time they had been very casual about the submarine danger, *Raven* had even anchored off the Palestine coast showing all her lights. No more. From this date forward all ships enforced a strict blackout and were escorted by French trawlers, torpedo boats or destroyers. Anti-submarine warfare was in its infancy, but the presence of the escorts would have made the submarine commander's job more difficult and dangerous. Merchantmen still sailed singly and unescorted, not until mid-1917 was an effective Mediterranean convoy system instigated.

The area around Port Said and Alexandria was not a prime hunting ground for the submarines. Hunting was simpler on the convoy routes in the middle of the Mediterranean. However, through 1917 and 1918 U-boats did visit the area. The U and UB-class submarines were armed with torpedoes and an 88 or 105-mm deck gun. The UC-class boats were primarily minelayers, but also were armed with torpedoes and a deck gun. It should be noted that none of these were lost whilst on patrol in the Eastern Mediterranean, except UB-66 as mentioned in Chapter 15.[454]

Submarine Sightings

Only four submarines were sighted and three of these attacked, during the many hours spent on patrol.

8 April 1918

At 08.40 Short N1581 was airborne from Alexandria, piloted by Capt Burling with AM H. Crisp as W/T Operator. The Short had two 65-lb bombs and a Lewis gun for Crisp's use. Twenty minutes later they were over an inbound convoy of eight ships, escorted by the gunboat *Ladybird* and sloop *Amaryllis* which was also flying a kite balloon.

U-boats *UC-34* and *UC-74* were Type UC-II mine laying submarines, capable of carrying 18 mines but also armed with two external torpedo tubes and a single internal stern tube, and an 88mm deck gun. Both were based in Pola from early 1917. Two identical UC-II class boats from Pola are seen here alongside *Amphitrite* an accommodation ship for German submarine crews at Cattaro, in the Adriatic Sea. *(navyphotos.uk)*

The first ship had just been picked up about half a mile from the outer buoy of the North channel when it was attacked by a submarine on its port beam. The track of the torpedo was seen and the periscope of the submarine observed on our starboard bow; the machine then dived at the submarine and dropped its first bomb from 200 ft. which fell on the port quarter of the submarine; the submarine turned sharply to starboard and continued in the opposite direction. The machine then turned sharply to the right and a second bomb dropped which fell just short of the periscope. The periscope disappeared suddenly and nothing more was seen of it. Signals were made to ships and trawlers to drop depth charges while we circled over the position where the submarine was last seen.[455]

The torpedoed ship was the steamer *Bengali* (5684 GRT, T. & J. Brocklebank, Ltd, Liverpool). It sank in a position 14 miles north of Alexandria, all the crew were rescued. The submarine was *UC-34* (*Oberleutnant* Horst Obermüller), based at Pola as part of the *Mittelmeer II Flotilla*. Obermüller sank another ship, the *Vasconia*, on 9 April and returned to Pola. *UC-34* survived to be scuttled outside Pola on 30 October, following surrender of Austro-Hungary.

8 May 1918

Little is known about this sighting and attack. Short N1639 had left Alexandria on a Submarine Patrol, piloted by Lt G.H. Willows with Lt S.C. Howes as observer. During the patrol, they sighted a submarine on the surface some 12 miles distant. Willows dived towards the submarine, which had seen them and quickly submerged. Two 65-lb bombs were dropped on the disturbance made by the diving submarine.

In addition to convoys, the Shorts flew escort to troopships as seen here on 12 October 1918. Top is HMAT *Ormonde* (1917, Orient Steam Navigation Co. Ltd.) and HMT *Kaiser-i-Hind* (1914, P&O). Both modern ships capable of 18 knots they each carried several hundred troops. As valuable targets they were always escorted, often by Japanese destroyers one of which can be seen in the distance under the Short's upper aileron. They were escorted in by two Shorts, N2791 and N2847, from Port Said, crews are unknown. *(EMK)*

The patrol only lasted one and a half hours, so they could not have been too far from Alexandria when sighting the submarine. Two U-boats were operating in the area in early May. *UB-51* was patrolling in the area of Cyprus, but *UC-74* (*Oberleutnant* Hans Adalbert von der Lühe) had been operating a short distance west of Alexandria, and was the probable target of the attack. *UC-74* returned safely to Pola and the *Mittelmeer II Flotilla*, and was to return to the area later in the year.

14 June 1918

This attack created quite a stir at the time, resulting in a claimed sinking and recommendations for decorations for those involved. So, let us take a careful look at the events.

On the morning of 14 June 1918 an inbound fast (15 knot) convoy of Australian troop transports, *Indarra, Canberra*,[456] and three P&O ships *Malwa, Caledonia*, and *Kaiser-i-Hind*, escorted by several Japanese destroyers, was approaching the outer buoy to the swept channel. In the approaches to Alexandria were many more ships, including five ML's, several trawlers and two sloops (one flying a kite balloon). Added to the protection were

two Shorts from Alexandria. At 06.11 Short N2904 took off piloted by Lt G. Waugh with AM T. Saunders W/T Operator, it was heavily loaded with a single 230-lb and two 65-lb bombs. A few minutes later Short N1639, Lt G.H. Willows and AM R.A. Chilman, followed with two more 65-lb bombs. By 06.45 both Shorts were over the convoy.[457]

The convoy had left Marseilles 'with reliefs and leave-expired men,' on 6 June. Arriving at Malta on 9 June for coal, water and supplies, sailing again two days later.

> Convoy was attacked by submarines on the 12th; several depth-charges let off and guns fired by Japanese escort. On the 14th attacked again by submarines outside the swept channel off Alexandria. This was the most severe attack of the series; it was said afterwards by the naval escort that there were three submarines seen attacking. The attack lasted about fifteen minutes, Japanese destroyer-escort firing guns and dropping depth-charges; three aeroplanes from Alexandria dropped depth-charges also, and other vessels from Alexandria were engaged as well. The attack, we heard afterwards, was expected, hence the reason for the aeroplanes and other vessels being out. I personally saw one submarine hit and blown up by aeroplane depth-charges (and Japanese), for the air was full of dirt and debris. No vessel in the convoy was hit. We heard afterwards that at least one submarine was sunk if not two – anyway they were not seen again about the place.[458]

There are similar accounts from the Chief Officer of *Indarra* and a passenger on the ship. The latter, Engineer Lt G.H. Ashton, Royal Indian Marine, is particularly interesting.

> I was on deck at the time, the morning being very calm and clear, when one of your machines sailed over, having spotted one of the enemy's submarines, who had by this time barely his periscope showing and was manoeuvring for position, the submarine's position at the time being about half a mile from the *Indarra* on the port quarter, and to my idea waiting for the ship to swing a little more or, failing that, to get a clear broadside shot at the following ship the *Canberra*. Your machine came over and dropped two bombs which I think caused damage and confusion. Immediately after the explosion from the bombs, the Japanese Destroyer "R" put about and on reaching the spot dropped two depth charges. I personally saw a portion of the submarine lifted clear out of the water from the result of the explosions...

Both Shorts observed a Japanese destroyer attacking with a single depth charge at approximately 06.50. Lt Willows, N1639, reported sighting the 'wash and periscope of a submarine steering approx NE away from the convoy, and about a quarter of a mile away from where the depth charge was seen to explode.' He immediately dived to attack, dropping his first bomb which 'was seen to explode not more than thirty feet ahead of the submarine.' His second bomb failed to explode.

Lt Waugh, N2904, was flying a little higher and saw the first machine drop a single bomb then fire the Very's light signal for 'Enemy submarine beneath me.' Waugh then reported seeing 'what was presumably the swirl of a submarine's propellers.' He then turned toward the swirl dropping the 230-lb bomb with a 2½ second delay fuse, 'which exploded about 70 feet ahead of the swirl.' He was unable to drop his remaining bombs 'owing to the proximity of a Japanese Destroyer "H" which had at once proceeded to the spot.'

Both Shorts remained over the convoy following their attacks. At 07.30, about 20 minutes after his attack, Lt Willows noted, 'Two upheavals were seen in approx position where submarine was last seen.' These were not observed by Lt Waugh. Both Shorts had returned to Alexandria by 09.00.

The Japanese destroyer was identified as "R" by Eng Lt Ashton, who was at sea level, both flyers identified it as "H". Each Japanese ship had a large letter painted in white on its bows to easy identification and reporting.[459] In this case the correct identification was probably R, the IJNS *Katsura* of the 10th Destroyer Division of the 2nd Special Squadron. The possibility of two Japanese destroyers cannot, however, be ruled out as the escort probably comprised eight destroyers. Destroyer H was *Sugi* of the 11th Destroyer Division. Both destroyers were Kaba-class, launched on successive days in February 1915, albeit from different shipyards.

The identity of the U-boat is more troublesome. During June 1918 only one U-boat was sunk in the Mediterranean, the *U-64*. But this boat was lost 17 June between Sardinia and Sicily. I have also been unable to identify any U-boat operating in the Alexandria or Port Said area at the time of the attack.

IJNS Katsura at Brindisi in July 1917. A *Kaba*-class destroyer she had been commissioned in 1915, one 120mm and four 76mm guns and two 450mm torpedo tubes, 30 knots. *(Courtesy Atsushi Yamashita)*

Nonetheless, Lt Col Risk was confident that the attacks had resulted in the sinking of a submarine. In a letter to the Headquarters, Naval Air Force Group, Egypt, dated 21 June, he wrote.

> From the evidence given by these Officers [from *Indarra*] as actual eye witnesses of the whole proceedings, it would appear that a legitimate claim might be advanced as to the actual destruction of this submarine.
>
> There have been previous cases in which bombs have been dropped in very close proximity to a submarine where one was inclined to claim that it had been destroyed. The submarine having been sighted, however, within the next 24 or 48 hours entirely negatived this. In this case, however, there has been no sign of the slightest submarine activity.
>
> Under these circumstances, I have the honour to request that the Pilots and Observers belonging to the Alexandria Seaplane Squadron, which carried out this attack may be credited with the destruction of this submarine. If this claim is allowed I would recommend the following Pilots and Observers for the under mentioned decorations:-
>
> > Lieut. George Waugh, RAF   ) Distinguished Flying
> > Lieut. George H. Willows, RAF) Cross.
> >
> > F-32649. Saunders, Thomas AM   ) Distinguished Flying
> > F-18932. Chilman, Reginald A. AM3) Medal.

Major General W.G.H. Salmond, Commanding Royal Air Force, Middle East, sent a brief message on the same day.

> Please convey my congratulations to the following Officers and men:-
>
> > Lieut. G. Waugh.
> > Lieut. G.H. Willows
> > A.M. W/T. Saunders.
> > A.M. W/T. Chilman.
>
> for their promptitude and action against the enemy submarine June 14th.
>
> (signed) W G H Salmond

The message was read at a parade at the Alexandria base a few days later. In his reply to General Salmond, Lt Col Risk concluded, 'Your message I am sure will encourage all ranks to further their efforts.' The matter of decorations proceeded much slower. Both Lt G. Waugh and Lt G.H. Willows were awarded the Air Force Cross in 1919.[460] There is no evidence that the two Airmen received any recognition.

2 November 1918

The Ottoman Empire signed an armistice on 30 October 1918, Austria-Hungary following in 3 November. The German U-boat flotilla at Pola began sailing for Germany on 30 October, scuttling or wrecking any submarines incapable of making the voyage. Some were still on patrol, including *UC-74* (*Oberleutnant* Hans Schüler) which was operating off the Egyptian coast.

On 2 November, Short N2792 from Port Said reported sighting a submarine, this could only have been *UC-74*. No attack was made, and no further details are available.

The submarine continued its patrol sinking two steamers (*Murcia* and *Surada*) off Port Said on 2 November. A third (*War Roach*) was mined on 4 November, *UC-74* had just laid 9 mines in front of an approaching convoy, the steamer was beached, later to be refloated and repaired. Finally, a small (38 tons) Italian sailing vessel *Stavnos* was sunk on 5 November. *UC-74* was now recalled and ordered to try to reach Germany but, having too little fuel, Schüler took his command to Barcelona where it was interned on 21 November 1918.

The Perils of Patrolling

If the vast majority of patrols were uneventful, engine problems could still make for some exciting moments. Lt Col Risk had complained from the start about the unreliability of the Sunbeam engines. The first engine problem on patrol occurred on 9 January 1918. Flt Lt G.D. Smith with LM Phillips had taken Short N1827 off from Port Said on an submarine patrol, and failed to return. Two subsequent searches, by Flt Lt E.M. King with Lt Ferguson on Short N1090, failed to locate the missing Short.

Smith provides his usual colourful story. 'My observer and I had been out on a rather long flight, through a bumpy atmosphere, and the constant jar and vibration put too great a strain on one of the copper petrol pipes, snapping it in two. Petrol fumes inhaled have just about the same effect as alcohol taken internally.' Feeling increasingly light headed and hilarious they headed towards Port Said, but the fuel loss forced them down in brackish Lake Menzala some thirty miles short. The water was only 18 inches deep, but the underlying mud was deep and sticky. Forced to wade ashore, slipping and sliding they were soon caked in mud. 'We were getting close to exhaustion when two Senegalese soldiers, discovering our plight, stripped off and rushed out to help us. We leaned on them over the last hundred yards. It took us about two days to get all that mud off.' He was able to telegraph the base from Damietta and the Short was eventually towed back to Port Said.

But on 11 April, Lt C.K.C. Dagg had a different problem, one that nearly cost his life. Flying Sopwith Baby N1126 from Alexandria conducting a convoy escort…

He noticed a disturbance in the water some distance away. He steered for this, although he knew he would not be able to get back to his base owing to a lack of petrol. He landed at 4.10 pm 4 miles north of the convoy and fired Very's lights. He was not picked up until 11 am the next morning by HMS *Rowan* [armed boarding steamer], after the seaplane had sunk. Lieut. Dagg keeping himself afloat by cutting the empty petrol tank adrift from the machine. He was in the water hanging on the tank for 17 hours.[461]

Captain King had his own problems with the Sunbeam engines. On 2 June he set out on a convoy escort with Lt Ferguson on Short N2811.[462] Reaching 150 feet the engine failed, King made a heavy landing and the undercarriage collapsed. The two airmen were rescued, and the wreckage towed in by *ML240*, although the machine was written off. His last flight before proceeding on extended leave home to New Zealand was on 16 July, on Short N2847 (Sunbeam 260-hp Maori). Together with Lt Dodds as observer, they left Port Said to escort a convoy. Forty minutes into the flight the engine failed. They landed safely and 'were towed in 12 miles to P.S. by trawler.'

Capt King was not the only sufferer from Sunbeam failure, as this account from Lt G.F. Hyams shows. Lt Hyams had spent some time flying Sopwith Babys from Westgate and Hornsea Mere before being posted to Egypt.

> At the beginning of August 1918, I went out to Alexandria in Egypt. I was flying Shorts and Sopwith Babys there. Our job there was to do anti-submarine patrols, and to escort incoming and outgoing convoys of troops, which were either arriving or going home. After I'd been there a short time, I went to Port Said, where we were doing exactly the same kind of work, with Sopwith Babies and Shorts, escorting vessels in and out, and doing reconnaissance patrols out at sea. In Alexandria, we had quite a good mess and living quarters. They were built on top of a flat roof of nitrate storage sheds in the harbour at Alexandria. At Port Said, things weren't nearly so luxurious; we had horrible cabins, built on the shore, and they were full of bugs most of the time.
>
> On my 20th birthday, 31st October 1918, I went out early in the morning with an observer in a Short to reconnoitre before a big convoy sailed for Salonika, and when we got about 80 miles out, the engine conked out, and we had to come down. I managed to get the engine going again, but by that time it was too rough to take-off. We had to get out, sit on the tip of the floats, and just wait. It midday when we came down, and we were out there all night. The sea got very rough indeed, and we were very seasick. During the night, we saw a Short circling all around, but it never spotted us! We had Very

flares, and I fired a lot of those. I sent a pigeon out, too. We used to carry pigeons; they were very useful, and used to take the messages straight back to base. Early the following morning, the sea became even rougher, and the machine suddenly sank by the tail and whipped over on its back, with just the bottom of the floats above the water. We scrambled out and sat on these things for a hell of a long time. That midday, after we'd given up hope of ever seeing anything again, we saw smoke on the horizon. I stood up on the bottom of the floats, with the other bloke holding me, and waved my shirt. By sheer luck, one of the lookouts spotted us, and eventually we were picked up by the Australian destroyer *Swan*, taken aboard, and finished up in Salonika. We waited there for a day or two, and then returned to Alexandria on an empty troop ship.[463]

He had been flying Short N2823 (Sunbeam 260-hp Maori), with 2Lt C.F. Standish as observer. N2823 had failed to return once before, on 23 August with Lt G.W. Morey and 2Lt L.T. Harris, it ran out of fuel returning from a convoy. The undercarriage was damaged, possibly when the machine was beached at El Arish. It was dismantled and returned to Port Said by rail. This time Hyams had damaged the tail float on landing. After restarting the engine they taxied SW for 1¾ hours at 4 knots, only stopping when the engine overheated. Trying to restart again the compressed air bottle was empty. The remainder of the story is told above.

These are just a few of the stories that could be told. But, since the area over which they patrolled was also patrolled by trawlers and other ships, no one appears to have been lost to the sea.

Short N2812 (260hp Sunbeam Maori) at the Port Said advance site. The original caption gives a date of 13 November 1918, adding that the pilot was Lt G.F. Hyams about to take Eng Lt Cdr Bloomfield from HMAS *Swan* for a flight. Swan had plucked Hyams and his observer 2Lt C.F. Standish from the sea on 1 November 1918 following a miserable night clinging the a float from their wrecked Short.

What Might Have Been

On 17 October 1918 Lt Col Risk wrote to the GOC, RAF, Middle East concerning some intelligence he had just received. This had come from Lt Cdr A.D. Cochrane, RN, who had recently arrived in Port Said having escaped from Yozgad Prison Camp.[464] He and six companions had escaped 7/8 August 1918 and after an overland march reached the Anatolian coast, close to Silifke, twenty three days later. Here they were stuck until they were able to steal a decrepit motor boat and sail for Cyprus on 12 September. Eventually they were transferred to Port Said where Cochrane was able to pass on his intelligence. During their period on the coast they had located a hostile seaplane station. From a map included with the letter, it appears to have been located close to modern Narlıkuyu. In 1918 it was a sheltered bay close to the coast road, with a good beach, a few huts and a supply of fresh water.[465]

BE2e B4596 was delivered to the Port Said Squadron (269 Squadron RAF) in early November 1918. It flew just one patrol of 1 hour 53 minutes on 9 November 1918. *(Paul Hare Collection)*

Risk added details of the latest visit to Cyprus, probably by a Friedrichshafen FF33 or FF49 from the *Wasserfliegerabteilung* station based at Mersina. On 12 October a seaplane had reconnoitred Karenina, Nicosia and Famagusta between 09.00 and 10.00, then returned to the north. This may also have been the last visit as the Mersina detachment was withdrawn at the end of October.

On 26 October Risk was informed:

> It may shortly be necessary to send your F.3 boat[466] to report to Palestine Brigade at Haifa or Beirut, in order to convoy 2 Bristol Fighters to Famagusta, Cyprus. Please inform me by wire what amount of petrol, and lubricating oil should be sent by sea to Cyprus for the return journey of the F.3 boat, and also to enable her if required to take part in a raid on the Turkish Seaplane Base referred to in your letter dated 17th instant.

Whilst further correspondence followed, the project was cancelled by RAF Headquarters, Middle East, on 1 November 1918. The Turkish Armistice had made the projected raid unnecessary.

Not With a Bang...

Captains King and Smith, and others, were due for extended home leave. A recent Air Ministry Order permitted Colonial Leave for long serving officers of the air services. Travel time was not included in the leave which only commenced when the officer reached his home. In Capt King's case the outward journey to New Zealand via Australia took almost six weeks. Leaving Suez on 19 July for Australia, aboard *Port Darwin*,[467] he travelled in company with Capt H.V. Worrall, a native of Melbourne. From Sydney he travelled by Union Steam Ship Company to Wellington, arriving at his family home in New Plymouth by mail train on 27 August.

Shortly before leaving Port Said, King was informed he had been awarded the Distinguished Flying Cross. He was to wait almost three years for the investiture at Wellington, on 22 March 1921, from His Excellency the Governor General of New Zealand, Admiral of the Fleet John Rushworth Jellicoe, 1st Earl Jellicoe. His DFC ribbon has the later (post 1919) diagonal stripes. Photographs taken during his leave (and during WW2) show him wearing the original horizontal striped ribbon, correct for the date of his award, which he must have acquired prior to leaving Egypt.

Before returning to Egypt he married Dorothy Saxton at St Mary's Church, New Plymouth on 17 September. They had become engaged in May 1914. They had a few weeks together before he had to return to Egypt. The exact date of his departure is not known, but the Armistice probably occurred whilst he was in transit. His first flight after returning to Port Said was on 12 December 1918.

Captain Smith had to travel to the west coast of North America. When his grant of leave came through he was informed that there was a ship, *War Dame*,[468] leaving for Boston two days later. He saw the captain of the ship and arranged to sail as a passenger. *War Dame* was no ocean greyhound and the passage took 28 days from Gibraltar to Boston, probably in a convoy. From Boston by train to Toronto then on to Vancouver, taking advantage of cheaper rail fares for serving soldiers in Canada. From Vancouver he took the ferry to Seattle catching the SS *Governor*, Pacific Coast Steamship Co, down the coast to San Francisco. Here Guy stepped off the ship 'into the arms of my father and sisters.' San Francisco newspapers took note of his arrival and he gave

Felixstowe F.3 probably N4360 flying the mail run from Port Said to Malta in the early post-war period. The ship just behind the port wing is the French *garde-côtes cuirassé Requin*, she left Port Said on 17 December 1918 and returned to Toulon. *Requin* was decommissioned late in 1919 and scrapped in 1921. (EMK)

several interviews, one in the *San Francisco Chronicle* being dated 7 August 1918. The memoir concludes at this point, but his Service Records show that he was back in Egypt by the end of February 1919.

Back at base, after the Armistice all the urgency must have gone from their lives, most of the officers and men counting down their days to demobilisation. Flying continued albeit at a reduced level.

A little out of the ordinary was a flight on 3 March 1919. Five Shorts escorted into Port Said the battlecruiser *New Zealand* with Lord Jellicoe aboard. Capt King was flying N2648 with Major Holmes, CO 269 Squadron, as passenger. The battle cruiser was taking Jellicoe to Australia and New Zealand to review the naval defences of the southern Dominions. It was possibly for this flight Eliot had *Kia-Ora* painted over the large fin surface of his favourite Short.[469]

Tensions had been rising in Egypt following the end of the war. A popular movement to gain full independence was gaining strength. On 8 March 1919 the British authorities arrested the leaders and exiled them to Malta. This led to widespread disturbances between 15–31 March. The base at Port Said was put on alert, and Capt King flew a reconnaissance on 16 March, with 2Lt Milne on *Kia-Ora*, around the Port Said and Damietta area. They found no disturbances to report. Eventually, the British declared limited independence to Egypt on 28 February 1922. The protectorate was abolished and the current Sultan of Egypt, Ahmed Fuad I, appointed King.[470] However, independence was illusionary as Britain retained control of the Canal Zone, Sudan, Egypt's external affairs, police forces, army, and railways.

As the rising calmed down, a return to routine and joy flights followed. Through April both Capts King and Smith were engaged flying the bi-weekly mail run to Alexandria. The mail was transferred to a Felixstowe F.3 from Kalafrana, Malta, for onward transmission and incoming mail flown to Port Said the following day.

Demobilisation was on the horizon, but the Sunbeam engine had one more trick to pull on Capt King. On April 23, he left Port Said with 2Lt Lummins as passenger, flying in his favourite Short N2648. 'Stopped at Damietta Light House & then had engine trouble 25 miles further on. Came down & fixed engine but waves damaged elevator so stayed night at Coastguards & were towed home next day by a tug.'

Following repairs, his final flight in Egypt was on N2648, on 19 May, was with Capt H.A. Buchanan-Wollaston, RN, who had recently been relieved as SNO Red Sea Patrol, as passenger. Eliot would have met the Captain whilst serving on *City of Oxford* in the Red Sea. Unfortunately, it was a one way trip down the Suez Canal to Ismailia. 'Floats both damaged on landing. ML towed m/c home on 21st after having brought new floats down to Ismailia.' The Captain probably took the train back to Port Said. It was not the way Captain King would have wished to end his flying career.

Captain Eliot Millar King, RAF, embarked on HMT *Kursk* at Suez on 10 July 1919. The ship was a troopship and his travel would not be as comfortable as a year earlier, but everybody was heading home and minor discomforts would be willingly endured. In Sydney, he again boarded the *Port Darwin* this time bound for New Zealand, and arrived in New Plymouth on 2 September 1919. He had been placed on the Unemployed List of the RAF from 22 August 1919.

Captain Guy Douglas Smith also left Port Said in May 1919. He spent some time in London, where he received his DSC from King George V at Buckingham Palace on Thursday 31 July 1919. He sailed from Liverpool on SS *Megantic* 12 Sept 1919, bound for Montreal. Then overland to Oakland, California. He was placed on the Unemployed List of the RAF from 28 September 1919.

There was also no longer any place for the Naval Wing, RAF, Egypt. The Alexandria Seaplane Squadron and Port Said Seaplane Squadron, having been combined, at least on paper, as 269 Squadron, RAF, was disbanded on 15 November 1919. Thus, effectively ending five years of floatplane operations— *l'escadrille de Port-Saïd*, East Indies and Egypt Seaplane Squadron, and RAF—in Egypt.

Short N2648 (260 hp Sunbeam Maori), Kia Ora, at the Port Said advance site. The de Lesseps statue is visible over the front of the floats. This was Flt Lt King's 'personal' Short after he returned from his long leave in New Zealand, he titled this image 'Arabs getting my Bus in.' *(EMK)*

Afterword

From the beginning Naval Aviation has had one huge advantage over land based aviation—its mobility. But that mobility was hard won and only slowly developed. True mobility required the development of the aircraft carrier, with a flight deck from which landplanes could take off and land upon. At the end of the Great War the first aircraft carriers were close to entering service. But that is another story.

The Middle East in World War 1 was a sideshow, but one that ultimately changed the world. Aviation in the Middle East developed both on land and upon the sea. Its geography of extended sea coasts, with transportation and population centres close to the coast, provided the ideal arena for naval aviation to demonstrate the advantages of mobility. By employing converted merchant ships, transporting small numbers of floatplanes, it was able to fly overland to look behind, and attack, the enemy far beyond the range of land based aeroplanes.

That said, it was only the fact that the area was considered a sideshow that permitted the obsolescent floatplanes to carry out the duties requested of them. Had they had to face modern aeroplanes in large numbers they would quickly have been forced from the sky. The few encounters they had with German designed two-seaters showed how vulnerable the floatplanes were.

The Nieuport, Short and Sopwith floatplanes had their origins in pre-war designs. The Nieuport and Sopwith were indeed developments of civilian sport planes. The Short 184 was the result of steady development dating back to 1912. All were designed for European conditions, their engines in particular struggled with the temperatures of a Middle Eastern summer. That they worked at all was due mainly to the unceasing work of the largely anonymous craftsmen and mechanics responsible for their care and maintenance.

The seaplane carriers were all conversions of civilian ships, cargo carriers or passenger ferries. The merchant ship conversions, *Aenne Rickmers/Anne, Rabenfels/Raven, Campinas* and *City of Oxford*, were makeshift. They were all slow, 10 knots on a good day, but were intended for long voyages. The floatplanes were carried on cargo hold hatch covers with, at best, canvas shelters to protect from the sun, wind, and seas. They were lifted on and off the ships using cargo derricks, intended to haul heavy loads of bulk cargo rather than delicate wood and fabric machines. The two converted passenger ferries, *Ben-my-Chree* and *Empress*, were provided with large metal hangars and derricks better adapted to their purpose. They were fast at 20 knots or more, but short ranged. Pre-war their coal bunkers could be recharged daily, in wartime service coal was stowed wherever space could be found, still they rarely sailed for more than three or four days at a time.

The pilots and observers were an equally ill assorted group. Some of the senior pilots had extensive pre-war experience. Most pilots were however wartime trainees. Training was, by modern standards, elementary and short. Eliot King, for example, in November 1915 was sent solo after less than two hours dual training and to a North Sea patrol squadron at Dundee with just 16½ hours total flying time. He had almost 138 hours when posted to Port Said in September 1916. Just six months later, in June 1916, RNAS training was becoming more organized, and FSL Popham was not sent solo until he had almost 10 hours dual. Even so, he was flying patrols from Killingholme with barely 50 hours flying time, and had 91 hours 40 minutes when posted to the EIESS.

For reasons explained earlier most of the observers in *l'escadrille de Port-Saïd* and the EIESS were Army officers, and nominally RFC. These were all locally trained and, until Childers established his observers school at Port Said, this was strictly on-the-job experience. The EIESS had a few naval observers, notably Sub Lt J.L. Kerry, RNVR, who had been trained as an observer at White City, Eastchurch, and Roehampton prior to his posting to Egypt. He was appointed Observer Lieutenant, RNAS, in April 1917.

Statistics quickly become wearisome, but a few numbers should be recorded.

During its time at Port Said *l'escadrille de Port-Saïd*, by the count of *Lieutenant de vaisseau* de l'Escaille, made 1072 flights. Of these over 300 were from the seaplane carriers *Aenne Rickmers/Anne, Rabenfels/Raven*, and *Campinas*. De l'Escaille also estimated the Nieuports spent over 500 hours over enemy territory.

The East Indies and Egypt Seaplane Squadron made around 550 flights from its seaplane carriers, *Ben-my-Chree, Anne, Raven, Empress*, and *City of Oxford*. These flights represent approximately 1000 hours over enemy territory.

Floatplanes damaged or destroyed could, in time, be rebuilt or replaced. Men lost in action could not.

L'ESCADRILLE DE PORT-SAÏD—RÔLE D'HONNEUR

Tué en action	QM Jean-Marie Le Gall, 28/1/15.
	2Lt Basil G.N.B. Partridge, 28/1/15.
	2Lt Horace M.C. Ledger, 22/12/15.
Prisonnier de guerre	QM Georges Marius Etienne Trouillet, 10/10/15.
	2Lt Sir Robert J. Paul, 10/10/15.
	LV Louis Marie J. Barthélémy de Saizieu, 22/12/15.

EAST INDIES AND EGYPT SEAPLANE SQUADRON—ROLE OF HONOUR

Killed in Action	Flt Lt J.T. Bankes-Price, 17/9/16.
	Lt N.W. Stewart, 23/1/17.
	Flt Lt M.C. Wood, 11/10/17.
Prisoner of War	Flt Cdr G.B. Dacre, 26/8/16.
	Flt Lt A.J. Nightingale, 2/12/16.
	Lt P.M. Woodland, 2/12/16.
	Flt Cdr A.W. Clemson, 11/10/17.
	Lt E.A. Newton, 11/10/17.
	Flt Lt C.G. Bronson, 28/1/18.
	Lt L.H. Pakenham-Walsh, 28/1/18.

Observer 2Lt K.L. Williams, although writing of his time with *l'escadrille de Port-Saïd* also spent much time in the air with the East Indies and Egypt Seaplane Squadron, adds an appropriate coda.

> On looking back one regrets the old 'buses and the fun that we used to get out of them; but when one remembers the frequent occasions on which the engine tried to cut out 20 or 30 miles inland, I think that we would have preferred the modern machine and engine, if we had had our choice.

* * *

APPENDIX 1

For the Family

I have been able to quote from several memoirs, articles, diaries, etc, in the main text, but the following were only intended for family consumption.

* * *

I am very grateful to James and Eileen Kerry (the Kerry Family Collection) for permission to reproduce a letter home written by Sub Lt James Leslie Kerry recounting his experiences on 26 March 1917 in the Laccadive Islands, and the photographs he mentions in the letter.

Kerry's letter is carefully worded due to censorship requirements. However, it describes his experiences with pilot FSL Guy Duncan Smith, flying Short 184, 8021, from *Raven* over Chetlat island in the Laccadives *(see Chapter 14)*. The letter has been transcribed without any editing.

Sub Lt James Leslie Kerry at the island base.

March 28th / 17

Dear Ones All,

Part of my letter to you this mail can be written by means of carbon paper—& part cannot! I want to tell you something about a very interesting experience which came my way Monday last. To save having to write the account twice, I am making two copies at one writing for wielding a pencil becomes more of an effort every day now as we proceed nearer the equator.

About 11.30 am on Monday a pilot & myself were ordered to fly in a big seaplane about 35 miles to a little coral island. By means of a compass we found our way without any trouble & we found the little place to be packed with palm trees & evidently populated. The island was, of course, surrounded by a coral reef on the outside edge of which the big swell was breaking. Inside the reef was a delightful looking lagoon with the green water perfectly smooth. We saw nothing likely to affect the war at the island. So we turned about and made for the ship.

Just as we turned the engine began to overheat & black smoke began to show itself. There was only one thing to do we did it—alighted. We came down about 5 miles from the island & taxied the machine slowly in that direction. We managed to cross the reef without trouble & as the engine continued to run slowly, we crossed the lagoon & approached the shore. As we drew near a canoe containing 12 dusky men put off from the shore & came to meet us. We did not know whether they were friendly or no, so I got the pilot to wave his white handkerchief, while I loaded the machine gun! Then I took a photo so as to have some sort of record in case we were made prisoners! When the canoe approached quite close to us, the occupants all stood up & salaamed violently. That was a good start, so we waved our hands & shouted "Salaam"  in return. By this time, we were about 50 yards from the shore & a dozen or more natives swam out some bringing with them some freshly cut cocoa-nuts which they had opened for us, and we were mighty glad of the nice refreshing drink.

The natives were not in the least frightened of the machine, and they clambered on to the floats & thoroughly inspected the machine & us. Presently the chief arrived & we managed to make him understand that we wanted an anchor. This was set for us & we anchored the machine about 20 yards from the shore. We examined the engine & found it very much overheated so we determined to let it cool down & try it again later on.

In the machine we had some water bottles, ships biscuits, & a tin of bully beef. I also had some Horlicks malted milk tablets, some beef lozenges, & a packet of chocolate. A crowd of about 200 natives had collected on the beach & were jabbering hard, but the shade of the palm trees looked so inviting that we decided to have lunch ashore. Accordingly, we pushed all the natives off the machine & made them take us ashore in a canoe. When we set foot on the beach they all came round us & we couldn't move. However, we pushed our way under the palm trees & sat down on the sand & began to open our provisions. But before we had time to start they produced two tiny little stools & we two sat down side by side & began our lunch before an audience of wonder-struck natives. They packed in all round us, some of them even climbed adjoining trees to get a view of us. Can you picture the scene? It is the most amazing lunch that I have ever partaken. We both had our flying helmets and goggles on & very dirty clothes.

Presently a dear old man appeared who was evidently the [chief] for they fetched a mat for him & he squatted down one yard in front of us solemnly watching us. Then they fetched us 6 eggs some water, more coconuts, & a syrupy kind of drink, all of which we enjoyed, & in return we gave them bully beef, biscuits & a few Horlicks tablets. The old chief liked the latter immensely! All this time these good fellows were chattering about us and chewing beetle nut which had the effect of making their tongues & teeth quite red. Presently we finished our repast & indicated we wanted to return to the plane, but no, they would not let us go. We had to follow the chief. About 100 yards further on in an opening under the palm trees was a stone building, the stone being coral-rock, and the roofing the leaves of the coconut palm trees plaited together. Inside was a plain room with two forms and a rough desk. We were introduced to somebody who salaamed & muttered something about "Koran" so we took him to be the priest & accordingly shook hands with him. He led us to a corner of the room where was an alms box & when we had made our contribution we were allowed to proceed. Back we went to the beach close to the seaplane, the crowd following. We boarded the machine & ran the engine but it was no good & we had to give it up as a bad job. We intimated to the natives that we wanted palm leaves to protect parts of the machine from the sun & these they got for us & proceeded to lash them on just as we showed them.

We knew that we should have to wait a few hours while the ship came to look for us, so back we went to the shore again & this time I took an aerial camera. I took a photo of the pilot in the centre of a group of natives, and he did a similar favour for me, & fortunately both these photos have come out splendidly, which is very good considering that any picture taken under a distance of 40 yards is out of focus with an aerial camera.

Presently there was much excitement for a number of natives came running towards us & pointing upwards. We guessed what it was. Sure enough there was another of our planes coming to look for us. She soon spotted our machine & returned to the ship. Then we retired under a little boat shed. We two sat on the gun'le, & the crowd gathered all round us waiting for the next thing. They understood no English, but the priest produced a piece of paper and we wrote on it thanking the inhabitants of the island for their kindness and signing our names. In return the priest wrote some mysterious stuff on the other half of the papers & his writing I have kept. Then I sang them a verse of "Rule-Britannia." Oh, you should have seen their expression. They were absolutely nonplussed! As they seemed to appreciate my effort, I switched on to "Married to a Mermaid" as an encore.[471] I have never sung to such a strange or big audience before & this time nobody burst out laughing. I thought you might like to know this. A lecture on the use of the camera followed, & the speed of the shutter somewhat amazed them. By this time, we could see our ships approaching so we concluded our entertainment by giving then a lesson in English but they were bad pupils or we were poor teachers.

They showed us how they worked with the cocoa-nut palms. How they climbed the trees like monkeys, how they obtained the nut itself, how they beat out the fibre on a stone & how they twisted the fibre up into ropes. They took us into one of their boat building sheds but they would not allow us into the village proper & we saw no females. Evidently the latter were not curious!

Now it was time to say good-bye so they knocked down more cocoa-nuts for us & patted us on the back. We shook them by the hand & they salaamed very properly. So we took our leave amidst much shouting & waving. Now don't you think that was an interesting experience?

Fortunately, the photos which I took came out splendidly & I will send you prints as soon as possible. I have already made the prints but I can't send them yet. They will make splendid souvenirs & are quite unique, I should think.

I am afraid I have only been able to give you a very rough idea of the events of the day but I hope it will interest you. You would have been charmed with the beauty of the scenery & the novelty & animation of the whole affair. And just to think that all this is to do with the great war!. Fortunately, the natives did not attempt to intern us or in any way molest us. They were perfectly honest & there was not the slightest attempt of pilfering. They did all they possibly could to befriend us & seemed quite sorry when we went. One old man patted me on the back and presented me with a prettily marked shell for which I shook him violently by the hand.

Well that is about all my story. You will be able to realize the surroundings more when you see the pictures, but when that will be, I can't say, except— as soon as possible.

Yours
Leslie.

The second 'For the Family' story was written post-war by Eliot Millar King. I express my grateful thanks to Adrienne Tatham, Eliot's youngest daughter, for permission to reproduce it.

AN IDEA OF A SEAPLANE RAID

by Eliot Millar King

At sea again and on the way to some 'stunt' or other! The burning question amongst the pilots and observers is 'Where is it to be?' Once well away from the harbour the C.O. calls us into his cabin and gives each one his instructions for the forthcoming flight. Four machines with such and such pilot and observer, will leave the ship at, say 4 am the ship having previously stopped at say, 20 miles off the hostile coast—and attack the ……… sheds at ……… We leave the cabin after a short discussion concerning the height to fly, in what order, etc, and prepare for dinner.

Everyone turns in early that night with instructions to be called at 3 am. The next one knows is a voice saying 'It's three o'clock, sir.' A grunt of 'All right, quartermaster' is the response and one is soon back in the Land of Forgetfulness. Again comes that irritating cry, 'It's three o'clock, sir!' It's beastly dark and at length getting up all one can see is the brilliant phosphorescence of the water. Somehow we find our own cabins—having slept on the deck outside and no lights allowed of course—and struggle into some clothes and then make tracks below to the wardroom where hot coffee is waiting. After that life seems more cheery and now all hasten aft to the hanger.

The air mechanics have been called some half hour previously and have already, under the supervision of the spare pilot, got one machine out of the hangar onto the quarter deck, where the wings are opened and the machine hooked onto the crane all ready to be hoisted over the side. The pilot comes aft, climbs in and starts up his engine to warm it up—(being water cooled)—the other pilots do the same with theirs inside the steel hanger. You know how loud one motor sounds inside a tin garage—just imagine four big aeroplane engines—each over 200hp and *without* silencers—running together and full out, and you will have an idea of the row inside the hanger for the next 10 minutes. The engines being all right they are switched off and now one has to wait awhile for the ship to get to the appointed spot. Dawn is now approaching and presently one can distinguish the hills of the coastline silhouetted against the sky, many miles away.

Soon a dark smudge shows up ahead, but it is no hostile submarine but the escort meeting us at the rendezvous. Arrived there the engine gong sounds, the ship comes to and immediately all is once more activity in the hanger. The crane suspending the first seaplane is swung outboard, the machine lowered onto the water and unhitched, the pilot starts up his engine and taxis about waiting for the others to join him. These do not take long and one after the other are quickly hoisted over and the last machines gone, the ship immediately gets up speed and recommences her old circling tactics, the destroyer escort meanwhile describing greater circles round her. Meantime the machines have taxied into a line abreast and as soon as this position is secured, the leader fires a light and he immediately gets up speed preparatory to rising off the water. The others following in their proper order and soon after a good deal of bumping on the slight swell due to the big load of bombs carried, all the machines are safely off and circling round to obtain a bit of height. The pilots all keep a watchful eye on the leader and presently as he changes course to the coastline, all the others do likewise and are soon swung out in a long line. The air is quite clear and now the sun is rising and showing up distinctly our objective—a big town—and very beautiful it looks from above too—all the houses with their red tiled roofs and white walls standing out in relief against the green vegetation which is here very plentiful, and then in the background rises the famous Lebanon range of mountains, here and there topped with snow.

Soon a column of smoke rising from the beach is discernable and others also appear at intervals—they are signals warning the enemy of the presence of hostile ships, etc. At length having arrived at our appointed target the first machine flies directly over it, releases some bombs, wheels and comes around again to repeat the process—the other machines doing likewise. One can see the bombs leave the machine and follow them with the eye until they hit, and there is a great explosion and the sound does not reach us for some considerable time after it has actually occurred, and then one feels the slight upward bump from the concussion. The bombing is pretty accurate as the machines have come down low purposely so as to make sure of the target, and there is not supposed to be any A.A. guns in the locality so only rifle fire is likely to be met with.

However, the raid must have been a complete surprise to the enemy as even rifle fire is very feeble and does no damage.

After dropping all the bombs and taking photos of their bursting and their work on the store sheds, etc. we turn back, our work finished, then fly back to the ship. At first we cannot see her owing to clouds being between us, so it is necessary to steer solely by compass for half an hour or so, and then coming down low, we see her, and one after the other, the machines come down and alight on the water near her, taxi up, hook on to the derrick and are hoisted inboard again. With the last 'bus in, the ship immediately gets under way again and is off back to port. After reporting their respective results, the pilots and observers have a good clean up and then brekker.

Next morning as we enter the harbour, there is much speculation as to whether a mail boat has arrived during our absence, even though everyone knew beforehand that she was not due for perhaps a week. And we away for how long? One day, two days, or perhaps three and yet it feels long and lots of things *might* have happened meanwhile. At any rate that is how it appears to us. And now, in our various messes, this raid will be the subject of discussion for weeks to come!

'What did you do in the Great War, Daddy'?

Finis (and a good job too!)

The story teller and the photographer.
Flight Sub-Lieutenant Eliot Millar King at RNAS Dundee in 1916 with Kodak in hand and Short 827 in background.

* * *

APPENDIX 2

Samson's Serials for EIESS Floatplanes

Whilst in command of the EIESS Samson made many improvements in the way the squadron operated and was administered. One that was not so successful were what I call Samson's Serials.

In *Ben-my-Chree*'s Log Books and Flight Reports some of the floatplanes were identified with simple serials, such as S.4 and B.3. The S stood for Short 184 and the B for Sopwith Schneider and Baby. They appear to have been used only on *Ben-my-Chree* and were dropped when Samson left the squadron. They can be seen, painted small, on the white part of the rudder marking of some machines.

For the Short 184 they ranged from S.1 to at least S.7, and for the Sopwiths B.1 to at least B.10. Identifying the actual serials represented remains a work in progress, it is not even clear if all the Samson Serials were actually allocated. Positive identification can only be made if the serial can be seen in a photograph, otherwise it is a matter of comparing written sources and making a judgement...

The use of n/k below indicates a paper record of use, but no identifiable serial.

Short 184

	S.1	n/k
	S.2	850 (*Ben-my-Chree* log book/Flight Report and Photo)
	S.3	8082 (*Ben-my-Chree* log book/Flight Report)
	S.4	8054 (Photo)
	S.5	8090—the 'Cut Short' (Photo)
	S.6	No paper record of use.
	S.7	8372 (*Ben-my-Chree* log book/Flight Report)

Sopwith Schneider and Baby

	B.1	3789 (*Ben-my-Chree* log book/Flight Report)
	B.2	n/k
	B.3	3790 (*Ben-my-Chree* log book/Flight Report and Photo)
	B.4	n/k
	B.5	8189 (*Ben-my-Chree* log book/Flight Report and Photo)
	B.6	No paper record of use.
	B.7	No paper record of use.
	B.8	Probably 8136 (Flight Reports).
	B.9	No paper record of use.
	B.10	Probably 8135 (Flight Reports).

APPENDIX 3

Nieuport *Hydroavions* serving with *l'escadrille de Port-Saïd*, 1914–1916

The following list and table of known *l'Aéronautique maritime* Nieuport Hydros has been put together with assistance from Michel Bénichou, Bernard Klaeylé and ARDHAN publications. However, any errors are the author's alone. Note: The following is an edited version of a Table that was published in *Cross and Cockade International*, V52N4, 2021. All dates are DD/MM/YY.

Serial	Type	Engine	Comments
N.3*bis*	VIH	100hp Gnome Double Ω	Arrived at the end of December. Found unsuitable for service and returned to France early in 1915. Possibly used to convert de Saizieu and Cintré to floatplanes. One crashed during training.
N.5	VIH		
N.7	XH Monoplace	80hp Le Rhône	Arrived Port Said on *Foudre* 30/11/14. Rebuilt at Port Said, possibly briefly used for training. Returned to France in March 1915.
N.11	XH	80hp Le Rhône	Arrived Port Said on *Foudre* 30/11/14. Crashed at sea close to west end of Lake Bardawil 4.12.15, engine and instruments were salvaged.
N.12	XH	80hp Le Rhône	Arriv ed Port Said on *Foudre* 30/11/14. Wrecked landing at Port Said 13/2/15. Parts were used to repair N.16, the remainder was returned to France in March 1915.
N.13	XH	80hp Clerget	Arrived Port Said on *Foudre* 30/11/14. Aboard *Minerva* 12/14 in the Gulf of Aqaba. 31/12/14, forced landed in Wadi Araba with engine trouble. Pilot *QM* Grall and observer Capt Stirling walked to coast and were rescued.
N.14	XH	80hp Le Rhône	Arrived Port Said on *Foudre* 30/11/14. 27/1/15, on recce from Port Said, forced down in sea by engine failure, crew (*QM* Le Gall and Lt Partridge) swam ashore and were killed by own forces. Nieuport recovered and repaired. 19/8/15, damaged in Gulf of Alexandretta. 10/10/15, Forced down in desert with engine trouble near Beersheba, *SM* Trouillet and Lt Paul PoW.

N.15	XH	80hp Clerget	Arrived Port Said on 3/1/15. 11/4/15, engine trouble returning from Beersheba, reached coast and was recovered. Damaged floats and controls June 1915, rebuilt. By Sept 1915 it was being used only for local patrols and training. To Malta 3/5/16 on *Anne*. To Argostoli May 1916.
N.16	XH	80hp Clerget	Arrived Port Said on 3/1/15. 16/2/15, capsized off Gaza, rebuilt with parts from N12. 21 May and 29 June 1915, damaged again. Sept 1915, being used only for local patrols and training. Converted at Port Said to 'standard' controls, crashed and destroyed during testing Nov 1915.
N.17	XH	80hp Clerget	Arrived Port Said on 3/1/15. 21/12/15, force landed near Beersheba with punctured fuel tank, after engine trouble, LV de Saizieu PoW, Lt Ledger killed when he fired on Turks after landing.
N.18	XH	80hp Clerget	Arrived Port Said April 1915. 10/6/13, nearly caught fire on take off due to leaking petrol. Major repairs required, back in service end of June. To Malta 7/4/16 on *Campinas*. To Argostoli May 1916.
N.19	XH	80hp Clerget	Arrived Port Said April 1915. No major problems or accidents. To Malta 7/4/16 on *Campinas*. To Argostoli May 1916.
N.20	XH	80hp Clerget	Arrived Port Said April 1915. Aboard *Montcalm* 5–27/5/15. *Raven*, 27/7/1915, float struts damage attempting take off. Last flight 31/12/1915 from *Raven*. To Malta on *Anne* 3/5/6, probably disassembled in hold. To Argostoli May 1916.
N.21	XH	100hp Clerget	Arrived Port Said Oct 1915. No record of shipborne operations. To Malta 7/4/16 on *Campinas*. To Argostoli May 1916. Crashed on take off at Argostoli in 1916.
N.22	XH	100hp Clerget	Arrived Port Said Oct 1915. Engine failed on first operation 7/11/15. 23/12/15, to Beersheba searching for N.17. 16/4/16 aboard *Anne*, capsized and sank on take off off Wadi Gaza (crew saved).
N.23	XH	100hp Clerget	Arrived Port Said Oct 1915. No major problems or accidents. To Malta 7/4/16 on *Campinas*. To Argostoli May 1916.
NB.1	XH	100hp Clerget	Arrived from Brindisi February 1916. 7–8/3/16, aboard *Anne*, combined operations with *Ben-my-Chree* and *Empress* on different parts of coast. 13/3/19, wrecked at Port Said, by SM Jeanblanc when landing with failed engine.

NB.2	XH	100hp Clerget	Arrived from Brindisi February 1916. 7–8/3/16 aboard *Anne*, combined operations with *Ben-my-Chree* and *Empress* on different parts of coast. 16/4/19 made the final flight of *l'escadrille de Port-Saïd* from *Anne* off Gaza, SM Grall and 2/Lt Finney. To Malta 3/5/16 on *Anne*. To Argostoli May 1916.
NB.3	XH	100hp Clerget	Arrived from Brindisi February 1916. No record of shipborne operations. To Malta 7/4/16 on *Campinas*. To Argostoli May 1916.

SPECIFICATIONS.
Nieuport Type XH Floatplane.

Main source: *L'Aéronautique maritime en 1914.* (ARDHAN)

Model	Nieuport XH *hydro biplace* (1914)
Engine	80 hp Le Rhône 9C (N.7, N.11, N.12 and N.14) 80 hp Clerget 7Z (N.13, N.15 to N.20) 100 hp Clerget* (N.21 to N.23, NB.1 to NB.3) [* probably 110 hp Clerget 9A or 9Z.]
Length	8,40 m
Span	12,50 m
Wing Area	24,00 m^2
Weight—Empty	450 kg
Weight—Loaded	850 kg
Speed	110 km/h max. (80–90 km/h cruise.)
Endurance—max.	5 hours (with war load 3 hours)
Armament	Nil. Adapted in service to carry up to two 105mm shell bombs (approx. 15 kg each) or two boxes of flechettes. Experimentally fitted at Port Said with a single Lewis gun. The pilot and observer usually had a single carbine and/or one or two pistols in the cockpit.

APPENDIX 4

Shorts and Sopwiths Serving With *EIESS*, January 1916–March 1918

The following list and table of known EIESS floatplanes has been put together from Operation Reports, Log Books and, the ever invaluable, Ray Sturtivant and Gordon Page, *Royal Navy Aircraft Serials and Units, 1911–1919*. With thanks to Peter Cowlan.

This Table does not include machines operated from *Ben-my-Chree* or *Empress* prior to service with EIESS. Nor does it include machines assembled but not used prior to 1 April 1918, for which see Appendix 5. All dates DD/MM/YY.

Short 166 / 184

Serial	Engine	Comments
\multicolumn{3}{c}{TYPE 166 (166 Short-built, remainder Westland-built)}		
166	200hp Canton-Unné	*Empress* Aegean. Received from *Ark Royal* 6/10/16. At Mudros by 28/12/16.
9755	200hp Canton-Unné	*Empress* Aegean. Received from *Ark Royal* 6/10/16. At Mudros by 23/11/16.
9758	200hp Canton-Unné	*Empress* Aegean. Received from Mudros 6/11/16. To A Sqd, Thasos 19/1/17.
9763	200hp Canton-Unné	*Empress* Aegean. Received from *Ark Royal* 28/12/16. To A Sqd, Thasos 19/1/17.
9764	200hp Canton-Unné	*Empress* Aegean. Received from *Ark Royal* 29/12/16. To *Ark Royal* 22/1/17.
\multicolumn{3}{c}{TYPE 184 (all Short-built, except as noted in Comments column)}		
846	225hp Sunbeam Mohawk	Arrived in Port Said aboard *Ben-my-Chree* 12/1/16. Lost 3/4/16 (at Port Said?).
849	225hp Sunbeam Mohawk	Arrived in Port Said aboard *Ben-my-Chree* 12/1/16. 11/2/16, lost off Sollum.

850	225hp Sunbeam Mohawk	Arrived in Port Said aboard *Ben-my-Chree* 12/1/16. Deleted (Del) 1917.
8004	225hp Sunbeam Mohawk	(S.E. Saunders) Delivered Port Said 12/4/16. Del 1917.
8018	240hp Sunbeam Gurkha	(S.E. Saunders) Delivered Port Said 6/2/17. Type D single-seat bomber. 9/10/17, blew up on landing with bombs onboard, Flt Lt Wood killed.
8019	240hp Sunbeam Gurkha	(S.E. Saunders) Delivered Port Said 10/3/17. Type D single-seat bomber. To RAF.
8020	240hp Sunbeam Gurkha	(S.E. Saunders) Delivered Port Said 10/3/17. Type D single-seat bomber. To RAF.
8021	240hp Sunbeam Gurkha	(S.E. Saunders) Delivered Port Said 26/2/17. 11/10/17, shot down over Adana. Flt Cdr Clemson wounded PoW, 2Lt Newton PoW.
8022	240hp Sunbeam Gurkha	(S.E. Saunders) Delivered Port Said 26/2/17. To RAF.
8031	225hp Sunbeam Mohawk	Arrived in Port Said on *Empress* 23/1/16. Crashed and WO 16/2/16 in Port Said harbour. *Empress'* first flight at Port Said.
8045	225hp Sunbeam Mohawk	Arrived in Port Said, from Mesopotamia, by 8/16. On *Raven* 24–28/8/16 (El Afule raid), no further record of use. Sturtivant records it still at Port Said in January 1917.
8051	225hp Sunbeam Mohawk	Delivered Port Said by 4/16. There are no records of its use, and Del by Nov 1916.
8054	225hp Sunbeam Mohawk	Delivered Port Said 18/4/16. Engine problems on 25/8/16, no further recorded flights.
8075	260hp Sunbeam Maori	Delivered Port Said by 5/16. Del Dec 1917.
8077	225hp Sunbeam Mohawk	Received from *Ark Royal*. *Empress* Aegean, between 19–26/6/16.
8078	225hp Sunbeam Mohawk	Received from *Ark Royal*. *Empress* Aegean, between 19–26/6/16.
8080	225hp Sunbeam Mohawk	Delivered Port Said by 5/16. 9/1/17, lost with *Ben-my-Chree* at Castellorizo.
8082	225hp Sunbeam Mohawk	Delivered Port Said by 1/5/16. Engine troubles at Aden 11/6/16, no further recorded flights.

8085	225hp Sunbeam Mohawk	Arrived in Port Said, from Mesopotamia, by 9/16. No recorded flights. Sturtivant records it still at Port Said in January 1917.
8087	225hp Sunbeam Mohawk	Delivered Port Said by 4/16. 23/5/16, crashed on take off and sank, *Ben-my-Chree*.
8088	225hp Sunbeam Mohawk	Shipped in *Empress* to Port Said. To Aegean on *Empress* 3/4/16. To *Ark Royal* June 1916, deleted.
8090	225hp Sunbeam Mohawk	Delivered Port Said by 4/16. The "Cut-Short". Crashed and WO landing 28/3/17, *Raven*.
8091	225hp Sunbeam Mohawk	Delivered Port Said by 4/16. 10/8/16, damaged in running fight with Rumpler from *FA300*. Del by Dec 1917.
8095	225hp Sunbeam Mohawk	Shipped in *Empress* to Port Said. To Aegean on *Empress* 3/4/16. Erected on *Ark Royal* and to *Empress* 17/6/16. At Mudros by 18/8/16.
8372	225hp Sunbeam Mohawk	(Phoenix Dynamo) Delivered Port Said by 7/16. 2/12/16, hit by AA fire and brought down overland. FSL Nightingale and Lt Woodland PoW.
8381	225hp Sunbeam Mohawk	(Frederick Sage) Shipped in *Empress* to Port Said. To Aegean on *Empress* 3/4/16. At Mudros by 18/8/16. Del 1919.
N1090	240hp Renault-Mercedes	Delivered Port Said by 9/17. To RAF.
N1091	240hp Renault-Mercedes	Delivered Port Said by 9/17. To RAF.
N1262	240hp Sunbeam Gurkha	(Robey) Delivered Port Said by 9/17. 30/10/17, severely damaged by German scout and crashed landing alongside *Raglan*. Del Dec 1917.
N1263	240hp Sunbeam Gurkha	(Robey) Delivered Port Said by 9/17. Del May 1918.
N1581	240hp Renault-Mercedes	Delivered Port Said by 11/17. To RAF,
N1582	240hp Renault-Mercedes	Delivered Port Said by 11/17. Night 27–28/1/18, shot down over Dardanelles, captured intact by Turks. Flt Lt Bronson and Lt Pakenham-Walsh PoW.
N1590	240hp Renault-Mercedes	(Frederick Sage) Delivered Port Said by 11/17. Crashed Port Said 28/3/18, deleted.

Serial	Engine	Comments
N1597	260hp Sunbeam Maori	(Frederick Sage) Delivered Port Said by 12/17. Port Said Seaplane Squadron from 2/18. To RAF.
N1639	240hp Renault-Mercedes	(Phoenix Dynamo) Delivered Port Said by 11/17. To RAF.
N1649	240hp Renault-Mercedes	(Phoenix Dynamo) Delivered Port Said by 11/17. Alexandria Seaplane Squadron from 1/18. To RAF.
N1679	260hp Sunbeam Maori	(Brush Electrical) Delivered Port Said by 1/18. Port Said Seaplane Squadron from 1/18. To RAF.
N1749	260hp Sunbeam Maori	(Phoenix Dynamo) Delivered Port Said by 2/18. Port Said Seaplane Squadron from 2/18. To *Empress* by 19/4/18.
N1784	260hp Sunbeam Maori	(Frederick Sage) Delivered Port Said by 2/18. Port Said Seaplane Squadron from 2/18. To *Empress* by 24/4/18.
N1827	260hp Sunbeam Maori	(Robey) Delivered Port Said by 12/17. Port Said Seaplane Squadron from 12/17. To RAF. Damaged 9/1/18, not used again.
N1839	260hp Sunbeam Maori	(Robey) Delivered Port Said by 2/18. Port Said Seaplane Squadron from 2/18. To RAF
N2822	260hp Sunbeam Maori	(Robey) Delivered Port Said by 2/18. Port Said Seaplane Squadron from 2/18. To *Empress* by 9/4/18.

Note: Several Short 184's were delivered to the EIESS, not erected but forwarded to other destinations, including N1580, N1583, N1664.

Sopwith Schneider / Baby

Serial	Engine	Comments
SCHNEIDER (all Sopwith-built)		
3721 *	100hp Gnome Monosupape	Arrived in Port Said aboard *Ben-my-Chree* 12/1/16. Del 1917.
3722 *	100hp Gnome Monosupape	Arrived in Port Said aboard *Ben-my-Chree* 12/1/16. Del 1916.
3727 *	100hp Gnome Monosupape	Delivered Port Said by 3/16. Del Jan 1917.

* *Schneiders 3721, 3722, and 3727 had wing warping and a small triangular fin. Later production machines had ailerons on upper and lower wings and an enlarged fin.*

3770	100hp Gnome Monosupape	Delivered Port Said by 4/16. 9/1/17, lost with *Ben-my-Chree* at Castellorizo.
3771	100hp Gnome Monosupape	Delivered Port Said by 7/16. Del Dec 1916.
3772	100hp Gnome Monosupape	To Aegean on *Empress* 3/4/16. To *Ark Royal* 29/12/16.
3773	100hp Gnome Monosupape	Arrived in Port Said on *Empress* 23/1/16. 16/3/17, damaged beyond repair, deleted.
3774	100hp Gnome Monosupape	Arrived in Port Said on *Empress* 23/1/16. Del 1917.
3775	100hp Gnome Monosupape	Arrived in Port Said on *Empress* 23/1/16. Del 1916.
3777	100hp Gnome Monosupape	Delivered Port Said by 4/16. 17/9/16, Combat with Rumpler C.I from *FA300*. FSL Man forced down on to sea, rescued. Schneider lost.
3778	100hp Gnome Monosupape	Delivered Port Said by 4/16. 9/1/17, lost with *Ben-my-Chree* at Castellorizo.
3779	100hp Gnome Monosupape	? A Report of Flight from *Ben-my-Chree* 14 Sept 1916. (According to Sturtivant—Del 25/4/16 at Calshot.)
3783	100hp Gnome Monosupape	? (According to Sturtivant—*Empress* Aegean, from *Ark Royal* Aug 1916. Ship's log books do not confirm.)
3786	100hp Gnome Monosupape	Delivered Port Said by 4/16. 2/7/16, FSL Man forced down on to sea. Schneider lost.
3787	100hp Gnome Monosupape	To Aegean on *Empress* 3/4/16. Erected on *Ark Royal* and to *Empress* by May 1916. To *Ark Royal* by Dec 1916. Del 1917
3788	100hp Gnome Monosupape	To Aegean on *Empress* 3/4/16. Erected on *Ark Royal* and to *Empress* by May 1916. To *Ark Royal* by Aug 1916.
3789	100hp Gnome Monosupape	Delivered Port Said by 5/16. 14/9/16, engine drops out on landing. Not repaired?
3790	100hp Gnome Monosupape	Delivered Port Said by 3/16. Del 1917.

BABY (all Sopwith-built)

8135	110hp Clerget 9Z	Delivered Port Said by 8/16. 17/9/16, Combat with Rumpler C.I from *FA300*. Flt Lt Bankes-Price shot down and killed.
8136	110hp Clerget 9Z	Delivered Port Said by 8/16. 25/8/16, Flt Cdr Dacre PoW after engine failure returning from El Afule.
8188	100hp Gnome Monosupape	Delivered Port Said by 5/16. 9/1/17, lost with *Ben-my-Chree* at Castellorizo.
8189	100hp Gnome Monosupape	Delivered Port Said by 5/16. Del 1917?
8202	100hp Gnome Monosupape	*Empress* Aegean. Received from *Ark Royal* 29/12/16. To A Sqd, Thasos 19/1/17.

SOPWITH BABY (all Blackburn-built; often referred to as the Blackburn Baby)

N1012	110hp Clerget 9Z	Delivered Port Said by 3/17? No recorded flights
N1014	110hp Clerget 9Z	Delivered Port Said by 3/17. On *Raven* by 10/3/17. *Brisbane*, and Colombo Detachment. Packed up 5/7/17 for return to Port Said. No further recorded flights. Del 1917?
N1016	110hp Clerget 9Z	Delivered Port Said by 3/17. Probably on *Raven* by 10/3/17. Colombo Detachment. Crashed and WO at Colombo, 4/7/17.
N1028	110hp Clerget 9Z	Delivered Port Said by 9/17. Crashed Port Said 19/4/18, deleted.
N1036	110hp Clerget 9Z	Delivered Port Said by 9/17. Damaged landing, sank 2/11/17.
N1038	110hp Clerget 9Z	Colombo Detachment by mid June 1917. Returned to Port Said July 1917. Damaged attempting take off, sunk by gunfire 2/11/17.
N1126	110hp Clerget 9Z	Delivered Port Said by 9/17. To RAF.
N1128	130hp Clerget 9B	Delivered Port Said by 9/17. To RAF.
N1129	130hp Clerget 9B	Delivered Port Said by 9/17. Damaged 2/11/17. Del Dec 1917.

FAIREY HAMBLE BABY (Parnall-built)

N1209	130hp Clerget 9B	Delivered Port Said by 9/17. To RAF.
N1210	130hp Clerget 9B	Delivered Port Said by 9/17. Engine failure, damaged landing, sunk by gunfire 2/11/17.

APPENDIX 5

Aircraft Serving with 64 (Naval) Wing, RAF, Egypt, from April 1918

The following list of machines employed by the Port Said and Alexandria Squadrons, RAF, has been compiled from weekly reports contained in AIR 1/455/15/312/35, and Sturtivant and Page, and may be incomplete. The two units were nominally 431 and 432 Flights, RAF, then 269 and 270 Squadrons, RAF, from 6 October 1918, but those designations do not appear in any of the reports. See Chapter 18. All dates are DD/MM/YY.

Serial	Engine	Comments
SHORT TYPE 827		
8649	150hp Sunbeam Crusader	From E Africa. Port Said by 29/6/18, being overhauled as a School machine. Stored on *City of Oxford* June to September 1918. Deleted (Del) at Port Said 22/12/18.
SHORT TYPE 184		
8019	240hp Sunbeam Gurkha	From EIESS. Port Said. One recorded patrol, 20/4/18 – beached and towed in. For deletion, 29/6/18. Del at Alexandria 10/18.
8020	240hp Sunbeam Gurkha	From EIESS. School machine only. Crashed Alexandria 15/7/18.
8022	240hp Sunbeam Gurkha	From EIESS. Port Said. Seven patrols, last 26/4/18. Del at Alexandria 10/18.
N1090	240hp Renault-Mercedes	From EIESS. Port Said and Alexandria. Last recorded patrol 2/6/18. Del 25/7/18.
N1091	240hp Renault-Mercedes	From EIESS. Port Said and Alexandria. Last recorded patrol 11/6/18. Del 1/5/19.
N1581	240hp Renault-Mercedes	From EIESS. Alexandria. Last recorded patrol 1/5/18. At Kalafrana, Malta, 27/8/19.
N1597	260hp Sunbeam Maori	From EIESS. Port Said. Last recorded patrol 17/10/18. Del 23/1/19.

N1639	240hp Renault-Mercedes	From EIESS. Port Said and Alexandria. Last recorded patrol 25/7/18. Del 22/3/19.
N1649	240hp Renault-Mercedes	From EIESS. Alexandria. No recorded patrols. Del 27/7/18.
N1679	260hp Sunbeam Maori	From EIESS. Port Said. Last recorded patrol 10/11/18. Del 1/5/19.
N1749	260hp Sunbeam Maori	From EIESS. Port Said. To *Empress* 19 April 1918.
N1756	260hp Sunbeam Maori	Alexandria. First recorded patrol 23/7/18. WO landing on road 23/7/18.
N1757	260hp Sunbeam Maori	Alexandria. First recorded patrol 23/5/18. Crashed and WO 1/8/19.
N1784	260hp Sunbeam Maori	From EIESS. Port Said. Last recorded patrol 21/4/18. To *Empress* 24 April 1918.
N1838	260hp Sunbeam Maori	From EIESS. Port Said. Last recorded patrol 6/5/18, crashed. Del 12/5/18.
N1839	260hp Sunbeam Maori	From EIESS. Port Said. Last recorded patrol 30/7/18, crashed, Del 1/8/18.
N2647	260hp Sunbeam Maori	Alexandria. First recorded patrol 15/9/18. At Kalafrana, Malta, 27/8/19.
N2648	260hp Sunbeam Maori	Port Said. No recorded patrols. At Kalafrana, Malta, 27/8/19.
N2649	260hp Sunbeam Maori	Port Said. 1919 only. At Kalafrana, Malta, 27/8/19.
N2653	260hp Sunbeam Maori	Port Said. No recorded patrols. At Kalafrana, Malta, 27/8/19.
N2791	260hp Sunbeam Maori	Port Said. First recorded patrol 6/7/18. Crashed and WO 13/10/18.
N2792	260hp Sunbeam Maori	Port Said. First recorded patrol 21/7/18. At Kalafrana, Malta, 27/8/19.
N2810	260hp Sunbeam Maori	Alexandria. First recorded patrol 19/5/18. At Kalafrana, Malta, 27/8/19.

N2811	260hp Sunbeam Maori	Port Said. First recorded patrol 29/5/18. Crashed and WO 2/6/18.
N2812	260hp Sunbeam Maori	Port Said. First recorded patrol 20/5/18. At Kalafrana, Malta, 27/8/19.
N2822	260hp Sunbeam Maori	From EIESS. Port Said. To *Empress* 9 April 1918.
N2823	260hp Sunbeam Maori	Port Said and Alexandria. First recorded patrol 20/10/18. Crashed and WO 31/10/18.
N2824	260hp Sunbeam Maori	Port Said. To *Empress* 14 April 1918. Last recorded patrol 19/4/18. UK by 20/12/18.
N2838	260hp Sunbeam Maori	Alexandria. First recorded patrol 3/6/19. At Kalafrana, Malta, 27/8/19 > 5/3/20.
N2847	260hp Sunbeam Maori	Port Said. First recorded patrol 6/7/18. At Kalafrana, Malta, 27/8/19.
N2848	260hp Sunbeam Maori	Alexandria. Crashed and WO 14/7/18.
N2849	260hp Sunbeam Maori	Port Said. First recorded patrol 6/7/18. At Kalafrana, Malta, 27/8/19.
N2903	260hp Sunbeam Maori	Port Said. First recorded patrol 4/6/18. Crashed and WO 16/6/18.
N2904	260hp Sunbeam Maori	Alexandria. First recorded patrol 14/6/18, see Chapter 18. Crashed and WO 1/8/19.
N2905	260hp Sunbeam Maori	Alexandria. First recorded patrol 9/6/18. Crashed and WO 30/7/18.
N2915	260hp Sunbeam Maori	Port Said. 1919 only. At Kalafrana, Malta, 12/8/19.
N2917	260hp Sunbeam Maori	Alexandria. First recorded patrol 29/6/18. Del 13/6/19.
N9066	260hp Sunbeam Maori	Port Said and Alexandria. 1919 only. At Kalafrana, Malta, 27/8/19.
N9067	260hp Sunbeam Maori	Alexandria. Probably not assembled.
N9076	260hp Sunbeam Maori	Alexandria. 1919 only. At Kalafrana, Malta, 27/8/19 > South Russia 1920.

		SOPWITH BABY (all Blackburn-built)
N1060	110hp Clerget 9Z	Port Said. First recorded patrol 29/5/18. Crashed and WO 3/6/18.
N1126	110hp Clerget 9Z	From EIESS. Port Said. Last recorded patrol 11/4/18, ran out of fuel, landed at sea and sank.
N1128	130hp Clerget 9B	From EIESS. Port Said. Last recorded patrol 17/10/18, forced landing and WO.
N2072	130hp Clerget 9B	Port Said. First recorded patrol 3/5/18. Hit dhow landing Port Said and WO 20/5/18.
N2073	130hp Clerget 9B	Port Said. First recorded patrol 19/5/18. Del 27/1/19.
N2131	130hp Clerget 9B	Port Said and Alexandria. First recorded patrol 21/5/18. Del 30/4/19.
N2132	130hp Clerget 9B	Port Said and Alexandria. First recorded patrol 17/5/18. At Kalafrana, Malta, 27/8/19.
		FAIREY HAMBLE BABY (Parnall-built)
N1209	130hp Clerget 9B	From EIESS. Port Said. Last recorded patrol 27/4/18. Del 11/18.
		ROYAL AIRCRAFT FACTORY BE2e
6802	90hp RAF1A	Port Said. First recorded patrol 1/11/18. Later, at Flying School, WO 1/4/19.
A1306	90hp RAF1A	Port Said. First recorded patrol 2/11/18. Later, at Flying School, WO 1/3/19.
B3657	90hp RAF1A	Port Said. First recorded patrol 1/11/18. X AD, Aboukir, 9/3/19.
B4596	90hp RAF1A	Port Said. First recorded patrol 9/11/18. X AD, Aboukir, 28/2/19.
		FELIXSTOWE F.3
N4360	2 x 375hp RR Eagle VIII	Alexandria from 8/10/18. No recorded patrols.

APPENDIX 6

L'escadrille de Port-Saïd, Known Pilots and Observers

Although a unit of *la Marine nationale*, *l'escadrille de Port-Saïd* relied on British Army officers as observers when operating from the seaplane carriers which were operated, commanded and crewed by the Royal Navy. Most flights from the base at Port Said were fully French crewed, employing enlisted men of *l'escadrille* as observers. Approximate comparative ranks:

LV	*Lieutenant de vaisseau*	(Flight Commander, RNAS)
EV	*Enseigne de vaisseau*	(Flight Lieutenant, RNAS
EV2	*EV de 2e classe*	(Flight Sub Lieutenant, RNAS)
SM	*Second maître*	(Petty Officer Mechanic, RNAS)
QM	*Quartier-maître*	(Leading Mechanic, RNAS)
Mot	*Matelot*	(Air Mechanic, RNAS)
Mecae	*Mécanicien d'air* (an enlisted rate usually preceded by rank, i.e. *Mot Mécanicien* or *QM Mécanicien*)	

Part 1: Pilots

Name	Joined *l'escadrille*	Comments
QM Raymond P. Bourgeois	Oct 1915	First operation from Port Said 20/10/15.
LV Alfred L.M. Cintré	Nov 1914	Initially an observer until qualified on Nieuport, Jan 1915. Also served at Antivari (in command) and Brindisi. *Chevalier de la Légion d'honneur* and *CdG*, Jul 1915. British DSC, Sept 1915.
Mot Pierre Collard	Feb 1916	Was operational with *l'escadrille de Brindisi*. At Port Said he only flew local patrols, later served at Argostoli.

LV Paul A.G. Delage	Nov 1914	Returned to Nieuport as Technical Director in February 1915.
LV Marcel V.A. Destrem	Nov 1914	Also served at Antivari and Argostoli. *Chevalier de la Légion d'honneur*, Nov 1915. *CdG*, August 1915. Recommended for British DSC, June 1916—lost in bureaucracy?
SM Charles A. Gramont	June 1915	Initially an observer until he completed his pilot training at Port Said. Flew his first operation 29/3/16.
QM Hervé Grall	Nov 1914	Promoted SM, 9/5/15. Also served at Argostoli. *Médaille militaire*, Feb 1915. *CdG*, two citations. British DSM, April 1915.
SM Emile A. Jeanblanc	Feb 1916	From *l'escadrille de Brindisi*. First operation from Port Said 8/3/16. Also served at Argostoli.
QM Jean-Marie Le Gall	Dec 1914	KIA, 28 Jan 1915 (Chapter 3).
LV Henry P. de L'Escaille	Nov 1914	*Officier de la Légion d'honneur*, July 1915. *CdG*, Nov 1915. British DSC, August 1915.
Mot Julien P.A. Levasseur	Nov 1914	Repatriated to France for medical reasons in March 1915. Also served at Antivari.
QM Henri J. Roussillon	Feb 1916	From *l'escadrille de Brindisi*. First operation from Port Said 7/3/16. Promoted SM, August 1916. Also served at Argostoli. Recommended for British DSM, June 1916—lost in bureaucracy?
LV Louis M.J. Barthélémy de Saizieu	Nov 1914	Initially an observer until qualified on Nieuport, January 1915. PoW, 22/12/15 (Chapter 4). *Chevalier de la Légion d'honneur*, Apr 1915. *CdG*, Feb 1916. British DSC, December 1915.
QM Georges M.E. Trouillet	Dec 1914	Promoted SM, 20/5/15. PoW, 10 Oct 1915 (Chapter 4). *CdG*, Aug and Nov 1915. British DSM, December 1915.

Part 2: British observers with *l'escadrille de Port-Säid*

Name	Unit	Comments
2Lt F.O. Baxter	2nd Rajput Light Infantry	Joined Egypt Detachment RFC (EDRFC) 19/1/15, to *l'escadrille* 30 Jan. Remained with EIESS.
2Lt C.A. Bourne	Royal Field Artillery (RFA)	His only flight with *l'escadrille*, 16/4/16, ended in a crash on take-off, with Grall as pilot. He later served as a pilot with 16 Sqn RFC in France.
2Lt J.M. Burd	RFA	Joined *l'escadrille* in early 1916. 1916 *Campinas*.
Capt E.L. Chute	Duke of Wellington's (West Riding) Regiment	Several flights with *l'escadrille*. In the first group of trainees in Childers' EIESS Observers School. Returned to the UK for training as a balloon observer. In 1917 was Commanding Officer of 12 Balloon Company, on the Arras front. Relinquished his commission on account of ill-health contracted on active service, 28/11/17.
2Lt Alfred D. Finney	RFA	Previous service as an observer with 2 Wing, RNAS, Imbros. Joined *l'escadrille* in Feb 1916. *Anne* March and April 1916.
Maj Herbert P. Fletcher	Middlesex Hussars	Senior observer working closely with Col Elgood. When Childers returned to UK he took over EIESS Observers School. Later returned to UK to train as a pilot. Died 3/8/16 after being struck by BE2c propeller. Fletcher was awarded the DSO and French *Croix de guerre* for his services with the *escadrille*.
Capt James R. Herbert	Duke of Cambridge's Hussars	Joined *l'escadrille* in Dec 1914. On leaving *l'escadrille* in Aug 1915, he set up the Frontiers District Administration in Sinai for Survey of Egypt.
2Lt H.G. Hillas	Duke of Wellington's Regiment	Joined EDRFC 18/1/15, to *l'escadrille* 28 Jan. Fletcher's assistant.

2Lt Horace M.C. Ledger	27th Punjabis	Joined EDRFC 20/1/15, to *l'escadrille* 30 Jan. KIA, 22/12/15 (Chapter 4). Awarded the French *Croix de guerre* in Oct 1915, with a second posthumous award.
2Lt Basil G.N.B. Partridge	2nd Rajput Light Infantry	Joined EDRFC 20/1/15, to *l'escadrille* 23 Jan. KIA, 28/1/15 (Chapter 3).
2Lt Sir Robert J. Paul	Egyptian Irrigation Department	Joined EDRFC 2/1/15, to *l'escadrille* 28 Jan. PoW, 10/10/15 (Chapter 4). Paul was awarded the French *Croix de guerre* for his services.
Capt Arthur J. Ross	Royal Engineers	Joined EDRFC 9/12/14, to *l'escadrille* probably only on temporary attachment. HMS *Minerva* 16–17/12/14. Later learned to fly, and as Major, Nov 1916 to April 1917 CO C Flight, Arabian Detachment, 14 Sqn, RFC. He was killed in a flying accident in Norfolk, 2/8/17.
Capt Walter F. Stirling	Royal Dublin Fusiliers	Joined EDRFC 30/11/14, to *l'escadrille* 21 Dec. 31/12/14, flying with QM Grall, crashed in Wadi Araba, north of Akaba, both walked out and rescued. Stirling later served with Lawrence.
Capt R.E. Todd	Royal Army Medical Corps	Flew several times from *Aenne Rickmers* and *Rabenfels*. Became Intelligence Officer *Raven* from June 1915.
Lt J.H.B. Wedderspoon	RFA	Joined *l'escadrille* in early 1916. 1916 *Campinas*. Remained with EIESS.
Capt L.B. Weldon	Royal Dublin Fusiliers	Intelligence Officer *Anne* 1915. Occasional flights from Port Said.
2Lt Kenneth L. Williams	2nd Rajput Light Infantry	Joined EDRFC 20/1/15, to *l'escadrille* 30 Jan. Was included in the first group of trainees in Childers' EIESS Observers School. Remained with EIESS.

Part 3: French *Observateurs d'aéronautique* with *l'escadrille de Port-Säid*

Name	Rank
Andreis, François Joseph Honoré	*Mecae*
Baille, Fernand Adolphe	*Mot*
Batier, Alfred	*Mot*
Blandin, Charles Louis Auguste François	*Mot*
Bus, Jean Modeste François	*Mot*
Capdegelle, Médard, Joseph	*Mot*
Carlavan, Marcelin	*QM*
Debart, Jean-François Eugène	*QM*
Duffaud, Edouard Joseph Ferdinand	*Mot*
Duval, Maurice	n/k
Falconet, Francisque Marius	*Mot → QM*
Galliano, Jean-Marie	*Mot*
Gazagne, Marcel	*Mot → QM*
Grelaud, Edouard	*Mot*
Guilbaud, Henri Auguste	*Mot*
Jamin, Eugène Félix Auguste	*QM*
Le Corf, Jean	*Mecae*
Machefaux, Georges	*Mot → SM*
Marchal, Victor Rigobert	*QM → SM*
Michel, François Auguste	*EV2R*
Michelet, Yves Marie Pierre	*QM*
Surre, François Eugène	*Mot → QM*
Venon, Jules Camille Marie	n/k

APPENDIX 7

East Indies and Egypt Seaplane Squadron, Known Pilots and Observers

Ranks shown in the following Tables are those worn when joining the EIESS, any promotions are noted in the Comments column. Award dates are from LG, *London Gazette*; EG, *Edinburgh Gazette*—and may be incomplete. All dates are DD/MM/YY.

Commanding Officers

Name	Comments
Sqn Cdr Cecil John L'Estrange Malone	Commanded *Ben-my-Chree* 23/3/15 to 14/5/16. Formed and commanded EIESS from 12/1/16 to 12/4/16. Ordered to UK 16/12/16. Promoted Wing Cdr 31/12/16 (in lieu of recommended DSC).
Wing Cdr Charles Rumney Samson	Commanded *Ben-my-Chree* 14/5/16 to 9/1/17. Commanded EIESS from 14/5/16 to 25/4/17. Returned to UK. Promoted Wing Captain 31/12/17. Awarded DSO, LG 20/10/14. Bar to DSO, LG 23/1/17.
Wing Cdr Charles Erskine Risk (Lt Col, RMLI)	Commanded EIESS from 25/4/17 to 31/3/18. As Lt Col commanded Naval Wing, RAF, Egypt, from 1/4/18. Awarded DSO, LG 3/6/19.

Part 1: Pilots (all RNAS)

Name	Comments
FSL G.S. Abbott *(Canada)*	*Empress* Aegean (Chapter 12). On detachment from No.2 Wing, RNAS, October 1916 to January 1917. Slightly wounded 30/11/16. Promoted Flt Lt 1/4/17.
FSL C.V. Arnold	Joined *Empress* in UK 31/7/15. Aegean until June 1917, transferred to Mudros Jan 1917. Promoted Flt Lt 1/10/16. Killed in flying accident at Chingford, whilst on UK leave, 16/8/17.
Flt Lt J.T. Bankes-Price *(Born Chicago, USA)*	Joined *Ben-my-Chree* in UK 3/5/15 as FSL. Promoted Flt Lt 1/1/16. KIA 17/9/16 (Chapter 9).

Flt Lt F.M.L. Barr	Joined EIESS May 1917. *Empress*. Promoted Flt Cdr 30/6/17. Returned to UK end of 1917 with poor record (ADM 273/4/1).
Flt Lt C.G. Bronson *(Canada)*	Joined EIESS April 1917. *Empress*, and *Raven*. PoW 28/1/18 (Chapter 16). Awarded *Croix de Guerre*, LG 22/2/18. DSC, LG 23/5/19.
Flt Lt J.C. Brooke	Joined EIESS May 1916. *Ben-my-Chree, Anne, Raven,* and *Empress*. Promoted Flt Cdr 31/12/16. Returned to UK July 1917. Awarded DSC, LG 20/7/17.
Flt Lt E.J.P. Burling	Joined EIESS October 1916. *Ben-my-Chree, Anne, Raven,* and *Empress*. Promoted Flt Cdr 31/12/17. Command of Alexandria Seaplane Base from December 1917. Transferred to RAF, remained in Egypt until end of war. Awarded MiD, LG 20/7/17. *Croix de Guerre*, LG 22/2/18. DFC, LG 2/7/18. DSC, EG 25/2/19.
Flt Lt A.W. Clemson	Joined EIESS May 1916. *Ben-my-Chree, Anne, Raven,* and *Empress*. Promoted Flt Cdr 31/12/16. HMAS *Brisbane* and Colombo Detachment, May-July 1917. PoW 11/10/17 (Chapter 15). Awarded DSC, LG 20/4/17.
FSL R.M. Clifford	Joined *Empress* in UK 16/11/15. *Raven* March/April 1916. Invalided to UK (malaria) early 1917.
Flt Cdr W.R. Crocker	Joined EIESS 25/2/16. Was to command the Island Base. Killed in railway accident at Port Said overnight 5/6 March (Ch. 5).
Flt Lt G.B. Dacre	Joined *Ben-my-Chree* in UK 22/3/15 as Flt Lt. *Ben-my-Chree, Anne,* and *Raven*. Appointed to command Port Said Island Base in March 1916 after Flt Cdr Crocker killed. Promoted Flt Cdr 30/6/16. PoW 26/8/16 (Chapter 9). Awarded DSO, LG 19/11/15 (Dardanelles). MiD 10/6/16
FSL M.G. Dover *(Canada)*	Joined EIESS June 1916. *Ben-my-Chree*. Hospitalised December 1916, and returned to UK. Promoted Flt Lt 30/6/17.
Flt Cdr C.H.K. Edmonds	Joined *Ben-my-Chree* in UK March 1915 as Flt Cdr. Promoted Sqn Cdr 30/6/16. Returned to UK May 1916. Returned to UK May 1916. Promoted Sqn Cdr 30/6/16. Early 1917, command of 6 Wing (Seaplane), at Otranto in Italy. Awarded DSO, LG 16/2/15 (Heligoland Bight). MiD 10/6/16.

Flt Lt T.H. England	Joined EIESS 25/2/16. *Ben-my-Chree*, and *Raven*. Promoted Flt Cdr 30/6/16. Returned to UK May 1917. Promoted Sqn Cdr 30/6/17. Awarded DSC, LG 24/10/16. MiD 10/9/16.
Flt Lt R.M. Field	Joined *Empress* in UK 8/7/15. Aegean until June 1916? With *Ark Royal* by 29/9/16. Promoted Flt Cdr 31/12/16. Promoted Sqn Cdr 31/12/17.
FSL F.C. Henderson *(Canada)*	Joined EIESS November 1916. *Ben-my-Chree*. To No.6 Wing, Otranto, March 1917. Promoted Flt Lt 1/4/17.
Flt Cdr P.L. Holmes	Posted to EIESS 18/11/17, arriving Port Said early December 1917. Promoted Sqn Cdr 1/1/18. Commanded Port Said Detachment from January 1918 (?). Transferred to RAF.
FSL E.M. King *(New Zealand)*	Joined EIESS September 1916. *Anne*, *Raven*, *Empress*, and *City of Oxford*. Promoted Flt Lt 1/10/17. Transferred to RAF. Long leave to New Zealand commencing July 1918. Returned to Egypt December 1918, remained until May 1919. Awarded *Croix de Guerre*, LG 22/2/18. DFC, LG 2/7/18.
Flt Lt H.deV. Leigh	Joined EIESS mid-1916. *Ben-my-Chree*, *Anne*, *Raven*, *Empress*, and *City of Oxford*. Promoted Flt Cdr 31/12/17. Transferred to RAF, joined *Empress* by May 1918. Awarded DFC, LG 2/7/18. MiD, LG 20/7/17.
FSL H.G.R. Malet *(Canada)*	*Empress* Aegean (Chapter 12). On detachment from *Ark Royal*, October 1916 to January 1917. Promoted Flt Lt 31/12/16. Promoted Flt Cdr 30/6/17. Returned to UK July 1917.
FSL W. Man	Joined EIESS April 1916. *Ben-my-Chree*, *Anne*, and *Raven*. Promoted Flt Lt 31/12/16. To No.6 Wing, Otranto, March 1917.
FSL A.S. Maskell	Joined *Ben-my-Chree* 22/3/15 as FSL. Promoted Flt Lt 25/6/15. Promoted Flt Cdr 31/12/16. Returned to UK in March 1917.
FSL G.W. Morey *(Australia)*	Joined EIESS January 1918. Transferred to RAF. Promoted Lt 1/4/1918.
FSL A.J. Nightingale *(Canada)*	Joined EIESS April 1916. *Ben-my-Chree*. Promoted Flt Lt 1/1/17. PoW 2/12/16 (Chapter 9).
FSL L.P. Paine	Joined EIESS early 1916. *Ben-my-Chree*, *Anne* and *Raven*. Returned to UK February 1917. Promoted Flt Lt 30/6/17. Awarded DSC, LG 22/6/17.

FSL G.A.A. Pennington	Joined EIESS January 1918. *City of Oxford*. May have transferred to RAF and joined *Empress* May 1918 for return to UK.
Flt Lt A.E. Popham	Joined EIESS 21/5/17. Posted to Colombo Detachment, returned to Port Said by 14/8/17. *Empress*, and *City of Oxford*. Transferred to RAF. Invalided (Colitis) to UK by 22/4/18. Awarded *Croix de Guerre*, LG 22/2/18. MiD 2/4/18.
Flt Cdr W.G. Sitwell	Joined EIESS, as new Chief Pilot for *Empress*, in February 1916. To *Ark Royal* in Aegean 7/10/16. Returned to UK early 1917. Awarded DSC, LG 30/4/18.
FSL G.D. Smith *(Living in USA)*	Joined EIESS April 1916. *Ben-my-Chree, Anne, Raven, Empress*, and *City of Oxford*. Promoted Flt Lt 1/4/17. Transferred to RAF. Long leave to California commencing July 1918. Returned to Egypt February 1919, remained until July 1919. Awarded DSC, LG 20/7/17. *Croix de Guerre*, LG 22/2/18.
FSL T.G.M. Stephens *(Canada)*	Joined EIESS January 1917. *Raven* and Empress. Promoted Flt Lt 1/4/17. Invalided to UK and Canada from February 1918, possibly due to problems with leg wound received 2/11/17 (Chapter 16). Awarded MiD, LG 20/7/17.
FSL G. Waugh	Joined EIESS February 1918. Transferred to RAF. Promoted Lt 1/4/18. Awarded AFC, LG 8/2/19. See Chapter 18.
Flt Cdr R. Whitehead	*Empress* Aegean (Chapter 12). On detachment from *Ark Royal*, October 1916 to January 1917. Returned to UK January 1917. Promoted Sqn Cdr 30/6/17, appointed No.6 Wing, Italy, Caproni Squadron.
Flt Lt M.C. Wood *(Canada)*	Joined EIESS May 1917. *Empress*. Promoted Flt Lt 30/6/17. KIA 11/10/17 (Chapter 15).
FSL H.V. Worrall *(Australia)*	Joined EIESS November 1916. *Ben-my-Chree, Raven* and *Empress*. Promoted Flt Lt 1/4/17. Transferred to RAF. Long leave to Australia commencing July 1918. May not have returned to Egypt. Awarded MiD, LG 20/7/17. *Croix de Guerre*, LG 22/2/18. DSC, LG 13/9/18 (Goeben). Bar to DSC, LG 18/2/19 (Gaza).
Flt Lt M.E.A. Wright	Joined *Ben-my-Chree* in UK 26/2/15 as FSL. *Ben-my-Chree, Anne*, and *Raven*. Promoted Flt Lt 7/5/15. Returned to UK 8/5/16. Awarded MiD 10/6/16.

Part 2: EIESS Known Observers

Naval officers employed as observers could be be either RNVR and/or RNAS. The rank of Observer (Obs) Sub Lt or Lt was introduced in April 1917 for observers who had completed a course of training in the UK. Whilst it was an RNAS rank, experienced RNVR observers could also be appointed as Obs Sub Lt or Obs Lt. In these cases their Service Records are unclear as to whether they remained RNVR or transferred to the RNAS. Military officers employed as observers were RFC (units given are pre-RFC), attached to RNAS. All Observers listed below completed at least one operational flight. Test flights are not included. Does not include observers from *l'escadrille de Port-Saïd*, unless transferred to EIESS.

Name	Unit / Service	Comments
2Lt F.O. Baxter	2nd Rajput Light Infantry	From *l'escadrille*. *Ben-my-Chree*, May 1916. No further recorded flights.
CPO J.W. Bell	RNAS	*Empress* Aegean. On detachment from No.2 Wing, May 1916.
2Lt W.W. Benn	Middlesex Yeomanry	Joined EIESS April 1916. EIESS Intelligence Officer and Chief Observer. Local appointment Temp Captain 8/7/16, "*Whilst employed as OC Mlt Observers Seaplane Squadron*", later confirmed and backdated to 5/4/16. Returned to UK after loss of *Ben-my-Chree*. Learnt to fly and served with RNAS, RFC and RAF in Italy. Awarded DSO, LG 4/6/17. DFC, LG 21/9/18.
2Lt C.A. Bourne	Royal Field Artillery (RFA)	From *l'escadrille*. *Empress*, March 1916. No further recorded flights. He later served as a pilot with 53 Sqn RFC in France. Transferred to '*American Air Service*' (AIR 76/47/200) September 1917.
Lt J.W. Brown	RFA	*Raven*, July 1916, Castellorizo. No further recorded flights.
2Lt J.M. Burd	RFA	From *l'escadrille*. *Ben-my-Chree*, Aden June 1916. No further recorded flights.
Lt R.E. Childers	RNVR	Joined *Ben-my-Chree* in UK, February 1915. Chief Observer, Returned to UK in March 1916. Transferred to RNAS, 2/4/17. Awarded DSC, LG 20/4/17.

Lt L. Clarke	6th Manchester Regt.	*Ben-my-Chree*, Aden June 1916. No further recorded flights.
2Lt A.D. Ferguson	Highland Light Infantry	Joined EIESS early 1917. Promoted Lieutenant 1/7/17. *Empress* and *City of Oxford*. Transferred to RAF in Egypt until end of 1918. Awarded *Croix de Guerre*, LG 22/2/18. MiD, LG 14/9/18.
CPO H.F. Goodwin	RNAS	*City of Oxford*, 24/2/18 and 22/3/18. No further recorded flights.
Ldg Mech E. Groucott	RNAS	*Raven*, Colombo 2 April 1917. No further recorded flights.
Sub Lt S. Hampton	RNVR	*Empress* Aegean 30/5/16. On detachment from No.2 Wing, September/October 1916. To UK from *Empress* Jan 1917.
Sub Lt A.W.C. Holcombe (Australia)	RNVR	*Empress* Aegean 30/5/16. On detachment from No.2 Wing, from August 1916. Promoted Obs Lt 1/4/17.
Obs Sub Lt S.C. Howes	RNAS	Joined EIESS early 1918. Appointed Alexandria Squadron by March 1918. Transferred to RAF in Egypt until June 1919. Promoted Lieutenant 1/4/18. Awarded AFC, LG 8/2/19.
Lt T.V. Hughes	RFA	Joined EIESS April 1916. *Anne*, 21–25/4/16. No further recorded flights.
Capt W.R. Kempson	RFA	Joined EIESS early 1917. *Raven*, *Empress* and *City of Oxford*. Transferred to RAF in Egypt until end of 1918. Awarded *Croix de Guerre*, LG 22/2/18. MiD 2/7/18.
Obs Lt R.C. Kennedy	RNAS	Joined EIESS early 1918. *City of Oxford*. Transferred to RAF, joined *Empress* April 1918.
Sub Lt J.L. Kerry	RNVR	Joined EIESS May 1916. *Ben-my-Chree*, *Anne*, *Raven*, and *Empress*. Appointed Obs Lieutenant, RNVR, 1/4/17. Posted to Eastchurch, UK, end of 1917. Transferred to RAF. Awarded DSC, LG 20/4/17.

2Lt E. King	King's Own Scottish Borderers	Joined EIESS May 1916. *Ben-my-Chree*, and *Raven*. May have left late 1916.
Capt R.E.C. Knight-Bruce	Royal 1st Devon Yeomanry	*Raven*, March/April 1917. His Service Record places him with 47 Squadron, RFC, in Salonika, from September 1916 to 18/8/17, although noted as being on an Aerial Gunnery course at Abassia, Egypt, in June 1917. His attachment to the EIESS is therefore a bit of a mystery.
Air Mech Lloyd	RNAS	*Ben-my-Chree*, 22/12/16. No further recorded flights.
Lt T.D. McLeish	RNVR	EIESS Photographic Officer from July 1917. *City of Oxford*, 16/3/18. No further recorded flights.
Lt W.C.A. Meade	RNVR	Joined EIESS May 1916. *Ben-my-Chree*, and *Raven*. Appointed Obs Lt, RNAS, 26/10/16. Posted Eastchurch, Armaments Officer, left Port Said August 1917. Returns to Port Said Dec 1917 as Lt, RNVR, Armaments Officer. Transferred to RAF. MiD, LG 20/7/17.
Lt V. Millard	Essex Regt.	EiESS from March 1916 to 31/3/18. *Anne*, *Raven*, *Empress* and *City of Oxford*. Promoted Captain by September 1917. Transferred to RAF in Egypt until July 1918. Awarded *Croix de Guerre*, LG 22/2/18. DFC, LG 2/7/18.
Lt E.A. Newton (Australia)	Camel Transport Corps	Joined EIESS mid-1917. *Empress*, Sept/Oct 1917. PoW 11/10/17 (Chapter 15).
2Lt L.H. Pakenham-Walsh	Cheshire Regt.	Joined EIESS February 1917. *Anne*, *Raven*, *Empress* and *City of Oxford*. Promoted Lieutenant 17/4/17. PoW 28/1/18 (Chapter 16). Awarded *Croix de Guerre*, LG 22/2/18. DFC, LG 8/2/19.
Air Mech A.J. Perry	RNAS	*City of Oxford*, 23/2/18, 1/3/18, and 18/3/18. No further recorded flights.

Ldg Mech Prince	RNAS	*Empress*, 11/10/17. No further recorded flights.
Lt A.P. Ravenscroft	RFA	Appears to have been attached to EIESS for *Raven*'s visit to Aden in July 1916.
Ldg Mech W. Robertson	RNAS	*City of Oxford*, 2/3/18, 4/3/18, and 20/3/18. No further recorded flights.
Sub Lt D.P. Rowland (*Canada*)	RNVR	*Empress* Aegean 6/10/16. On detachment from No.2 Wing, from October 1916.
Lt W.L. Samson	RNVR	Joined EIESS October 1917. *Ben-my-Chree*. Left EIESS March 1917. Awarded MiD 31/1/17. DFC, LG 7/2/19.
2Lt A.K. Smith	Highland Light Infantry	Joined EIESS April 1916. *Anne*, *Ben-my-Chree*, and *Raven*. Last recorded flight 10/8/16 from *Raven* practice spotting for monitor *M.21*.
Lt N.W. Stewart	7th Royal Scottish	Joined EIESS mid-1916. *Ben-my-Chree*, *Anne*, and *Raven*. KIA 23/1/17 (Chapter 13).
Mid H.D. Thornton	RNVR	Appointed *Empress* April 1915 to September 1916. *Empress* Aegean (Chapter 12). Single flight 3 May 1916.
Lt Lord Torrington	RNVR	*Empress* Aegean. On detachment from *Ark Royal*, May 1916.
Lt J.H.B. Wedderspoon	RFA	From *l'escadrille*. *Ben-my-Chree*, Aden June 1916. Later trained as a pilot, KIA on 6 April 1917 with 27 Sqn, RFC, in France.
Lt E. Williams (*le petit*)	East Yorkshire Regt.	May have come from *l'escadrille*. Was included in the first group of trainees in Childers' EIESS Observers School. *Empress*, 7/3/16. No further recorded flights.
2Lt Kenneth L. Williams (*Long*)	2nd Rajput Light Infantry	From *l'escadrille*. Was included in the first group of trainees in Childers' EIESS Observers School. Promoted Lt on 1/12/16. Remained with EIESS until early 1917.
Lt P.M. Woodland	RNVR	Joined EIESS August 1916. *Ben-my-Chree*. PoW 2/12/16 (Chapter 9).

APPENDIX 8

Summary of Shipboard Operations of *l'escadrille de Port-Saïd*, 1914–1916

Shipboard Operations represent less than half of the total operational flights made by *l'escadrille de Port-Saïd*. Many more flights, local recconaissances of Northern Sinai and submarine patrols, were made from Port Said. Most of these were flown by the French pilots with observers drawn from the *mécanicien* of *l'escadrille* itself. Details of these flights have not been located, although a list of known French *Observateur d'aéronautique* is included in Appendix 6. Note: The following is an edited version of a Table that was published in Cross and Cockade International, V52N4, 2021. Main source: TNA AIR1 files. Secondary sources: ARDHAN, Cronin and CCI Nieuports.

Ship	Dates	Pilots / Observers	Nieuport Serials	Area of Operations
		1914		
Doris	12 Dec	*LV* Destrem / Capt Herbert	N/K	El Arish and Gaza. Attempt to fly to Beersheba – engine trouble 20 miles inland forced to return
	14–17 Dec	*LV* Destrem / Capt Herbert	N/K	Lake Bardawil, Gaza, Jaffa and Haifa. First successful flight to Beersheba, 15 Dec.
Minerva	9–13 Dec	*LV* Destrem / Capt Stirling	N/K	Akaba and locality. 31 Dec—N.13 crashes in Wadi Araba, Grall and Stirling walk out and rescued.
	16–18 Dec	*Mot* Levasseur / Capt Ross	N.12?	
	22–24 Dec	*QM* Grall / Capt Stirling	N/K	
	29–31 Dec	*QM* Grall / Capt Stirling	N.13	

	1915			
Doris	1–3 Jan	*Mot* Levasseur N/K	N/K	El Arish and Gaza. Attempts to reach Hebron and Beersheba.
Aenne Rickmers	15 Jan	*LV* de Saizieu Capt Todd	N/K	Lake Bardawil. Reconnaissance of sandspit with intent of establishing a temporary ALG for the Egypt Detachment, RFC.
	17–22 Jan	*LV* de Saizieu, *Mot* Levasseur Capt Todd, Capt Herbert	N.15, N.16	Kossaima, El Arish, Rafah, Khan Yunis, Gaza, and Beersheba.
	26–30 Jan	*QM* Grall, *QM* Trouillet Capt Todd, Capt Herbert	N.11, N.15	El Arish. 28 Jan: Recovered N.14, forced down with engine failure on reconnaissance from Port Said previous day. Crew (*QM* Le Gall and Lt Partridge) swam ashore and were killed by own forces.
	1–8 Feb	*QM* Grall, *QM* Trouillet Lt Paul, 2Lt Hillas	N.15, N.17	El Arish, Gaza, Beersheba.
	14–21 Feb	*QM* Grall Capt Todd, Lt Paul	N.17, N/K	El Arish, Gaza, Beersheba.
Rabenfels	7–13 Feb	N/K N/K	N.12 (7–12), N.16 (13 only)	El Arish.
	15–16 Feb	N/K N/K	N.16	El Arish. 16 Feb: N.16 capsized immediately after take off. Machine badly damaged but salvaged, and rebuilt. Pilot and observer OK.
	20–27 Feb	*LV* de Saizieu, *QM* Trouillet Lt Paul, 2Lt Hillas	N.14, N.15	Rafa, El Arish. Flights to El Auja and Jebel Libni.

Ship	Date	Crew	Aircraft	Notes
Aenne Rickmers	24 Feb / 3 March	LV Destrem, QM Grall, Capt Todd, Lt Paul.	N/K	Gaza. Unsuccessful attempts to reach Beersheba and Jaffa.
Rabenfels	5–14 March	QM Trouillet, N/K	N.15, N.16	Gaza and Jaffa. Flight to Beersheba, 8 March.
Aenne Rickmers	5–11 March	LV Destrem; Lt Paul	N.11 and/or N.17	Smyrna. *Aenne Rickmers* torpedoed 11 March. Returned to Port Said following repairs and refit early July.
Rabenfels	19 March / 4 April	LV de l'Escaille, LV de Saizieu, QM Trouillet, Capt Herbert, 2Lt Hillas	N.14, N.15. N.16	Smyrna, Mudros, Tenedos.
Rabenfels	7–13 April	LV Cintré, Capt Todd, Lt Paul	N.15, N.16	El Arish, Gaza, Beersheba. Spotting for French ship *Saint-Louis*.
Rabenfels	15–20 April	QM Gral, Capt Todd, Lt Paul	N.11, N.16	El Arish, Gaza. Spotting for French ship *Saint-Louis*.
Rabenfels	20–30 April	LV Cintré, Capt Todd, Lt Paul	N.18, N.20	Mersina, Gulf of Alexandretta. Reconnaissances of railway and bridges.
Rabenfels	5–10 May	QM Grall, 2Lt Ledger	N.11, N.19	Ashkelon, Haifa, Jaffa, El Arish. Railway reconnaissances. Spotting for French ship *Jeanne d'Arc*.
Rabenfels	19–26 May	LV Cintré, 2Lt Ledger	N.16, N.18	El Arish, Gaza, Beersheba. Reconnaissance Jaffa, Ludd, Ramleh. Bombs dropped on new camp at Ramleh.
Montcalm	10–27 May	LV Destrem, Lt Baxter	N.20	Red Sea and Gulf of Akaba.

Ship	Date	Crew	Aircraft	Operations
Rabenfels	29 May / 9 June	SM Grall, Lt Paul, 2Lt Ledger	N.11, N.19	Haifa, Beirut, Gulf of Alexandretta, Chicaldere Bridge, Gaza, El Arish.
Hardinge	4–16 June	LV Cintré, Major Fletcher	N.16, N.18	Red Sea. Mowila, Dibbah (Dhuba), Jeddah.
	23–30 June	LV Cintré, 2Lt Hillas	N.16, N.18	Red Sea. Series of coastal reconnaissance flights, north to south, from Dibbah to Rabegh.

Rabenfels renamed *Raven II* on 12 June 1915.

Ship	Date	Crew	Aircraft	Operations
Raven	13–23 June	LV Destrem, SM Trouillet, Capt Herbert, Capt Todd	N.15, N.20	El Arish, Jaffa, Beirut, Haifa..
	28 June / 10 July	LV de Saizieu; 2Lt Ledger	N.17, N.19	El Arish, Jaffa, Beirut. Bombs dropped at El Arish. Gaza, Jaffa and Ramleh reconnaissances completed.
Aenne Rickmers	8–9 July	N/K	N/K	Submarine search and patrol, Aboukir Bay.
	20–25 July (approx)	SM Trouillet, Capt Herbert	N/K	Delaman Chai, Makri, Bay of Marmaris (Turkish coast above Rhodes). Local reconnaissances, bombed barracks at Makri, submarine search.
Raven	14–28 July	LV Destrem, SM Grall, Major Fletcher, 2Lt Hillas	N.15, N.20	El Arish, Jaffa, Haifa, Beirut, Tripoli, Latakia. The flights were mostly local reconnaissances. Bombed Samaria and Tul Keram. The ship attacked a shore camp with rifles(!) and captured two schooners.
	3–14 Aug	LV Cintré, Lt Paul, 2Lt Ledger	N.18, N.19	El Arish, Jaffa, Ashkelon, Rafa, Gaza.

Aenne Rickmers renamed *Anne* on 5 August 1915.

Ship	Dates	Crew	Aircraft	Operations
Anne	12–23 Aug	*LV* Destrem, *SM* Grall, Major Fletcher, Capt Herbert	N.11, N.20	El Arish, Haifa; Bay of Tarsus, Gulf of Alexandretta (with *Raven*, 18–20 Aug); Beirut. 15 Aug: *Anne* captures a small schooner and hoists it aboard!
Raven	18–24 Aug	*LV* de Saizieu, *SM* Trouillet, Lt Paul, 2Lt Ledger	N.14, N.17	Bay of Tarsus, Gulf of Alexandretta (with *Anne*, 18–20 Aug); Jaffa. To reconnoitre the railway and bomb Chicaldere bridge with machines from *Anne*.
Anne	30 Aug / 4 Sept	*LV* de Saizieu, *SM* Trouillet, Major Fletcher, 2Lt Hillas	N.16, N.17	al flights, but one inland to Nazareth.
Anne	14–16 Sept	Helping French ships rescue Armenians from Musa Dagh. No flying operations.		
Anne	19–26 Sept	*LV* Destrem, *SM* Grall, Lt Paul, 2Lt Ledger	N.11, N.20	El Arish, Ruad, Jaffa. Local flights to check on camps and railway. Coastal flights adjacent to Ruad island to search for troops and gun batteries.
Anne	9–14 Oct	*LV* de Saizieu, *SM* Trouillet, Lt Paul, 2Lt Ledger	N.14, N.17	Gaza, Ruad, Jaffa. Bombed trenches on coast near Ruad. A return to Beersheba after almost five months. 10 Oct: N.14 lands 12 miles from the coast returning from Beersheba, Nieuport destroyed by Arabs. Trouillet and Paul PoW.
Anne	19–24 Oct	*LV* Destrem, *QM* Bourgeois, Major Fletcher, 2Lt Ledger	N.19, N.20	El Arish, Ashkelon, Jaffa, Gaza. Flights to reconnoitre railway, including Ramleh and Beersheba.

	5–13 Nov	*LV* de Saizieu, *SM* Grall Major Fletcher, 2Lt Ledger	N.11, N.22	Gaza, Jaffa.
Anne	6–12 Dec	*LV* Destrem, *QM* Bourgeois Major Fletcher, Capt Chute	N.19, N.23	Ruad, Gulf of Alexandretta. Determines that Ruad is unsuitable as *an aeroplane station*. Survey of the railway between Mersina and Seihun.
	15–17 Dec	*LV* Destrem, *QM* Bourgeois Capt Chute	N.23	Acre, El Arish. Local coastal patrols.
	21–24 Dec	*LV* de Saizieu, *SM* Grall 2Lt Ledger, 2Lt Williams	N.17, N.22	Gaza and Beersheba. 21 Dec: N.17 17 came down with punctured fuel tank, returning from Beersheba. 2Lt Ledger killed when he fired on Turks after landing, *LV* de Saizieu PoW.
Raven	28 Dec to 6 Jan 1916	*LV* Destrem Major Fletcher	N.20, N.23	Haifa, Jaffa.
1916				

From 12 January 1916 *l'escadrille de Port-Saïd* became part of the East Indies and Egypt Seaplane Squadron. See Appendix 9 for all 1916 Shipboard Operations.

APPENDIX 9

Summary of Shipboard Operations EIESS, 1915–1918

Unlike *l'escadrille de Port-Saïd*, the majority of operations by the East Indies and Egypt Seaplane Squadron took place from the squadron's seaplane carriers. Only a small number of submarine patrols were flown from Port Said itself. An almost complete set of operation reports for the EIESS have survived and the following Table has been built from this material.

On 12 January 1916 *l'escadrille de Port-Saïd* became part of the East Indies and Egypt Seaplane Squadron. The remaining operations of *l'escadrille de Port-Saïd* from *Anne* and *Campinas* are included below.

Main source: TNA AIR 1 Files. Secondary sources: Diaries and Ship's Log Books. Machines: Short = Short 184. Sch = Sopwith Schneider. Baby = Sopwith Baby. FHB = Fairey Hamble Baby.

Ship	Dates	Pilots / Observers	Machines	Comments
		1916		
Anne	11–23 Jan	SM Grall, QM Bourgeois Capt Chute, 2Lt K.L. Williams *(unless otherwise stated, all Williams are K.L. 'Long' Williams)*	Nieuport: N.19, N.22	11–12 Jan. Beirut, El Min (Tripoli). Local flights, looking for evidence of submarines and mines. Sheltered from bad weather in Famagusta harbour, 15–19 Jan. 22–23 Jan. Local check up El Arish and railway Ramleh to Tul Keram.
Ben-my-Chree	20 Jan	Flt Cdr Edmonds, Flt Lt Dacre Lt Childers	Short: 846, 849	Reconnaissances of Kossaima & El Auja, and Gaza & Lake Bardawil.

Ship	Date	Crew	Aircraft	Mission
Ben-my-Chree	30/31 Jan	Flt Cdr Edmonds, Flt Lt Dacre, Flt Lt Wright, Lt Childers	Short: 850	Reconnaissance to Beersheba.
Anne	31 Jan / 1 Feb	*LV* Destrem, Major Fletcher	Nieuport: N.23	Gulf of Alexandretta. Reconnaissance of railway
Ben-my-Chree	11 Feb	Flt Cdr Edmonds, Flt Wright, Lt Childers, 2Lt Williams	Short: 849	Reconnaissance of Bardia & Sollum. Short 849 lost, Edmonds and Childers rescued.
Anne	11 Feb	*SM* Grall, Capt Chute	Nieuport: N.22	Gaza. Beersheba and El Auja.
Ben-my-Chree	7 March	Flt Cdr Edmonds, Flt Lt Wright, Lt Childers, 2Lt Williams	Short: 846, 850	Reconnaissances of Beersheba and Gaza regions.
Empress	7 March	Flr Cdr Sitwell, Flt Lt Field, Lt Bourne, Lt E. Williams	Short: 8088, 8381	Reconnaissances of Ludd, Ramleh and El Falujeh.
Anne	7–8 March	*SM* Jeanblanc, *SM* Roussillon, Capt Chute, 2Lt Finney	Nieuport: NB.1, NB.2	El Arish. Local reconnaissance and Kossaima, El Auja, Rafa.
Campinas	24–31 March	*LV* Destrem, *QM* Gramont, Lt Burd, Lt Wedderspoon	N.19, N.23	Castellorizo, Ruad, Beirut.
Raven	31 March to 4 April	Flt Cdr Edmonds, Flt Lt Bankes-Price, Flt Lt England, Flt Lt Wright, FSL Clifford, Lt Millard	Short: 850 Sch: 3721, 3722, 3727, 3775, 3790	Aden
Empress	Detached to Aegean, 3 April 1916. Returns to Port Said, 11 April 1917. See Chapter 12.			

Anne	16 April	*SM* Grall, *SM* Roussillon 2Lt Bourne, 2Lt Finney	Nieuport: N.22, NB.2	Gaza, El Arish. Final flight to Beersheba, N.22 crashed on take off. Local flights Gaza-Rafa and El Arish. **Final operations of *l'escadrille de Port-Saïd*.**
Anne	21/25 April	Flt Lt Wright Lt Hughes, 2Lt Smith	Short: 8004, 8054	Castellorizo, local coastal reconnaissances.
Raven	25 April	FSL Clifford	Sch: 3774	Reconnaissance El Arish to Bir Mazar
Ben-my-Chree	17/18 May	Wing Cdr Samson, Flt Lt Bankes-Price, Flt Lt England, FSL Man 2Lt Baxter	Short: 8082, 8084 Baby: 8188, 8189	Reconnaissances El Arish to Gaza. Spotting for Monitors *M.15* and *M.23*.
Ben-my-Chree	22/23 May	Wing Cdr Samson, +? 2Lt Benn, +?	Short: 8087 Sch: 3790	Reconnaissances Jaffa. 23 May: Short 8087, crashed on take off and sank.
Ben-my-Chree	26/27 May	Wing Cdr Samson, Flt Lt Bankes-Price, Flt Lt England, Flt Lt Wright 2Lt Baxter, 2Lt Benn, 2Lt Smith	Short: 850 Sch: 3774, 3790	Reconnaissances and bombing, Jaffa, Ramleh, Gaza, and El Arish.
Ben-my-Chree	2–16 June	Wing Cdr Samson, Flt Lt Bankes-Price, Flt Lt England, Flt Lt Wright, FSL Paine Lt Wedderspoon, 2Lt Benn, 2Lt Clarke	Short: 850, 8082 Sch: 3789, 3790 Baby: 8189	Aden and Red Sea.
Raven	7 June	Flt Lt Brooke Lt Stewart	Short: 8090	Reconnaissance of El Arish and area.
Raven	22 June	FSL Man	Short: n/k Sch: 3786	Bombed camps at El Arish.

Raven	1–9 July	Flt Lt Brooke, Flt Lt Dacre, FSL Man Lt Brown, Lt Ravenscroft, 2Lt King	Short: 8090, 8091 Sch: 3786 Baby: 8189	El Arish reconnaissance. Haifa reconnaissance and bombed camps. 2 July: FSL Man (Schneider 3786) forced down on to sea, rescued. Sopwith sunk. Castellorizo, local coastal reconnaissances.
Ben-my-Chree	6–11 July	Wing Cdr Samson, Flt Lt Bankes-Price, Flt Lt England, Flt Lt Maskell, FSL Paine Sub Lt Kerry, 2Lt Benn, 2Lt Smith, *LV* Picard	Short: 850 (S.2), 8054 (S.4) Sch: 3789 (B.1), 3790 (B.3)	Reconnaissances and bombing, El Arish, Ruad and Beirut.
Ben-my-Chree	12–21 July	Wing Cdr Samson, Flt Lt Bankes-Price, Flt Lt England, Flt Lt Maskell Sub Lt Kerry, Capt Benn, Lt Millard, 2Lt Smith, 2Lt Williams, 2Lt Woodland	Short: 850, 8054, 8372 Sch: 3774, 3789, 3790	Submarine Patrols from Port Said
Ben-my-Chree	23/24 July	Wing Cdr Samson, Flt Lt England, Flt Lt Maskell Capt Benn, Lt Stewart, 2Lt Smith	Short: 8054, 8372 Sch: 3771, 3774	Reconnaissance to Beersheba, and coast between Khan Yunis and El Arish.
Ben-my-Chree	25/26 July	Wing Cdr Samson, Flt Lt Bankes-Price, Flt Lt England, Flt Lt Maskell Capt Benn, Lt Stewart, 2Lt Smith	Short: 8054, 8372 Sch: 3771, 3774	Reconnaissances from El Arish to Haifa, including inland to Nazareth, and Nablus.
Raven	28 July to 7 Aug	Flt Lt Clemson, FSL Smith Lt Millard, 2Lt Williams	Short: 8075, 8091	Red Sea. Akaba—photo mosaic. Mowila and Wejh (Sherm Wej).

Ship	Date	Crew	Aircraft	Operation
Anne	5/6 Aug	Flt Lt Dacre, FSL Man, Sub Lt Kerry, Lt Stewart	Short: 8090	Reconnaissances Gaza to Rafa.
Anne	10 Aug	Flt Lt Dacre, FSL Man, Lt Stewart	Short: 8090, Sch: 3777	Mersina Bay. Spotting for *Pothuau* (Fr), bombing railway station and shed.
Raven	10 Aug	Flt Lt Brooke, Flt Lt Clemson, 2Lt Smith, 2Lt Williams	Short: 8075, 8091	To spot for monitor *M.21*, Bardawil, Bir el Mazar. Short 8091 in running fight with Rumpler from FA300. Severely damaged but reached *Raven* and recovered. Crew Brooke and Smith unhurt.
Ben-my-Chree	14/15 Aug	Wing Cdr Samson, Flt Cdr England, Flt Lt Bankes-Price, Flt Lt Maskell, FSL Dover, Lt Woodland, Capt Benn, 2Lt King	Short: 8054, 8080, 8372, Sch: 'B.8'	Reconnaissances of Haifa Valley to El Afule and Nazareth, Jaffa-Ludd-Ramleh, coast from Gaza to El Arish.
Ben-my-Chree	24–30 Aug	Wing Cdr Samson, Flt Cdr Dacre, Flt Cdr England, Flt Lt Bankes-Price, Flt Lt Maskell, FSL Dover, Lt Woodland, Capt Benn, 2Lt King	Short: 8054, 8080, 8372 (S.7), Baby: 8135, 8136, 'B.10'	Squadron bombing raids on railway at El Afule, and Wadi el Hesi viaduct. 25 Aug: Flt Cdr Dacre PoW after engine failure returning from El Afule. Baby 8136.
Anne	25/26 Aug	Flt Lt Brooke, FSL Man, 2Lt Williams	Short: 8091, Sch: 3777	Also: *Ben-my-Chree*: Nahr el Kebir, Adana and railway bridge, shipping near Mersina. *Anne*: Reconnaissances Eski Adalia, Adalia and Fineka Bay to search for submarine base.
Raven	25–28 Aug	Flt Lt Clemson, Flt Lt Leigh, FSL Paine, FSL Smith, Sqn Cdr Malone, Sub Lt Kerry, Lt Millard	Short: 8045, 8075, Baby: 8189	

Ship	Dates	Crew	Aircraft	Notes
Anne	1 Sept to 26 Oct	Flt Lt Clemson, FSL Smith, Sqn Cdr Malone, Lt Millard, 2Lt Williams	Short: 8004, 8075	Red Sea. *Anne* returned to Suez to coal and replenish supplies 19/20 Sept. Leaving one Short with *Northbrook*, later transferred to *Dufferin*. Met *Raven* 26 Oct, see below. (*Anne* appears to have undergone a long refit following this cruise as there are no records of her until January 1917.)
Ben-my-Chree	13/14 Sept	Wing Cdr Samson, Flt Cdr England, Flt Lt Bankes-Price, Flt Lt Leigh, Capt Benn	Short: 8372 Sch: 3779, 3789 Baby: 8135	Reconnaissances of Gaza and Bir Saba (Beersheba). Spotting for monitors. 14 Sept: 3789 engine drops out on landing. Samson OK, Baby not repaired?
Ben-my-Chree	16/17 Sept	Wing Cdr Samson, Flt Cdr England, Flt Lt Bankes-Price, Flt Lt Maskell, FSL Man, FSL Nightingale, Sub Lt Kerry, 2Lt King	Short: 8080, 8372 Sch: 3770, 3777, 3778 Baby: 8135	Spotting for monitors and reconnaissance of El Arish. 17 Sept: Combat with Rumpler C.I from *FA300* (*Ltn* Bulow and *Ltn* Hesler). Flt Lt Bankes-Price (Baby 8135) shot down and killed. FSL Man (Sch 3777) forced down on to sea with engine failure, rescued.
Raven	26 Oct to 2 Nov	Sqn Cdr Malone, plus 2 n/k, 2 n/k	Short: 8004, 8075	Red Sea. Met with *Anne* and took aboard 2 Shorts and Malone, with 2 POs & 11 air mechanics. No record of any flights.

Ben-my-Chree	2/5 Nov	Flt Cdr England, Flt Lt Maskell, FSL Nightingale, FSL Dover Lt Woodland, Capt Benn, Lt Stewart	Short: 8072, 8080, 8372	Adalia and Castellorizo.
Ben-my-Chree	1–3 Dec	Flt Cdr England, Flt Lt Clemson, FSL Dover, FSL Nightingale Sub Lt Kerry, Capt Benn, Lt Woodland	Short: 8080, 8372 Sch: 3778	Reconnaissances from Haifa to Gaza. 2 Dec: Short 8372 hit by AA fire and brought down overland. FSL Nightingale and Lt Woodland PoW.
Raven	1–21 Dec	Flt Lt Burling, FSL Man, FSL Worrall Lt Stewart, 2Lt Williams	Short: 8004, 8075	Red Sea. Yenbo.
Ben-my-Chree	22 Dec	Wing Cdr Samson, Flt Cdr England, Flt Lt Clemson, FSL Henderson, FSL Smith Lt Meade, Lt W.L. Samson, Sub Lt Kerry, Capt Benn, AM Lloyd	Short: 8080, 8090	Reconnaissances from Haifa to Gaza.
Ben-my-Chree	26/27 Dec	Wing Cdr Samson, Flt Cdr England, Flt Lt Brooke, Flt Lt Clemson, Flt Lt Maskell, FSL Henderson, FSL Smith Lt Meade, Lt W.L. Samson, Capt Benn	Short: 8080, 8091 Sch: 3770, 3778 Baby: 8188	Squadron bombing raids on Chicaldere railway bridge.
Raven	27 Dec	Flt Lt Burling, FSL King Lt Stewart, 2Lt Williams	Short: 8004, 8075	

1917

Ship	Dates	Personnel	Aircraft	Notes
Ben-my-Chree	8/9 Jan	Wing Cdr Samson, Flt Cdr England, Flt Lt Clemson, Flt Lt Maskell, FSL Smith, FSL Worrall Lt Meade, Lt W.L. Samson, Sub Lt Kerry, Capt Benn	Short: 8080 Sch: 3770, 3778 Baby: 8188	Castellorizo. *Ben-my-Chree* shelled and sunk, 9 January 1917. All machines lost. Minor injuries only.
Anne	16/26 Jan	Flt Lt Burling, FSL King, FSL Paine Lt Stewart, 2Lt Williams	Short: 8004, 8075	Red Sea. Capture of Wejh, 23/24 January 1917. 23 Jan: Lt Stewart shot and killed by rifle fire whilst over Wejh.
Anne	26 Feb to 1 March	Flt Cdr Clemson, Flt Lt Leigh, Flt Lt Smith Sub Lt Kerry, 2Lt Pakenham-Walsh	Short: 8021, 8022	Reconnaissances of Haifa, Beirut and Damascus.
Raven	10 March to 12 June	Wing Cdr Samson, Flt Cdr Clemson, Flt Cdr England, Flt Lt Burling, FSL/Flt Lt Smith, FSL Stephens Obs Lt Kerry, Capt Knight-Bruce, Lt Kempson, Lt Meade	Short: 8018, 8019, 8021, 8090 (S.5) Baby: N1014, N1016?	Aden, Indian Ocean, Laccadive, Maldive and Chagos Islands, Colombo. 8018 and 8019 both flown as two-seaters. Wing Cdr Samson leaves EIESS whilst at Colombo, 8 May 1917. Wing Cdr C.E. Risk had assumed command of EIESS at Port Said on 25 April. 16 May @ Colombo, *Raven*'s log book: Discharged Sopwith N1014 [+N1016?] & motor car & motor boat to depot. Flight Commander Clemson, Flight Sub Lieutenant Stephens, 1 PO & 8 mechanics discharged to depot.—Colombo Detachment.

Anne	21/23 March	Flt Cdr Brooke, Flt Lt Leigh Lt Millard	Short: 8020	Reconnaissances of Haifa, Nazareth and railway at El Afule. 8020 flown as two-seater. ***Anne*'s final operation with EIESS and RNAS.**
Empress	17 April	FSL Worrall 2Lt Ferguson	Short: 8020	Deir-el-Belah, submarine search. 8020 flown as two-seater.
Empress	12/13 May	Flt Cdr Brooke, Flt Lt Leigh, FSL King Lt Millard, 2Lt Ferguson	Short: 8004, 8020, 8075	Beirut—bombing and submarine search. 13 May: Short 8020 flown solo, Flt Cdr Brooke, with 500-lb bomb.
Empress	22/23 June	Flt Lt Worrall, Flt Lt Barr, FSL Bronson, FSL King Wing Cdr Risk, Lt Millard, 2Lt Ferguson, 2Lt Pakenham-Walsh	Short: 8004, 8019, 8020, 8075	Bombing railway junction at Tul Keram. 8019, 8020, both flown as two-seaters.
Empress	30 June	Flt Lt Leigh, Flt Lt Wood 2Lt Ferguson, 2Lt Pakenham-Walsh	Short: 8019, 8020	Search for submarine and ship sending SOS 30 miles NW of Port Said. Ship was moored in harbour. 8019, 8020, both flown as two-seaters.
Empress	12 July	Flt Lt Wood, Flt Lt Worrall Wing Cdr Risk, Lt Obs Kerry	Short: 8019, 8020	Reconnoitre for mines and submarines in approaches to Port Said. Ship was moored in harbour. 8019, 8020, both flown as two-seaters.

Empress	14–16 July	Flt Lt Bronson, Flt Lt Burling, Flt Lt Leigh, Flt Lt Smith, Flt Lt Wood, Flt lt Worrall, FSL King Wing Cdr Risk, Obs Lt Kerry, Capt Kempson, Lt Millard, 2Lt Pakenham-Walsh	Short: 8004, 8018, 8019, 8020	Bombed cotton factories NW of Adana. Submarine patrol on approach to Port Said. 8018, 8019, 8020, all flown as two-seaters.
Empress	16–18 Aug	Flt Lt Bronson, Flt Lt Burling, Flt Lt Worrall, FSL King Wing Cdr Risk, 2Lt Pakenham-Walsh	Short: 8018, 8019, 8004, 8022	Beirut – Bombing and submarine search. Both 8018 and 8019 flown solo, carried 36 16-lb bombs.
Empress	17/21 Sept	Flt Lt Bronson, Flt Lt Burling, Flt Lt Wood, Flt Lt Worrall, FSL King Wing Cdr Risk, Obs Lt Kerry, Capt Kempson, Capt Millard, 2Lt Newton, 2Lt Pakenham-Walsh	Short: 8021, 8022, N1090, N1263	(At Ismailia) W/T and spotting practice with monitor *M.31*.
Empress	26/28 Sept	Flt Lt Bronson, Flt Lt Burling, Flt Lt Worrall, FSL King Capt Millard, 2Lt Pakenham-Walsh	Short: 8018, 8019, 8021, N1090	Beirut—bombing, submarine search, spotting for *Grafton*. 8018 (King solo) carried 2 65-lb & 16 16-lb bombs. 8019 (Burling solo) failed to start.

Ship	Date	Crew	Aircraft	Notes
Empress	8–12 Oct	Flt Cdr Clemson, Flt Lt Leigh, Flt Lt Popham, Flt Lt Stephens, Flt Lt Wood Capt Kempson, 2Lt Newton, Ldg Mch Price	Short: 8018, 8019, 8021, N1091	Bombing Adana and Chicaldere railway bridge. 9 Oct: Wood killed when 8018 (solo) blew up on landing with bombs onboard. 3 65-lb & 16 16-lb. Also 9 Oct: 8019 (Popham solo) also 3 65-lb & 16 16-lb. Radiator punctured and engine overheated. 11 Oct: Clemson and Newton, shot down, Short 8021 at Adana. Clemson wounded PoW, Newton PoW.
Operations in support of Third Battle of Gaza				
City of Oxford	29 Oct to 2 Nov	Flt Lt Burling, Flt Lt Leigh, Flt Lt Smith Capt Kempson, Capt Millard, 2Lt Ferguson	Short: 8019, N1090, N1091, N1262	Spotting off Wadi el Hesi and Gaza for *Raglan*, *Ladybird*, and *Aphis*. 8019 flown as two-seater. 30 Oct: N1262, with Burling and Kempson, attacked by 'Halberstadt' Kempson wounded. Machine salved by *Raglan*.
Raven	1/2 Nov	Flt Lt Bronson, Flt Lt Worrall 2Lt Pakenham-Walsh	Short: 8022, N1263	Spotting for *Requin* (Fr), *Grafton*, and *Raglan*, off Wadi el Hesi. N1263 to *Raglan*. **Raven's final operation with EIESS and RNAS.**
Empress	1–3 Nov	Flt Lt King, Flt Lt Leigh, Flt Lt Popham, Flt Lt Smith, Flt Lt Stephens	Baby: N1028, N1036, N1038, N1129 FHB: N1209, N1210	Tulkeram, bombing Jiljulie(?) railway bridge. Haifa, bombing oil factory. 2 Nov: N1036, N1038, and N1210 machines lost, N1129 damaged, but recovered. Leigh and Stephens injured, other pilots OK.

City of Oxford	3–9 Nov	Flt Lt Burling, Flt Lt King, Flt Lt Popham, Flt Lt Smith Capt Kempson, Capt Millard, 2Lt Ferguson, 2Lt Pakenham-Walsh	Short: 8019, N1090, N1091, N1263	Spotting off Wadi el Hesi and Gaza for *Raglan*, *M.15*, *M.22*, *M.29*, *M.31*, *Requin* (Fr), *Ladybird*, and *Aphis*. 8019 flown as two-seater.
1918				
Empress	24 Jan to 4 Feb	Flt Lt Bronson, Flt Lt Leigh, Flt Lt Worrall Capt Kempson, Lt Ferguson, Lt Pakenham-Walsh	Short: N1090, N1581, N1582, N1590	Mudros. Night Bombing *Goeben/Yavuz*. Night 27/28 Jan: N1582 shot down and captured by Turks. Bronson and Walsh PoW.
City of Oxford	13 Feb to 29 March	Flt Cdr Leigh, Flt Lt King, Flt Smith, FSL Pennington Capt Millard, Lt Ferguson, Ob Lt Kennedy, CPO Goodwin, Ldg Mch Robertson, AM Low, AM Perry	Short: 8020, 8022, N1263, N1639	Red Sea. Operations in southern Red Sea (No visit to Aden). 8020 flown as two-seater. **Final Seaplane Carrier Operation of the East Indies and Egypt Seaplane Squadron, RNAS.**

Port Said. A bathing party at the island base posing with a Short and an Egyptian labourer. CPO Stone standing on the right. *(Stone Family Collection)*

RAF Port Said Babys 1918

Lieutenant Geoffrey W. Morey, RAF, poses on one of the handful of Blackburn Baby's remaining in service with the Port Said Squadron, RAF, in 1918. Taken at the Out Station, Port Said, the lighthouse can be seen in the background, in the latter half of 1918. Morey was originally from Adelaide, which explains the Australian flag on the fuselage aft of the roundel. The serial is not fully visible but may be N2132. *(From the estate of S/Ldr Frank Barry Willmott, RAAF.)*

Another victory for an Egyptian dhow. Blackburn Baby N2072 at the small boat dock at the northern end of the island base. The Baby hit the dhow whilst landing in the harbour at Port Said on 20 August 1918, pilot Lt C.H. Biddlecombe, RAF, was unhurt. There is no information about the condition of the dhow, but the Baby was written off. *(From the estate of S/Ldr Frank Barry Willmott, RAAF.)*

APPENDIX 10

Some Comparative Ranks

Royal Navy Rank	RNAS Rank †	RFC/RAF Rank	French Naval Rank	German Army	Ottoman Army ‡
Captain	Wing Capt	Colonel	*Capitaine de vaisseau*	*Oberst*	*Miralay*
Commander	Wing Commander	Lt Colonel	*Capitaine de frégate*	*Oberstleutnant*	*Kaymakam*
Lt Commander	Squadron Commander	Major	*Capitaine de corvette*	*Major*	*Binbaşi*
Lieutenant	Flight Commander	Captain	*Lieutenant de vaisseau*	*Hauptmann*	*Yüzbaşi*
	Flt Lieutenant	Lieutenant		*Oberleutnant*	*Mülazım-ı-evvel*
Sub-Lieutenant	Flight Sub-Lieutenant	2nd Lieutenant	*Enseigne de vaisseau*	*Leutnant*	*Mülazım-ı sani*
Chief Petty Officer	CPO Mechanic 1st, 2nd or 3rd Grade	Flight Sergeant	*Premier-maître*	*Feldwebel*	*Başçavuş*
Petty Officer 1st Class	Petty Officer Mechanic	Sergeant	*Maître-principal*	*Vizefeldwebel*	*Çavuş*
Petty Officer 2nd Class			*Second-maître*		
Leading Seaman	Leading Mechanic	Corporal	*Quartier-maître*	*Obergefreiter*	*Onbaşi*
Able Seaman	Air Mechanic 1st/2nd Grade	Air Mechanic 1st Class	*Matelot breveté*	*Gefreiter*	*Nefer*
Ordinary Seaman	Aircraftsman	Air Mechanic 2nd/3rd Class	*Matelot*	*Soldat*	

The Table opposite is intended only as a guide to ranks encountered in this book and does not include senior ranks. I have used the Royal Navy and Royal Naval Air Service as the base and extrapolated from there. Ranks do not always have direct equivalents, especially when comparing navy to army, nation to nation, and enlisted ranks. So, some of the comparisons are approximate.

---

† Extract from Admiralty Circular Letter C.W.13964/14 – 26.6.14,
Royal Naval Air Service—Organisation (TNA ADM 1/8378)

Rank in the Royal Naval Air Service.

Officers of the Royal Naval Air Service will be graded in the following ranks, and will take rank and command accordingly:—

Wing Captain with the relative rank of Captain, RN.

Wing Commander with the relative rank of Commander, RN.

Squadron Commander (when in command) with relative rank of Lieutenant Commander, RN.

Squadron Commander (when not in command) with the relative rank of Lieutenant over 4 years' seniority (but senior to all Flight Commanders).
On attaining 8 years' seniority in the relative rank of Lieutenant these Officers will rank with Lieutenant Commanders, RN.

Flight Commander with the relative rank of Lieutenant, RN over 4 years' seniority.

Flight Lieutenant with the relative rank of Lieutenant, RN.

Flight Sub-Lieutenant with the relative rank of Sub-Lieutenant, RN.

Grades.

All ratings enrolled in the Royal Naval Air Service will be graded in one of the following [see Table left] ratings. Men who, for special reasons, may not be so graded, will continue in all respects under the conditions of the general naval service as regards pay, advancement, &c.

---

‡ Based on *Handbook of the Turkish Army 1916* and
Notes on the Turkish Army 1915

Notes on pronunciation: Modern Turkish uses the Latin alphabet with a few additional characters and some unique pronunciations. It is a phonetic language, meaning words are generally pronounced exactly as they are spelled. Stress typically falls on the last syllable of a word.

i is pronounced like the ee in see, but ı (without the dot) is pronounced like the u in but

c is like the j in jungle, ç is like the ch in cherry

s is the same as in English, but ş is like sh in ship

ö is like e in her or the German ö, while ü is like u in universe or the French u

g is hard, as in the word get, but ğ is silent and lengthens the preceding vowel

BIBLIOGRAPHY

ARCHIVES

ENGLISH

The National Archives

Service Records, various ADM188, ADM273 and AIR 76 files.

ADM 53—Log Books for HMFA *Rabenfels*, HMS *City of Oxford*, *Doris*, *Hardinge* and *Warrior*. Sadly, the Log Books for *Aenne Rickmers / Anne* have not survived. Additional Log Books (including *Ben-my-Chree*, *Empress*, *Raven*) were consulted online—see below.

ADM 137/364 to 368, 545, and 546 – Weekly Reports of the Vice Admiral Eastern Mediterranean Squadron, Salonika and the Aegean Sea, February to December 1916.

ADM 137/3667 – Proceedings of Court Martial for the Loss of HMS *Ben-my-Chree*.

ADM 186/600 Report, ADM 186/601 Plates, ADM 186/602 Maps. Committee appointed to investigate the attacks delivered on, and the enemy defences of the Dardanelles Straits: The Mitchell Report.

AIR 1/117/15/40/35, Movement of a flight of RFC aeroplanes to Egypt, 14 July 1914 – 30 Jan 1915.

AIR 1/271/15/226/119, HMS *Anne* reports 26 Feb to 1 March 1917.

AIR 1/271/15/226/120, Weekly patrol reports from the British Adriatic Force, British Aegean Squadron, and R.N.A.S. seaplane squadrons operating from Malta and the East Indies and Egypt station, with summaries.

AIR 1/271/15/226/121, Operations Against *Goeben* & *Breslau*, 20–29 January 1918.

AIR 1/279/15/226/129, Training of Observers, March-April 1916.

AIR 1/361/15/228/10, Concluding Report, RNAS Operations Against *Goeben*, 27–28 January 1918.

AIR 1/436/15/284/1, HMS *Empress* Employment and Correspondence, March 1915 to December 1918.

AIR 1/436/15/289/1/ Telegrams and report – *Raven* at Aden, March 1917.

AIR 1/455/15/312/35, EIESS and RAF Anti-Submarine Patrols, Egypt, Feb-Nov 1918.

AIR 1/648/17/122/393, RNAS Operations, Aden, April 1916.

AIR 1/649/17/122/409, Report on RAF Mediterranean, 8 September 1918.

AIR 1/649/17/122/420, RNAS re-organization, Eastern Mediterranean, January–March 1916.

AIR 1/649/17/122/422, Report on RNAS Units Eastern Mediterranean Squadron, 26 Feb to 27 March 1916.

AIR 1/651/17/122/452, Recommendations of Officers and Men of the French Seaplane Squadron Port Said [also EIESS], 10 June 1916 to 10 November 1917.

AIR 1/654/17/122/503, Report on the RNAS in Eastern Mediterranean, Oct–Dec 1915.

AIR 1/660/17/122/620, Summary of Work Egypt Seaplane Squadron, 18–27 May 1916.

AIR 1/665/17/122/714, Work of Short seaplanes with HMS *Roberts* at Rabbit Islands.

AIR 1/665/17/122/722, *l'escadrille de Port-Saïd*, Observers Reports, January 1915 to March 1916.

AIR 1/667/17/122/743, Summary of Reports of Operations, HMS *Raven*, May/June 1917.

AIR 1/1168/204/5/2589. Egypt Detachment RFC (30 Squadron), 18 May 1915 to 14 October 1915. Contains the War Diary, November 1914 to November 1915, and Flight Reports, November 1914 to August 1915.

AIR 1/1706/204/123/64, HMS *Raven II*, reports April to August 1916.

Air 1/1706/204/123/65, HMS *Raven II*, reports August to December 1916.

AIR 1/1707/204/123/67, HMS *Raven II*, reports March to Nov 1917.

AIR 1/1707/204/123/68, HMS *Ben-my-Chree*, reports Jan to May 1916.

AIR 1/1707/204/123/69, HMS *Ben-my-Chree*, reports Aug to Dec 1916.

AIR 1/1708/204/123/70, HMS *Ben-my-Chree*, reports Jan 1916.

AIR 1/1708/204/123/71, EIESS Reports, Jan and Feb 1916.

AIR 1/1708/204/123/73, HMS *Anne*, reports Jan 1916 to March 1917.

AIR 1/1708/204/123/74, HMS *Empress*, reports March 1916 to Feb 1918. Excludes Aegean operations.

AIR 1/1708/204/123/75, HMS *City of Oxford*, reports October 1917 to March 1918.

AIR 1/1709/204/123/76, Summary of Work Done, all EIESS ships, January 1916 to March 1918.

AIR 1/1710/204/123/79, Port Said and Alexandria Seaplane Sqns, Submarine Patrols, Dec 1917 to Nov 1918.

AIR 1/1711/204/123/86, Re-numbering of RNAS Units, March to April 1918.

AIR 1/1711/204/123/89, Organisation *Anne* and *Raven II*, 1915–16.

AIR 1/1712/204/123/95, Turkish Seaplane Station Asia Minor Coast.

AIR 1/1712/204/123/98, Organisation of EIESS, January to March 1916.

AIR 1/1713/204/123/103, Attack on Submarine off Alexandria, 14 June 1918.

AIR 1/1713/204/123/104, EIESS Organisation of Seaplane Squadron Base, 1916.

AIR 1/1713/204/123/109, State of Aircraft Available Alexandria and Port Said, June 1918.

AIR 1/1720/204/123/190, HMS *City of Oxford*. Intelligence and Operation Reports, Feb and March 1918.

AIR 1/1720/204/123/192, Personnel and Aircraft Requirements Seaplane Squadrons, June to December 1917.

AIR 1/2283/209/75/4, RFC [and RNAS] Organisation Egypt, December 1915 to May 1916.

AIR 1/2284/209/75/8, HMS *Raven II* – Operations in Gulf of Akaba, Red Sea, July to August 1916. [Includes photographs—see also WO 319/1.]

AIR 1/2285/209/75/13, HMFA *Raven II* and HMFA *Anne* reports Jan 1915 to March 1916. Including reports from HMS *Hardinge* and *Montcalm*.

AIR 2/38, Photographic Report—6 Wing, RNAS, 1915–1918.

AIR 20/611, Escorting Convoys, Port Said Squadron, 64 Wing, RAF Egypt, 1918.

AIR 27/191, 14 Squadron RFC/RAF Operational History, 1915–1939.

CAB 37/124/13, HMS *Doris*, Report of Proceedings Off Syrian Coast, 14th to 27th December, 1914.

WO 95/4401, War Diary, Administrative Commandant Port Said, January to June 1916.

WO 319/1, Aerial survey of 'Akaba', Palestine by HMS *Raven II*, August 1916.

Fleet Air Arm Museum

A.E. Popham Log Book and Diary.

G.B. Dacre Diary, *Ben-my-Chree*, 1915-1916.

—, A Prisoner in Turkey (YEORN 1987/374/1)

Imperial War Museum

IWM Documents.471a. Robert Erskine Childers, HMS *Ben-my-Chree* Diary, Vol.3, 26 December 1915 to 10 March 1916.

IWM Documents.011534. Midshipman Parkes-Buchanan Log, HMS *Minerva*, November 1914 to March 1915.

IWM Documents.12080. Private Papers—Samson, Charles Rumney, CMG, DSO, AFC.

—, Flying log-books, 1911 – end of 1918.

—, Samson's copy of EIESS reports, 18 May 1916 to 27 January 1917, in 4 Vols.

FRENCH

La Bibliothèque nationale de France (Gallica)

bnf.fr/ark:/12148/cb45152879b. *L'organisation de l'Aéronautique maritime / Capitaine de Frégate Louis Marie Alix de Carné et Capitaine de Vaisseau Jean Constantin*, ca.1920.

bnf.fr/ark:/12148/cb451513144, *Organisation et opérations de l'aviation maritime française en Egypte et en Syrie / Lieutenant de Vaisseau* de Saint-Maurice, 1923.

bnf.fr/ark:/12148/bpt6k9759930x, *La défense du canal de Suez / Lieutenant de Vaisseau* Lavigne (*École supérieure de guerre navale*. 1935–36).

Le Service Historique de la Marine, Château de Vincennes (SHM Vincennes)

SHM Vh 13, Reports of *lieutenant de vaisseau* Henri P. de L'Escaille.

GERMAN

German Military Archive, Freiburg

BA-RH61/1805; Hesselberger, Joseph; Report of the operation against the Island of Kekova 3 November 1916 for Marshal Liman von Sanders; 8 November 1917.

BA-RH61/1805; Schmidt-Kolbow, Karl; Report of the artillery activity during the operation against Meis Island; Smyrna; 23 January 1917.

PUBLISHED SOURCES

ENGLISH

Books

Anon, Air Publication 125, *A Short History of the Royal Air Force* (Air Ministry, revised 1936).

Anon, *Handbook of Aircraft Armament*, Admiralty, Air Department 1916 (IWM and The Naval & Military Press, reprint 2016).

Anon, *Report of the Committee Appointed to Investigate the Attacks Delivered on and the Enemy Defences of the Dardanelles Straits*, aka. The Mitchell Report, 1919 (Admiralty, Naval Staff, Gunnery Division, CB 1550)

Capt W.W. Benn, *In The Side Shows* (Hodder and Stoughton, 1919).

W.H.D. Boyle (Admiral of the Fleet, the Earl of Cork and Orrery), *My Naval Life 1886-1941* (Hutchinson, 1942).

J.M. Bruce, *Short 184 - Windsock Datafile 85* (Albatros Publications, 2001).

J.M. Bruce, *Short 184 - Profile Number 74* (Profile Publications, 1966).

Ian Burns, *Ben-my-Chree - Woman of My Heart, Isle of Man Packet Steamer and Seaplane Carrier* (Colin Huston, 2008).

Ian Buxton, *Big Gun Monitors* (Seaforth, 2008).

Dick Cronin, *Royal Navy Shipboard Aircraft Developments, 1912-1931* (Air Britain, 1990).

Lt Col P.G. Elgood, *Egypt and the Army* (Oxford U P, 1924).

Edward J. Erickson, *Palestine, The Ottoman Campaigns of 1914-1918* (Pen & Sword, 2016)

Major J. Everidge, RAF, *History of No.30 Squadron RAF. Egypt and Mesopotamia 1914 to 1919* (Naval and Military Press / IWM, 2009). This is a reproduction of a typescript dated May 1919, Air Ministry Library 11517.

Capt Cyril Falls, *Military Operations, Egypt & Palestine*, 3 volumes (HMSO, 1928-1930).

—, *Military Operations, Macedonia*, Vol.1 (HMSO, 1933).

Stuart Hadaway, *Pyramids and Fleshpots, The Egyptian, Senussi and Eastern Mediterranean Campaigns, 1914-16* (Spellmount, 2014).

—, *From Gaza to Jerusalem, The Campaign for Southern Palestine 1917* (Spellmount, 2015).

David Hobbs, *The Royal Navy's Air Service in the Great War* (Seaforth, 2017).

C.E. Hughes, *Above and Beyond Palestine* (Benn, 1930).

Philip Jarrett, *Frank McClean - Godfather to British Naval Aviation* (Seaforth, 2011).

Wing Commander C.G. Jefford, *Obervers and Navigators and other non-pilot aircrew in the RFC, RNAS and RAF* (Airlife, 2001).

H.A. Jones, *The War in the Air*, Vol.5 (Oxford UP, 1935).

Jean Labayle-Couhat, *French Warships of World War 1* (Ian Allan, 1974).

Mark Lax, Mike O'Connor and Ray Vann, *Wings Over Mesopotamia* (Cross and Cockade International, 2017).

Russell McGuirk, *The Sanusi's Little War* (Arabian Publishing, 2007).

David Méchin, *Oriental Adventures of the French Air Force, 1914-1918* (Aeronaut Books, 2022).

Jacques Mortane, *Raoul Lufbery and Marc Pourpe* (Schiffer, 2023)

David Murphy, *The Arab Revolt 1916-1918* (Osprey, 2008).

Dimitar Nedialkov, *The Air Component of the Kingdom of Bulgaria in the Great War* (Albatros MDV Publishing, 2018). Published in a joint Bulgarian-English format.

Henry Newbolt, *Naval Operations*, Vols.4/5 (Longmans Green, 1928 / 1931).

Ole Nikolajsen, *Ottoman Aviation 1909-1919* (Ole Nikolajsen, Updated Edition 2012).

Colin Owers, *Shorts Aircraft of WW1*, 3 Vols (Aeronaut Books, 2022).

Alan Palmer, *The Gardeners of Salonika* (Andre Deutsch, 1965).

Nicholas G. Pappas, *Near Eastern Dreams - The French Occupation of Castellorizo 1915-1921* (Halstead Press, Sydney, NSW, 2002).

Nicholas G. Pappas and Nicholas C. Bogiatzis, *An Island in Time - Castellorizo in Photographs 1890-1948* (Halstead Press, Sydney, NSW, 2015).

Leonard Piper, *The Tragedy of Erskine Childers* (Hambledon, 2003). Also known as *Dangerous Waters: The Life and Death of Erskine Childers*. This biography is the only one that attempts to detail Childers' wartime service, devoting almost sixty pages to it.

Gérard Pommier, *Nieuport 1875-1911, A Biography of Edouard Nieuport* (Schiffer, 2002).

C.R. Samson, *Fights and Flights* (Benn, 1930).Ray Sanger, *Nieuport Aircraft of World War One* (Crowood Press, 2002).

Ray Sturtivant and Gordon Page, *Royal Navy Aircraft Serials and Units, 1911-1919* (Air Britain, 1992).

Yigal Sheffy, *British Intelligence in the Palestine Campaign, 1914-1918* (Routledge, 2014).

Col W.F. Stirling, *Safety Last* (Hollis and Carter, 1953).

Vice Admiral C.V. Usborne, *Smoke On The Horizon* (Hodder and Stoughton, 1933).

Various, *Lawrence of Arabia & Middle East Air Power* (Cross and Cockade International, 2016)

Alan Wakefield and Simon Moody, *Under The Devil's Eye* (Pen & Sword, 2011).

Colonel A.P. Wavell, *The Palestine Campaigns* (Constable, 1928).

Capt L.B. Weldon, *Hard Lying* (Herbert Jenkins, 1925). A new 2023 edition, with a useful biography of Weldon, is available from Eland Publishing Ltd.

Naval Intelligence Staff Handbooks

Handbook of Asia Minor, Vol.1, 1919.

Handbook of Asir, 1916.

Handbook of Bulgaria, 1920.

Handbook of Macedonia, 1920.

Handbook of Syria (Including Palestine), 1920.

Handbook of the Hejaz, 1917.

Handbook of Turkey in Europe, 1917.

Handbook of Yemen, 1917.

Articles and Papers

Anon, Three Months on the Syrian Coast (HMS *Doris*), *The Naval Review*, 1915, pp.621-636.

John Baldry, British Naval Operations against Turkish Yaman 1914-1919, *Arabica*, Tome XXV, Fascicule 2, June 1978.

Captain W.H.D. Boyle, Naval Operations in the Red Sea, 1916-1917, *The Naval Review* (The Naval Society), Vol.XIII No.4, November 1925.

—, Naval Operations in the Red Sea, 1917-1918, *The Naval Review* (The Naval Society), Vol.XIV No.1, February 1926.

Bernard de Broglio, Footnote to a Sideshow, *The '14-'18 Journal*, 2018 Vol.2, The Australian Society of World War One Aero Historians.

Ian Burns, Aden Flight, Aviation in Aden 1915-1919, *Cross and Cockade International*, V46N4, 2015.

Ian Burns and Gunter Hartnagel, Castellorizo and the Loss of HMS *Ben-my-Chree*, *Cross and Cockade International*, V48N1, 2017.

Ian Burns, Floatplanes Over The Desert – The Adventures of *l'escadrille de Port-Saïd*, 1914-1916, *Cross and Cockade International*, V52N2/3/4, 2021.

Mark Connelly, The British Campaign in Aden, 1914-1918, *Journal of the Centre for First World War Studies*, 2:1 (2005), pp.65-96.

A.E. Ferko, Hans Henkel – The Desert Lion, *Cross & Cockade (US)*, V13N1, Spring 1972.

Dr Brian P. Flanagan (Editor), The Ottoman Air Force in the Great War – The Reports of Major Eric Serno, *Cross & Cockade (US)*, V11N2/3/4, 1970.

Dieter H.M. Gröschel and Jürgen Ladek, Wings Over Sinai and Palestine, The Adventures of Flieger Abteilung 300 'Pascha', *Over The Front*, V13N1, 1998.

Dieter H.M. Gröschell and Elimor Makevet, Flieger-Abteilung 301 and the German Aerial Force in Palestine in WWI, *Cross and Cockade International*, V52N3/4 and V53N1, 2021/22.

Thierry Le Roy, The Port-Said Squadron, The First Squadron of the French Naval Aviation, 1914-1916, *Over The Front*, V11 N4, 1996. Note: This is an abridged version of the *Avions* (qv) article.

Lt W.C.A. Meade, Our Seaplane Adventure, *The Wide World Magazine*, Vol.41, No.245, September 1918, pp.414-420.

David Nicolle, 'Ertugrul' the Blériot and 'Osmanli' the Deperdussin, *Air Enthusiast 96*, Nov/Dec 2001, pp.46-55.

Mike O'Connor and Mike Napier, 14 Squadron, A First War History, *Cross and Cockade International Journal*, V42N1 to V42N3.

Timothy D. Saxon, Anglo-Japanese Naval Cooperation, 1914-1918, *Naval War College Review*, Vol.53, No.1 (Winter 2000), pp. 62-92

Guy Duncan Smith and Ian Burns, The Aerial Crusader, *Cross and Cockade International Journal*, V44N2/3, 2013.

Lt K.L. Williams, Aviation in Egypt in 1915 – By An Observer in the French Seaplane Squadron, Port Said, *The Aeroplane*, 24 Sept 1919, pp.1198-1200.

FRENCH

Books

Paul Chack, *On Se Bat Sur Mer* (Les Editions de France, 1926). English translation by Commander L.B. Denman, RN, *The Entente Upon The Seas* (Vaillant-Carmanne, 1928).

Lieutenant de vaisseau Georges Douin, *L'attaque du canal de Suez* (Librarie Delagrave, 1922).

Commandant Bernard Frank, *Corsaires du XXe Siècle* (Flammarion, 1956).

Docteur Charles Héderer, *L'ile du Chateau-Rouge* (Société d'Éditions Géographiques, Maritimes et Coloniales, Paris, 1924)

Robert Feuilloy and Lucien Morareau, *L'aéronautique maritime dans la Grande Guerre* (ARDHAN, 2019).

Capitaine de vaisseau A. Thomazi, *La guerre navale dans la Méditerranée* (Payot, 1929).

Articles and Papers

Maud Jarry, L'Aéronautique navale naquit avec la Foudre, *Le Fana de l'Aviation*, No.377/378, Avril/Mai 2001.

Thierry Le Roy, L'Escadrille de Port-Saïd, Premier Escadrille De L'Aviation Maritime Française, 1914–1916, *Avions*, Nos. 34–36, Janvier/Février/Mars 1996.

Baron Marc de Villiers du Terrage, *Les Aérostiers Militaires en Egypte, Campagne de Bonaparte, 1798–1801* (Imprimerie G. Camproger, 1901). A paper presented at the International Aeronautical Congress in Meudon and read at the *Société de Navigation aerienne* on 28 March 1901.

GERMAN

Books

Hans Henkel (as Henkelburg), *Als Kampfflieger am Suez Kanal* (August Scherl, Berlin, 1917). English translation by Michael A. O'Neill, *Flames in the Desert* (BfP Books, 2023).

Vizeadmiral Eberhard von Mantey, *Der Krieg zur See 1914–1918, Der Kreuzerkrieg, Band 3 – Die deutschen hilfskreuzer* (E.G. Mittler und Sohn, Berlin, 1937).

TURKISH

Books

Mustafa Aydemir (editor), *Ben Bir Türk Zabitiyim (I am a Turkish Officer)* (Denizler Kitabevi, Istanbul, 2007).

Online Resources

academia.edu/70398646/Ottoman_German_1918_Naval_Raid_on_Imroz_Island_Stinging_Turkish_Critique_1927

albindenis.free.fr/Site_escadrille/page_recherches A site devoted to French Aviation Services of WW1.

naval-history.net/OWShips-LogBooksWW1

netmarine.net/tradi/marins14-18/index — A site devoted to Gallantry Awards, French Navy of WW1.

trove.nla.gov.au/work/26989887 — The Akabarbarian—HMS Minerva—Christmas 1915.

ENDNOTES

1. *L'Association pour la Recherche de Documentation sur l'Histoire de l'Aéronautique Navale* (ARDHAN) (www.aeronavale.org) is devoted to researching, writing and publishing the history of the French Naval Air Arm – *l'Aéronavale*.

2. The Urabi Revolt was a nationalist uprising in Egypt from 1879 to 1882 seeking to end British and French influence over the country. It was led by and named for Colonel Ahmed Urabi.

3. Modern Türkiye, but contemporary usage Turkey is employed throughout the book.

4. Third Supplement (Number 29010) to the *London Gazette* of 15th of December, 1914.

5. The preceding is necessarily brief. For a full, and accessible, overview of Egypt and Britain see Stuart Hadaway, *Pyramids and Fleshpots, The Egyptian, Senussi and Eastern Mediterranean Campaigns, 1914–16* (Spellmount, 2014).

6. Marc de Villers, *Les Aérostiers Militaires en Egypte*, for all the quotes regarding the balloon flights, except endnote 7.

7. Abd al-Rahman al-Jabarti, *Tarikh muddat al-faransis bi-Misr* (The History of the Period of the French Occupation in Egypt), 1798. 1975 translation by Shmuel Moreh, *Al-Jabarti's Chronicle of the French Occupation* (Markus Wiener Publishers, 1997).

8. Accounts vary as to its capacity, ranging from 4500 to 5000 ft^3.

9. *The Egyptian Gazette*, 4 March 1890. General Dormer was then commander-in-chief of the British troops in Egypt. The village was Shibin El Qanater, now part of El Obour City, about 30 km NE of Cairo.

10. Jasper Kemmis in *The Sphinx*, 25 December 1909. *The Sphinx* was an English language illustrated weekly published during the 'season' (December to April) between 1892 and 1947. Mr Kemmis was probably Jasper Harold Kemmis, born in Dublin in 1886, who appears to have been a jobbing journalist and editor. He certainly had some familiarity with the aviation scene in Britain, having attended the Doncaster Meeting in October 1909.

11. www.thefirstairraces.net/meetings.php.

12. Jasper Kemmis from the 12 February 1910 edition of *The Sphinx*.

13. Jasper Kemmis, *ibid*.

14. Edited from the 19 February 1910 edition of *The Sphinx*.

15. *Flight*, 27 December 1913.

16. *The Aeroplane*, 15 Jan 1914.

17. Védrines returned to Paris still desiring to fight M. Quinton with French Army revolvers at ten paces. The affair made headlines in the International press for several weeks, but experts in dueling protocol eventually decided that there was no cause for attempted bloodshed.

18. Alternatively, *Osmanli* per David Nicolle. However, according to Nikolajsen, *Osmanli* a two-seater Deperdussin was delivered to Turkey on 15 March 1912, on the same day as a single seater, and served until shot down over the Black Sea on 5 May 1915. A second two-seater Deperdussin *Prens Celaleddin* was delivered on 10 July 1912, this was used on the attempted Cairo flight but crashed at Jaffa on 11 March 1914.

19. Probably a new machine similar to the one that Roland Garros had flown across the Mediterranean from St Raphael (Frejus) to Bizerta on 23 September 1913. A non-stop flight lasting 7 hours 53 minutes. See Mortane, *Raoul Lufbery and Marc Pourpe*, p.66.

20. Edited from *Flying*, March 1914, p.49.

21. Olivier was Louis Olivier, *Aéro-Club de France* certificate 556 issued on July 28, 1911, records indicate that he was killed in a flying accident at Melun, France, on 2 Sept 1913. However, it seems that whilst he was badly injured, his passenger, Desvaux de Lyf, was killed. There are several photographs of the 30 January crash, and more about Olivier, and the Bogos prize, in Phil Jarrett's *Fine Flights and First-Class Scandals*, *Aeroplane Monthly*, April 2005.

22 These details from *The Aeroplane* 22 January 1914. However in the 16 April edition it states, 'a distance of 1,375.5 miles, was covered in a net flying time of 16 hrs. 18 mins.—an average speed of 83 m.p.h'. Either way, a fine achievement.

23 'HW' (E. Howard-Williams), *Something New Out of Africa* (Pitman, 1934), p.186.

24 For the complete story of this flight see Philip Jarrett, *Frank McClean – Godfather to British Naval Aviation* (Seaforth, 2011). For more about the Short S.80 see Colin Owers, *Shorts Aircraft of WW1, Vol.1* (Aeronaut Books, 2022), pp.44–47.

25 Possibly Giuseppe 'Beppe' Leonardi, *Aéro-Club de France* Brevet 935 (1912), Italian Brevet 122.

26 The two ships served with the Grand Fleet as HMS *Agincourt* and HMS *Erin*, respectively.

27 Edward J Erickson, *Palestine, The Ottoman Campaigns of 1914–1918* (Pen & Sword, 2016), pp.8/9.

28 The Men of Anzac, edited by C.E.W. Bean, *The Anzac Book* (Cassell, 1916), p.72.

29 Colonel A.P. Wavell, *The Palestine Campaigns* (Constable, 1928). There are many later printings, the latest in 2016, text and pagination appears unchanged throughout. Later editions have updated authorship to Field Marshal Earl Wavell.

30 All the same, traffic dropped significantly over the wartime period from about 20 million tons in 1914 to a low of 9 million in 1917, rising to about 14.1 million in 1918.

31 1st Camel Regt, formed Jan/Feb 1915 in Syria. Based initially at Beersheba, to Baghdad in Sept 1915, possibly returning to Palestine 1916/17.

32 See Chapters 3 and 7.

33 The Turkish divisions mentioned by Col Wavell were under strength. In the Hejaz nominal strength was 10000 men, Asir 6000, and Yemen 22000. Actual effectives were much lower. *Notes on the Turkish Army, 1915*, War Office.

34 Boué de Lapeyrère was probably the first air minded admiral in any navy. He had been involved with the creation of *l'Aéronautique maritime* as early as 1910/11 and remained a supporter.

35 *Vice-amiral* Dartige du Fournet *Commandant la 3ème Escadre* during 1915, relieved by *vice-amiral* Frédéric Moreau in 1916.

36 All Royal Navy ships should be understood to carry the prefix HMS (His Majesty's Ship) or HMAS (His Majesty's Australian Ship). Ships of the Royal Indian Marine (1892–1934) were prefixed RIMS, although during the war they were more frequently referred to as HMS. The Kingdom of Italy, *Regia Marina*, used RN (*Regia Nave* – Royal Ship); the French navy, *Marine nationale*, did not employ a prefix. Allied navies Anglicised *Dai-Nippon Teikoku Kaigun* (Imperial Japanese Navy) to a prefix, IJNS. Similarly, German and Austrian navy ships were prefixed SMS (*Seiner Majestät Schiff*). The Ottoman Navy does not appear to have used a prefix.

37 Wemyss was knighted in the 1916 New Years Honours list and promoted Vice Admiral in December 1916.

38 See Chapter 15.

39 For the history of French Naval Aviation in WW1 see, Robert Feuilloy and Lucien Morareau, *L'Aéronautique maritime dans la Grande Guerre* (L'ARDHAN, 2019). RNAS strength from Hobbs, *The Royal Navy's Air Service*.

40 Wing-warping was controlled by the feet, whilst the rudder and elevator were operated by a control column. A fore and aft movement of the control column operated the elevators, a sideways motion the rudder.

41 For more on the Nieuport monoplanes see, Ian Burns, Floatplanes Over the Desert: The Adventures of *l'escadrille de Port-Saïd, 1914–1916*, Part 1, *Cross and Cockade International* Journal, V52N2.

42 Although often written as Henri, the spelling of Henry is confirmed by his birth certificate.

43 De l'Escaille writing in *Revue des forces aériennes*, July 1930, pp.785–803.

44 Carné, *L'organisation de l'Aéronautique maritime*, p.17.

45 Whilst Britain bought a major shareholding in 1875 the *Compagnie universelle du canal maritime de Suez* remained, at heart, French, with a French president until 1956.

46 *Revue des forces aériennes, op cit.*

47 The Hervieu design, a predecessor to the Bessonneau which Hervieu also helped design, was for a single machine and could be quickly erected by a team of four men. Each hangar was 40 ft wide × 43 ft deep and 12 ft high.

48 *Revue des forces aériennes, op cit.*

49 De l'Escaille's reports for *l'escadrille de Port-Saïd* are held in *Le Service Historique de la Marine*, Château de Vincennes. They are quoted extensively in this and the following chapter.

50 2/Lt Kenneth L Williams, 2nd Rajput LI, had joined the Egypt Detachment, RFC, on 20 January 1915 and transferred to *l'escadrille de Port-Saïd* on 30 January. This and following quotes are from his article *Aviation in Egypt, 1915*, The Aeroplane, 24 September 1919, pp.1198/1200.

51 The French Navy used 100 mm guns, the French Army 105 mm. Both shells weighed ±15 kg, with an explosive charge of 2 kg or less. In comparison, the '20 lb' Hales Bomb weighed around 18.5 lb / 9 kg with a charge of 4.5 lb / 2 kg.

52 From Elgood, *Egypt and the Army*, pp.118/9.

53 Midshipman Maurice John Parkes-Buchanan Log (IWM Documents.011534).

54 ADM 137/1091: SNO Egypt to CO *Doris*, Sailing Orders, 13 Dec. 1914. Quoted in Yigal Sheffy, *British Military Intelligence in the Palestine Campaign*, p.50.

55 Accounts can be found on the Web, but the most readable is contained in Vice Admiral Usborne's *Smoke On The Horizon*, Chapter II.

56 *TB.63* (1885), 125 feet, 75 tons, 2 3-pdr and 3 14-inch torpedoes, 19½ knots as built. Served in the Mediterranean throughout the war, Lieutenant in Command Thomas Dunne. The Nieuport was presumably towed to Port Said as open deck space on the torpedo boat was very limited.

57 Stirling, *Safety Last*, p.55–60.

58 Umm al-Rashrāsh existed until 1949 when an abandoned British police post was captured by Israeli Defence Forces. The area subsequently being renamed Eilat.

59 Stirling, *op cit.*

60 *London Gazette*, 10 April 1915, p.3551. For gallant behaviour on reconnaissance in a Hydroplane at Akaba about the 6th January, 1915.

61 For a full account of the Egypt Detachment, RFC, see Ian Burns, *Wings Over The Sinai, Cross and Cockade International* Journal, V53N4, 2022, pp.242–265.

62 A ship's 'tonnage' can be very confusing, a merchant ship can have as many a five different values. Generally, for a merchant ship the Gross Tonnage is most often quoted. This is the total cubic enclosed space of the ship – engines, accommodation, freight holds, etc. – at 100 cubic feet to the ton. For warships, the normal measurement is Displacement Tonnage, the total weight of the ship and everything on board.

63 *Deutsche Dampfschiffahrts-Gesellschaft* Hansa (German Steamship Company Hansa).

64 Captain Weldon, Egyptian Survey Dept, was initially attached to General Staff Intelligence working out of Headquarters in Cairo, he was posted to *Aenne Rickmers* to work for Major Elgood at Port Said. This and subsequent extracts are from his memoir *Hard Lying*.

65 A reference to Harry Tate's famously chaotic Music Hall sketches.

66 AIR 1/2285/209/75/13. Herbert was an observer with *l'escadrille de Port-Saïd*.

67 Anjer has not been identified.

68 This and the preceding quote from the third Supplement to the *London Gazette*, 20 June 1916.

69 *London Gazette*, Friday, 10 September 1915, p.9065.

70 *Ibid.*

71 *Saint Louis* (1899) one of three *Charlemagne*-class pre-dreadnoughts, 4–305 mm guns in two turrets, plus 10–138.6 mm, 8–100 mm and 20–47 mm guns. At the time she was flagship of the 3ème Escadre.

72 Williams, *op cit.*

73 *Montcalm* (1900), a *Gueydon*-class armoured cruiser, 2–194 mm guns in single turrets, 8–164.7 mm guns in casements and numerous lighter guns.

74 *Requin* (1877) an ironclad barbette battleship, completely reconstructed and rearmed in 1901 she was reclassified as *garde-côtes cuirassé*. She carried 2–274 mm, 6–100 mm and 10–47 mm guns. She served along the Syrian coast until the end of the war, and will re-enter our story several times.

75 Essentially, a build up of deposits, often due to poor quality coal and/or salt from sea water, in the boiler condenser tubes causing a reduction in power. The only solution is to replace the tubes.

76 *Swiftsure* and *Triumph* (1904) two identical pre-dreadnoughts ordered by Chile but purchased for the Royal Navy before completion; Two twin 10-inch turrets and 14 7.5-inch guns. *Euryalus* (1901) a Cressy-class armoured cruiser, two 9·2-inch and 12 6-inch guns.

77 One of four identical *38m torpilleurs* built by Schneider et Cie, Chalons-sur-Saône, France, in 1907 for the Ottoman navy; *Demirhisar, Sultanhisar, Sivrihisar,* and *Hamidabad*. Two 37 mm QF guns and three 450 mm torpedo tubes.

78 Weldon, *Hard Lying*, pp.25–90.

79 *London Gazette*, 10 April 1915, p.3551. For gallant behaviour on reconnaissance in a Hydroplane at Akaba about the 6th January, 1915.

80 See George H. Cassar, *Reluctant Partner – French Participation in the Dardanelles Expedition* (Helion, 2019), pp.105–112.

81 One of de l'Escaille's reports indicates that several sets of landing gear were available, having been brought out with *l'escadrille*.

82 There was already a *Raven* in the Royal Navy, an old gunboat (1882) being used as a training ship in the UK. *Raven II* will not be used again, except in direct quotes.

83 Both Jenkins and Kerr were awarded the DSC for 'services in command of a seaplane-carrying vessel, on the East Indies and Egypt Station' (LG 20/7/17).

84 See Chapter 2.

85 *d'Entrecasteaux* (1898) protected cruiser, 2–240mm, 12–138.6mm, and numerous smaller calibre; *d'Estrées* (1899) third-class cruiser, 2–138.6mm, 4–100mm, and 8–47mm.

86 *Jeanne d'Arc* (1903) armoured cruiser, 2–194 mm in single turrets, 14–138.6 mm guns in casements and numerous lighter guns. Some of the 138.6 mm guns were removed pre-war when the ship was serving as a training ship and may not have been reinstalled. *Jeanne d'Arc* was flagship of *3ème Escadre* from April 1915 until 30 March 1916.

87 Modern Çakaldere. The bridge still exists, it is located approximately 1.5 kilometres east of the village, and is in regular daily use.

88 *Irmak* translates as river, and *Jeihan* is one of several names (including the Pyramus of antiquity) applied to the Ceyhan.

89 One of the first records of photographs by *l'escadrille*. For more information see Chapter 5.

90 *Raven*'s report, AIR 1/2285/209/75/13.

91 In antiquity, Strabo in his *Geographica* (ca. 7BCE), mentioned a freshwater sea spring on to which the people let down from the water-fetching boat an inverted, wide-mouthed funnel made of lead, the upper part of which contracts into a stem with a moderate-sized hole through it; and round this stem they fasten a leathern tube, which receives the water that is forced up from the spring through the funnel. Now the first water that is forced up is sea-water, but the boatmen wait for the flow of pure and potable water and catch all that is needed in vessels prepared for the purpose and carry it to the city.

92 Weldon, pp.104/5.

93 See Chapter 7.

94 *London Gazette*, 21 December 1915, p.12795.

95 Decauville Railway – a light narrow gauge system employing prefabricated track lengths of 600 mm gauge. This was probably the similar German *feldbahn* system.

96 *London Gazette*, 21 December 1915, p.12794.

97 Gramont appears to be the correct family name, although records also use Gramant and Grammont.

98 Following quotes are from AIR 1/654/17/122/503. Papers concerning the RNAS, Eastern Mediterranean, Oct–Dec 1915.

99 Rear Admiral Rosslyn Erskine Wemyss, responsible for the successful re-embarkation of troops from Anzac and Suvla.

100 HMS *Ark Royal, Ben-my-Chree*, and *Empress*, also two Balloon Ships.

101 AIR 1/1708/204/123/71. *Ben-my-Chree* Reports Jan and Feb 1916.

102 Malone Service Records, ADM 273/2/10.

103 AIR 1/1713/204/123/104, Organisation of Seaplane Squadron Base, 1916.

104 Elgood, *Egypt and the Army*, p.119, fn.2.

105 For the story of *Ben-my-Chree* see Ian Burns, *Ben-my-Chree – Woman of My Heart* (Colin Huston, 2008).

106 *The Aeroplane*, 4 July 1912, p.45. For more details see Colin Owers, *Shorts Aircraft of WW1, Vol.1, Early Types*, pp.18–23.

107 *Ben-my-Chree* appears to have brought three Short 184, 846, 849 and 850, and two Sopwith Schneider, 3721 and 3722, floatplanes from Galliopoli in her hangar.

108 Promoted Commander, RNR, 12 April 1917.

109 Leonard Piper, *The Tragedy of Erskine Childers*, provides a very plausible account on pp.139/144.

110 R.E. Childers, *Ben-my-Chree* Diary.

111 Rear Admiral Wemyss, although appointed on 1 January, did not arrive in Port Said, aboard his flagship HMS *Euryalus*, until 23 January and took up command two days later.

112 For details of RNAS (and RFC) observer training see Jefford, *Observers and Navigators*.

113 AIR 1/279/15/226/129. Training Observers March-April 1916.

114 For more details of Short Brothers early RNAS aircraft see Colin Owers, *Shorts Aircraft of WW1, Vol.1* (Aeronaut Books, 2022), Chapter 2.

115 *Canadian Aviation*, June 1936 – February 1937.

116 Squadron Leader T.D. Hallam, *The Spider Web* (Blackwood, 1919), pp.95/96. Shipboard operations were slightly different, see Eliot Millar King's account in Appendix 1.

117 The Diary of Lt G.B. Dacre, RNAS. Fleet Air Arm Museum, RNAS Yeovilton. Also the source of all following quotes by Dacre.

118 ADM 101/335. HMS *Ben-my-Chree*, Medical Officer's Report, March 1915 to May 1916.

119 Group Captain G.E. Livock, *To the Ends of the Air* (HMSO, 1973), p.32.

120 Hyams later served with 269 Squadron at Alexandria and 1918/19.

121 AIR 1/1712/204/123/98. Notes on Organization of Seaplanes and Seaplane-Carrying Ships, Jan-March 1916.

122 Writers were primarily clerical, being responsible for legal, pay, and career issues for a crew. Chief Writer ranks equivalent to a CPO.

123 Frederick Cusden was promoted CPO on 1 August 1916. DCM – *London Gazette*, 20 July 1917.

124 Probably a Goerz Anschutz, a folding camera popular with press photographers because of its light weight and speed of operation. It would have been provided with 13x18 cm (approx 5x7 inch) film back.

125 AIR 1/1713/204/123/104.

126 Cecil Eldred Hughes, a Lieutenant in the Somerset Light Infantry and Wedgwood Benn's brother in law, was appointed Assistant Intelligence Officer in late 1916 and continued to serve in that role with the RAF until 1919. His memoir *Above and Beyond Palestine* was originally published privately in Cairo, 1923, it was republished in 1930 by Ernest Benn (W.W. Benn's older brother).

127 Flt Cdr C.H.K Edmonds, and FSL G.D. Smith.

128 Probably, Captain Wedgwood Benn, of whom more anon.

129 One of several army privates employed as batmen.

130 Probably CPO Levy Natkiel who had also served at Yarmouth before being posted to Port Said.

131 Diary of Flt Lt A.E. Popham. Popham had arrived in Port Said on 21 May.

132 From this point forward any use of Short should be understood to refer to the Short 184. Any other model will be identified as such.

133 *Ben-my-Chree* log book (ADM 53/35187) says 'Pilot & Observer were picked up by HMT [trawler] Charlsin.' *Charlsin*, ex-*Esteburg*, was a captured German trawler, it was sunk by *UC-74* off Mersa Matruh, 30 September 1917.

134 For a full account see Russell McGuirk, *The Sanusi's Little War* (Arabian Publishing, 2007).

135 This was the final recorded flight for Lt E. *'le petit'* Williams, all future references to Williams refer to 2Lt K.L. *'long'* Williams.

136 The quote from Sitwell's log book is drawn from description of Bonhams' Auction, Lot 275, 22 November 2011.

137 *TDM250* was built in 1904, she displaced only 90 tons and carried an armament of two 37-mm guns and three 356mm (14 inch) torpedo tubes. Converted for anti-submarine work, with the aft torpedo tubes were replaced by a naval *soixante-quinze* and several depth charges, *TDM250* was a familiar escort for ships of the EIESS.

138 See AIR 1/665/17/122/722, *Campinas* Reports, March 1916.

139 See Chapter 2.

140 Weldon, p.118/9.

141 *Als Kampfflieger am Suez Kanal,* pp.24/6. Published in 1917 and written under the thin pseudonym Ltn Hans Henkelburg. The translation and interpretation is mine.

142 See Chapter 11.

143 Weldon, p.125/7.

144 Short: 846, 849, and 850. Sopwith Schneider: 3721, and 3722.

145 The Khedivial Mail Steamship & Graving Dock Co. was formed in 1898 to operate ships and docks owned by various departments of the Egyptian Government. The dry dock itself can be dated back to around 1860.

146 *The Aeroplane* 22 July, 29 July, and 5 August 1914.

147 See Chapter 3.

148 See Ian Burns, *Wings Over The Sinai - The Egypt Detachment RFC, 1914–1915, Cross and Cockade International* Journal, V53 N4, Winter 2022.

149 For a concise and contemporary account see G.P. Neumann, translated by J.E. Gurdon, *The German Air Force in the Great War* (Hodder and Stoughton, 1921), pp.243–265. Also Serno per fn.150 below. For modern accounts see fn.153 and 154 below.

150 Dr Brian P. Flanagan (Editor), *The Ottoman Air Force in the Great War – The Reports of Major Eric Serno,* Cross & Cockade (US), V11 N2/3/4, 1970. Major Serno from the German Military Mission, had been appointed General Inspector of the Ottoman Turkish Air Force in 1915 until 1918. Issus is modern İskenderun.

151 Neumann, *op. cit.*, pp.243–258. *Hptm* Hellmuth Felmy should not to be confused with *Obltn* Gerhard Felmy, his younger brother, a pilot with FA300 from September 1916 to August 1917.

152 A.E. Ferko, Hans Henkel—The Desert Lion, *Cross & Cockade* (US), V13N1, 1972. Dittmar, a German NCO pilot rated as *Mulazim-i-evvel* (Lieutenant) in Turkish service, had been serving at Gallipoli, but was seconded to Palestine for the purpose of establishing *FA300*'s base.

153 Dieter H.M. Gröschell and Jürgen Ladek, Wings Over Sinai and Palestine, *Over The Front*, V13N1, Spring 1998.

154 WO 95/4401.

155 A Parabellum MG14. FA300's Pfalz E.IIs were fitted with a Spandau MG08.

156 *Als Kampfflieger am Suez Kanal, op. cit.* Hans Henkel, born in Frankfurt in 1892, began studying at the *Technische Universität* Berlin after completing his secondary school education, then discovered his love of flying and trained as a pilot at the Albatros flying school in Johannisthal, gaining his German Certificate 486 on 16 August 1913. In the same month he joined the Field Artillery Regiment No. 63. After the start of the war, he flew with the *FFA 2* on the Western Front before joining *FA 300*.

157 Lynden-Bell Papers, IWM, quoted in Yigal Sheffy, *British Military Intelligence in the Palestine Campaign*, p.190.

158 For full account of *Flieger-Abteilung 300*, see Dieter H.M. Gröschell and Jürgen Ladek, Wings Over Sinai and Palestine, *Over The Front*, V13N1, Spring 1998. For *Flieger-Abteilung 301*, see Dieter H.M. Gröschell and Elimor Makevet, Flieger-Abteilung 301 and the German Aerial Force in Palestine in WWI, *Cross and Cockade International*, V52N3/4 and V53N1, 2021/22.

159 The Reports of Major Eric Serno – *op. cit.*, and Ole Nikolajsen, *Ottoman Aviation*, Chapter 7.

160 This is probably a reference to the EIESS as a detachment of 14 Squadron, RFC, did not arrive in country until November 1916.

161 A Lt Fazel (sic) was noted as being under instruction at the Bristol School, Salisbury Plain, in *Flight*, 19 October 1912 (p.936), along with several other Turkish officers, but there is no record of him qualifying for a Royal Aero Club certificate. The Turkish officers were recalled on the outbreak of First Balkan War.

162 The Damascus *Tayyare Istasyonu* (Aircraft Station) between August 1916 and November 1917 refurbished 13 Rumpler C.Is which were taken over in a non-flyable condition from the German unit *FA300*. Once completed these aircraft were issued to Turkish units in Arabia and Iraq. Ole Nikolajsen, *Ottoman Aviation*, Chapter 7.

163 For a full account of RFC/RAF operations, including details of the Handley Page 0/400, see *Lawrence of Arabia and Middle East Air Power, Cross and Cockade International*, 2016.

164 Ole Nikolajsen, *Ottoman Aviation*, Chapter 8, p.120.

165 See Colin Owers, *Shorts Aircraft of WW1, Vol.1, Early Types*, pp.18–23.

166 Sir Frederick Sykes, *From Many Angles* (Harrap, 1942), p.170.

167 Richard Bell Davies, *Sailor in the Air* (Peter Davies, 1967), p.75.

168 The Vickers 2-pdr was taken on board on 16 May 1916, and its installation can be seen on photographs. However, the 12-pdr HA does not appear on any photographs although *Ben-my-Chree*'s log book confirms it was test fired on 30 May 1916.

169 Wedgwood Benn, *In The Side Shows* (Hodder & Stoughton, 1919), p.16.

170 C R Samson, *Fights and Flights* (Benn, 1930), p.296. Like most memoirs of the period, one has to be careful with quotes as the story was often more important than the facts.

171 Samson, pp.306–308.

172 Benn, *op. cit.*, p.124.

173 *Espiegle* (1901), a *Cadmus*-class sloop, was typical of the older warships doing useful work in this theatre of the war. Built at the turn of the century, she epitomized the late Victorian gunboat. A handsome ship having a fine schooner bow complete with figurehead, and three tall masts intended to carry auxiliary sails, she packed six 4-inch guns into a 1070 ton, 200 foot long hull. *M.15* and *M.23* were from a numerous class of small monitors built during the war; *M.15* to *M.33*, the latter is now a Museum Ship at Portsmouth Historic Dockyard. They were armed with a single 9.2-inch gun (*M.15* to *M.28*) or two 6-inch guns (*M.29* to *M.33*).

174 Samson, *op. cit.*, p.302.

175 The Rumplers of *FA300* had left El Arish earlier, but would return.

176 Samson, *op. cit.*, p.302.

177 Short 8087 was a total loss, together with all contents.

178 This is the earliest recorded use of Samson Serials – see Appendix 2.

179 Gröschell and Ladek, Wings Over Sinai and Palestine, confirms the loss of a Rumpler, one damaged (later repaired), and damage to Pfalz wings. There is a photograph of the Rumpler on p.14 of the article. The surviving Rumplers bombed Port Said that evening in retaliation.

180 16-lb Bombs. Almost all EIESS flight records refer only to 16-lb, not 20-lb, bombs. These were HERL 16-lb, although some 20-lb HE Hale bombs may have been used.

181 Samson's Log Book for this period is a small notebook with scribbled entries. There are no serials and, although flight durations are recorded, no take off or landing times.

182 According to the Flight Reports (AIR 1/1707/124/123/68) on 24 July both Samson and England were flying S.4/8054 at the same time on different routes. Two Shorts, 8054 and 8372 were aboard, Samson probably flew 8372 which would become his favourite machine.

183 Benn, *op. cit.*, p.120.

184 Ole Nikolajsen, *Ottoman Aviation*, Chapter 12.

185 AIR 1/1708/204/123/73. Short 8090, with FSL Man and observer Lt Stewart.

186 AIR 1/1706/204/123/65, source for all quotes 10 August 1916.

187 Gröschell and Ladek, *op. cit.*, p.21.

188 *Dacre's PoW Diary, A Prisoner in Turkey* (YEORN 1987/374/1, Fleet Air Arm Museum.), 'It is because I have to kill time that I have written this little book, hoping that it will not be taken from me.' It was not and is the source for all of Dacre's quotes in this chapter.

189 The Lewis gun 'hoppers' used by the EIESS were of the 47 round variety. There is no documentary or photographic evidence for the use of the 97 round magazine drum.

190 Guy Duncan Smith, *The Aerial Crusader*, written whilst back home in California during a long leave in 1918. He was a great story teller and, as it was intended for family only, I have had to be selective in the use made of his memoir. Frustratingly, it names no names and provides no dates, but was laid out chronologically. Suffice to say the quotes used can, mostly, be supported by official reports, but there is always room for a good story. Unless otherwise indicated all of Smith's subsequent quotes are from this source. For his full story see The Aerial Crusader, *Cross and Cockade International* Journal, V44N2/3.

191 The bridge at Deir Sineid will feature again in Chapter 16.

192 ADM 273/4/22, Service Records, G B Dacre.

193 AIR 1/1707/204/123/69, Bankes-Price's Report of Flight.

194 Samson, *Fights and Flights*, p.331.

195 *Fights and Flights*, p.332.

196 *Samson's Operation Reports.*

197 Short 8372 was built by Phoenix Dynamo, one of a batch of 12 machines, and was not fitted with a Scarff ring mounting. A photos of 8368 (first Phoenix Dynamo machine), and 8383 (a Short built by Frederick Sage), show a pillar mounting in the observer's cockpit and what may be a second pillar on the starboard side of the cockpit. 8372 was the first but not the last EIESS Short to carry two Lewis guns.

198 Gröschell and Ladek, p.25, mention 'two aircraft dropping sixteen 20 lb bombs from an altitude of 100m,' apparently causing no damage.

199 Samson's Operation Report.

200 AIR 1/1707/204/123/69, Maskell's Report of Flight.

201 *Fights and Flights*, p.336.

202 Benn, *In The Side Shows*, p.133.

203 William 'Bill' Samson, who after serving with 3 Wing at Gallipoli had been with the RNAS in East Africa. He had joined *Ben-my-Chree* in October 1916.

204 Mann Egerton Co were building ten similar 184s, 9085–9094, as their Type B as Samson was modifying 8090. It is unlikely either was aware of the other's work.

205 *Fights and Flights*, p.341. *Anne* was still in a refit, and would shortly be off once more for a cruise to the Red Sea.

206 See Chapter 4.

207 C.E. Hughes, *Above and Beyond Palestine*, p.48.

208 The Divisions comprised four (39th) or three (40th) Infantry Regiments with some supporting cavalry and artillery. Nominally, each infantry Regiment comprised three Battalions with an establishment of 1081 officers and men. However, each Battalion was estimated only to have 250 to 350 effectives. *Handbook of the Turkish Army, 1916*, pp.194/5.

209 See Ian Burns, Aden Flight, Aviation in Aden 1915–1919, *Cross and Cockade International*, V46N4, 2015.

210 AIR 1/1706/204/123/64. Flt Cdr Edmonds' report on Air Operations at Aden. Reference for all *Raven*'s operations. All quotes for Raven's visit are from this source.

211 Walton was Aden Infantry Brigade Commander, Nov 1915 – Nov 1917.

212 The Aden operations are one of the few times 20-lb bombs are mentioned. Most Flight Reports from EIESS operations mention only 16-lb.

213 Place names often do not match, or appear on, period and/or modern maps. A map prepared aboard *Raven* locates Sharaj between Waht and Dard and a little further east. It also has different names for places shown on an official map from the period. The map accompanying this chapter attempts to reconcile the major differences, and shows place names as used in the text.

214 AIR 1/1706/204/123/64. Clifford's Report of Flight.

215 AIR 1/657/17/122/567. Request for Retention of Aeroplanes at Aden.

216 All Flight Reports employ only Samson's Serials which will be followed in the text. Shorts, S.2/850 and S.3/8082; Schneiders, B.1/3789 and B.3/3790, and Baby B.5/8189. See Appendix 2.

217 C.R. Samson, *Fights and Flights*, p.305.

218 AIR 1/1707/204/123/68. Operation reports: HMS *Ben-my-Chree*. Reference for all *Ben-my-Chree*'s operations.
219 Short S.3/8082 had engine problems which were not repairable on board.
220 Port Sudan was built between 1905 and 1909 to replace the old harbour at Suakin.
221 Samson, *Fights and Flights*, p.311. The 'Senior Naval Officer' was probably Commander L.N. Turton, Royal Indian Marine Ship *Northbrook* (1907, six 4.7-inch guns).
222 HMS *Fox* was an old (1893) *Astraea*-class 2nd class protected cruiser armed with two 6-inch and eight 4.7-inch guns, she was the flagship of the Red Sea Patrol, Captain William Henry Dudley Boyle (later Admiral of the Fleet, Twelfth Earl of Cork and Orrery). RIMS (Royal Indian Marine Ship) *Hardinge* (1900, six 4.7-inch guns), and *Dufferin* (1904, six 4.7-inch guns), were armed transports specially designed and built for the RIM. HMS *Perth* was an armed boarding steamer (three 4.7-inch guns).
223 Samson, *Fights and Flights*, p.313.
224 James Goldrick, *Coal and the Advent of the First World War at Sea*, War in History, V21N3, July 2014.
225 This would be a Thornton Pickard A Type camera, as detailed in Chapter 5.
226 An original copy exists at the National Archives, WO 319/1.
227 AIR 1/2284/209/75/8. *Raven* – Operations in Gulf of Akaba.
228 Killing Private George William Simpson and injuring 12 men of the 1st Garrison Battalion, Sherwood Foresters, also killing Private R G Winters, RE. Six civilians are also believed to have been killed when one of the bombs struck a local café.
229 Surgeon Probationer C Herington, RNVR; George Garge (?), Syrian; M Fernandez, G Frangoulis, D Ghlias, D Eskovo, Mohammed Ali and El Said Fahmy firemen. All were transported to British Hospital. Surgeon Herington was eventually evacuated to the UK.
230 The British had been planning to develop a manganese mine below the escarpment at Um Bogma before the war, Ras Abu Zenimeh to be developed as the port. Lt Gen Sir George MacMunn, *The History of the Sikh Pioneers*, p.484/5, says in part 'On the 24th of April [1916], the battalion [23rd Sikh Pioneers] was embarked for Abu Zenimeh, some seventy miles down the Gulf of Suez, and for Tor, some sixty miles further on, in some attempt to close roads by which mines might be brought to the Gulf and curb Turkish intrigue among the Bedouin, but especially to protect the railway that ran inland for nine miles from Abu Zenimeh towards the valuable manganese mines.'
231 AIR 1/1708/204/123/72. *Anne* Red Sea reports, Sept–Oct 1916.
232 Hughes, *Above and Beyond Palestine*, pp.101/2.
233 Naval Operations in the Red Sea, 1916–17, *The Naval Review*, 1925, Issue 4, p.652. Although Anonymous, the author is believed to have been Captain W.H.D. Boyle.
234 The fort was restored in 2018 as an important pilgrim fort on the Haj route from Egypt dating from the mid-17th century.
235 They were eventually permitted to land at Rabegh on 17 November 1916. See *Lawrence of Arabia and Middle East Air Power*, pp.23–33.
236 The infantry battalion strengths are unknown, but were likely very similar to those in the Yemen, ie; 250–350 effectives. *Handbook of the Turkish Army*, 1916, p.199, and Murphy, *The Arab Revolt 1916–1918*, p.27.
237 Lt (later Major) Herbert Garland, Army Ordnance Corps. 'He taught me how to be familiar with high explosive. Sappers handled it like a sacrament, but Garland would shove a handful of detonators into his pocket with a string of primers, fuse, and fusees and jump gaily on his camel for a week's ride to the Hejaz railway.' T.E. Lawrence, *Seven Pillars of Wisdom*.
238 Lawrence, *Seven Pillars of Wisdom*, Chapter XX. Dakhil-Allah el Kadhi, later changed allegiance and guided Lawrence on one of his raids.
239 See Murphy, *The Arab Revolt 1916–1918*, pp.33–39, for a more complete account with excellent maps.
240 AIR 1/1706/204/123/65. *Raven*, Aug–Dec 1916.
241 This time *Raven* had started out with 671.5 tons of coal, returning with 480 tons.

242 For a complete and absorbing account of the French occupation see Nicholas G. Pappas, *Near Eastern Dreams*.

243 Saint-Salvy (b. 9 March 1872) was a career naval officer, and gunnery specialist, entering the *l'Ecole Navale* in October 1890. Promoted *Lieutenant de Vaisseau* in June 1903, he served in the Mediterranean Fleet and various school appointments. At the beginning of 1916 he was serving on the old cruiser *Pothuau* based at Port Said. Promoted *Capitaine de Frégate* in June 1917, he remained *gouverneur de l'île de Castellorizo* until July 1917. In 1920/1 he was the French naval agent in Constantinople. He retired in 1933 as a *Capitaine de Vaisseau*. Saint-Salvy retired to a family property in Albi, Tarn department, where he died on 5 June 1960.

244 Light field guns for naval landing parties.

245 Governor Saint-Salvy's log of orders, January-February 1916. *Service historique de la Marine*.

246 AIR 1/1708/204/123/73, HMS *Anne*, Report of Flights, January 1916 to March 1917; Weldon, *Hard Lying*, pp.119–122.

247 AIR 1/1706/204/123/64, Operation Reports HMS *Raven II*, April to August 1916; Diary of Lt G.B. Dacre, RNAS, Fleet Air Arm Museum.

248 Ioannis Yeorgiou Lakerdis (1880–1944). He was appointed mayor before the French departed. Under Italian rule he became increasingly unpopular and by the early 1940's had left the island to live out his days on Rhodes. Pappas, *Near Eastern Dreams*, and Biographical note on Ioannis Lakerdis, 3 November 1920 (*Service historique de la Marine*, Vincennes).

249 Héderer, *L'Ile du Chateau-Rouge*, fn p.78, 'The same day [20 January 1917], the island of Kekova, occupied by militiamen from Megiste, was attacked with guns and by plane. It had to be evacuated.'

250 *CdV* A. Thomazi, *La Guerre Navale dans La Méditerranée* (Payot, Paris,1929), pp.115/6.

251 The regional military command was the *21nci Kolordu* (XXI Army Corps) under *Miralay* (Colonel) Nureddin Bey based at Aydin. The Corps was responsible for the area between Muğla and Antalya. On the operational level, *Binbaşı* (Major) Cemal Bey served as district commander, and commanded the 3rd Battalion of the 135th Regiment.

252 Joseph Hesselberger, *Report for Marshal Liman von Sanders*. German Military Archive, BA-RH61/1805.

253 A battery normally comprised four guns, it is not clear whether the full battery was transported to the coast.

254 Karl Schmidt-Kolbow, Report of the artillery activity during the operation against Meis Island. German Military Archive, BA-RH61/1805. Modern road D635 traces the route followed from Burdur to Elmali.

255 Ertuğrul, *Ben Bir Türk Zabitiyim*, pp.60–71. His memoirs were written at the request of Kemal Ataturk in 1934, and are very Turko-centric. His dates are corrected to the Gregorian calendar, thus 23 December 1915 becomes 5 January 1917. His route from Denzeli to Elmali cannot be accurately traced but modern road E87 to İstanos (Korkuteli) is a close approximation, he would then have followed Schmidt-Kolbow's route south.

256 Schmidt-Kolbow's dates follow the European (Gregorian) calendar and are left as reported. Baindyr is about 2 kilometres east of Antiphilo and 4 kilometres NE of the battery sites. Actual overland distances are much greater due to mountain ridges across the routes.

257 Romanos was subsequently sentenced to death by the Turkish army commander, but released following the payment of a large bribe. He was decorated by a later Governor of Castellorizo for *les nombreaux services qu'il rendus à la Cause Française pendent la guerre Européenné*. Pappas, *Near Eastern Dreams*, fn pp. 67 and 93.

258 The Royal Navy's equivalent to a half day's holiday. All but essential ship's work is cancelled, and the crew is free to relax or make and mend clothing.

259 ADM 137/3667, Loss of HMS *Ben-my-Chree*. Report of *Contre-Amiral* de Spitz.

260 ADM 137/3667. Unless otherwise cited all officers reports are from this source. (*qv*. Chapter 13.)

261 Maskell had been promoted Flight Commander with effect from 31 December 1916. However, it is probable that the news had not yet reached him as he is still referred to as Flight Lt in the official records.

262 Lt Cdr A.L. Braithwaite, RNVR (Rtd), letter to magazine *Sea Breezes*, March 1966.

263 Benn, p.149. Although the boats came under shell fire there is no evidence that the boats and men were deliberately targeted by the guns. They just happened to be in the general target area.

264 Sub Lt J.L. Kerry, edited diary entries.

265 Braithwaite, *op. cit.*

266 Héderer, *op. cit.*, fn p.78, refers to Emin Aga as 'a certain *bimbashi* named Emin Agha, a convicted criminal and creature of the Germans.'

267 Ertuğrul, *op. cit.*, pp.82–88. Héderer, *op. cit.*, fn pp.76/78.

268 Nikolajsen, *Ottoman Aviation 1909–1919*. Albatros C.III s/n *AK.25, AK.34, AK.35* and *AK.36* were serving with *5.TyBl* at this time. None of the surviving Turkish records of *5nci Ty Bol* confirm the story. However, I am inclined to believe it as both sides provide essentially the same details.

269 Report on Condition of *Ben-my-Chree*, 9 May 1918. *Capitaine de Corvette* Le Camus, Governor of Castellorizo, July 1917-July 1919. Service historique de la Marine, Vincennes.

270 *Handbook of Macedonia* (Naval Intelligence Division, Admiralty, 1920).

271 Except for the British Naval Mission to Serbia, 1914–1916. A story ably told by Charles E.J. Fryer, *The Royal Navy on the Danube* (East European Monographs, Columbia UP, 1988). Also, by one who was there, Commander C.L. Kerr, *All In The Day's Work* (Rich & Cowan, 1939), Chapters XVII to XXI.

272 The above is necessarily a brief overview. See Palmer, *The Gardeners of Salonika*, or Wakefield and Moody, *Under The Devil's Eye*, or Jon Lewis, *The Forgotten Front: The Macedonian Campaign 1915–1918* (Helion, 2023), or the Official Histories, for a complete account.

273 Aegean Squadron from September 1917.

274 Henry Newbolt, *Naval Operations* (Longmans, 1928), Vol.4, p.128.

275 1st Detached Squadron – Crete and Anatolian Coast between Samos and Rhodes.
2nd Detached Squadron – Dardanelles Area.
3rd Detached Squadron – Salonika Area.
4th Detached Squadron – Smyrna, Khios and Samos Area.
5th Detached Squadron – Doro [Cape Kafireas] – Salonika Patrol and East Coast of Greece.
6th Detached Squadron – Stavros and the Bulgarian Coast.

276 Edited from H.A. Jones, *The War in the Air* (Oxford, 1935), Vol.5, pp.370/1.

277 See David Méchin, *Oriental Adventures of the French Air Force* (Aeronaut, 2022), pp.37/8.

278 ADM 137/350, Salonika and the Aegean Sea Weekly Reports 20 February to 29 April 1916, and ADM 137/365, Weekly Reports 30 April to 17 June 1916. If not otherwise identified all quotes are from various ADM 137 files listed in Sources. Especially useful are the Weekly Operation Reports, Air Service Headquarters, Mudros.

279 ADM 137/365, with additional input courtesy David Mechin and Paschalis Palavousis.

280 Nedialkov, *The Air Component of the Kingdom of Bulgaria*, pp.49, 50 and 68.

281 See www.theaerodrome.com/forum – *FA and Jasta's in Macedonia,* and www.frontflieger.de/o-navi.html. There is also *Haupt* Heydemarck's *War Flying in Macedonia*, but that essentially begins in 1917, when he was posted to *Feldflieger-Abteilung 30*.

282 *FFA30* had been transferred from the Galicia to Serbia in September 1915, first to Wershetz (Vršac), then Leskovac. *FFA30* was renumbered *Flieger-Abteilung (FA)30* on January 11, 1917.

283 There is a small airstrip, *Xanthi* Neos Zygos, used by the Aeroclub of Xanthi that may be located at the site of the original Maswakli airfield.

284 The *Abercrombie*-class of big gun monitors – *Abercrombie, Havelock, Raglan,* and *Roberts* – were each fitted with a twin 14-inch gun turret designed and built in the USA for a Greek battlecruiser *Salamis* being built in Germany. Not delivered prior to the war the guns and turret assemblies were sold to Britain. The ships were originally given names of American Civil War leaders, *Raglan* was to be *Robert E. Lee*, the US Government did not appreciate to compliment and all were given the names of British generals.

285 Reports for 3 May 1916, ADM 137/365, pp.88/102.

286 Ferejik, a railway junction about 25 kilometres east of Dédé Agatch and 17 kilometres inland.

287 Dunning is better remembered for his trial landings on the forward flight deck of *Furious* in August 1917. He was killed attempting his third landing.

288 George Master Byng, 9th Viscount Torrington. He was attached to *Ark Royal* but served aboard *Empress* in May 1916. On 2 December 1916 the Henry Farman F.27, 3905, from A Flt Thasos, he was flying in was shot down over Bulgaria whilst attacking the seaplane base at Gereviz (Xanthi). Torrington and his pilot Flt Lt E.J. Cooper were captured and made PoW.

289 The flight reports for May 1916 survive in ADM137/365 and are quoted here and below.

290 Many of the locations mentioned in the following reports have been since reclaimed from a flood plain and now rest under the Mila-Bodrum Airport.

291 HMS *Wolverine* (1910), one 4-inch and 3 12-pdr, two 21-inch torpedo tubes, part of the numerous 5th Destroyer Flotilla. Served throughout the Gallipoli Campaign and in 1916 was part of an anti-submarine force in the Aegean. She may have been a regular escort for *Empress*.

292 HMS *Earl of Peterborough* (1915), Lord Clive-class monitor, single twin 12-inch gun turret taken from decommissioned pre-dreadnought battleships. There were eight in the class only *Earl of Peterborough* and *Sir Thomas Picton* served in the Mediterranean theatre.

293 *Grafton*, log book, ADM 53/43385.

294 The quote from Sitwell's log book is drawn from description of Bonhams Auctions, Lot 275, 22 November 2011.

295 *Naval Operations, op. cit.*, p.183.

296 The whole Greek situation is far to complicated to even attempt to review here. *Naval Operations, op. cit.*, Chapter V, contains a very useful summary – from a naval point of view. Alternatively, see the books listed in fn.272 above.

297 Vice Admiral Sir Cecil F. Thursby since 19 June 1916.

298 ADM 137/545, pp.241–245.

299 The *Kilkis* and *Lemnos* were ex-US Navy pre-dreadnoughts *Mississippi* and *Idaho*; *Georgios Averof* a modern (1910) armoured cruiser; *Spetsai* and *Hydra* two coast defence ships; *Helle* a modern (1912) protected cruiser. *Georgios Averof* survives as a museum ship at Palaio Faliro, Athens to this day.

300 There is a surviving ancient stone bridge at modern Sarikemer, probably the one bombed on 26 July 1916.

301 HMS *Grafton* (1892), Edgar-class protected cruiser, two 9.2-inch and ten 6-inch guns. Refitted for shore bombardment early in 1915, the 9.2 inch guns were replaced with 6 inch guns, and anti-torpedo bulges fitted.

302 Dranli was a small Muslim village in the foot hills of the Bunar Dagh. To the southwest, a kilometre away across a ridge, was another Muslim settlement, Lungor. Both were abandoned by the end of the war and a new village, Ofrynio, established about three kilometres south of Dranli.

303 See Some Experiences in Macedonia, *The Naval Review*, Vol.VII No.3, August 1919, pp.370/6.

304 ADM 137/545, pp.504/5.

305 Stilwell flew a few times from *Ark Royal*, then transferred to UK in command of South Shields Air Station from 2 April 1917 until March 1919. He was promoted Squadron Commander on 1 January 1918. Field was promoted Flt Cdr in December 1916 and Squadron Cdr the following December. He remained with *Ark Royal*, but by the end of 1917 was noted as CO of Suda Bay (ADM 273/4/90). See fn.314 below.

306 ADM 137/545, p.523. Note: The reference to Gulf of Ruphani probably means Rendina or Orfano.

307 Graham Mottram, Norman Woodhead – Naval Aviator, *Cross and Cockade (UK)* Journal, V9N2, 1978, p.80.

308 About 1.5 kilometres west of Orfano.

309 Cyril Falls, *Military Operations Macedonia*, Vol.1, pp.192–196.

310 9758 had arrived from Mudros on 6 November and had been kept busy since then.

311 HMS *Edgar* (1890), lead ship of the Edgar-class protected cruisers, refitted for bombardment as her sister *Grafton*.

312 Buxton, *Big Gun Monitors*, p.234.

313 *Controller*, hired trawler, Admiralty No 298. Built 1913, 201grt, Granton-reg GN.79.

314 A detachment of Short 184s was formed at Suda Bay sometime in 1916/17. It appears to have been administered from *Ark Royal* and not as a separate Flight of 2 Wing. With the formation of the RAF it eventually became 439 Flight, then part of 264 Squadron. It was commanded by Sqd Cdr / Major R.M. Field, ex-*Empress*.

315 *Ben-my-Chree*'s crew on 9 January 1917 comprised 30 officers (7 RNAS; 2 RN; 15 RNR; 5 RNVR; 1 Military – Benn) and 216 other ranks (46 RNAS; 68 RN, RFR, RNR; 104 T.124; 2 Military). There were seven wounded and none killed. ADM 137/3667.

316 Hughes, *Above and Beyond Palestine*, pp.156/161.

317 Lt Cdr A L Braithwaite, RNVR (Rtd), letter to magazine *Sea Breezes*, March 1966.

318 ADM 137/3667. The Royal Navy operates on the principal that the Captain bears ultimate responsibility for everything that happens on his ship. Thus, in almost every case, the Captain will have to prove he was not negligent at a court martial.

319 'Highly satisfactory' in Royal Navalese is high praise indeed.

320 Document #118, dated 7 February 1917. Paul G. Halpern (editor), *The Royal Navy in the Mediterranean, 1915–18* (Navy Records Society, 1987).

321 Letter dated 6 May 1917, written aboard *Raven* then at Colombo.

322 Boyle, *My Naval Life*, p.102.

323 Lawrence, *Seven Pillars of Wisdom*, Book II, Chapter XXIII (pp.148/9, Cape 1955 edition).

324 *Fox*, *Espiegle*, *Suva*, *Hardinge*, and *Anne*.

325 Boyle, *op. cit.*, and *The Naval Review*, 1925-4, Naval Operations in the Red Sea, 1916–1917.

326 *El Kahira* (1892) of the Khedivial Mail Steamship & Graving Dock Co.

327 An X-lighter, which were designed as basic, shallow-draft powered landing craft.

Many, completed in 1915, were sent out to the Mediterranean to see service at Gallipoli. Later they were used throughout the Middle East as transports and water or fuel tankers.

328 *Nord Caper*. Trawler built in Dunkirk 1907. 418 tons, 47m length. Requisitioned as an *Aviso auxiliaire* in December 1914. Armament one 65 mm forward and a 47 mm aft; crew two officers and 25–30 *matelot*. On 7 November 1915, captured a Turkish schooner (crew 43) near the southern tip of Crete – ship and crew were awarded the *Croix de Guerre* for this action. Returned to fishing in 1919. Requisitioned by *Kriegsmarine* in 1940 as *Vorpostenboot V1607*. Survived the war and once again returned to fishing. Scrapped in 1953.

329 AIR 1/271/15/226/119. HMS *Anne*, 26 February – 1 March 1917.

330 Photographs show that there was one Lewis on mounting in the observer's cockpit, and one on a pillar mounting firing over centre section.

331 *Above and Beyond Palestine*, pp.184–191.

332 Hughes waxes lyrically, at length, about the city. British troops, when they marched in a year and a half later, were not so impressed.

333 *London Gazette*, 21 April 1917, Supplement: 30029, Page: 3820.

334 Carrier Skeleton Tubular 112 lb. Could be used for bombs between 50-lb and 112-lb.

335 Report of Experiments Carried out at the Isle of Grain during April 1916. AIR1/4316/15/278/2.

336 AIR 1/1720/204/123/190, Intelligence and Operation Reports, Feb-March 1918.

337 *Wachtfels* (1913; 135m/443 ft, 5809 tons, 11 knots) was a younger, bigger, half-sister to *Rabenfels/Raven*.

338 To learn more of this remarkable cruise, see Richard Guilliatt and Peter Hohnen, *The Wolf* (Free Press, 2010). To read about *Wölfchen*, see Peter M Grosz, *FF33E* (Windsock Datafile 73, 1999).

339 C E Hughes, *Above and Beyond Palestine*, Chapter VIII.

340 *Raven*'s log book.

341 AIR 1/1707/204/121/67, *Raven*, Aden, Indian Ocean, March to June 1917. The source for many of the quotes in this chapter.

342 *Odin* (1901) was another of the ubiquitous Cadmus-class sloop identical to *Espiegle*. See fn.173 (Chapter 8).

343 Details from *Odin*'s log book ADM 53/53206, and *Naval Operations*, Vol.4, pp.214/5.

344 A Collectorate of India is a district administered by a Collector (of revenue or taxes) as the Chief Executive.

345 For a full account see Sub Lt Kerry's account in Appendix 1.

346 AIR1/1707/204/121/67, *op. cit.*

347 When and where the couple met is not recorded, but they had one daughter before divorcing in December 1923. Samson remarried in 1924 to Miss Winifred Reeves, they had a son and a second daughter.

348 AIR 1/1707/204/121/67, report dated 24 April 1917.

349 Hughes, *op. cit.*

350 AIR 1/1707/204/121/67, report dated 10 June 1917.

351 Hughes, *op. cit.*

352 A Town-class cruiser built in Australia. *Brisbane* spent most of her war in eastern waters.

353 Hughes, *op. cit.*

354 See *The Kipling Journal*, Vol.73, No.290, June 1999, pp.24/31.

355 Kipling's *A Flight Of Fact* was first published in *Nash's Magazine* and the *Pall Mall Magazine* for June 1918. Meade's account appeared in *The Wide World Magazine*, Vol.41, No.245, September 1918, pp.414/420. Guy Duncan Smith also expanded on his letter in his unpublished memoir. Whilst Kipling's tale is by far the most entertaining, Meade's account is probably the most accurate.

356 Guy Duncan Smith's memoir.

357 *Diana* (1895) was an Eclipse-class protected cruiser, a sister to *Doris* and *Minerva* from Chapter 3.

358 Smith made no mention of a shed, but all other sources, and photos, clearly indicate a covered shed was built.

359 *Raven*'s log book.

360 RNAS Operations Report, No.33, 1–16 May 1917, p.19.

361 Diary of Flt Lt A.E. Popham, the main source for information about the Colombo Detachment (FAAM). Popham had arrived in Port Said on 21 May, informed he was to proceed to Colombo on 2 June and left Port Said on 4 June. Landed at Bombay on 16 June, he completed his journey to Colombo by rail and steamer. He arrived on 23 June. Popham was an opionated, almost priggish, young man and his diary makes interesting reading.

362 David Stevens, *In All Respects Ready* (OUP Australia, 2014), p.279.

363 Popham diary.

364 SS *Somali* (1901) built for P&O. Not to be confused with the Deutsche Ost-Afrika Linie *Somali* (1889), supply ship for SMS *Königsberg*, sunk in the Rufigi Delta 1915.

365 *Exmouth* (1901) a pre-Dreadnought battleship: *Kinfauns Castle* (1899) had been an Armed Merchant Cruiser, briefly used as a seaplane carrier, during 1914/15 before being employed as a troopship.

366 For more on Aden, see Ian Burns, Aden Flight, Aviation in Aden 1915–1919, *Cross and Cockade International*, V46N4, Winter 2015.

367 He had been promoted Captain, RM, during the course.

368 AIR 1/1720/204/123/192, reports dated 9 June and 4 July 1917.

369 Short 184 – 8004, 8018, 8019, 8020, 8021, 8022, and 8075. Sopwith Baby – probably N1014, N1016 and N1038 at Colombo.

370 ADM 137/1230: C-in-C East Indies to Secretary, Admiralty, No.1002/1155, 4 Dec 1916. Quoted in Sheffy, *British Military Intelligence in the Palestine Campaign*, p.192.

371 The staff comprised 65 German Officers and just nine Turkish officers, the highest Turkish rank was *Binbaşı* (major).

372 Mustafa Kemal Atatürk after 1921.

373 For a full account see Falls, *Military Operations, Egypt and Palestine*, Wavell, *Palestine Campaign*, or Hadaway, *From Gaza to Jerusalem*.

374 *Above and Beyond Palestine*, pp.203/5.

375 AIR 1/1708/204/123/74. *Empress* Reports, March 1916–February 1918 (excludes Aegean operations). Sturtivant and Page note that Short 8004 was fitted with the 240-hp Gurkha

376 *War in the Air*, Vol.5, pp.231/2.

377 AIR 1/1708/204/123/74.

378 Eliot Millar King, Pilot's flying log book, ARC2002–128, Box 1, Item 5. Puke Ariki, Taranaki Research Centre, New Zealand. King's log book is most unusual as it often contains substantial notes on his operational flights.

379 AIR 27/191.

380 Smith does not name the crew, but in his report Wing Cdr Risk draws attention to them, although subjected to vigorous infantry fire, descended to 400 feet in order to make certain of carrying out the mission as ordered.

381 AIR 1/1708/204/123/74. Report dated July 16th, 1917.

382 *UB-66* was last sighted off Famagusta on 12 January and is believed to have sunk the steamer *Windsor Hall* some 70 kilometres NW of Alexandria on 17 January 1918. *UB-66* did not return to base, possibly sunk by the sloop *Campanula* on 18 January, all crew members were lost.

383 AIR 10/307, RNAS Bomb Dropping Mirror Instructions.

384 The Report of Flight shows two 65 lb bombs but King only records dropping one as does Risk's report (AIR 1/1708/204/123/74). Most likely it was a typing error.

385 Commander Drury, commanding *Empress*, report in AIR 1/1708/204/123/74.

386 From Clemson's Service Records, ADM 273/4/46.

387 Benn, *In The Side Shows*, p.81.

388 Wemyss returned to London and, in September, was appointed Deputy First Sea Lord. In January 1918 he replaced Sir John Jellicoe as First Sea Lord and Chief of the Naval Staff, holding the post until November 1919. He was promoted full Admiral in February 1919 and Admiral of the Fleet in November 1919.

389 See fn.327 (Chapter 13).

390 Newbolt, *Naval Operations*, Vol.5, pp.80/1.

391 AIR 1/1708/204/123/75. *City of Oxford*. Report of Operations, source for most *City of Oxford* quotes.

392 For more on *Managem* and *Veresis* see Weldon, *Hard Lying*.

393 Newbolt, *Naval Operations*, Vol.5, p.79.

394 Newbolt, *Naval Operations*, Vol.5, p.79. For this and later work off the Wadi el Hesi *Requin* was awarded the *fourragère* (a braided cord awarded as a decoration to the colours of a military unit) in the colours of the ribbon of the *Croix de Guerre*. The citation dated 7 July 1919 reading in part: '*Les 1er et 2 novembre 1917 et les 6, 7, 8, 9 novembre 1917, devant le Wadi el-Hessi (Nord de Gaza), a brillamment participé, avec la division navale anglaise aux attaques de l'armée alliée contre les positions ennemies, en dépit d'un feu violent d'artillerie qui a mis hors de combat une partie notable de son équipage et malgré des attaques d'avions et la menace de sous-marins.*'

395 Operation Gratitude, 12 January 1945, a very effective strike on SE French Indochina, by aircraft from the carriers *Enterprise* (CV-6) and *Independence* (CVL-22), Task Group 38.5 from Task Force 38. The stern section was re-floated in June 1956 and was under tow to breakers in Hong Kong but was lost in heavy weather.

396 N1028, N1036, and N1038 were armed with two Lewis guns, the remaining Baby's having a single Lewis.

397 Except as noted, all *Empress* reports from AIR 1/1708/204/123/74.

398 Flt Lt Stephens' wound did not heal and he was invalided to UK and on to Canada from February 1918. Flt Lt Leigh quickly recovered from his injuries.

399 His log book says N1090, but the Report of Flight has the serial altered by hand to N1091, confirmed by the final Report of Operations.

400 Entry in King's log book.

401 William Thomas Massey, Official *Correspondent* of London Newspapers with the Egyptian Expeditionary Force, *How Jerusalem Was Won* (Charles Scribner, 1920), pp.101/2.

402 Joe Bull, *One Airman's War* (Banner Books, 1997), p.86.

403 Mediterranean Division of the German Imperial Navy. SMS *Goeben* (1911), ten 280 mm, twelve 150 mm and twelve 88 mm guns, 25.5 knots. SMS *Breslau* (1912), twelve 105 mm guns, 27.5 knots. *Breslau* was rearmed in 1917, all the 105 mm guns were removed and eight 150 mm guns installed – one fore and aft and three on each beam.

404 There are many books about the escape of *Goeben* and *Breslau*. Two I find particularly interesting are Georg Kopp, *Two Lone Ships* (Hutchinson, 1931), Kopp was a wireless operator on *Goeben* an offers a first hand view of events. Redmond McLaughlin, *The Escape of 'The Goeben'* (Charles Scribner, 1974), a balanced account mercifully free of the political infighting of many books. For a modern general history I can recommend, Paul G. Halpern, *The Naval War in the Mediterranean, 1914–1918* (US Naval Institute Press, 1987). Note: *Goeben* and *Breslau* will continue to be used instead of their Turkish names, as in most contemporary English language reports and accounts.

405 *Kapitänleutnant* Herman Baltzer, quoted in the Mitchell Report, Chapter 15, pp.411/7.

406 Both were *Lord Nelson*-class ships, commissioned in 1908, four 12 inch (in two turrets) and ten 9.2 inch guns, 18 knots.

407 *Kapitänleutnant* Baltzer, *op. cit.*

408 *Basra, Samsun* (1907), built in France, one 65 mm and six 47 mm guns, two torpedo tubes, 28 knots on trials, but 17 knots by 1915. *Muavanet-i Milliye,* and *Nümune Hamiye* (1910), built in Germany, two 75 mm and two 57 mm guns, three torpedo tubes, 32 knots on trials, but 26 knots in 1912.

409 All times have been corrected where necessary to British Eastern Mediterranean fleet local times.

410 Flt Lt J.W.B. Grigson, War Experiences, RAF Staff College, Third Course, 1924. AIR 1/2387/228/11/42.

411 *Kapitänleutnant* Baltzer, *op. cit.*

412 See *War in the Air*, Vol.V, Chapter VII, pp.410–414, for an account. AIR 1/271/15/226/121 and AIR 1/361/15/228/10, contain a complete *Narrative of Operations Against "Goeben".*

413 *Kapitänleutnant* Baltzer, *op. cit.*

414 *War in the Air*, Vol.V, p.413.

415 AIR 1/1708/204/123/74. *Empress* Operation Reports, March 1916-Feb 1918. Report of Operations, dated 4 February 1918.

416 AIR 1/271/15/226/121, p.12.

417 AIR 1/271/15/226/121, p.13. A letter from Bronson was dropped on Imbros by a 'German seaplane' on 2 March confirming they were safe and PoW.

418 *London Gazette*, 23 May 1919, Supplement: 31354, Page: 6445.

419 AIR 1/271/15/226/121, p.14.

420 Naval Operations in the Red Sea, 1917–19, *The Naval Review*, 1926, Issue 1, pp.48–56. See fn.27, Chapter 9.

421 The 21st Independent Division was thought to comprise four Regiments, with a total effective strength of under 6000. *Handbook of the Turkish Army, 1916*, p.199.

422 The steam launch had been requisitioned from the pilgrim quarantine station on Kamaran Island, fitted with a Maxim machine gun and manned by a piratical group of seamen under CPO William Henry Duke. It was used to patrol the coast between Kamaran Island and Midi, a distance of some 80 kilometres, essentially to intercept dhows attempting to smuggle arms into Yemen for the Turkish. It was an area of shallow water and numerous low lying islands. Duke was awarded the DSM on 7 June 1918, and MID on 31 December 1917, for the period he commanded *Kamaran*. For more details see Cork and Orrery, *My Naval Life*, p.106.

423 Naval Operations in the Red Sea, *op. cit.*

424 Log book of HMS *Clio*, another of the ubiquitous Cadmus-class sloops. In addition her sisters *Odin* and *Espiegle* were also involved on various dates at Loheia.

425 *The Red Sea Pilot*, Hydrographic Office, Admiralty, 1900, p.307.

426 AIR 1/1708/204/123/75, *City of Oxford* (Oct 1917 to March 1918) and AIR 1/1720/204/123/190, *City of Oxford* Reports (Feb–March 1918). These together with details from Flt Lt King's log book provide the principal sources for the account of the work carried out by *City of Oxford*.

427 The map S.W. Arabia, Sheet 2, SANAA, published in 1915 by the Ordnance Survey for the Geographical Section, General Staff, was so unreliable that maps were prepared on board *City of Oxford* based on observations and photographs. Probably drawn by Leading Mechanic W. Robertson, a first draft became available on 26 February and was continually being revised.

428 Unfortunately, Guy Duncan Smith's unpublished memoir takes too many flights of fancy to be a reliable source for this period and, accordingly, will be used sparingly.

429 Lt Thomas McLeish, RNVR, was a well regarded photographic officer who had been with the EIESS since June 1917. He remained with the RAF Seaplane Squadron in Egypt until the end of the war.

430 King's log book.

431 Possibly confirming that the cells could be converted to carry four 16-lb bombs, as a two-seater 8020 had just two bomb cells. See Chapter 13.

432 *Naval Operations in the Red Sea*, op. cit.

433 AIR 1/1720/204/123/190, op. cit.

434 Leading Mechanic W. Robertson was awarded a Mentioned in Despatches 28 June 1918. I have not located any awards for PO S. Lay or Leading Mechanic J. Stokes.

435 15 Short flights and 12 Sopwith flights.

436 Samson's Operation Reports.

437 AIR 1/1710/204/123/79. Summary of Anti-Submarine Patrols, December 1917 to March 1918.

438 Burling was promoted Flight Commander on 31 December 1917.

439 On 24 May 1918, so his RAF rank was Captain.

440 AIR 1/455/15/312/35. (EIESS and RAF) Anti-Submarine Patrols, Egypt Feb-Nov 1918.

441 Mentioned in Chapter 16. Royal permission to wear the medals was published in the *London Gazette* 22 February 1918 for the RNAS recipients, and 18 April 1918 for the military observers.

442 *Contre-Amiral* Georges F.C. Varney, had replaced *Contre-Amiral* Spitz as *Commandante Division Navale de Syrie*.

443 Founded by Rothschild interests the *Société Anonyme de Gérance et d'Armement* was created in 1919 for the management of French state-owned ships.

444 Flt Lt Stephens had been invalided Home on 22 February 1918, and Flt Lt Popham would be invalided Home on or about 10 April 1918.

445 FSL G.J. Pilgrim and FSL G. Waugh both arrived in Port Said at the end of March 1918.

446 The Kite Balloon Base proposed for Port Said was never completed.

447 AIR 1/649/17/122/409: 'The whole of the first 12 F.3 seaplanes which formed the original contract in Malta dockyard have now been turned out, and these have been disposed as follows; – 4 to Adriatic, 5 to Malta seaplane station, while one has been deleted and two are undergoing large repairs in the Dockyard.'

448 AIR 1/649/17/122/409: 'A temporary base for aeroplanes has been established on the Marsa sports ground, and two D.H.9 aeroplanes are stationed here for this purpose. These machines are being fitted with special "9A" large area wings, and the special F.V. flotation devices described in Air Technical Order No. 505 of February 1918, for use over the sea.' No DH9s were delivered to the Naval Wing, RAF, Egypt before the end of the war.

449 13th Sloop Flotilla, comprising at various times: *Amaryllis, Bryony, Cornflower, Gardenia, Ivy, Lily, Mallow, Magnolia, Nigella, Valerian,* and *Verbena*. How many were fitted for Kite Balloons is not known.

450 Mex Road (Al Max) runs East-West behind the location of the Alexandria Seaplane Station. There were some clear areas close by, perhaps the location of the Balloon Base.

451 The Imperial Japanese Navy's 2nd Special Squadron was based in Malta from 16 April 1917. It comprised a cruiser as flagship, and up to twelve modern destroyers. They mainly escorted troopships from Marseille and Malta to Salonika and Alexandria. British officers rated Japanese naval professionalism as significantly higher than that of the French or the Italians.

452 AIR 1/455/15/312/35, EIESS and RAF Patrols, Egypt Feb-Nov 1918. *Empress* had four 260-hp Sunbeam Maori engined Shorts aboard when she left Port Said; N1749, N1784, N2822, and N2824. She also carried away Flt Cdr Leigh, FSL Pennington, and Obs Lts Kennedy and Copley, the number of "E" ratings is unknown.

453 AIR 1/1713/204/123/109, Aircraft available Alexandria and Port Said. June 1918. AIR 1/1710/204/123/79, a report of State of Aircraft at Port Said for week ending 29th June 1918.

454 For more information consult www.uboat.net/wwi/boats/ or Wilson & Kemp, *Mediterranean Submarines*.

455 AIR 1/455/15/312/35, Captain Burling's report.

456 Both *Indarra* and *Canberra* had been hired in November 1917 to bring Australian reinforcements to Europe.

457 Unless otherwise noted, all reports are from AIR 1/1713/204/123/103, Attack on Submarine, 14 June 1918, Alexandria.

458 Captain M.M. Osborne, Master, *Indarra*. For *Indarra*'s experiences in the preceding month, including several submarine attacks and the loss of two ships from the convoy, see Australian Official History, Vol.IX, *The Royal Australian Navy, 1914-1918*, pp.424-426.

459 Not all have been identified, but the following are believed to be correct. H—*Sugi*; I—*Sakaki*; J—*Matsu*; O—*Ume*; P—*Kusunoki*; Q—*Kaede*; R—*Katsura*; V—*Momo*.

460 *London Gazette*, 8 February 1919, p.2049.

461 AIR 1/455/15/312/35. Report for the Month of April, 1918. *Rowan* was an impressed Laird Line steamer converted to an armed boarding vessel.

462 All details are from King's log book.

463 Extract from a longer account at voicesinflight.webs.com/gordonhyams.htm.

464 On 4 September 1915 HM Submarine *E7* was attempting a passage through the Dardanelles into the Sea of Marmora when the starboard propeller became entangled in the anti-submarine nets at Nagara. Despite all efforts the submarine was unable to break free and to avoid being depth charged to destruction the Cochrane surfaced the submarine to abandon ship and then scuttled the submarine. All of the crew survived and were made Prisoners of War. However four of the crew died in captivity.

465 AIR 1/1712/204/123/95, Turkish Seaplane Station Asia Minor Coast. All material used comes from this file.

466 Malta built Felixstowe F.3 N4360 had been delivered to Alexandria by 6 October 1918. The F.3 was the first of the second contract to be completed, several other boats joining it at Alexandria post war.

467 Commonwealth & Dominion Line Ltd, a reefer built by Workman Clark & Co Ltd, Belfast, launched in 1918 and immediately hired as a transport.

468 *War Dame* was a War Standard Ship built to a standard design a quickly as possible, similar in idea to the Liberty Ships of a later war. She was built by Mitsubishi Zosen Kaisha, Nagasaki, in 1917.

469 *Kia ora* is a Māori language greeting which means, *be well/healthy*. It has become a universal greeting, farewell and expression of thanks in New Zealand.

470 Fuad became Sultan of Egypt on 8 October 1917 upon the death of his brother Hussein Kamel. Fuad died on 28 April 1936 to be succeeded by his son Farouk.

471 *Married to a Mermaid* dates from around 1750, possibly to the music of *Rule Britannia!* by Thomas Arne. In the late 19th Century it became a popular Music Hall piece, where the chorus echoing that of *Rule Britannia!* ensured its popularity. In its original form, it tells the story of a young sailor who falls overboard and is married to a mermaid, and later rises from the sea and says goodbye to his comrades and messmates.

INDEX

A

Abbott, FSL G.S., RNAS (P), 305, 307, 308, 491
Abu Aweigila, 90, 135
Abu Zenimeh, 243-244
Acre, 32, 39, 84, 178, 182, 504
Adalia (Antalya), 203, 211-212, 264, 509, 511
Adana, 21, 24, 31, 80-81, 160-161, 163, 204-205, 359, 364-367, 375-378, 381, 477, 509, 514-515
Aden, 36-37, 98, 112, 177, 223-234, 236, 329, 332-334, 351, 409-411, 477, 495-496, 498, 506-507, 512, 516
Afion Kara Hissar, PoW camp, 201, 203
Aircraft *(individual aircraft serial numbers are not indexed)*
Aircraft, British
    AIRCO (DeHavilland) DH.1A, 149
    AIRCO (DeHavilland) DH.2, 150, 160
    AIRCO (DeHavilland) DH.4, 403, 408
    AIRCO (DeHavilland) DH.9, 150, 442
    Bristol Fighter, 150-151, 160, 257
    Bristol M.1 Monoplane, 150-151
    Bristol Scout, 285, 363
    Felixstowe F.3 flyingboat, 439, 441, 443, 457, 458, 485
    Martinsyde, 149, 160, 325
    Royal Aircraft Factory BE2a/BE2c/BE2e, 35, 59, 149-150, 160, 187, 363, 446, 457, 485, 488
    Royal Aircraft Factory BE12, 149-150, 160
    Royal Aircraft Factory RE8, 150
    Short seaplanes, 95, 99-100
    Short S.38, Sommer Pusher Biplane, 99, 165
    Short S.80, 28
    Short Type 166, 284, 290-291, 301-302, 305-309, 476
        Comparison with Type 184, 305
    Short Type 184
        Description, 103-106
        Armament, 104-105, 168-169, 327
        Sunbeam engines, 95, 102, 104, 106, 121, 177, 305, 328, 337-338, 356, 394, 433, 446-448, 454-456, 459-460, 476-479, 482-484
        Model D single-seat bomber, 325, 327-328, 360, 363, 368, 376, 423, 445, 477
        'Cut Short' (8090), 214, 332-333, 337-338, 472, 478
    Short Type 320, 405
    Sopwith 1½ Strutter, 403
    Sopwith Camel, 346, 406, 408
    Sopwith Schneider/Baby, 99-100
        Description, 106-107
        Armament, 107, 171-172, 206-207, 291, 348, 386
        Blackburn Baby, 107, 386, 388, 481, 485
        Fairey (Parnall) Hamble Baby, 107, 355, 372, 386, 389, 447, 481, 485
    Vickers FB.16, 150
Aircraft, French
    Antoinette, 15, 19
    Bleriot, 15, 18, 21-6, 28, 42, 59
    Borel, 20, 22
    Caudron, 41, 92, 210, 285
    FBA flying boat, 136, 298
    Henry Farman, 27, 28, 59
    HF.27, 285
    Maurice Farman, 15-16, 102
        MF.7 / Longhorn, 59
        MF.11 / Shorthorn, 59, 285
    Morane(-Saulnier), 20, 22, 26-27
    Nieuport
        Monoplane, 21-23, 40, 52
        Floatplane, 28, 41-47, 473-475
        Description, 41, 44-45, 91-92, 144-145, 475
        Armament, 46-48, 138, 475
        Le Rhône and Clerget engines, 41, 44-46, 66, 90, 92, 145, 475
        Scout, 159
        10/12 two-seater, 285, 301, 306
Aircraft, German
    AEG C.IV, 155, 157, 161-162, 393
    AEG G.II, 286
    Albatros B.I, 158, 160-161
    Albatros C.I, 158-159, 286-287
    Albatros C.III, 157-161, 279, 286-287, 308
    Albatros D.III/D.V, 153, 157, 162, 382
    DFW C.V, 161
    Fokker E type, 153, 158, 287
    Friedrichshafen FF33/FF.49, 161, 286, 288, 329-330, 457
    Gotha, 287-288
    Grade monoplane, 15-19
    Halberstadt Scout (D.II), 160, 382, 515
    LVG B.II, 285-286
    Otto C.I, 285
    Pfalz A.II, 158
    Pfalz E.II, 152-155, 180
    Rumpler 4A Fethi, 148
    Rumpler B.I, 158, 160
    Rumpler C.I, 141, 152-156, 159, 161-163, 175, 180, 189, 210, 393, 480-481, 510
        Description, 154-155
    Rumpler C.IV, 157
    Sablatnig SF.5, 161
Akaba, 32, 36, 49, 52-56, 58, 62, 73-74, 77, 98, 159-160, 237-241, 489, 499, 501, 508
Aleppo, 24, 32, 148-149, 163

Alexander, Colonel, GSO Aden, 233-234, 334
Alexandretta (İskenderun), 32, 49, 80, 85, 94, 148-149, 160, 378,
Alexandretta, Gulf of, 38, 51, 80-81, 83, 85, 98, 128, 215, 220, 364, 473, 501-504, 506
Alexandria, 5-6, 14, 16, 23, 25, 28, 52, 59, 77, 79, 83, 128, 131-132, 150, 166, 327, 407, 414, 432-433, 439-446, 448-456, 459-460, 479, 482-485, 492, 496.
  Seaplane Base, 432, 440-444
    Kamaria Port, 432, 441
  Kite Balloon Base, 440, 443-445
Allenby, General Edmund H.H., 36, 358, 379, 384
Allied Army Units
  Aden Brigade, British Indian Army, 224, 227
  Aden Field Force, 226, 334
  ANZAC (Australian and New Zealand Army Corps), 31, 34-35, 94, 100, 208, 352
  Armée d'Orient, 283
  British Salonika Force/Army, 283
  Egyptian Expeditionary Force (EEF), 36, 109, 139, 149, 325
  Mediterranean Expeditionary Force, 35, 93, 247
Allingham, Lt H.C., RNR, 269
Al Zareeb Castle, 246-247, 253
Amanus mountains, 31-32, 151, 153, 160, 204
Amman, 161-163
Antalya (*see* Adalia)
Anatolia (Asia Minor), 22, 143, 256, 265, 281, 285, 287, 295, 299, 311, 457
Antiphilo (Kaş), 256, 259-260, 263-264, 269
Antioch, 85, 179
Arab 12, 24, 37, 88, 110-111, 115-116, 118, 120, 126, 130, 158-160, 162, 202, 214, 229-230, 234, 233-235, 239, 241, 243, 246-250, 252, 254, 313-317, 323, 354-355, 409-414, 418-422, 424, 428-429, 460, 503 (*see also* Bedouin)
Arab Revolt, 36-37, 177, 205, 223, 229, 233-236, 243-247, 252, 313, 322, 409-410, 413, 430
Argostoli, 139, 474-475, 486-487

Arnold, FSL C.V., RNAS (P), 102, 289-290, 292, 294, 298-299, 302, 305, 491
Asir, 36-37, 409-410, 430 (*see also* Idrisi)
Ashkelon, 50, 85, 128, 133, 179, 186, 190, 199-200, 382, 385, 394-395, 397, 501-503
Athlit, 189-191, 319
Attrill, CPO Charles, 4, 118, 140
Australian Flying Corps (AFC)
  1 Squadron, AFC 149-150, 382 (*see also* 67 Squadron, RFC)
Ayas, 81-82, 216

B

Baghdad Railway, 20, 31-32, 80, 151
Balsan, Jacques, 15, 17-19
Baltzer, *Kapitänleutnant* Herman, 400, 403, 408
Bankes-Price, FSL J.T., RNAS (P), 101, 123, 172, 174, 179-181, 186, 190, 193, 200, 204, 207-211, 227-229, 231, 233, 235, 463, 481, 491, 506-510
Bardawil, Lake, 48, 50, 62, 68-69, 87, 90, 125, 188, 473, 499-500, 505, 509
Bardia, 129-130, 506
Barnier, Joseph, 21
Barr, Flt Lt F.M.L., RNAS (P), 361-362, 492, 513
Baxter, 2Lt F.O., 2nd Rajput L.I. (O), 73-74, 171, 174, 488, 495, 501, 507
Bedouin, 35, 56, 61-62, 69, 91, 253
Beersheba, 32, 35, 46, 48-51, 62-63, 65, 72, 76, 82, 86-91, 98, 126-127, 133-134, 140-141, 150, 153, 156, 203, 207, 242, 358-359, 379-380, 390, 473-474, 499-501, 503-504, 506-508, 510
Beirut, 21, 24-25, 32, 39, 84, 126, 138-139, 180-182, 258, 319-324, 359-361, 368-374, 457, 502-503, 505-506, 508, 512-514
Bell, CPO J.W., RNAS (O), 290, 292, 294-295, 495
Bell-Davies, Admiral Richard, RN, VC, 2, 166

Benn, Capt W. Wedgwood, Middlesex Hussars (O), 167-170, 173-174, 180-181, 183, 185-186, 189, 193, 196-199, 203-205, 207, 212-213, 216-219, 229-231, 234-235, 237, 268-269, 273, 278, 312, 327, 378, 495, 507-512
Bennison, AB Edward A, 275
Berthold, *Oberleutnant* Fritz (O), 154, 175, 189
Beschik Lake (Lake Volvi), 301
Betts, Cdr E.A.A., RN, 171
Bican, *Mulazim* Hüseyin (O), 188
Billett, Ldg Seaman William, 275
Bir el Abd, 48-49, 62-63, 69, 90, 154, 183
Bir Lahfan, 71
Bir el Mazar, 184, 188, 190, 208, 507, 509
Bir Mabieuk, 62-63
Bir el Murra, 71-72, 135
Blon, Hubert Le, 15, 18-19
Boblen (Akropotamos), 307-309
Bogdanov, *Kapitan* Nikifor, 285
Bonnier, Marc, 21-22
Borton, Lt Colonel A.E., RFC, 361
Boué de Lapeyrère, *amiral*, 38, 42, 136, 139
Bourgeois, *Quartier-maître* Raymond P. (P), 91, 126, 486, 503-505
Bourne, 2Lt C.A., RFA (O), 133, 140, 488, 495, 506-507
Boyle, Capt W.H.D., RN, 39, 234-235, 244, 247-248, 252, 313-315, 409, 411, 429-430
Braithwaite, Lt A.L. RNVR, 269, 273, 278, 310
Brandes, *Kapitänleutnant* Iwan, 333
Brindisi, 67, 91-92, 452, 474-475, 486-487
Bristol Flying School, 158, 201
Bronson, Flt Lt C.G., RNAS (P), 361-363, 368-372, 374, 383-385, 405-407, 433, 463, 478, 492, 513-516
Brooke, Flt Lt J.C., RNAS (P), 122, 177, 188-189, 193, 197, 200, 216-218, 260-261, 360, 492, 507-509, 511, 513
Broome, Cdr Henry F. Chevalier, Viscount, RN, 404

Brown, Captain Francis Clifton, RN, 307
Brown, Lt J.W., RFA (O), 177-178, 260, 262, 495, 508
Buchanan-Wollaston, Captain H.A., RN, 39, 409, 414, 459
Budrum (Bodrum), 292-294, 299
Bulgaria, 20, 100, 147, 281-289, 297
Bulgarian Aviation
  1-bo Aeroplanno Otdelenie (AO), 285-286
  2-po AO, 285
Bülow-Bothkamp, *Leutnant* W. von, 209-210
Bunar Dagh (Mt Pangaion), 281, 302
Burd, 2Lt J.M., RFA (O), 137, 229, 233, 488, 495, 506
Burdur, 266
Bureir, 190, 200, 213
Bureika, 247
Burling, Flt Lt E.J., RNAS (Capt, RAF) (P), 218, 248-250, 252-254, 313, 315, 319, 332, 341, 351, 354, 363-364, 366-370, 372-373, 381-385, 390-391, 394-395, 398, 432-433, 437-439, 449, 492, 511-512, 514-516

C

Cajagzi (Chai Aghizi), 301-302, 307-308
Carden, Admiral Sackville Hamilton, 38, 43, 52
Castellorizo, 84, 142, 179, 203, 211-212, 255-280, 310-311, 477, 480-481, 495, 506-508, 511-512
  Navlakas cove, 277-278
Caters, Baron Pierre de, 13-14
Cemal Bey, Binbaşı, 265, 269-270
Chesme (Çesme), 76, 296
Ceyhan river, 31, 81-82, 205, 217, 219
Chagos Archipelago, 330, 333, 335-336, 333-339, 512
Chauvel, Major Gen Henry George (Harry), 208, 210
Chevilliard, Maurice, 28
Chicaldere railway bridge, 31-32, 80, 82, 98, 128, 211, 215-221, 273, 375-377, 502-503, 511, 515

Childers, Lt R.E., RNVR (O), 99, 101-103, 113, 125-127, 129-135, 167-169, 462, 488-489, 495, 498, 505-506
Chilman, AM R.A., RAF (O), 451, 453
Chryssidy, Sub Lt S., Hellenic Naval Air Service (O), 408
Chustan Island (*see* Long Island)
Chute, Capt E.L., Duke of Wellington's (West Riding) Regt. (O), 126, 133-135, 488, 504-507
Cintré, *Lieutenant de vaisseau* Alfred L.M. (P), 43, 45, 49, 63, 67, 71-75, 78, 80, 91, 473, 486, 501-502
Clarke, Lt L., 6th Manchester Regt (O), 229, 234, 496, 507
Clemson, Flt Lt A.W., RNAS (P), 188, 193, 197, 211-212, 216, 218, 220, 237-239, 243-244, 246-247, 268, 275-276, 319-325, 332, 342, 348-350, 375, 377-378, 386, 463, 477, 492, 508-512, 515
Clifford, FSL R.M., RNAS (P), 102, 105, 227-228, 492, 506-507
Cochrane, Lt Cdr A.D., 457
Colombo, 120, 329, 338-339, 342, 345-351, 353, 372, 481, 496, 512
Colombo Detachment (*see* EIESS)
Compagnie d'aérostiers, 9
Constantine, King of Greece, 282
Constantinople (İstanbul), 1, 6, 13, 20-25, 29, 31, 52, 59, 88, 129, 148, 153, 178, 201, 258, 401, 408, 436-437
Conté, Nicolas-Jacques, 9-10
Coutelle, *colonel* Jean-Marie Joseph, 9
Crete, 309
Crisp, AM H., RAF (O), 449
Crocker, Flt Cdr W.R., RNAS, 111, 492
Cusden, PO Frederick, 113-114
Cyprus, 21, 126, 160-161, 166, 178, 204, 260, 280, 364, 450, 457

D

Dacre, Flt Lt G.B., RNAS (P), 101, 106, 110-112, 125, 127-128, 177, 187, 191, 194, 199-203, 260-264, 463, 481, 492, 505-506, 508-509

Dagg, Lt C.K.C., RAF (P), 454-455
Damascus, 24-25, 32, 36, 88, 148-149, 158-159, 162-163, 203, 319-325, 409
Dardanelles, 29-30, 38, 43, 76-78, 93-94, 100-101, 113, 134, 144, 165-167, 283, 400-408, 478, 492.
Nagara Point, 401, 403-407
(*see also* Gallipoli)
Dartige du Fournet, *amiral* Louis, 68, 83, 297, 299
Daucourt, Pierre, 20-22
Dead Sea, 52, 55-56
Dedeagatch (Alexandroupolis), 100, 281, 289-290
Delage, *Lieutenant de vaisseau* Paul A.G. (P), 41, 45, 49, 63, 67, 487
Dera'a, 159, 161-162
Destrem, *Lieutenant de vaisseau* Marcel V.A., 40, 43, 45, 49-54, 63, 67, 73, 76-77, 81-83, 87, 91, 128, 137-140, 487, 499, 501-504, 506
Dhow, 39, 138, 180, 182, 185, 199-200, 205, 345, 409, 411, 430, 485
Deir El Belah, 326, 361, 363, 379, 384, 398, 513
Deir Sineid, 200, 381-383, 385, 390-392, 394-385, 399
Dittmar, *Offizierssstellvertreter* Edgar, 153, 162
Djemal Pasha, 62
Dover, FSL M.G., RNAS (P), 190, 193, 199, 213, 492, 509, 511
Drama, 287
Dranli, 302-304, 307-308
Drury, Lt Cdr Edward D., RNR, 101, 377-378, 437
Duke, CPO W.H., RN, 411
Duleidela, 246-247
Dunning, Flt Lt E.H., RNAS (P), 285, 291
Duray, Arthur, 15

E

East Indies and Egypt Seaplane Squadron (EIESS), 2, 37, 91
Formation, 93-98
  Royal Navy Egypt Order No.39, 96-97

545

East Indies and Egypt Seaplane Squadron (continued)
  Role of the EIESS, 94, 98, 150, 187, 211, 221, 359, 381, 431
  Island Base, 109-123, 187, 199, 310, 328, 353-354, 433, 440
    Buildings and Hangars, 110-112, 115-117, 353-354, 440
    Clerical staff, 111, 113
    Intelligence Office, 113, 169-170
    Photographic Office, 113-115
  Squadron attack on El Afule junction, 191-200
  Anti-submarine patrols, 183, 211, 326-327, 363, 408, 431-433 (see also RAF Egypt)
  Alexandria Detachment, 432-433 (see also RAF Egypt)
  Colombo Detachment, 342, 348-350, 353, 372, 481, 492, 494, 512
  Port Said Detachment, 431-433, 439, 445
  Wing Cdr Risk's report on state of EIESS, February 1918, 433
  Establishment, 28 March 1918, 438
Edmonds, Flt Lt C.H.K., RNAS (P), 101, 125, 127, 129-130, 134, 174, 227-229, 492, 505-506,
Egypt, 1, 5-7, 34-39, 52, 62, 84, 88, 93-95, 98, 100-101, 129, 139, 147, 149-150, 159, 165-167, 224, 226-227, 229, 233, 236, 247, 280, 283-285, 346, 357, 379, 405, 410, 435, 439, 443, 448, 455, 457, 458-460, 462, 492-494, 496-497
  Ottoman Egypt, 1, 5
  British Egypt, 5-7
  Napoleon's invasion, 9-11
  Early balloon flights in, 9-13
  Early flights in, 13-28
    Heliopolis Aviation Meeting, 1910, 15-20
    Long distance flights to Cairo, 20-25
    Pyramid flights, 26-27
    Nile flights, 27-28
  Defence of Egypt and Suez Canal, 34-36, 187

El Afule, 84-85, 185, 189-194, 196-197, 199, 205, 212, 215, 319, 325, 477, 481, 509, 513
El Atn, 411-412, 430
El Arish, 36, 38, 48-49, 51, 59, 62, 64, 71-72, 80, 83, 87, 98, 125-126, 133-135, 140-142, 150, 153-154, 171-172, 175, 177, 179-180, 182-184, 187, 190, 205, 208-211, 214, 242, 260, 325, 359, 456, 499-510
El Auja, 62, 65, 86, 89-90, 98, 125, 127, 133-135, 500, 505-506
El Falujeh, 133, 186, 190, 214, 506
Elgood, Lt Col Percival G., 48, 60, 96-97, 488
El Khalasa, 133-134
El Magdhaba, 135
El Qels, 50, 62, 64
El Shellal, 140, 183, 190
El Sirr, plain of, 71, 134
Elias, Dr, 148
Elias, *Hauptman* Hermann, 163
Eleusis, 297-299
England, Flt Lt T.H., RNAS (P), 111, 172, 174, 180-186, 190, 193-194, 200, 203-204, 207-208, 212-213, 216, 221, 227-229, 231-233, 268, 319, 332, 334, 342, 493, 506-512
En Nezle (El Nuzle), 390-391
Errington, Sub Lt A.G., RN (O), 291
Ertuğrul, *Topçu Yüzbaşi* (Artillery Captain) Mustafa, 263, 265-267, 269-271, 274-275, 279
Escadrille de Port Said (l'escadrille), 41-92, 103, 128, 136-146
  Formation and role, 41-43
  Base, 44, 108-109
Escaille, *Lieutenant de Vaisseau* Henry J.P. de l' (P), 42-46, 49, 52, 54, 58, 63-64, 67, 69-70, 77-78, 91-92, 95, 97, 103, 106, 136, 139-140, 143-146, 154, 462, 487
Euringer, *Oberleutnant* Richard (P), 154, 175, 189
Evitt, Mr H.C.J., Gunner RN, 272, 275

F

Fakhri Pasha, General, 37, 248-249, 251-252

Falkenhayn, *General der Infanterie* Erich von, 358
Felmy, *Hauptmann* Hellmuth, 151
Famagusta, 126, 166, 178, 180-181, 205, 221, 364, 377, 457, 505
Fazil, *Yüzbaşı* (P), 158-159
Ferejik, 289-290
Ferguson, 2Lt A.D., HLI (O), 319, 326, 360-362, 381, 383, 390-391, 393, 395, 397-398, 405-406, 413, 419-422, 425, 427, 429, 433-435, 437-439, 454-455, 496, 513-513, 515-516
Fethi, *Yüzbaşı*, 24, 148
Field, Flt Lt R.M., RNAS (P), 102, 133, 289, 292, 294-295, 299, 303, 305, 493, 506
Fineka, 137, 509
Finney, 2Lt Alfred D., RFA (Lt, RAF) (O), 134, 140-142, 475, 488, 506-507
Fitzmaurice, Cdr Raymond, RN, 317
Flechettes, 47-48, 193-194, 228, 230-231, 475
Fletcher, Major Herbert P., Middlesex Hussars (O), 74, 81, 83-84, 87, 89, 128, 167, 488, 502-504, 506
French Air Force (*Armée de l'Air*)
  Groupe de Bombardement d'Orient, 285
  MF.98M, 283
French Navy (*Marine nationale*), 38-39, 96-97, 86, 486
  3ème Escadre (Levant), 38, 68, 73, 92, 96-97, 136, 297
  3ème Escadre de Ligne, 297
French Naval Aviation (*Aéronautique maritime*), 1-2, 28, 41-43, 135, 139, 473, 475
  Escadrille de Brindisi, 67, 91-92, 474-475, 486-487
  Escadrille de Port Said (see separate listing)
Fricks, *Kapitänleutnant* Freiherr von, 76

G

Gallipoli, 1, 31, 35, 58, 76-80, 93-94, 100-101, 112, 119, 124, 146-147, 164, 224, 281, 283, 289, 352, 399 (*see also* Dardanelles)

Gaskell, Lt Robert A., RNR, 60, 70
Gaunt, Admiral Ernest F.A., RN, 39, 353, 379
Gaza, 34, 36, 50, 62, 65, 72-73, 76, 80, 86-90, 126-128, 133-134, 140-141, 149-150, 153, 157, 161, 173-174, 183-185, 187, 202-203, 207, 211-214, 239, 325, 328, 352, 356, 358-359, 368, 372, 379, 431, 474-475, 494, 499-504, 505-507, 509-511
   Gaza, 1st and 2nd Battles of, 34, 36, 149, 325-327
   Gaza, 3rd Battle of, 34, 150, 157, 161, 359, 379-399, 515-516
Genoa, 309, 313, 326, 352
Gereviz, Lake Boru, 286, 288
German Air Force *(Deutsche Luftstreitkräfte)*
   Feldflieger-Abteilung 1 (FFA1), 286
   Feldflieger-Abteilung 30 (FFA30), 286-287
   Feldflieger-Abteilung 57 (FFA57), 286
   Feldflieger-Abteilung 66 (FFA66), 286
   Feldflieger-Abteilung 69 (FFA69), 286
   Flieger-Abteilung (FA) 300 'Pascha', 141, 151-157, 159, 161-163, 180, 183, 189, 210-211, 242-243, 382, 393, 478, 480-481, 509-510
   Flieger-Abteilung (FA) 301, 157, 382, 393
   Flieger-Abteilung (FA) 302, 161
   Flieger-Abteilung (FA) 305, 157, 161-162
German Navy *Wasserfliegerabteilung*, 161, 286, 288, 457
Gerrard, Wing Commander E.L., RNAS, 283
Gobron, Jean, 15, 17
Gömbe, 266
Goodwin, CPO H.F., RNAS (O), 418, 429, 496, 516
Goss, Temporary Surgeon Lt L.S., RN, 276
Grace, Captain Edgar, RN, 301-302, 304, 307, 373
Grade, Hans, 15-16, 18-19

Grall, *Quartier-maître* Hervé (P), 40, 45, 55-58, 65, 67, 69, 76-77, 81, 83, 87, 89-91, 126, 133-134, 140-142, 473, 475, 487-489, 499-507
Gramont, *Second-maître* Charles A. (P), 91, 137, 487, 506
Greece, 61, 101, 255, 258, 281-283, 297-299, 326
   Greek IV Army Corps, 282
Grey, C.G., 22-23
Grigson, Flt Lt J.W.B., RNAS, 403
Groucott, Ldg Mech E., RNAS (O), 349, 496
Guillaux, Maurice, 28
Gumuljina (Komotini), 285-286

H

Habil, 425-428
Hacıkırı, 31, 152, 157
Haifa, 39, 51, 84-85, 98, 177-178, 182-183, 185, 189-191, 194, 200, 212, 215, 260, 319, 325-326, 359, 377, 387-389, 457
Hakkı, Yüzbaşı Ismail, 24-25
Hampton, Sub Lt S., RNVR (O), 302-303, 305, 496
Harris, 2Lt L.T., RAF (O), 456
Hasani Island, 246, 313
al-Hashimi, Emir Hussein ibn Ali, Sharif of the Holy Cities of Mecca and Medina, 37
   Sons, 37
      Emir Ali, 37, 248-249
      Emir Abdullah, 37
      Emir Faisal, 37, 248-249, 252, 313, 317
      Emir Zeid, 37, 248-249
Hauvette-Michelin, Gabriel, 15, 18
Hayes-Sadler, Admiral Arthur, RN, 297, 299, 401
Heemskerke, *Hauptmann* Hans-Eduard von, 152
Héderer, Docteur Charles, 276
Hejaz, 32, 34, 36-37, 39, 52, 150, 157-159, 244, 252, 409
Hejaz Railway, 32, 52, 98
Heliopolis, 13, 15-22, 26-28, 59
Henderson, FSL F.C., RNAS (P), 220, 493, 511
Henkel, *Leutnant* Hans, 141, 153-156

Herbert, Capt James R., Duke of Cambridge's Hussars (O), 49-50, 62-63, 70, 78, 81-83, 85, 488, 499-503
Hervey, Flt Cdr Lionel A., RNAS (P), 408
Hesler, *Leutant* Friedrich von (O), 209-210, 510
Hesselberger, *Oberleutnant* Joseph, 264-265, 269-270, 279
Hillas, 2Lt H.G., Duke of Wellington's Regt (O), 65, 75, 78, 83-85, 488, 500-503
Hodeida, 410
Holcombe, Sub Lt A.W.C., RNVR (O), 298, 305-306, 308, 496
Holmes, Sqn Cdr P.L., RNAS (Major, RAF), 432, 435, 438-439, 459, 493
Homs, 163, 180, 203, 321
Howes, Obs Sub Lt S.C., RNAS (Lt, RAF) (O), 438, 449, 496
Hughes, Lt C.E., Somerset Light Infantry, 115-117, 221, 321, 324, 326, 331, 334, 359, 435, 437
Hughes, Lt T.V., RFA (O), 142, 257-259, 496, 507
Hull, Surgeon H.R.B, 106
Hussein ibn Ali al-Hashimi, Emir, Sherif of Mecca and Medina, 37, 410
Hyams, FSL G.F., RNAS (Lt, RAF) (P), 107, 455-456

I

Idrisi Arab forces, 412, 418, 420, 430
Idrisi, Sayyid Muhammad ibn Ali al (aka, the Idrisi), 36, 410-411, 413, 415, 425-427, 429-430
Idrisi, Sayyid Mustafa el, 411-412, 416, 419, 424, 430
Ilkucan, *Yüzbaşı* Salim, 25
Imbros, 114, 283-285, 289, 381, 401-406, 433, 488
   Kephalo Bay, 289-290, 373, 403
   Kusu Bay, 403
Iralti (Eratino), 287
Isle of Man Steam Packet Company, 1, 98, 312
Ismailia, 6, 59, 62-63, 65, 96-97, 191, 215, 372, 383, 413, 436, 459, 514

J

al-Jabarti, Abd al-Rahman, 11
Jackson, Admiral Thomas, RN, 379, 391, 398
Jaffa, 21, 24, 39, 51, 72, 76, 85, 126-128, 133, 173-174, 182, 186, 190, 200, 203, 213, 215, 359, 363, 386-387, 499, 501-504, 507, 509
Jeanblanc, *Second-maître* Emile A. (P), 92, 134-135, 474, 487, 506
Jebel Abs, 425, 428
Jebel al Milh, 414-425, 429-430
Jeddah, 36-37, 39, 75, 177, 229, 234-236, 243, 247, 313, 502
Jeihan Irmak 81, 204-205 (see also Ceyhan River)
Jenkins, Lt John, RNR, 60, 80, 82, 189, 227, 342
Jerusalem, 21-25, 34, 36, 85, 88, 127, 161, 179, 210, 359
Jizán, 409-411
Jordan Rift Valley, 52, 161

K

Kalamaki, 137, 260, 263
Kalymnos island, 292
Kamaran island, 429
Kantara, 21, 48, 62-63, 154
Karagach, 259
Karatash Burnu, 204, 364, 376-377
Karjani, 302, 307-308
Kasaba, 266
Katia, 35-36, 63, 69, 189
Kavala, 281-282, 284, 287
Kekova Island, 137, 259, 264-265
Kazaviti (Kavamiti, Prinos), Cape (see Thasos)
Kemal, Mustafa (Atatürk), 358
Kemal, *Yüzbaşı*, 24
Kemmis, Jasper, 20
Kempson, Capt W.R., RFA (O), 319, 332, 363-364, 367, 372, 375-377, 381-385, 390-391, 394-395, 398, 405, 434-435, 438-439, 496, 512, 514-516
Kennedy, Obs Lt R.C., RNAS (O), 413, 421-425, 428, 438-439, 496, 516
Kennedy, Captain T.W.B., RN, 291
Kephalo (see Imbros)
Kerr, Lt John, RNR, 77, 79-80

Kerry, S/Lt J.L., RNVR (O), 4, 180, 182, 190, 200, 208-209, 249, 268, 272, 275-276, 319-325, 332, 337-339, 341, 343, 347, 351, 363-364, 366-367, 372, 462, 464-468, 496, 508-514
Kesme Burnu, 205
Khan Yunis, 62, 90, 133, 173-174, 183-184, 359, 500, 508
Khartoum, 11-12, 20, 26-28
Khios, 76, 78, 296, 301
Kilner, Sqn Cdr C.F., RNAS (P), 291
King, 2Lt E., KOSB (O), 177, 190, 193, 200, 203-205, 208, 260-261, 497, 508-510
King, FSL E.M., RNAS (Capt, RAF) (P), 4, 218, 313, 316-317, 319, 360-363, 368-370, 372-373, 386-391, 394-396, 398, 413, 417-418, 420-428, 431, 434-435, 438, 454-455, 457-460, 462, 468-471, 493, 511-516
Konya, 20-21, 24, 163
Kossaima, 62, 86, 94, 98, 125, 135, 500, 505
Kressenstein, *Oberst* Kress von, 62
Krieken, *Major* Frederik Schueler van (*Binbaşı* Şule), 279
Kum Kale, 78
Knight-Bruce, Capt R.E.C., Royal 1st Devon Yeomanry (O), 332, 337-339, 346, 497, 512

L

Laccadive (Lakshadweep) Islands, 330, 334-338, 340, 464, 512
Chetlat island, 337, 464
Lakerdis, Ioannis Yeorgiou, 263
Laroche, *Madame* Raymonde de, 15, 18-19
Lahej, 223-230, 232, 351
Latham, Hubert, 15, 18-19
Lawrence, T.E., 56, 58, 159-160, 313, 489
Lay, PO S., RNAS, 429
Lebanon, 179, 203, 320, 323-324, 470
Ledger, 2Lt Horace M.C., 27th Punjabis (O), 46, 69, 81-82, 87, 89-91, 463, 474, 489, 501-504

Le Gall, *Quartier-maître* Jean-Marie (P), 40, 45, 63-64, 463, 473, 487, 500
Leigh, Flt Lt H. de V., RNAS (P), 200, 207, 211, 319, 325, 360, 363-364, 367, 375-378, 381, 383, 386-389, 405, 413-416, 421-423, 425, 427, 429, 435, 438-439, 493, 509-510, 512-516
Lejh, 245, 252, 313
Lemnos, 77, 284
 Mudros Bay, 76, 77-78, 100, 130, 143, 147, 283-284, 289-290, 292-293, 296, 299, 301, 305, 308-309, 400-408, 437, 476, 478, 491, 501, 516
Leros, 292, 295
 Port Lakki, 292, 295-297, 299
Lesbos (see Mytilene)
Levant, 35
Levasseur, *Matelot* Julien P.A. (P), 43, 45, 51, 54, 62, 67, 70, 487, 499-500
Levisi, 258-259
Linberry, Cdr T.J., RN, 75
Lith, 409
Lloyd, Air Mech, RNAS (O), 497, 511
Loheia (Loheiya), 409-416, 418-419, 423, 425, 428-430
Long Island (Chustan Island), 76, 326
Ludd, 127-128, 133, 186, 190, 501, 506, 509
Lufbery, Raoul, 26
Lynden-Bell, General Sir Arthur, 156

M

Ma'an, 52-54, 62, 159-162
Macedonia, 151, 281-282, 286-287
MacGregor, Lt Cdr Donald Priaulx, RN, 404
Mackenzie, Lieutenant R.J.H.L., RE, 12
Maeandere river, 295, 300
 Derin Guel, 300
 Lade Island, 300
Makri (Fethiye), 144, 258-262, 502
Maldive Islands, 328-330, 335-336, 338-340, 342, 345, 347, 351, 364, 512

Ari Atoll, 340-341, 345
Felidu Atoll, 340-341
Malé, 336, 340-341, 345
Mulaku Atoll, 340
Nilandhu Atoll, 345-346
    Fiale island, 345-347
South Malé Atoll, 340
Malet, FSL H.G.R., RNAS (P), 305-306, 308, 493
Malone, Sqn Cdr C.J. L'Estrange, RNAS (P), 94-95, 98, 101, 114, 129, 143, 147, 167, 193, 197, 238, 243-273, 491, 509-510,
    Pre EIESS, 99
    Post EIESS, 247-248
Malta, 37-39, 42, 45, 135, 139, 143, 145-146, 373, 435, 439-443, 451, 458-459, 474-475, 482-485
Man, FSL W., RNAS (P), 173, 177-178, 187, 193, 197, 200, 208-209, 211-212, 248, 250, 252-254, 261-263, 480, 493, 507-511
Marchal, *Second-maître* Victor Rigobert (O), 43, 69, 490
Marcon, Lt Cdr Reginald E., RN, 101
Marduna Island, 245, 315
Maritsa river (Evros), 281, 289
Marmaris, 258-259, 502
Marmora, Sea of, 401
Martin, PO J.A., RN, gunlayer, 185
Maskell, FLt Lt A.S., RNAS (P), 101, 107, 122, 180-184, 186, 190, 193, 205, 208, 220, 268, 271, 493, 508-512
Maswakli, 286-287
Maxwell, General Sir John, 34-36, 48-49, 55, 60, 63, 66, 68, 78, 92, 96, 128, 143
McClean, Francis, 28
McLeish, Lt T.D., RNVR (O), 423, 438, 497
McMahon, Sir Henry, High Commissioner for Egypt, 7, 37, 63
Meade, Lt W.C.A., RNVR (O), 216, 268, 332-333, 339-344, 347, 497, 511-512
Mecca, 36-37, 159, 236, 243, 247, 253, 313, 409-410
Medina, 36-37, 158-159, 161, 246-249, 251-252, 313, 409

Mendelia, Gulf of, 293, 300
    Asin Kalesi Bay, 293
    Dalian, 294-295
    Kuluk, 294-295
    Milas, 294
    Sari Chai, 294
Mersina, 80-81, 160-161, 187-188, 457, 501, 504, 509
Mesopotamia, 31-32, 80, 120, 149, 331, 349, 358, 477-478
Métrot, René, 15, 17-19
Midhat Nuri, *Mülazım* (P), 160
Midi, 425-428
Millard, Lt V., Essex Regt (O), 193, 199, 227-228, 237-238, 241, 243, 246-247, 325, 360-364, 367, 372, 374, 381, 383, 391, 395-396, 398, 413, 416-419, 421-423, 425, 428, 434-435, 437-439, 497, 506, 508-510, 513-516
Milos island, 127, 136, 296-299, 301
Mitla Pass, 62-63
Monnaque, *Capitaine de Frégate*, 186
Moreau, *amiral* Frédéric Paul, 92
Morey, Lt G.W., RAF (P), 438, 456, 493
Morrison, Flt Lt J.S.F., RNAS (P), 291
Mowila (Al Muwaileh), 74-75, 239, 241, 245, 313, 502, 508
Mudros (*see* Lemnos)
Munro, General Charles, 93
Murray, General Sir Arthur, 35-36, 96, 139, 156, 187, 247, 358
Musa Dagh, 83-85, 503
Myra, 268, 279
Mytilene, 25, 99, 283
    Port Iero (Gulf of Yera), 100, 296, 300

N

Nablus, 133, 186, 203, 508
Nagara Point (*see* Dardanelles)
Nahr el Kebir, river, 180-181, 203-205, 509
Nahr Iskanderun, river, 127, 133, 186, 362
Nakhl (Mubarak), 249-252
Nazareth, 84, 194, 198, 212, 319, 325-326, 503, 508-509, 513
Nekhl, 62, 150, 243

Nerger, *Fregattenkapitän* Karl August, 329, 333
Newton, Lt E.A., Camel Transport Corps (O), 372, 375-378, 463, 477, 497, 514-515
Nicolle, Midshipman E.J., RNR, 278
Nightingale, Flt S/Lt A.G., RNAS (P), 208-200, 211, 213-214, 463, 478, 493, 510-511
Nile, river, 11, 20, 28, 35
Noman Island, 313, 317

O

Ocean Salvage Company, 280
Olivier, Louis, 27-28
Orfano (Orfani), 302-303, 306-308
Orfano (Rendina), Gulf of, 281, 283, 285-286, 307
Osmaniye, 82
Ottoman Empire/government, 1, 5, 7, 9, 27, 30, 32, 109-110, 129, 138, 163, 255, 263, 282, 285, 410, 433, 437, 454, 457
Ottoman (Turkish) Air Force (*Osmanli Hava Kuvvetleri*), 24-25, 41, 148, 154, 157-163, 188, 285
    3ncü Tayyare Bölük (Aircraft Company, *Ty Bol*), 157-163, 248
    4ncü Ty Bol, 160-163, 188
    5nci Ty Bol, 287, 279
    14ncü Ty Bol, 161-163
Ottoman (Turkish) Army, 30-39, 48-50, 52, 54-58, 59-66, 68, 70, 73-74, 78, 80, 83-85, 88, 90-91, 93, 149-151, 169, 179-183, 185, 187-189, 191, 200-203, 205, 210-213, 215, 224, 229- 236, 239-241, 243, 246-252, 256, 259-260, 265, 269, 272, 279, 283, 301, 306, 309, 313, 316, 325-327, 334, 358-359, 365, 372, 379-380, 384, 390, 394, 409-415, 418-423, 425-426, 430, 518-519
    4th Army, 148, 161, 163
    Yemen Army Corps, 224
    Yildirim Army Group (*Yıldırım Ordular Grubu*), 358
Ottoman (Turkish) Navy, 29-30, 39, 76, 100, 399-408

P

Pakenham-Walsh, 2Lt L.H., Cheshire Regt (O), 319, 383-385, 392-393, 397-398, 433, 437, 463, 478, 497, 512-516
Paine, FSL L.P., RNAS (P), 180, 193, 196, 199-200, 229, 231, 234, 313, 493, 507-508, 512
Palestine, 31-34, 38-39, 49, 59-60, 68, 77, 79-80, 86, 88, 98, 125, 133, 148-149, 150-152, 157-158, 161-163, 176-177, 179, 205, 212, 227, 260, 319, 334, 358-359, 362-363, 368, 379, 381, 400, 439, 448, 457
Palmer, Lt Cdr E.M., RN, 333
Parker, Assistant Paymaster J.M., 278
Parkes-Buchanan, Midshipman M.J., 49, 52
Partridge, 2Lt Basil G.N.B., 2nd Rajput Light Infantry (O), 63-64, 463, 473, 489, 500
Paul, 2Lt Sir Robert J., Egyptian Irrigation Dept (O), 65, 71-72, 76, 80-82, 85, 87-89, 463, 473, 489, 500-503
Peirse, Admiral Sir Richard H., 38, 49, 63, 64, 69, 76-77, 94, 103
Pennington, FSL G.A.A., RNAS (P), 413, 419, 421-422, 438, 493, 516
Perim Island, 224, 227, 233
Perry, Air Mech, RNAS (O) A.J., 417, 422, 424, 497, 516
Pharmako Island, 296, 300
Picard, *Lieutenant de Vaisseau*, 179, 182, 508
Piræus, 280, 298-299
Popham, Flt Lt A.E., RNAS (P), 120, 349-350, 352, 375-377, 386-389, 391, 393, 395, 397-398, 433-435, 462, 494, 515-516
Port Iero (*see* Mytilene)
Port Lakki (*see* Leros)
Port Said, 1, 4, 5-6, 22, 42-43, 45, 48-49, 51-52, 54-55, 58-70, 73, 75-80, , 85-87, 89-94, 100-104, 125-126, 128, 130, 132, 134-147, 149, 167, 169, 171, 173-175, 177, 179, 182-183, 187-190, 194, 199-201, 203, 205, 207, 210-216, 220-221, 223, 229, 236-237, 247-248, 254, 257, 259-260, 263-264, 268, 278, 289, 309-310, 312-313, 319, 325-328, 331-333, 346-347, 349-351, 353, 359-363, 367-368, 371-375, 377-378, 381, 383-386, 389, 395, 397-399, 405, 408-409, 413-414, 424, 428, 431-433, 437, 439-443, 445-448, 450, 452, 454-460, 462, 473-487, 489, 492-494, 497, 499-501, 505-508, 512-514
  Naval base, 38-39, 96-98, 108, 139, 166
  Seaplane bases, 44, 108-113, 115-124, 440, 442
  Kite Balloon Base, 440, 443
  Bombing raids on, 154, 242-243
Port Sudan (*see* Sudan)
Porto Lagos, 285-286, 288-289, 291
Pourpe, Marc, 22, 26-27
Pozantı, 21, 31, 152-153, 160
Prince, Ldg Mech, RNAS (O), 375, 378, 498

Q

Quishon, Valley of, 194

R

Rabegh, 159, 243-249, 251, 502
Rafa, 32, 36, 62, 90, 133, 172-174, 183-185, 211, 358-359, 399, 500, 502, 506-507, 509
Railways
  Baghdad Railway, 20, 31-32, 80, 151
  Bulgarian, 290
  Feldbahn system, 32, 157
  Hejaz Railway, 32, 36, 39, 52, 98, 159
  Narrow gauge (1.05m), 32, 153, 319, 321, 380
  Standard gauge (4 ft 8½ in / 1.435 m), 32, 153, 319, 380
Ramleh (Er-Ramleh, Ramla), 85, 89, 98, 126, 128, 133, 153, 157, 163, 174, 186, 190, 203, 213, 361, 363, 501-503, 505-507, 509
Ravenscroft, Lt A.P., RFA (O), 177, 261, 498, 508
Rayak, 32, 153, 162-163, 319-322, 324-325
Rebeur-Paschwitz, *Vizeadmiral* Hubert von, 400-401, 404
Redington, Ward Room steward S., 276
Red Sea, 1, 11, 32, 36-39, 52, 55, 73-75, 87, 98, 144, 177, 179, 188, 211, 215, 222-224, 227, 234, 236-237, 242-245, 248, 254, 313, 317, 326-327, 331, 334, 379, 399, 409-411, 413-414, 424, 430, 433, 459, 501-502, 507-508, 510-512, 516
Red Sea Patrol (*see* Royal Navy)
Riemsdijk, Frederick van, 15, 17
Rhodes, 255-256, 259-260, 278, 300, 502
Ridley, Chief Writer, RN, 113, 213, 273
Risk, Wing Cdr C.E. RMLI (P), 346, 352-353, 355, 361-363, 368-372, 374, 381, 431-435, 437-439, 491, 512-514
Pre EIESS, 352
RAF, Egypt, 439, 446, 453-454, 457, 491
Robeck, Admiral John M. de, 283-284
Robertson, Ldg Mech W., RNAS (O), 425, 429, 498, 516
Robinson, Eng Lt G., RNR, 271, 312
Romani, 69, 187
Romani, Battle of, 36, 183, 187, 208
Romanos, Abbot Kyrillos, 268
Ross, Capt Arthur J., RE (O), 54, 489, 499
Rougier, Henri, 15, 17-19
Roussillon, *Quartier-maître* Henri J. (P), 92, 134, 140, 142, 487, 506-507
Roux, J, 20-23
Rowland, Sub Lt D.P., RNVR (O), 305, 308, 498
Royal Air Force (RAF), 2, 98, 109, 123, 150, 159, 161, 167, 203, 226, 248, 312, 346, 386, 415, 428, 432-434, 436, 438-439, 445, 453, 457, 460, 477-479, 481, 492-497
Royal Air Force, Egypt, 436-460, 482-485
  Organization – Egypt, 436-439
  Anti-submarine and Convoy patrols, 431-433, 440, 443, 445-456

Royal Air Force (continued)
  Naval Wing, Royal Air Force, Egypt, 436, 439, 445, 460, 482, 491
    Alexandria Detachment / Squadron, 432-433, 439-446, 448-456, 459-460, 479, 482-485, 492, 496
    Port Said Detachment / Squadron, 432-433, 437, 439-443, 445-448, 452, 454-460, 482-485,
  64 (Naval) Wing, RAF, 439-440, 482
    431 and 432 (Seaplane) Flights, 439, 482
    269 Squadron, RAF, 439, 457, 459-460, 482
    270 Squadron, RAF, 439
  Machines available, June 1918, 446-448, 482
  Report on RAF in the Mediterranean, August 1918, 439-445
Royal Engineers, balloon detachment, 11-12
Royal Flying Corps (RFC), 28, 48, 54, 123, 133, 139, 149-151, 161, 180, 187, 209, 211, 226-227, 247, 285, 325, 327, 359, 361-363, 380, 383, 431, 433, 436, 437, 462, 488, 495, 498
Royal Flying Corps, Egypt,
  Egypt Detachment (EDRFC), 35, 59, 62-63, 65, 67, 73, 98, 149-150, 488-489, 500
  5th (Army/Corps) Wing, 149-150, 325, 361
  40th (Army) Wing, 150
  14 Squadron, 149-150, 157, 159-160, 180, 183, 239, 325, 363
  17 Squadron, 149, 285
  47 Squadron, 346, 497
  67 Squadron (1 Squadron, AFC), 149-150, 382
  111 Squadron, 150-151
  113 Squadron, 150
  142 Squadron, 150
  144 Squadron, 150
  145 Squadron, 150
Royal Indian Marine (RIMS), 74-75, 246, 250, 451

Royal Naval Air Service (RNAS), 1-2, 41, 65-66, 93, 95, 98-99, 101-103, 106, 109, 111, 115, 118, 123, 139-140, 146, 154, 156, 165, 167, 177, 187, 199-201, 210, 227, 237-238, 248, 257-260, 277, 283-287, 301, 305-306, 309, 322, 325, 348, 352, 359, 361, 381, 386, 430, 405, 414, 433-437, 462, 471, 486, 488, 491, 495-498, 513, 515-516, 518-519
  East Indies and Egypt Seaplane Squadron (see separate listing)
  Eastchurch Naval Air Station/ Flying School, 99, 165, 462, 496-7
  Eastchurch Squadron, 165
  2 Wing, RNAS, 283-285, 287, 289, 301, 306, 408, 488, 491, 495-496, 498
    A Flight (later Squadron), 285, 287
    D Flight (later Squadron), 301, 303, 306-307
    G Squadron, 408
  3 Wing, RNAS, 283
  RNAS to RAF, 434-439
    Admiralty Fleet Order (AFO) No.1391, 434-436
Royal Navy (RN), 37-39, 49, 51, 62, 70, 73-74, 78, 80, 96-100, 102, 104, 106, 128, 143, 147, 165, 172, 200, 236, 247, 249-250, 260, 276, 281-283, 298, 301, 311, 342, 346, 371, 379, 400, 413, 438, 457, 459, 486, 518-519
  Eastern Mediterranean Squadron/Station, 38, 43, 93-94, 147, 281-284, 289, 297, 309
    6th Detached Squadron, 301, 304, 307
  East Indies and Egypt Station, 38, 69, 94-96, 98, 147, 166, 311, 353, 356, 361, 379
  Red Sea Patrol, 37, 39, 234, 248, 409-411, 414, 430, 459
Ruad island, 83-84, 137-138, 180, 204, 264, 503-504, 506, 508

S

Sadık, *Mülazım*, 24

Saizieu, *Lieutenant de vaisseau* Louis M.J. Barthélémy de (P), 45-46, 62-63, 67, 70, 77-78, 81-85, 87, 89-91, 167, 463, 473-474, 487, 500-504
Sakelarov, *Kapitan* Dimitar, 285
Şakir Fevzi, *Mülazım*, 148-149
Salih Rifat, *Mülazım* (O), 160
Salamis, 298-299
Salonika, 3, 35, 92, 136, 149, 281-286, 287, 307, 346, 401, 439, 455-456, 497
Salmond, Major General W.G.H., RAF, 453
Samos island, 285, 295, 300
Samaria, 133, 186, 213, 502
Samson, Wing Cdr C.R., RNAS (P), 78, 147, 165-175, 179-183, 185-187, 189-191, 193-200, 203-210, 212-, 214-216, 218, 220, 226, 229, 231-235, 237, 268-278, 310-312, 332-334, 339, 342, 346, 352, 431, 472, 491, 507-512,
  Pre EIESS, 164-165
  Court Martial, 310-312
  Marriage, 338
  Post EIESS, 346
Samson, Lt W.L., RNVR (O), 220, 268, 498, 511-512
Sana'a, 36, 416
Sanders, *General der Kavallerie* Otto Liman von, 163, 265, 358
Sandwell, Flt Cdr A.H., RNAS (P), 105
Sarrail, *général* Maurice P.E., 282, 297
Saunders, AM T., RAF (O), 451, 453
Scalanuova, Gulf of, 292, 296
Scarlett, Wing Captain F.R., RNAS, 284
Schmidt-Kolbow, *Major* Karl, 265-267, 269-270, 274-275
Sedat, *Yüzbaşı* Hüseyin, Ottoman Navy, (P), 160, 188
Seihun (Seyhun, Seihan) river, 31, 204, 504
Senussi, 129-130, 132, 150, 410
Serbia, 282-283
Sheikh Othman, 223, 226, 231
Sheikh Said, 224, 233

Ships—
Ships, Merchant vessels
  Aenne Rickmers (see British Seaplane Carriers)
  Ben-my-Chree (IoMSPC) (see British Seaplane Carriers)
  Campinas (Compagnie des Chargeurs Réunis) (see French Seaplane Carriers)
  City of Oxford (see British Seaplane Carriers)
  Crosshill, 79
  Empress (South Eastern & Chatham Railway Company) (see British Seaplane Carriers)
  Engadine (South Eastern & Chatham Railway Company) (see British Seaplane Carriers)
  Hardi (Suez Canal Company tug), 69
  Khyber (P&O), 135
  Port Darwin, 457, 460
  Rabenfels (see British Seaplane Carriers)
  River Clyde, 58, 77
  Riviera (South Eastern & Chatham Railway Company) (see British Seaplane Carriers)
  Somali, 350
  Turritella (see German Raider Iltis)
  Uganda (British India Steam Nav. Co. Ltd), 146
  Wachtfels (see German Raider Wolf)
  War Dame, 458
  Wathfield, 295
Ships, Warships, British
  Battleships and battlecruisers
    Africa, 165
    Agamemnon, 130, 400, 403
    Exmouth, 297, 351
    Hannibal, 166, 311
    Hibernia, 165
    London, 99
    Lord Nelson, 400-401
    Swiftsure, 38, 61-62, 64, 76-77
    Triumph, 76
  Cruisers
    Diana, 338, 346
    Doris, 49-51, 58, 80, 149
    Edgar, 307, 309
    Euryalus, 76, 191, 227
    Fox, 39, 234, 245, 247-248, 313, 315-316, 409-412, 414, 425, 428
    Grafton, 296, 301-304, 306-307, 372-374, 379, 384-385, 390-391, 514-515
    Lowestoft, 290-291
    Minerva, 49, 51-52, 54-58, 73, 238, 473, 489, 499
    Warrior, 59
  Destroyers
    Comet, 379
    Jed, 300
    Lizard, 401, 403-404
    Savage, 52
    Scourge, 52, 301
    Staunch, 379, 398
    Tigress, 403-404
    Wolverine, 295-296
  Monitors
    Abercrombie, 289-290, 307
    Earl of Peterborough, 296-297, 307
    M.15, 171-172, 208, 395-396, 398, 507, 516
    M.18, 304, 306, 308
    M.21, 188, 509
    M.22, 516
    M.23, 171-172, 507
    M.28, 291, 303-304, 403-404
    M.29, 297, 391, 516
    M.30, 379
    M.31, 208, 248, 372, 397, 514, 516
    M.32, 304, 306, 308
    M.33, 297, 300
    Raglan, 289-290, 379, 381-385, 390-391, 394-395, 401-404, 478, 515-516
    Sir Thomas Picton, 301, 304, 307
  Seaplane carriers
    Aenne Rickmers / Anne, 59-65, 70-71, 76-77, 79-80, 87, 89-92, 95-98, 112, 126, 128, 133-136, 140, 142-143, 147, 154, 166-167, 177, 187-188, 191, 193, 199-200, 211, 215, 223, 227, 236, 242-247, 257-260, 264, 310, 313-319, 321, 324-326, 328, 352, 356-358, 431, 462, 474-475, 488-489, 492-494, 496-498, 500-510, 512-513
      Capture, 59
      Conversion, 59, 79
      Manning, 60-61, 79, 147
      Renaming Aenne Rickmers to Anne, 80
      Bombing attacks on, 141, 242
      Damascus Flight, 28 February 1917, 319-325, 512
      Post EIESS history, 326
    Ark Royal, 102, 147, 283-284, 289-290, 292, 296, 305, 308-309, 403, 476, 480-481, 493-494, 498
    Ben-my-Chree, 1-2, 37, 93-95, 98-103, 106-107, 109, 113-119, 124-128, 130, 132-135, 140, 143, 146-147, 165-166, 171-190, 191-194, 197, 199-222, 223, 226, 229-236, 260, 264, 268-280, 283-284, 309-312, 356, 462, 472, 474-481, 491-498, 505-512
      History and conversion to seaplane carrier, 98-99
      Armament, 99, 166, 171, 175, 271
      Early operations (pre-EIESS), 99-100
      Intercepts Turkish blockade running dhows, schooners, and tugs, 180-182, 184-185, 199-200
      Bombing attacks on, 175-176, 183, 215
      Aden and the Red Sea, 223, 229-236
      Loss at Castellorizo, 268-280
      Salvage, 280
    City of Oxford, 223, 356, 371, 381-383, 386, 389-391, 393-399, 409, 431, 433, 436-437, 446, 459, 462, 482, 493-494, 496-498, 515-516
      History and conversion to seaplane carrier, 356-358

City of Oxford (continued)
   The Final Cruise, 413-430
   Operations over the Jebel Al Milh, 416-425
   Post EIESS history, 436-437
Empress, 93-95, 98, 120, 133-134, 146-147, 313, 325-327, 328, 352-353, 356, 359-364, 367-372, 374-378, 381, 383, 386-389, 396, 399, 409, 431, 433, 436-437, 439, 446, 462, 474-481, 483-484, 491-498, 506, 513-516
   History and conversion to seaplane carrier, 100-101
   Early days at Port Said, 102, 104-105
   Aegean operations, 281-309, 405-408
   Post EIESS history, 437, 439
Engadine, 98-101
Manxman, 405
Rabenfels / Raven, 59-61, 65, 67, 71-73, 77-83, 92, 95-96, 98, 100, 112, 124, 136, 143, 146-147, 166-167, 177-178, 181, 188-189, 191, 193, 199-200, 203, 211, 215-216, 218, 221, 223, 226-230, 236-242, 247-254, 260-263, 275, 313, 319, 325, 327-329, 353, 356-357, 381, 383-385, 390, 409, 431, 448, 462, 464, 474, 477-478, 481, 489, 492-498, 501-504, 506-512, 515
   Capture, 59
   Conversion, 59, 61
   Manning, 60, 80, 147
   Renaming Rabenfels to Raven II, 80
   Survey of Akaba, 237-239
   Bombing attacks on, 242-243
   Indian Ocean cruise, 331-351
   Post EIESS history, 385
Riviera, 98, 100

Submarine
  E.14 (Lt Cdr G.S. White), 406-408
Miscellaneous vessels
  Amaryllis (anti-submarine sloop), 449
  Aphis (gunboat), 379, 383, 395, 515-516
  "*Charasin*", aka *Chirlsin* (armed trawler), 132
  Clio (sloop), 412, 430
  El Kahira (water-distilling ship), 317
  Espiegle (sloop), 171-172, 208, 313, 317, 412, 430
  Dufferin (Royal Indian Marine), 234, 246-248, 250, 252, 510
  Hardinge (Royal Indian Marine), 74-75, 234, 237-239, 245, 247, 313-315, 502
  Kamaran (armed steam launch), 411, 430
  Ladybird (gunboat), 379, 383, 395, 397, 449, 515-516
  Managem (armed yacht), 383
  Northbrook (Royal Indian Marine), 246, 510
  Odin (sloop), 333, 430
  Perth (armed boarding steamer), 234, 412
  Rononia (armed trawler), 209
  Rowan (armed boarding steamer), 455
  Southland (troopship), 100
  Suva (armed boarding steamer), 248-250, 313, 430
  TB63 (torpedo boat), 51
  TB81 (torpedo boat), 165
  Trawler 298 (armed trawler), 309
  X-lighter, 313, 317, 379
Ships, Warships, Australian
  Brisbane (cruiser), 342, 348-349, 351, 481, 492
Ships, Warships, French
Battleships
  Provence, 297
  Saint Louis, 72-73, 80, 501
Cruisers
  Bruix, 21

Amiral Charner, 38-39, 85, 133, 256, 258, 448
Jeanne d'Arc, 80-81, 83, 130, 256, 501
Montcalm, 73-74, 87, 91, 474, 501
Pothuau, 187-188, 211, 334, 337-338, 340, 342, 509
Destroyers, 38, 133, 171, 177, 297, 379, 400, 445, 448
  Arbalète, 183-186, 191, 199, 204, 372, 375, 379
  Coutelas, 326, 360, 379, 386, 388-389
  Dard, 179, 181-182, 186, 207-208, 211-212, 264
  Fauconneau, 379
  Hache, 143, 189, 191, 361-362, 375, 378-379, 386
  Pierrier, 214-215, 268, 273-274, 360
  Voltigeur, 171, 173, 187-188, 363, 379
Seaplane Carriers
  Campinas, 98, 136-139, 143, 145, 257, 297-298, 462, 474-475, 488-489, 505-506
  Foudre, 40-45, 85, 473
Miscellaneous vessels
  Alexandra (armed trawler), 280
  Ariane II (armed yacht), 211, 257, 268, 273-274
  Cachalot (armed trawler), 263
  Canada II (armed trawler), 211
  Cydnus (armed tug), 84
  Laborieux (armed tug), 84, 177, 260, 262-263
  Maroc (armed trawler), 215
  Nord Caper (armed trawler), 215, 258-259, 319, 325
  Paris II (armed trawler), 191, 278-280
  Requin (garde-côtes cuirassé), 74, 379, 383-385, 390-397, 458, 515-516
  Surmulet (armed trawler), 263
  Torpilleur de Défence Mobile 250 (torpedo boat), 134-135, 141, 154, 268, 273

553

Ships, Warships, German
  Battlecruisers
    Goeben, 29, 379, 396, 399-408, 433, 494, 516,
  Cruisers
    Breslau, 29, 399-404
  Submarines 39, 129, 138, 144, 181, 183, 258-259, 262, 469-470 (anti-submarine, *see* EIESS and RAF)
    U-21 (Hersing), 39, 133, 258, 448
    UB-43 (Obermüller), 373
    UB-66 (Wernicke), 368
    UC-34 (Obermüller), 449
    UC-38 (Wendlandt), 398
    UC-74 (von der Lühe / Schüler), 449-450, 454
  Raiders
    Iltis, 333
    Wolf, 329-330, 333-334, 348
Ships, Warships, Greek
  Georgios Averof (cruiser), 298
  Helle (cruiser), 298-299
  Kilkis (battleship, ex-USS *Idaho*), 298
  Lemnos (battleship, ex-USS *Missisippi*), 298
Ships, Warships, Italian
  Francesco Ferrucio (armoured cruiser), 39
  Giuseppe Garibaldi (armoured cruiser), 39
Ships, Warships, Japanese, 39, 445, 450-452
  Katsura (destroyer), 452
  Sugi (destroyer), 452
Ships, Warships, Ottoman (Turkish)
  Battleships and battlecruisers
    Reshadiye, 29
    Sultan Osman-i-Evvel, 29
    Torgut Reis, 408
    Yavuz Sultan Selim (*see* Goeben)
  Cruiser
    Midilli (*see* Breslau)
  Destroyer
    Basra, 401, 404
    Muavanet-i Milliye, 30, 401
    Nümune Hamiyet, 401
    Samsun, 401

  Miscellaneous vessels
    Avnillah (Ironclad), 39, 138, 368, 370
    Demirhisar (torpedo boat), 30, 76, 78, 135
    Sultanhisar (torpedo boat), 76
Ships, Warships, Russian
  Askold (cruiser), 52
Silifke, 21, 160-161, 189, 457
Simvolon Dagh, 281, 302, 307
Sinai, 32, 34-36, 48-49, 52, 55, 59, 61-63, 65, 68, 134, 150, 154, 325, 488, 499
Singer, Adam Mortimer, 15-16
Sırrı, *Mülazım-ı Evvel*, 35
Sitwell, Flt Cdr W.G., RNAS (P), 102, 133, 292-296, 299, 305, 494, 506
Smith, Lt A.K., Highland Light Infantry (O), 142, 175, 180-181, 183, 185-186, 188-189, 257-259, 498, 507-509
Smith, Midshipman F.R., RNR, 269, 271
Smith, Flt Lt G.D., RNAS (Capt, RAF) (P), 193, 195-199, 216-217, 237-239, 241, 243, 246, 268, 273, 275, 278, 319, 332, 337, 339-347, 351, 363-364, 367, 381-383, 386-389, 392, 397-398, 413-414, 418, 421-422, 424-425, 428-429, 433-435, 437-438, 454, 457-460, 464, 494, 508-512, 514-516
Smith, CPO S.J., RNAS, 118, 140
Smyrna (Izmir), 76-78, 143, 279, 281, 283, 287, 295, 379, 501
Snepp, Capt J.W., RMLI, 55-56
Sokia (Söke), 295
Sollum, 128-132, 476, 506
Spelterini, 'Captain' Eduard, 8, 13
Spitz, *Contre-Amiral* Henri de, Commandant Division Navale de Syrie, 38, 179, 211-212, 257, 258, 264, 268-269, 273, 311-312
Stalter, *Oberleutnant* Karl (O), 154-155
Standish, 2Lt C.F., RAF (O), 456
Stavros, 284-286, 289-291, 301-309, 373
Steen, Lt C.H.C., RNR, 191

Stephens, FSL T.G.M., RNAS (P), 332, 337-338, 341, 350, 376-377, 386-389, 494, 512, 515
Sterling, Type 52 W/T set, 170, 207, 372
Stewart, General J.M., 334, 351
Stewart, Lt N.W., 7th Royal Scottish (O), 183-187, 218, 248-250, 253-254, 313, 316, 463, 498, 507-509, 511-512
Stirling, Capt Walter F., Royal Dublin Fusiliers (O), 52, 54-58, 473, 489, 499
Stokes, Ldg Mech J., RNAS, 429
Stone, CPO Charles Henry, RNAS, 4, 118-119, 174, 517
Stone, Lt Cdr R.G., RN, 308
Struma river, 281-283, 285-287, 301-303, 306
Suakin, 11-12
Subar, 227-229, 233-334, 351
Souchon, *Konteradmiral* Wilhelm, 29-30, 399-400
Suda Bay, 309
Sudan, 5, 11, 34-35, 74, 141, 410, 430, 459
  Port Sudan, 74, 234, 236, 247, 414
Suez, 6, 52, 54-55, 58, 62, 73-75, 146, 177, 187, 223-224, 227, 234, 236-237, 242-243, 246-248, 252, 254, 313, 317, 325, , 331-332, 348, 351, 381, 383, 413-414, 429, 457, 460, 510
Suez Canal, 5-6, 21, 34-36, 59, 68, 86, 98, 100, 120, 149-150, 154, 222-223, 229, 358, 372, 400, 459
  Defence of, 32, 42-43, 48-49, 52, 59, 61-67,72, 74, 187
  Turkish attack on, 62-65, 148-149
Suez Canal Company (*Compagnie Universelle du Canal de Suez*), 5-6, 44, 69, 91, 100, 135-136
Sykes, Col F.H., RFC, 165, 283
Syria, 9, 20-21, 32-34, 38, 52, 60, 62, 68, 88, 93-94, 144, 157, 161, 179, 191, 205, 227, 311, 358

T

Tahinos Lake, 302, 306
Taif, 37, 236

Talat, *Yüzbaşi* Lütfi, 76
Tarsus, 24, 80-82, 128, 204, 503
Tartus, 21, 83-84
Taurus mountains, 20-21, 24, 31-32, 148, 151-153, 157, 160, 204, 358-359
Tel el Sheria, 87, 90, 134, 358, 390
Tel Kale, 180, 204
Templer, Major J.L.B., RE, 12
Tenedos, 76, 78, 164, 406, 501, 517
Tersana Island, 258-262
Thornton, Mid H.D., RNVR (O), 289-290, 498
Thornton Pickard A Type aerial camera, 113-114, 132, 170
Thasos, 282, 284-287, 291, 301, 305, 308-309, 476, 481
  Cape Kazaviti, 284-285
Todd, Capt R.E., RAMC (O), 62, 67, 70-73, 76, 80, 489, 500-502
Toprakkale Junction, 32, 82, 128
Torrington, Lt Lord, RNVR (O), 292-294, 498
Trabaud, *Lieutenant de Vaisseau* Albert, Governor of Ruad, 83-84, 180
Trouillet, *Quartier-maître* Georges M.E. (P), 40, 45, 65, 67, 77-78, 81-85, 87-89, 463, 473, 487, 500-503
Tubaun, 185, 190
Tul Keram, 89, 126-128, 133, 213, 359, 361-363, 502, 505, 513
Turkey (Türkiye), 84, 147, 211, 258, 268, 278, 281, 283, 502
Turkish Air Force, Army, Navy (*see* Ottoman)
Tuzla farm (or *chiftlik*), 302-303, 307

U

Umrashash, 55, 58, 238
Urmek island, 414

V

Varney, *amiral* Georges, 38, 433-434
Védrines, Jules, 21-23
Venizelos, Eleftherios, Prime Minister of Greece, 282
Vickerey, Major Charles, 313, 315
Villacoublay, 40, 42
Volos Island, 137, 260

W

Wadi Araba, 52, 54-56, 74, 473, 489, 499
Wadi el Hesi, 200, 381, 383-385, 389-390, 393, 396-397, 509, 515-516
Wadi Gaza (el Ghuzze), 86-87, 89-90, 133, 140, 474
Wadi Gharandal, 52-54, 244
Wadi Imleih, 90, 133
Wadi Ithm (Yetham), 52-54, 56-57, 73-74
Wadi Mukhsheib, 63
Walker, Lt C.F.T., RNVR, 328, 423, 438
Walton, General William Crawford, GOC Aden, 227-229, 233
Ward, Private Frank, RMLI, 56
Waugh, FSL G., RNAS (Lt, RAF) (P), 438, 451-453, 494
Wedderspoon, Lt J.H.B., RFA (O), 137-139, 229, 232, 234, 489, 498, 506-507
Weldon, Capt L.B., Royal Dublin Fusiliers, 48, 60, 71, 77, 79, 81, 84, 87-88, 90, 96-97, 126, 140, 143, 258-260, 489
Wells, 'Professor' Rufus Gibson, 13
Wejh (Sherm Wej, Al Wajh), 232, 241, 245-246, 248, 252-254, 313-317, 508, 512
Wemyss, Rear Admiral Rosslyn, RN, 38, 93-95, 97, 139, 146, 166, 186, 191, 208, 210, 220, 227-238, 268, 311-312, 353, 356, 379
Whitehead, Flt Cdr R., RNAS (P), 305, 494
Whitmore, CPO, RN, 140, 319
Wiedemann, *Unteroffizier* (first monteur [fitter]), 155
Williams, Lt E. 'le petit', East Yorkshire Regt (O), 102, 105, 133, 498, 506
Williams, 2Lt K.L. 'Long', 2nd Rajputs (O), 46-47, 70, 76, 90, 126, 129, 134, 188, 193, 197, 200, 218, 227, 237-239, 243, 246, 248-250, 252-253, 313, 317, 463, 489, 498, 504-506, 508-512
Willows, Lt G.H., RAF (P), 449, 451-453
Witt, Warrant Officer, RN, 319
Wood, Flt Lt M.C., RNAS (P), 363-364, 367, 372, 375-377, 463, 477, 494, 513-515
Woodhead, FSL Norman, RNAS (P), 305
Woodland, Lt P.M., RNVR (O), 190, 193, 199, 212-214, 463, 478, 498, 508-509, 511
Worrall, FSL H.V., RNAS (Capt, RAF) (P), 248-251, 253-254, 268, 319, 326, 360-363, 367-369, 372, 374, 383-384, 405, 434-435, 437-438, 458, 494, 511-516
Wright, Flt Lt M.E.A., RNAS (P), 101, 127-129, 134, 142, 171, 175, 227-229, 234, 257-259, 494, 506-507

X

Xanthi, 285-288, 291
Xeros, Gulf of, 281

Y

Yemen, 36-37, 223-224, 409, 411, 419, 425, 430
Yenbo, 74, 158, 243, 245-252, 511
  Sherm Yenbo, 245, 248-249, 252
Yeşilköy, 24-25, 160-161
Yozgad PoW camp, 214, 457

Z

Zahle, 322, 324
Zeki, *Çavuş* Ismail (P), 160
Zohrah, 413, 415-416, 421-422, 424-425, 429

www.ingramcontent.com/pod-product-compliance
Lightning Source LLC
Chambersburg PA
CBHW081151070526
44583CB00021B/2793